VOLUME SIXTY SEVEN

ADVANCES IN
MARINE BIOLOGY

Advances in Cephalopod Science: Biology, Ecology, Cultivation and Fisheries

ADVANCES IN MARINE BIOLOGY

VOLUME SIXTY SEVEN

ADVANCES IN MARINE BIOLOGY

Advances in Cephalopod Science: Biology, Ecology, Cultivation and Fisheries

Edited by

ERICA A.G. VIDAL
Center for Marine Studies
University of Parana (UFPR)
Parana, Brazil

ELSEVIER

AMSTERDAM • BOSTON • HEIDELBERG • LONDON
NEW YORK • OXFORD • PARIS • SAN DIEGO
SAN FRANCISCO • SINGAPORE • SYDNEY • TOKYO
Academic Press is an imprint of Elsevier

Academic Press is an imprint of Elsevier
32 Jamestown Road, London NW1 7BY, UK
Radarweg 29, PO Box 211, 1000 AE Amsterdam, The Netherlands
The Boulevard, Langford Lane, Kidlington, Oxford, OX5 1GB, UK
225 Wyman Street, Waltham, MA 02451, USA
525 B Street, Suite 1800, San Diego, CA 92101-4495, USA

First edition 2014

ISBN: 978-0-12-800287-2
ISSN: 0065-2881

CONTRIBUTORS TO VOLUME 67

Caroline B. Albertin
Department of Organismal Biology and Anatomy, University of Chicago, Chicago, Illinois, USA

José P. Andrade
CCMAR-CIMAR L.A., Centro de Ciencias do Mar do Algarve, Campus de Gambelas, Universidade doAlgarve, Faro, Portugal

Alexander I. Arkhipkin
Falkland Islands Fisheries Department, Stanley, Falkland Islands

Diogo D. Barcellos
Laboratório de Ecossistemas Pesqueiros (LabPesq), Universidade de São Paulo, Instituto Oceanográfico Praça do Oceanográfico, Butantã, São Paulo, SP, Brazil

Isobel Bloor
School of Ocean Sciences, Bangor University, Menai Bridge, Anglesey, United Kingdom

Kathrin S.R. Bolstad
Institute for Applied Ecology New Zealand, Auckland University of Technology, Auckland, New Zealand

Heather E. Braid
Institute for Applied Ecology New Zealand, Auckland University of Technology, Auckland, New Zealand

Felipe Briceño
Institute for Marine and Antarctic Studies (IMAS), University of Tasmania, Hobart, Tasmania, Australia

Claudia Caamal-Monsreal
Unidad Multidisciplinaria de Docencia e Investigación, Facultad de Ciencias, Universidad Nacional Autónoma de México, Yucatán, Mexico

Maria E. Chimal
Unidad Multidisciplinaria de Docencia e Investigación, Facultad de Ciencias, Universidad Nacional Autónoma de México, Yucatán, Mexico

Nicola Downey
Department of Ichthyology and Fisheries Science, Rhodes University, Grahamstown, and Bayworld Centre for Research & Education, Constantia, Cape Town, South Africa

Eric Edsinger-Gonzales
Molecular Genetics Unit, Okinawa Institute of Science and Technology, Okinawa, Japan

Aaron B. Evans
Institute for Applied Ecology New Zealand, Auckland University of Technology, Auckland, New Zealand

Rita M. Franco-Santos
Center for Marine Studies, University of Parana (UFPR), Parana, Brazil

Dirk Fuchs
Freie Universität Berlin, Institute of Geological Sciences, Branch Paleontology, Berlin, Germany

Pedro Gallardo
Unidad Multidisciplinaria de Docencia e Investigación, Facultad de Ciencias, Universidad Nacional Autónoma de México, Yucatán, Mexico

Maria Gasalla
Fisheries Ecosystems Laboratory, Oceanographic Institute, University of São Paulo, São Paulo, Brazil

Ian G. Gleadall
International Fisheries Science Unit, Graduate School of Agricultural Sciences, Tohoku University, Sendai, Japan

Michaël Gras
Université de Caen Basse-Normandie, Institut de Biologie Fondamentale et Appliquée Department, UMR BOREA: Biologie des ORganismes et des Ecosystèmes Aquatiques, Esplanade de la paix, CS 14032, and BOREA, UMR CNRS7208, IRD207, UPMC, MNHN, UCBN, Caen, France

Bret Grasse
Monterey Bay Aquarium, Monterey, California, USA

Corey P. Green
Department of Environment and Primary Industries, Fisheries Victoria, Queenscliff, Victoria, Australia

Angel Guerra
Instituto de Investigaciones Marinas (CSIC), Vigo, Spain

Manuel Haimovici
Institute of Oceanography, Federal University of Rio Grande, CEP, Rio Grande, Brazil

Jennifer Hofmeister
Caldwell Laboratory, Department of Integrative Biology, University of California, Berkeley, California, USA

Henk-Jan T. Hoving
GEOMAR, Helmholtz Centre for Ocean Research Kiel, Kiel, Germany

José Iglesias
Oceanographic Center of Vigo. Spanish Institute of Oceanography, Subida a Radio Faro, Pontevedra, Spain

Heather Judkins
Department of Biological Sciences, University of South Florida St. Petersburg, St. Petersburg, Florida, USA

Jesse T. Kelly
Institute for Applied Ecology New Zealand, Auckland University of Technology, Auckland, New Zealand

Hideaki Kidokoro
Japan Sea National Fisheries Research, Institute, Fisheries Research Agency, Suido-cho, Niigata, Japan

Noussithé Koueta
UMR BOREA, MNHN, UPMC, UCBN, CNRS-7028, IRD-207, IBFA Université de Caen Basse-Normandie, Esplanade de la Paix, Caen cedex, France

Vladimir V. Laptikhovsky
CEFAS, Lowestoft, Suffolk, United Kingdom

Charles Le Pabic
UMR BOREA, MNHN, UPMC, UCBN, CNRS-7028, IRD-207, IBFA Université de Caen Basse-Normandie, Esplanade de la Paix, Caen cedex, France

Evgenia Lefkaditou
Helenic Centre for Marine Research, Aghios Kosmas, Hellinikon, Athens, Greece

Marek R. Lipiński
Department of Ichthyology and Fisheries Science, Rhodes University, Grahamstown, South Africa

Silvia A.P. Lourenço
IPMA Div. RP, Departamento do Mar e Recursos marinhos, Lisboa, Portugal

José E.A.R. Marian
Departamento de Zoologia, Instituto de Biociências, Universidade de São Paulo, Sao Paulo, Brazil

Maite Mascaró
Unidad Multidisciplinaria de Docencia e Investigación, Universidad Nacional Autónoma de México, Puerto de Abrigo s/n, Sisal, Yucatán, México

Jennifer Mather
Psychology Department, University of Lethbridge, Lethbridge, Alberta, Canada

Natalie A. Moltschaniwskyj
School of Environmental & Life Sciences, University of Newcastle, Ourimbah, New South Wales, Australia

Ryuta Nakajima
Department of Art and Design, University of Minnesota Duluth, Duluth, Minnesota, USA

Mike Navarro
Integrative Oceanography Division, and Center for Marine Biodiversity and Conservation, Scripps Institution of Oceanography, University of California, San Diego, La Jolla, California, USA

Owen C. Nichols
School for Marine Science and Technology, University of Massachusetts – Dartmouth, Fairhaven, and Center for Coastal Studies, Provincetown, Massachusetts, USA

Cristina Pascual
Unidad Multidisciplinaria de Docencia e Investigación, Facultad de Ciencias, Universidad Nacional Autónoma de México, Yucatán, México

João Pereira
Instituto de Investigação das Pescas e do Mar (IPIMAR), Lisboa, Portugal

Jose Angel A. Perez
Centro de Ciências Tecnológicas da Terra e do Mar Universidade do Vale do Itajaí, Itajaí, Santa Catarina, Brazil

Uwe Piatkowski
GEOMAR, Helmholtz Centre for Ocean Research Kiel, Kiel, Germany

Graham J. Pierce
Oceanlab, University of Aberdeen, Newburgh, Aberdeenshire, United Kingdom, and CESAM & Departamento de Biologia, Universidade de Aveiro, Aveiro, Portugal

Cristina Pita
Oceanlab, University of Aberdeen, Newburgh, Aberdeenshire, United Kingdom, and CESAM & Departamento de Biologia, Universidade de Aveiro, Aveiro, Portugal

Jorge E. Ramos
Institute for Marine and Antarctic Studies, Marine Research Laboratories Taroona, Nubeena Crescent, Taroona, Tasmania, Australia

Amanda Reid
Australian Museum Research Institute, Sydney, New South Wales, Australia

Michael Roberts
Rhodes University, Grahamstown, and Oceans & Coasts Research, Victoria & Alfred Waterfront, Cape Town, South Africa

Jean-Paul Robin
Université de Caen Basse-Normandie, UMR BOREA: Biologie des ORganismes et des Ecosystèmes Aquatiques, Esplanade de la paix, CS 14032, 14032 Caen, France, and UMR BOREA, UMR CNRS7208, IRD207, UPMC, MNHN, UCBN, 14032 Caen, France

Paul G.K. Rodhouse
British Antarctic Survey, Cambridge, United Kingdom

Almendra Rodriguez
El Colegio de la Frontera Sur, Colonia Casasano, Cuautla, Morelos, Mexico

Clyde F.E. Roper
Smithsonian Institution, National Museum of Natural History, Washington, District of Columbia, USA

Carlos Rosas
Unidad Multidisciplinaria de Docencia e Investigación, Facultad de Ciencias, Universidad Nacional Autónoma de México, Yucatán, Mexico

Katina Roumbedakis
Laboratório Sanidade de Organismos Aquáticos. Universidade Federal de Santa Catarina, Florianopolis, Brazil

Kazuhiro Sadayasu
Marine Fisheries Research and Development Center, Fisheries Research Agency, Yokohama, Kanagawa, Japan

Mitsuo Sakai
National Research Institute of Far Seas Fisheries, Shizuoka, Japan

Warwick H.H. Sauer
Department of Ichthyology and Fisheries Science, Rhodes University, Grahamstown, South Africa

Susumu Segawa
Tokyo University of Fisheries, Minato-ku, Tokyo, Japan

Michael Vecchione
NMFS National Systematics Laboratory, National Museum of Natural History, Washington, DC, USA

Erica A.G. Vidal
Center for Marine Studies, University of Parana (UFPR), Parana, Brazil

Roger Villanueva
Institut de Ciències del Mar (CSIC), Barcelona, Spain

James Wood
Mounts Botanical Garden, West Palm Beach, Florida, USA

José C.C. Xavier
Institute of Marine Research, Department of Life Sciences, University of Coimbra, Coimbra, Portugal, and British Antarctic Survey, NERC, Cambridge, United Kingdom

Lou Zeidberg
Department of Ecology and Evolutionary Biology, University of California, Los Angeles, California, USA

CONTENTS

SERIES CONTENTS FOR LAST FIFTEEN YEARS*

Volume 35, 1999.

Creasey, S. S. and Rogers, A. D. Population genetics of bathyal and abyssal organisms. pp. 1–151.

Brey, T. Growth performance and mortality in aquatic macrobenthic invertebrates. pp. 153–223.

Volume 36, 1999.

Shulman, G. E. and Love, R. M. The biochemical ecology of marine fishes. pp. 1–325.

Volume 37, 1999.

His, E., Beiras, R. and Seaman, M. N. L. The assessment of marine pollution—bioassays with bivalve embryos and larvae. pp. 1–178.

Bailey, K. M., Quinn, T. J., Bentzen, P. and Grant, W. S. Population structure and dynamics of walleye pollock, Theragra chalcogramma. pp. 179–255.

Volume 38, 2000.

Blaxter, J. H. S. The enhancement of marine fish stocks. pp. 1–54.

Bergström, B. I. The biology of Pandalus. pp. 55–245.

Volume 39, 2001.

Peterson, C. H. The "Exxon Valdez" oil spill in Alaska: acute indirect and chronic effects on the ecosystem. pp. 1–103.

Johnson, W. S., Stevens, M. and Watling, L. Reproduction and development of marine peracaridans. pp. 105–260.

Rodhouse, P. G., Elvidge, C. D. and Trathan, P. N. Remote sensing of the global light-fishing fleet: an analysis of interactions with oceanography, other fisheries and predators. pp. 261–303.

Volume 40, 2001.

Hemmingsen, W. and MacKenzie, K. The parasite fauna of the Atlantic cod, Gadus morhua L. pp. 1–80.

*The full list of contents for volumes 1–37 can be found in volume 38

Kathiresan, K. and Bingham, B. L. Biology of mangroves and mangrove ecosystems. pp. 81–251.
Zaccone, G., Kapoor, B. G., Fasulo, S. and Ainis, L. Structural, histochemical and functional aspects of the epidermis of fishes. pp. 253–348.

Volume 41, 2001.
Whitfield, M. Interactions between phytoplankton and trace metals in the ocean. pp. 1–128.
Hamel, J.-F., Conand, C., Pawson, D. L. and Mercier, A. The sea cucumber Holothuria scabra (Holothuroidea: Echinodermata): its biology and exploitation as beche-de-Mer. pp. 129–223.

Volume 42, 2002.
Zardus, J. D. Protobranch bivalves. pp. 1–65.
Mikkelsen, P. M. Shelled opisthobranchs. pp. 67–136.
Reynolds, P. D. The Scaphopoda. pp. 137–236.
Harasewych, M. G. Pleurotomarioidean gastropods. pp. 237–294.

Volume 43, 2002.
Rohde, K. Ecology and biogeography of marine parasites. pp. 1–86.
Ramirez Llodra, E. Fecundity and life-history strategies in marine invertebrates. pp. 87–170.
Brierley, A. S. and Thomas, D. N. Ecology of southern ocean pack ice. pp. 171–276.
Hedley, J. D. and Mumby, P. J. Biological and remote sensing perspectives of pigmentation in coral reef organisms. pp. 277–317.

Volume 44, 2003.
Hirst, A. G., Roff, J. C. and Lampitt, R. S. A synthesis of growth rates in epipelagic invertebrate zooplankton. pp. 3–142.
Boletzky, S. von. Biology of early life stages in cephalopod molluscs. pp. 143–203.
Pittman, S. J. and McAlpine, C. A. Movements of marine fish and decapod crustaceans: process, theory and application. pp. 205–294.
Cutts, C. J. Culture of harpacticoid copepods: potential as live feed for rearing marine fish. pp. 295–315.

Volume 45, 2003.
Cumulative Taxonomic and Subject Index.

Volume 46, 2003.

Gooday, A. J. Benthic foraminifera (Protista) as tools in deep-water palaeoceanography: environmental influences on faunal characteristics. pp. 1–90.

Subramoniam, T. and Gunamalai, V. Breeding biology of the intertidal sand crab, Emerita (Decapoda: Anomura). pp. 91–182.

Coles, S. L. and Brown, B. E. Coral bleaching—capacity for acclimatization and adaptation. pp. 183–223.

Dalsgaard J., St. John M., Kattner G., Müller-Navarra D. and Hagen W. Fatty acid trophic markers in the pelagic marine environment. pp. 225–340.

Volume 47, 2004.

Southward, A. J., Langmead, O., Hardman-Mountford, N. J., Aiken, J., Boalch, G. T., Dando, P. R., Genner, M. J., Joint, I., Kendall, M. A., Halliday, N. C., Harris, R. P., Leaper, R., Mieszkowska, N., Pingree, R. D., Richardson, A. J., Sims, D.W., Smith, T., Walne, A. W. and Hawkins, S. J. Long-term oceanographic and ecological research in the western English Channel. pp. 1–105.

Queiroga, H. and Blanton, J. Interactions between behaviour and physical forcing in the control of horizontal transport of decapod crustacean larvae. pp. 107–214.

Braithwaite, R. A. and McEvoy, L. A. Marine biofouling on fish farms and its remediation. pp. 215–252.

Frangoulis, C., Christou, E. D. and Hecq, J. H. Comparison of marine copepod outfluxes: nature, rate, fate and role in the carbon and nitrogen cycles. pp. 253–309.

Volume 48, 2005.

Canfield, D. E., Kristensen, E. and Thamdrup, B. Aquatic Geomicrobiology. pp. 1–599.

Volume 49, 2005.

Bell, J. D., Rothlisberg, P. C., Munro, J. L., Loneragan, N. R., Nash, W. J., Ward, R. D. and Andrew, N. L. Restocking and stock enhancement of marine invertebrate fisheries. pp. 1–358.

Volume 50, 2006.

Lewis, J. B. Biology and ecology of the hydrocoral Millepora on coral reefs. pp. 1–55.

Susanne P. Eriksson, Bodil Hernroth, and Susanne P. Baden. Stress Biology and Immunology in *Nephrops norvegicus*. pp. 149–200.
Adam Powell and Susanne P. Eriksson. Reproduction: Life Cycle, Larvae and Larviculture. pp. 201–246.
Anette Ungfors, Ewen Bell, Magnus L. Johnson, Daniel Cowing, Nicola C. Dobson, Ralf Bublitz, and Jane Sandell. *Nephrops* Fisheries in European Waters. pp. 247–314.

Volume 65, 2013.
Isobel S.M. Bloor, Martin J. Attrill, and Emma L. Jackson. A Review of the Factors Influencing Spawning, Early Life Stage Survival and Recruitment Variability in the Common Cuttlefish (*Sepia officinalis*). pp. 1–66.
Dianna K. Padilla and Monique M. Savedo. A Systematic Review of Phenotypic Plasticity in Marine Invertebrate and Plant Systems. pp. 67–120.
Leif K. Rasmuson. The Biology, Ecology and Fishery of the Dungeness crab, *Cancer magister*. pp. 121–174.

Volume 66, 2013.
Lisa-ann Gershwin, Anthony J. Richardson, Kenneth D. Winkel, Peter J. Fenner, John Lippmann, Russell Hore, Griselda Avila-Soria, David Brewer, Rudy J. Kloser, Andy Steven, and Scott Condie. Biology and Ecology of Irukandji Jellyfish (Cnidaria: Cubozoa). pp. 1–86.
April M. H. Blakeslee, Amy E. Fowler, and Carolyn L. Keogh. Marine Invasions and Parasite Escape: Updates and New Perspectives. pp. 87–170.
Michael P. Russell. Echinoderm Responses to Variation in Salinity. pp. 171–212.
Daniela M. Ceccarelli, A. David McKinnon, Serge Andréfouët, Valerie Allain, Jock Young, Daniel C. Gledhill, Adrian Flynn, Nicholas J. Bax, Robin Beaman, Philippe Borsa, Richard Brinkman, Rodrigo H. Bustamante, Robert Campbell, Mike Cappo, Sophie Cravatte, Stéphanie D'Agata, Catherine M. Dichmont, Piers K. Dunstan, Cécile Dupouy, Graham Edgar, Richard Farman, Miles Furnas, Claire Garrigue, Trevor Hutton, Michel Kulbicki, Yves Letourneur, Dhugal Lindsay, Christophe Menkes, David Mouillot, Valeriano Parravicini, Claude Payri, Bernard Pelletier, Bertrand Richer de Forges, Ken Ridgway, Martine Rodier, Sarah Samadi, David Schoeman, Tim Skewes, Steven Swearer, Laurent Vigliola, Laurent Wantiez, Alan Williams, Ashley Williams, and Anthony J. Richardson. The Coral Sea: Physical Environment, Ecosystem Status and Biodiversity Assets. pp. 213–290.

PREFACE

It is my pleasure to introduce this special volume of *Advances in Marine Biology* devoted to current topics in cephalopod research to a wider readership. Cephalopods—squid, octopuses, cuttlefish, and nautiluses—are highly evolved marine mollusks with diverse life histories and a wide range of habitat preferences. These rapidly growing animals have relatively short life cycles and die after a single reproduction period. Combined with a high production-to-biomass ratio, these opportunistic features earn them a place as important components in the trophic structure of marine ecosystems and as the objects of valuable capture fisheries.

Cephalopods also are noticeably sensitive to environmental variations. This has become of special concern in a time of changing ocean climate and, by itself, warrants research that adds to our knowledge of these attractive invertebrates with impressively large brains, strikingly familiar eyes, and intriguingly complex behaviours.

The idea for this publication was born while undertaking preparations for the Cephalopod International Advisory Council (CIAC) Symposium held in Florianópolis, Brazil in October–November 2012. Every 3 years, CIAC Symposia gather the world's cephalopod researchers in both formal and informal settings designed to encourage the open exchange of ideas and cross-discipline collaboration.

A tradition of CIAC Symposia is our pre-conference workshops. These are led by recognized experts in their field and treat topics of particular interest to attendees. For CIAC 2012, 15 topics initially were proposed and 4 of these, receiving together more than 50% of the votes from the cephalopod scientific community, were selected for presentation.

While preparing the 2012 Symposium, for which I served as Chair, I was struck by the idea that our organization would benefit from introducing the outputs of the workshops to a broader audience. I felt that this best would be achieved by publishing our workshop proceedings in a high-quality volume. This idea grew and eventually became one of the goals of my chairmanship. I thus am very happy that this has been realized with the publication of the present volume that contains the four review papers from the CIAC 2012 workshops.

Each paper reflects the perspective of recognized leaders in their field, together with the contributions of colleagues and students who participated actively in the workshops. In each case, the result is an authoritative review that combines the expertise of multiple 'contributing authors' representing many research groups distributed over five continents.

The first topic is cephalopod culture. Owing to their very high growth rates, short life cycles, and important markets, cephalopods are one of the more attractive candidates for marine aquaculture. This paper summarizes state-of-the-art culture methodologies for four species chosen as biological models. It addresses problems associated with transportation, water quality, nutritional requirements, and digestive physiology. Opportunities for, and present limitations to, commercialization are analyzed and research priorities are outlined.

The next review deals with fisheries management, particularly its effect on cephalopod population dynamics. This paper describes management approaches to cephalopod fisheries and underlines the essential role of forecasting in sustaining profitable harvest of these short-lived species. The need to enhance our understanding of the relationship between the environment and both population dynamics and trophic dynamics at all phases of the life cycle is stressed.

The third contribution summarizes current knowledge of deep-sea cephalopods, for which information naturally is sparse, as these animals remain exceedingly difficult to collect and observe *in vivo*. The authors provide a valuable survey of what thus far is known about their diversity, biogeography, systematics, trophic ecology, and aspects of life cycles. Their reproductive and feeding strategies, in particular, are quite different from those of their much more familiar near-surface 'cousins'. Recommendations are made for filling in the gaps in our knowledge of these mysterious and charismatic cephalopods.

The last paper reviews the morphological, physiological, ecological, and behavioural changes that occur during each of four 'transitions' identified in cephalopod life history: fertilization to release of eggs; egg hatching to paralarvae; paralarvae to sub-adults; and sub-adults to adults. Of broad application to several areas of cephalopod research, it analyzes the implications of each 'transition' and its consequences, particularly at the population level.

These four contributions are testament to the impressive progress that has been made in cephalopod science over the last few decades. Overall, it is my earnest hope that these reviews give readers a comprehensive understanding of the present state of cephalopod biology, ecology, cultivation, and fisheries. At the same time—and of immediate importance—they provide a detailed outline of contemporary challenges in cephalopod science that usefully frame areas of future research.

Finally, on a personal note, it has been a very rewarding journey from the pleasant weekend of discussions during the workshops to completion of this volume. I am very grateful to the coordination, commitment, and understanding of the lead authors during its preparation, and also to all co-authors for their valuable participation and contributions. I extend my special thanks to Paul G. K. Rodhouse for his encouragement throughout this work and for his always thoughtful and positive comments. Many thanks also are owed to José Eduardo A. R. Marian for his enthusiastic support, friendship, and hard work during the organization of the symposium. It gives me great pleasure to thank the referees (anonymous or not) for their valuable and constructive comments: Sigurd von Boletzky, Pedro Domingues, Ángel Guerra, Vladimir Laptikhovsky, Julio Portela, Warwick Sauer, and Elizabeth Shea. I save my deepest gratitude for my husband, José Guilherme Bersano Filho, our children, Lyanna and Adriano, and my parents, Fernando and Thereza (in memorian), for their substantial contribution of unwavering love and support.

Erica A.G. Vidal

CHAPTER ONE

Cephalopod Culture: Current Status of Main Biological Models and Research Priorities

Erica A.G. Vidal[*,1], Roger Villanueva[†], José P. Andrade[‡], Ian G. Gleadall[§], José Iglesias[¶], Noussithé Koueta[||], Carlos Rosas[#], Susumu Segawa[**], Bret Grasse[††], Rita M. Franco-Santos[*], Caroline B. Albertin[‡‡], Claudia Caamal-Monsreal[#], Maria E. Chimal[#], Eric Edsinger-Gonzales[§§], Pedro Gallardo[#], Charles Le Pabic[||], Cristina Pascual[#], Katina Roumbedakis[¶¶], James Wood[||||]

[*]Center for Marine Studies, University of Parana (UFPR), Parana, Brazil
[†]Institut de Ciències del Mar (CSIC), Barcelona, Spain
[‡]CCMAR-CIMAR L.A., Centro de Ciencias do Mar do Algarve, Campus de Gambelas, Universidade doAlgarve, Faro, Portugal
[§]International Fisheries Science Unit, Graduate School of Agricultural Sciences, Tohoku University, Sendai, Japan
[¶]Oceanographic Center of Vigo. Spanish Institute of Oceanography, Subida a Radio Faro, Pontevedra, Spain
[||]UMR BOREA, MNHN, UPMC, UCBN, CNRS-7028, IRD-207, IBFA Université de Caen Basse-Normandie, Esplanade de la Paix, Caen cedex, France
[#]Unidad Multidisciplinaria de Docencia e Investigación, Facultad de Ciencias, Universidad Nacional Autónoma de México, Yucatán, Mexico
[**]Tokyo University of Fisheries, Minato-ku, Tokyo, Japan
[††]Monterey Bay Aquarium, Monterey, California, USA
[‡‡]Department of Organismal Biology and Anatomy, University of Chicago, Chicago, Illinois, USA
[§§]Molecular Genetics Unit, Okinawa Institute of Science and Technology, Okinawa, Japan
[¶¶]Laboratório Sanidade de Organismos Aquáticos. Universidade Federal de Santa Catarina, Florianopolis, Brazil
[||||]Mounts Botanical Garden, West Palm Beach, Florida, USA
[1]Corresponding author: e-mail address: ericavidal2000@yahoo.com.br

Contents

Advances in Marine Biology, Volume 67
ISSN 0065-2881
http://dx.doi.org/10.1016/B978-0-12-800287-2.00001-9

Abstract

A recent revival in using cephalopods as experimental animals has rekindled interest in their biology and life cycles, information with direct applications also in the rapidly growing ornamental aquarium species trade and in commercial aquaculture production for human consumption. Cephalopods have high rates of growth and food conversion, which for aquaculture translates into short culture cycles, high ratios of production to biomass and high cost-effectiveness.

However, at present, only small-scale culture is possible and only for a few species: the cuttlefish *Sepia officinalis*, the loliginid squid *Sepioteuthis lessoniana* and the octopuses *Octopus maya* and *O. vulgaris*. These four species are the focus of this chapter, the aims of which are as follows: (1) to provide an overview of the culture requirements of cephalopods, (2) to highlight the physical and nutritional requirements at each phase of the life cycle regarded as essential for successful full-scale culture and (3) to identify current limitations and the topics on which further research is required.

Knowledge of cephalopod culture methods is advanced, but commercialization is still constrained by the highly selective feeding habits of cephalopods and their requirement for large quantities of high-quality (preferably live) feed, particularly in the early stages of development. Future research should focus on problems related to the consistent production of viable numbers of juveniles, the resolution of which requires a better understanding of nutrition at all phases of the life cycle and better broodstock management, particularly regarding developments in genetic selection, control of reproduction and quality of eggs and offspring.

Keywords: Aquaculture, Broodstock, Cephalopod, Digestive physiology, Embryonic development, Experimental model, Grow-out, Nutrition, *Octopus maya*, *Octopus vulgaris*, Paralarvae, *Sepia officinalis*, *Sepioteuthis lessoniana*, Water quality

1. INTRODUCTION

Cephalopods have been reported as experimental models since long ago in many research areas, particularly in neuroscience and physiology (Gilbert et al., 1990), and ongoing research increasingly relies on the maintenance of healthy animals in the laboratory. However, the history of cephalopod culture is not that old (except, perhaps, for the paper by Grimpe, 1928), with major advances beginning around five decades ago (Choe and Ohshima, 1961, 1963; Itami et al., 1963) and laying the foundations for the current interest in the culture of these molluscs.

A more recent spark for improving cephalopod culture arose mainly from the biomedical research community as their attention became drawn to the highly developed nervous system of cephalopods, along with their advanced vision and other organs as subjects for experimentation (Hanlon, 1987; Koueta and Boucaud-Camou, 1999). However, experimental studies have gone far beyond the initial surge in curiosity, and cephalopods *per se* have been an important means by which crucial research problems have been solved (e.g. Budelmann, 1995; Budelmann and Bleckmann, 1988; Fiorito and Scotto, 1992; Nyholm and McFall-Ngai, 2004; Shashar et al., 1998; Villanueva et al., 1995) and new ideas presented (e.g. models for tissue regeneration and observations on beak attachment as a new direction in improving the integration of prosthetic limbs with the human body; Chiao et al., 2011; Fox et al., 2013; Laschi et al., 2012; Miserez et al., 2008; Rohrbach and Schmidtberg, 2006). These have all contributed to enhancing knowledge of broad areas of cephalopod biology, physiology, life cycles, immunology, and behaviour. Other stimuli for cephalopod culture are the rapidly growing global industry for ornamental cephalopods, mainly octopus and cuttlefish, and the worldwide trend in commercial aquaculture production of food organisms.

As a consequence of this increasing attention on cephalopods, there have been periodic updates during the past few decades on progress in cephalopod culture, which has been reviewed (Boletzky, 1974, 2004; Boletzky and Hanlon, 1983; Hanlon, 1987, 1990; Oestmann et al., 1997; Shvetsova, 1977) and discussed (García García and García García, 2011; Iglesias et al., 2004, 2007; Vidal et al., 2002a; Villanueva et al., 1995; Walsh et al., 2002) and the aquaculture potential of some species evaluated (Lee et al., 1994; Rosas et al., 2013; Sykes et al., 2006b; Vaz-Pires et al., 2004). The publication of a book on cephalopod culture represents a major recent advance, collating the details of husbandry techniques for the most relevant species (Iglesias et al., 2014). From delicate planktonic forms to robust benthic adults, a variety of species and life phases have been maintained, reared, and cultured, albeit with different degrees of success. Most are coastal cephalopods due to the ease of collecting and maintaining them and the consequent availability of much information on their life cycles and bioecology.

Cephalopods have many features considered favourable and attractive for aquaculture. Among those deserving attention are their short life cycles (mostly between 4 and 18 months) and the fact that eggs can be obtained readily since many species, particularly octopus and cuttlefish, will promptly lay eggs when introduced to captivity. The fecundity of some species is quite high and comparable to fish species (Saville, 1987), resulting in the production of large numbers of planktonic paralarvae, while other species show low fecundity but produce large eggs and developmentally advanced benthic hatchlings. These directly developing hatchlings are miniatures of the adults, born with a highly developed nervous system and innate hunting skills. The growth rate of some cephalopods is impressive, ranging across 5–13% of body wet weight (WW) d^{-1} (Lee et al., 1994), and combined with high food conversion rates of 30–50% (Lee, 1994; O'Dor and Wells, 1987) they rate as one of the most attractive groups of marine invertebrates considered suitable for aquaculture. In addition, cephalopods are well appreciated as a high-protein source (75–85% protein dry weight (DW)) and for the fact that 80–85% of their total body weight is edible for human consumption (Lee, 1994). This represents a significant impact on profitability and marketability, particularly when compared with only 40–45% for crustaceans and 40–75% for teleost fishes (Kreuzer, 1984). The implications of these biological features for commercial culture are obvious, as they translate into short culture cycles with high ratios of production to biomass and consequently high cost-effectiveness. However, profitability is also dependent on other factors, such as demand, markets and high trade value. It is therefore important to consider the markets likely to sustain potential commercial ventures involving cephalopod aquaculture.

The current demand depends almost exclusively on wild-caught stocks, but it seems doubtful that sea fisheries can continue to meet this demand. The decline in many finfish stocks has increased fishing pressure on cephalopods, expanding their commercial importance and consequent rapid rise in capture statistics over recent decades (Pierce and Portela, 2014; Rodhouse et al., 2014). Annual global cephalopod landings have increased steadily since 1950 from a total of ~0.5 million t annually to a peak of over 4 million t in 2007, followed by a recent decline to ~3.5 million t in 2010 (FAO, 2011; Pierce and Portela, 2014). Thus, long-term trends seem to indicate that the boom years of cephalopod fisheries may have come to an end because of overfishing. This tendency follows the widespread perception that fisheries production in general is declining worldwide (Hilborn, 2012; Rodhouse et al., 2014; Worm et al., 2009).

Nevertheless, there is a well-established and lucrative market for cephalopods in European and Asian countries. Most European countries are net importers of cephalopods and Spain and Italy are on the top of the importers' list (imports over 5 and 4 million t, respectively). During 1976 and 2009, Japan imported more cephalopods than any other country (~7 million t), attaining the largest total net import–export deficit of any country in the world (Pierce and Portela, 2014). Recent cephalopod trade statistics emphasize the limited availability of octopus in 2011 (Globefish, 2012), and figures for 2012 indicate that Japan will further increase its octopus imports. For squid and cuttlefish, imports are also increasing (Pierce and Portela, 2014), so clearly, there is a growing demand for cephalopods in Europe and in Japan and its export markets. If world cephalopod fisheries continue to show a decline in landings, the increased demand will tend to drive prices up, which certainly will represent the most efficient stimulus for cephalopod aquaculture development as perhaps the only plausible alternative to supply the demand. However, compared to fish, shrimp and other molluscs, cephalopod aquaculture is in its infancy and at present there is no commercial-scale culture.

Production is constrained largely by the highly selective feeding habits of cephalopods, which are carnivores and thus need a protein-based diet with high-quality lipids rich in essential fatty acids, phospholipids and cholesterol. Satisfying their voracious appetites requires large quantities of food, preferably live feed. Owing to their high metabolic and feeding rates (20–50% WW d^{-1}; Hanlon, 1987; Lee, 1994), if their nutritional requirements are not met, they can quickly starve during early development, and from an early age they will resort to cannibalism (see, e.g. Ibáñez and Keyl, 2010). The early developmental stages are particularly vulnerable since they also require live feed of specific size, shape and—as they are visual hunters—mobility, to sustain their exorbitant feeding rates (~80% of WW d^{-1}; Hanlon, 1987; Segawa, 1990). The rearing of hatchlings and production of juveniles on a commercial basis have been severely constrained by these drawbacks, along with insufficient knowledge of their feeding habits and food types, causing high mortality rates during early life in captivity. This underscores the need either to find a reliable live-feed supply for hatchlings or to develop cost-effective artificial feeds. As with other commercial marine aquaculture species, focusing on nutritional improvements is one strategy to shorten the time to achieve commercial production.

Small-scale cephalopod culture is already possible for a few species, such as the bigfin reef squid, *Sepioteuthis lessoniana* Lesson, 1839; the European cuttlefish, *Sepia officinalis* Linnaeus, 1758; the Mexican four-eyed octopus, *Octopus maya* (Voss and Solis, 1966); and (as the ongrowing of subadults in sea cages) the common octopus, *Octopus vulgaris* Cuvier, 1797 (Figure 1.1). All of these selected species except *O. vulgaris* produce large hatchlings without a vulnerable planktonic paralarval phase. They also each

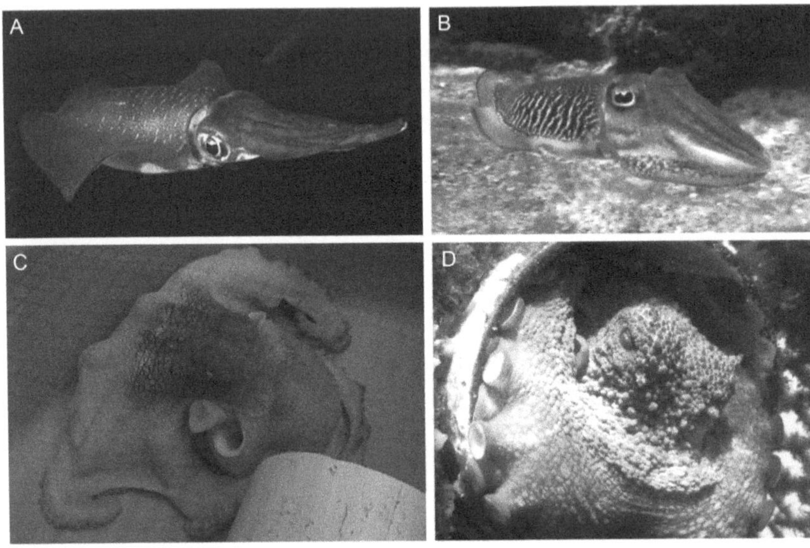

Figure 1.1 The four species chosen as experimental models for this chapter. (A) *Sepioteuthis lessoniana*, (B) *Sepia officinalis* at the aquarium of the Musée Océanographique de Monaco, (C) *Octopus maya* (original image) and (D) *Octopus vulgaris* inside an artificial shelter held for ongrowing in a sea cage in the Canary Islands. *(A) Photo courtesy of J. Forsythe, (B) photo courtesy of P. Bustamante, and (D) photo courtesy of J. Estefanell.*

have a wide range of distribution (with the exception of *O. maya*) and are also important commercial fishing resources, with well-established market channels. As a consequence, they have been used extensively as cephalopod culture models around the world, and comprehensive information is available on their biology and life cycles, which is why they are the focus of this chapter. However, for anyone wishing to experiment in starting up small-scale culture or to design the transition from laboratory-based to commercial-scale production, this essential information is widely dispersed throughout the literature. The last comprehensive review was published more than 30 years ago (Boletzky and Hanlon, 1983), so compiling and interpreting recent advances in culture techniques are necessary steps in taking basic and applied cephalopod research further to overcome its diverse array of challenges. To this end, this chapter results from a workshop held at the 2012 CIAC (Cephalopod International Advisory Council) Symposium on *Cephalopod Culture* and its main goals are to (1) summarize the salient features of requirements for culturing cephalopods (such as water quality, nutrition and disease treatment and prevention), (2) review the current status of information critical for culturing the four species chosen as biological models and (3) clearly identify current limitations and critical research areas.

2. STATUS OF CEPHALOPOD CULTURE

The terminology follows that proposed by Boletzky and Hanlon (1983). *Maintenance* means the holding of wild-caught late juveniles, subadults or adults at the same developmental stage for variable periods, albeit with some growth occurring to a more advanced stage. *Rearing* applies to growing individuals over a certain period of time, for example from eggs to juveniles or adults, but without achieving a second generation. *Culture* (sensu stricto) refers to growing a cephalopod at best from hatchling to the complete life cycle, until egg laying and production of viable hatchings of the first filial captive generation (F_1). Nevertheless, the term culture also may be used in general, referring to maintenance, rearing and culture.

2.1. Transportation

Transportation is a critical bottleneck in the transfer of cephalopods from the wild to long-term captivity. The challenges arise from housing an active aquatic animal in a small environment over an extended period of time with little direct control over conditions. Cephalopods have a high metabolism and heightened sensitivity to deteriorating water quality, and many species will often release ink. Transporting cephalopods and their eggs therefore requires careful thought and preparation to ensure optimum survival (Hanlon, 1990).

However, cephalopods can be successfully shipped if their needs are met. It is important to start with healthy non-senescent specimens, pack them appropriately, minimize shipping time and stress and acclimate the animals appropriately upon arrival. Successful transportation methods fulfil part of the obligations for ethical responsibility and care of cephalopods. This is becoming increasingly important for research, culture and exhibitory purposes regardless of geographic location.

2.1.1 Animal assessment

Visual inspection and selection of healthy animals and eggs are very important because injuries caused by capture and handling eventually lead to bacterial infection and can be lethal (Hulet et al., 1979). Attempting to transport animals of less than perfect condition is likely to be a waste of time and resources. When eggs are collected in the field, they should be half on the way to hatching (around late gastrulation stages), when the embryos are robust and metabolic rate is relatively low, minimizing impact on water quality and therefore reducing the likelihood of mortality. If collected at very early stages, the instability of water parameters during transport may cause abnormalities and death. Eggs at later stages may hatch prematurely during transportation and handling as they also have higher metabolic rates, causing rapid deterioration of water quality (Hanlon, 1990).

Wild-caught cephalopods should be acclimated to captivity in aquaria prior to shipping (Bower et al., 1999; Ikeda et al., 2004). Healthy cephalopods should have a good appetite, be free of skin lesions and exhibit no unusual behaviour. They should not be fed prior to shipping to avoid compromising water quality while confined (Hanlon, 1990; Ikeda et al., 2004). Mortality during transport can be avoided by proper aeration or by supplying pure oxygen.

2.1.2 *Packing*

Only a single juvenile or adult should be packed per bag for shipping. Reasons for separating animals include inking, aggression, cannibalism, stress, reducing the biological load and preventing a single mortality from compromising water quality for multiple animals. The use of PVC tubes or individual mesh bags reduces mortality during transport for octopuses. In contrast, many embryos or egg capsules can be packed per bag, but the total number should be split between two or more bags to increase the chances of success.

There are two commonly practised shipping methods for adult cephalopods, based on the animal size and the specific shipping conditions (shipping time and animal bioload to sea water volume ratio). In the first method, the shipping bag is filled to 30–50% of its volume with sea water and the remaining volume with pure oxygen. The reservoir of oxygen optimizes water quality but allows water turbulence during shipping. This turbulence can cause increased stress and issues with air entrapment in the mantle cavity. To avoid this, smaller cephalopods can be shipped without the oxygen reservoir, but the shipping water should be saturated with oxygen prior to shipping (to around 20 mg L^{-1}), and the shipping bag should be closed tightly. This method solves the air entrapment problem but greatly reduces the available oxygen.

Depending on the geographic destination, water temperature should be maintained as close as possible to the norm for the species. Cephalopods are generally shipped in Styrofoam fish boxes or coolers. Ice packs or heating packs may be required depending on the species (Hanlon, 1990). Shipping cephalopods slightly below their normal optimum temperature will lower their metabolic rate and allow the shipping water to hold more oxygen.

For the transport of large active cephalopods, the use of a sedative should be considered, such as magnesium chloride (6.8 g L^{-1}, approx. 73 mM, as recommended by Bower et al., 1999) or magnesium sulphate (cf. 5 g L^{-1} of the heptahydrate, 20 mM, used with the firefly squid; Gleadall, 2013b).

Bags should be packed tightly or otherwise secured in the shipping box so that they do not move during transport. It is important to double bag the animals in case one of the bags becomes damaged in transit. The shipping box should be clearly labelled: "this side up" and "live animals".

2.1.3 Shipping

Over short distances, such as between field and laboratory, cephalopods can be transported in buckets, live bait wells or coolers. For octopus species, a secure lid is essential or individuals can be placed inside net bags or PVC. For squid and cuttlefish, the largest practical container should be used (Bower et al., 1999; Ikeda et al., 2004). Siphons or turkey basters can be used to remove any ink released. To deal with the release of copious quantities of ink, it is important to carry extra sea water for water changes depending on the duration of transportation. Oxygen can be supplied via an air diffuser from either an air pump or an oxygen bottle.

Temperature fluctuations should be minimized, for example, by keeping onboard containers in the shade when transporting by boat and using the air conditioner when transporting by car. When using a commercial freight company for long-distance shipping, overnight delivery is best. Be sure to obtain a tracking number and follow the package's progress online and, as required, have all necessary permits and documentation prepared ahead of time. For international shipping, it is useful to use a customs broker with live animal experience to avoid potential delays.

2.1.4 Receiving

Care of cephalopods upon arrival depends on the circumstances. A very slow drip or partial water change acclimation is preferable (Ikeda et al., 2004). Water quality parameters (especially pH, dissolved oxygen (DO) and temperature) should be monitored on a regular basis. The length of acclimation depends on the differences (in temperature, pH, DO and salinity) between shipping water and water conditions at the final destination (Hanlon, 1990).

For closed systems, the shipping water should not be added with the animals. However, if the water quality is extremely poor upon receiving animals, it is better to promptly remove the cephalopod and add it to the new system directly. Due to high metabolism, the presenting of food during acclimation is a helpful practice for smaller specimens.

2.2. Water quality requirements

Water quality is of critical importance when keeping cephalopods in laboratories, aquaculture facilities and public aquaria, so careful monitoring is essential for successful culture, particularly for early stages of development (Boletzky and Hanlon, 1983). Without careful monitoring, water quality can deteriorate quickly and compromise animal health. It should be noted, however, that different developmental stages of the life cycle have different tolerance limits and this should always be pondered with care (Kinne, 1971).

The animals, tanks, equipment (to maintain specific conditions), water flow and aeration should be checked first thing in the morning and at the end of the afternoon. Water quality parameters should be monitored daily and should be kept as similar as possible to natural sea water (see, e.g. Forsythe et al., 1991; Koueta and Boucaud-Camou, 1999; Oestmann et al., 1997). Trace elements, such as strontium and calcium, should be kept close to natural sea water values (Hanlon et al., 1989). Irrespective of the sea water system used, water flows should be strong enough to maintain water quality and sustain the best reproduction results for broodstock but very gentle for tanks holding paralarvae (Vidal et al., 2002a). Water and tanks should be kept clean from leftovers, faeces and other debris by removing them by water siphoning. Also, to avoid problems with cleaning, sand substrates (e.g. for *Sepia* spp.) should not be used (Forsythe et al., 1994).

2.2.1 Temperature

Cephalopods are poikilothermic organisms, so their metabolism accelerates as temperatures rise. Temperature is the most influential water parameter in altering the life cycle of cephalopods in captivity (Forsythe et al., 2001) because it directly influences egg development, yolk absorption rate, growth rate, feeding rate and lifespan (Domingues et al., 2001a, 2002; Forsythe et al., 2002; Grigoriou and Richardson, 2004, 2008; Mangold and Boletzky, 1973; Vidal et al., 2002b). It is therefore open to manipulation to obtain the objectives required during culture. For instance, if the goal is to culture cephalopods for human consumption, the temperature can be elevated within the optimal temperature range of each species, maximizing growth over a shorter time period to boost production. However, if the goal is to maximize the lifespan of a certain species for public display or laboratory experiments, a cooler water temperature will promote longevity (Forsythe et al., 2001).

The optimal temperature range is generally species-specific and stage-specific and is the range within which individuals can maximize their growth, fitness and fecundity (Noyola et al., 2013b), so it is also an important consideration for the success of the grow-out period and sustained cephalopod culture (Delgado et al., 2011). A basic rule of thumb is that survival is optimized if the temperature (as well as other culture parameters) is close to that of the natural environment at the capture site (Boyle, 1991). If widely distributed, the temperature tolerance range will therefore depend on the natural geographic location of a given population. This is well illustrated by *O. vulgaris*, which has at least two genetically divergent populations in the Mediterranean (Fadhlaoui-Zid et al., 2012) and four off the Brazilian coast (Moreira et al., 2011). For populations in Europe, the recommended temperature is between 10 and 20 °C (Vaz-Pires et al., 2004). Growth increases at higher temperatures within this range but death ensues above 23 °C (Aguado Giménez and García García, 2002; García García et al., 2009). However, in the Caribbean tropics, the populations of *O. vulgaris* in Bonaire are reported to thrive at sustained temperatures as high as 28 °C (Putnam and Edmunds, 2011).

System temperature not only influences juveniles and adults but also greatly affects cephalopod embryonic development (Boletzky, 1983, 1987; Boletzky and Hanlon, 1983; Caverivière et al., 1999; Choe, 1966a): warmer water temperatures are associated with shorter incubation periods and vice versa. *Sepia officinalis* embryos should be incubated at 18–25 °C, with incubation time possibly doubling at 15 °C and embryos not developing at temperatures below 12 °C (Sykes et al., 2006b). High temperatures will decrease incubation time and increase the food requirements of paralarvae to support their high metabolic demand (Vidal and Boletzky, 2014). Thus, rearing temperature can be used to extend or reduce developmental time and overall lifespan (Forsythe and Hanlon, 1988; Villanueva, 1995).

2.2.2 Salinity

The ambient salinity of most tropical seas is around 34–37 psu, although coastal (estuarine) waters can range from 5 to 30 psu (Moe, 2009). Cephalopods are stenohaline and salinities outside the range of 27–37 psu can be fatal, so salinity is a limiting factor in the distribution of most species (e.g., Vaz-Pires et al., 2004), with only two known exceptions: the squid *Lolliguncula brevis*, an estuarine species tolerating salinities as low as 16 psu (Hendrix et al., 1981), and the sepiid *Sepiella inermis*. Therefore, for the majority of species, the potential for freshwater influx (e.g. from nearby rivers or subsurface seeps or in regions subject to heavy rain) needs to be considered when planning an open culture or maintenance system (Vaz-Pires et al., 2004). In culture, research indicates a working salinity range of 27–35 psu for *O. maya*, *O. vulgaris*, *S. officinalis* and *S. lessoniana* (Boletzky, 1983; Boletzky and Hanlon, 1983; Farías et al., 2009; Nabhitabhata et al., 2005b). Maintaining the salinity above 30 psu is recommended (Berger, 2010; Chapela et al., 2006), along with the slow addition of deionized (DI) or reverse osmosis (RO) water if salinity rises above 35 psu (Walsh et al., 2002).

2.2.3 pH

Like many animals, cephalopods are sensitive to pH, because of the impact on aerobic performance (see, e.g. Pörtner and Zielinski, 1998), and should be kept within the range of open sea water: between 7.8 and 8.2 (Boletzky and Hanlon, 1983; Boyle, 1991). Hanlon (1990) pointed out that a pH above 7.6, as well as high levels of dissolved oxygen, is necessary to accommodate the high metabolic demands of constant swimming in teuthoid squid. Eggs and paralarvae are extremely sensitive to pH changes. Lacoue-Labarthe et al. (2011) studied the combined effects of low pH and high CO_2 partial pressure on the metabolism of embryos and paralarvae of *Loligo vulgaris* and observed changes in the bioaccumulation of metals. The addition of sodium bicarbonate can be used to adjust low-pH conditions (Boletzky and Hanlon, 1983), but this should be done very slowly as rapid changes in pH can be detrimental to animal health. Elevated pH can be reduced by slowly adding sea water or DI/RO freshwater to the system (while carefully monitoring salinity) or by increasing tank aeration (Moe, 2009). However, the best solution for conditions of low pH is to gradually replace tired water with fresh sea water.

2.2.4 Dissolved oxygen

Oxygen is the most important of all the dissolved gases for water quality because it is essential for respiration, oxidizing the nutrients that release the energy necessary for locomotion, reproduction, feeding and other vital activities (Cerezo Valverde and García García, 2005). Oxygen consumption can be affected by metabolism, ambient temperature and body weight.

Higher temperatures induce higher oxygen consumption (Wells et al., 1988) and decreased oxygen solubility (Weiss, 1970). Smaller animals have a relatively higher mass-specific oxygen consumption than larger animals, and both feeding and movement lead to large increases in oxygen consumption (Cerezo Valverde and García García, 2004; Grigoriou and Richardson, 2009; Melzner et al., 2007; Wells et al., 1983).

Parra et al. (2000) evaluated the respiration rates of *O. vulgaris* late embryos and hatchlings at 20 °C and observed that oxygen consumption rates increase threefold at hatching, because of the high energy demands of jet propulsion. They also observed that the oxygen consumption of a medium-sized egg mass is approximately twice that of the female brooding them. In general, then, DO levels for most cephalopods should always be near saturation, although benthic octopods are more tolerant to lower oxygen levels (Boletzky and Hanlon, 1983).

2.2.5 Nitrogenous wastes

Like most aquatic invertebrates, cephalopods are ammonotelic, releasing their nitrogen end products mainly as ammonia (Boucher-Rodoni and Mangold, 1985; Katsanevakis et al., 2005). The harmful inorganic nitrogenous compounds are unionized ammonia (NH_3), nitrite (NO_2^-) and nitrate (NO_3^-). Total ammonia nitrogen (TAN) includes both the ionized form, ammonium (NH_4^+), and the unionized form, which occur in an equilibrium determined by water temperature, salinity, pH and pressure. Ammonia management deserves special attention in closed systems as its unionized form is considerably toxic even at very low concentrations. The levels of TAN, nitrite and nitrate must be monitored carefully to avoid reaching lethal levels (Boletzky and Hanlon, 1983). Numerous publications suggest the importance of keeping unionized ammonia $<0.1\ mg\ L^{-1}$, nitrate $<20.0\ mg\ L^{-1}$ and nitrite $<0.1\ mg\ L^{-1}$ in cephalopod culture (Boletzky and Hanlon, 1983; Vidal and Boletzky, 2014). Segawa and Hanlon (1988) observed that, except at hatching, oxygen consumption and ammonia excretion rates increase linearly with increasing body weight. This emphasizes the importance of carefully monitoring captive bioload to tank volume ratio as cephalopods grow in captivity. Increase in and breakdown of nitrogenous compounds can also decrease pH, causing additional animal stress and health concerns (Moe, 1993). Removing harmful nitrogenous wastes from the sea water is critical in maintaining cephalopod culture. This is the main function of biological filtering, but if levels still rise above the recommended minima, replacement of the sea water is the only option.

2.2.6 Organic wastes

All animals produce various organic wastes that must be removed, primarily because of the decomposition of the organic wastes of digestion, which can quickly cause water quality to deteriorate by encouraging microbial growth and add stress to captive animals. Organic matter should not represent a problem if it is quickly removed from the system. Solid particulate organic matter sinks and can be removed efficiently by siphoning. Suspended particulate matter and excreted waste can be physically removed by mechanical filtration using sand, gravel, floss, filter pads or filter bags, helping to reduce variations in pH and also levels of nitrogenous compounds (Vidal et al., 2002a). The type of mechanical filtration should be selected based on specific culture needs. However, protein skimmers and chemical filtration such as activated carbon should be considered to effectively reduce dissolved and particulate organic matter. Protein skimmers are also very effective for ink removal. Filtration to remove suspended particles should occur before sterilization, as particulate organic matter is the main substrate for bacteria in recirculation systems (Appleford et al., 2003).

2.2.7 Disinfection and sterilization

The delicate skin of cephalopods is characterized by a single layer of columnar epithelial cells. Substantial damage caused by capture, transport, handling and constant contact with the tank environment (bumping into walls and abrasion of skin) can produce small wounds in the epidermis followed by bleeding. Bacteria in sea water (see Section 2.3.1) can enter these wounds leading to infection and death (Hanlon, 1990). Thus, disinfection and sterilization are very important to efficiently remove potentially harmful organisms such as bacteria, microalgae, fungi, protozoans and viruses. Such procedures have proven effective for maintaining healthy cephalopods from paralarvae through to adults (Hanlon, 1990; Vidal et al., 2002a).

Sterilization can be accomplished by either UV or ozone. Ozone sterilization requires careful monitoring because ozone leakage into the culture system can harm and kill cephalopods, particularly eggs and paralarvae. UV sterilizers are usually safe, as long as they are securely housed to protect both animals and maintenance staff from the harmful effects of UV light on the retina and skin.

Additionally, careful cleanliness protocols are advisable. Examples include, in addition to daily siphoning, weekly cleaning of airlifts, air stones and water outlet filters, using a scrubber under running tap water. In order to assure proper disinfection, the cleaning process must be followed by immersion of these circuit components for 24 h in Atlantol 914 (Atlantol, Belgium), rinsing under running tap water, and immersion for 15 min in Virkon S (DuPont Animal Health Solutions, Europe). Finally, all components should be abundantly rinsed under running tap water. This cleaning procedure must be applied to nets and all other materials used in animal handling and will contribute to decrease contamination rates, pathologies and the risk of spreading disease.

2.2.8 Avoidance of toxic fixtures

Special care must be taken to avoid contact between cephalopods and toxic materials (e.g. PVC glue and silicone for use in sea water) when assembling tanks. Toxic materials include many plastics and paints, particularly if they give off an odour; and inorganic metals such as copper plumbing, well known to be toxic to cephalopods (Establier and Pascual, 1983; Hanlon and Forsythe, 1985; Paulij et al., 1990a). As a precaution, prior to introducing cephalopods, systems should be rinsed thoroughly with tap water for at least 24 h and then with running sea water.

2.3. Diseases and parasites

Most pathogenic agents described in the literature are gram-negative *Vibrio* bacteria, fungi, parasites and viruses, and they will be briefly presented here.

2.3.1 Bacterial infection

Cephalopods are susceptible to bacterial infections of the skin, as a result of secondary infection of wounds by opportunistic pathogens (Castellanos-Martínez and Gestal, 2013), particularly *Vibrio* spp., which are primarily located on the epidermis and produce ulcers. In severe cases, the circulatory system and reproductive organs can be affected (Sangster and Smolowitz, 2003), leading to death within a few days. Cases of *Vibrio* spp. infections have been described in *Loligo pealei*, *O. vulgaris*, *O. joubini*, *O. briareus*, *O. maya*, *L. brevis*, *S. officinalis*, *Sepia pharaonis* and *S. apama* (Cruz et al., 2008; Farto et al., 2003; Ford et al., 1986; Gamboa, 2011; Hanlon et al., 1984; Harms et al., 2006; Leibovitz et al., 1977; Reimschuessel et al., 1990; Sangster and Smolowitz, 2003; Scimeca, 2012). This bacterium is common in coastal waters and has been found in higher densities associated with tank walls in cephalopod culture (Elston and Wood, 1983; Sangster and Smolowitz, 2003) compared to natural conditions (Ford et al., 1986). This highlights the possibility that infection is indirect and occurs secondary to wounding, especially if injury is caused by collisions or abrasion in a captive environment (Harms et al., 2006; Hulet et al., 1979; Sangster and Smolowitz, 2003; Sherrill et al., 2000).

Enzymatic analyses of *S. officinalis* tissues indicate the presence of strong immune activities in the skin (C. Le Pabic, personal observation). Therefore, reducing stress and skin damage is paramount when keeping cephalopods in long-term captive conditions (Harms et al., 2006; Sangster and Smolowitz, 2003). This can be achieved by the use of specific holding techniques, such as circular tanks separated from water conditioning facilities to prevent friction and resulting epidermal damage (Hanlon, 1990) and maintain optimal water quality to suppress bacterial outbreak.

2.3.2 Fungal infection

Reports of fungal infections are rare in cephalopods and treatments have not been evaluated (Harms et al., 2006). *Cladosporium* sp. has been reported in an unidentified octopus but was not described in detail (Scimeca, 2012). It was also reported in an adult female *S. officinalis*, with local bacterial infection associated with subsequent ulceration (Harms et al., 2006). These infections appear on the epidermis as described for *Vibrio* spp. Similar to bacterial infections, most fungal infections in cephalopods are secondary, as a result of trauma or a compromised immune system (Harms et al., 2006).

2.3.3 Macroscopic parasites

Cephalopods play an important role in parasite transmission, since they can be primary hosts for protozoans, dicyemids and crustaceans and can also serve as a reservoir of secondary or tertiary intermediate hosts for digeneans, cestodes, acanthocephalans and nematodes (González et al., 2003; Hanlon and Forsythe, 1990; Hochberg, 1990; Overstreet and Hochberg, 1975; Pascual et al., 1996). Most do not seem to cause major problems in cephalopods, but under culture conditions, they could proliferate and cause long-term health issues, and the effects of parasites on growth, reproduction and survival of cephalopods are yet to be characterized (Pascual et al., 1996).

The most frequent parasites of cephalopods belong to the coccidian *Aggregata* (Apicomplexa: Aggregatidae), commonly found in the digestive tract (Dobell, 1925; Gestal, 2000; Hochberg, 1990; Pascual et al., 1996). Severe infections may extend to the connective and epithelial tissues of the mantle and gills (Mladineo and Bočina, 2007). To date, a total of 10 species have been described infecting a wide range of cuttlefish, squid and octopus species (Gestal et al., 2010). *Agreggata* spp. are intracellular parasites, transmitted by ingestion of decapod crustaceans, which are the intermediate hosts (Gestal et al., 2007). These protozoans are a dominant epidemiological agent in wild and cultured cephalopods, causing a disease called coccidiosis (Gestal, 2000). Histopathological damage has been observed in affected tissues (Gestal et al., 2002a), and coccidiosis can negatively affect gastrointestinal function through the malfunction and/or reduction of enzymes required for efficient absorption, a phenomenon known as malabsorption syndrome (Gestal et al., 2002b). Although coccidiosis is not a leading cause of death, it is likely that malabsorption syndrome impairs the development and growth of octopuses (Gestal et al., 2002b).

2.3.4 Viral infection

The potential pathogenic effect of viruses in aquaculture is a new area of research. Recent studies have been conducted on *O. maya* to determine the presence of the white spot syndrome virus (WSSV) that adversely affects the shrimp aquaculture industry. Artisan fisheries commonly use crustaceans as bait for *O. maya*, raising the possibility that *O. maya* could potentially act as a vector for this virus. Although no infection was observed in the wild, laboratory experiments have demonstrated direct transfers of WSSV from *O. maya* to *Litopenaeus vannamei* and vice versa (Rodríguez-Canul et al., 2012). *Octopus maya* hosts WSSV in its gills and digestive gland, causing no apparent symptoms, and may, therefore, be a paratenic host for this viral agent (Rodríguez-Canul et al., 2012). Additional research is necessary to determine how WSSV could impact the long-term culture of *O. maya*.

2.3.5 Treatment of diseases

There have been few studies of treatments for cephalopod pathogens. The available information was recently summarized by Scimeca (2012), identifying six antibiotics and one antiprotozoal treatment. Only one study has described the pharmacokinetic parameters and intravenous, oral or bath administration of the fluoroquinolone antimicrobial enrofloxacin on *S. officinalis* (Gore et al., 2005). Of particular interest is a potentially practical and efficient use of enrofloxacin by oral administration (10 mg/kg in live shrimp). Enrofloxacin appears to have a very fast elimination rate and a very low distribution in cuttlefish tissues compared to fish, birds, reptiles and mammals, emphasizing the fact that pharmacokinetic studies in cephalopods are in their infancy.

2.4. Nutritional requirements and digestive physiology

The characteristic fast growth rate of cephalopods sets high requirements for food ingestion, digestion and assimilation. A number of factors influencing cephalopod growth have been identified (Forsythe and Van Heukelem, 1987; Koueta and Boucaud-Camou, 2003; Koueta et al., 2002), but the determinant factor for cephalopods raised by aquaculture is the quality and efficiency of nutrition and digestion.

It is well known that cephalopods have high requirements for protein and relatively low requirements for high-quality lipids (Miliou et al., 2005; Rosas et al., 2013): the cephalopod diet must consist of over 60% protein and ~4% lipids. Studies on *S. officinalis* and *O. maya* have demonstrated that it is possible to feed cephalopods with native protein (as opposed to that processed by, e.g. heating), frozen or fresh (Domingues, 1999; Domingues et al., 2001b, 2002, 2004; García-Garrido et al., 2010; Petza et al., 2006; Quintana et al., 2011; Rosas et al., 2008, 2013). Experiments with these two species have demonstrated that their digestive physiology is adapted to process native, which is promptly absorbed by the digestive gland (Boucaud-Camou and Boucher-Rodoni, 1983; Boucaud-Camou et al., 1985; Boucher-Rodoni et al., 1987; Koueta et al., 2000, 2006; Martínez et al., 2012).

The last review on cephalopod digestion was presented by Boucaud-Camou and Boucher-Rodoni (1983) more than 30 years ago. Budelmann et al. (1997) provided details of the histomorphologic and functional characteristics of the digestive tract, demonstrating its complexity. Further research has clarified that the digestion process starts externally at the prey, where salivary enzymes (mostly chymotrypsin) are injected after perforation by the beaks or salivary papilla (Messenger and Young, 1999). Although the exact biochemical mechanisms in different species are not fully known, it is considered that predigestion is probably limited to loosening muscle attachments (Boucaud-Camou and Boucher-Rodoni, 1983). Partially digested food is then ingested and enters the crop, in octopus, or goes to the stomach, in cuttlefish and squid, where digestive enzymes from the digestive gland initiate digestion. Enzyme-bound soluble nutrients pass from the crop to the stomach in octopus or directly to the stomach in cuttlefish and squid, where fibrillar proteins and other macromolecules are degraded until a liquid chyme is formed. The chyme is then separated by the caecum to be transported to the digestive gland or to form faecal pellets (Boucher-Rodoni et al., 1987; Budelmann et al., 1997). Once in the digestive gland, nutrients are dissolved and absorbed by pinocytosis in the digestive gland cells, where intracellular digestion occurs. This process can take from 4 to 8 h, depending on the size of the animal and temperature (Boucher-Rodoni et al., 1987; Martínez et al., 2011, 2012; O'Dor et al., 1984). Once digestion ends, the digestive gland cells are renewed in preparation for a new digestion process.

The activity of digestive enzymes in *O. vulgaris* was reviewed by Boucaud-Camou and Boucher-Rodoni (1983). Different proteases were identified, including alkaline proteinases (such as chymotrypsin and trypsin, mainly from the anterior salivary glands) and cathepsin-like enzymes (from the gastric juice). Many questions were raised from those studies, including whether or not acidic proteases were present in the gastric juice or if different species of cephalopods would have different digestive enzymes. Most information on digestive enzymes in cephalopods has been obtained from studies on *S. officinalis*, *O. vulgaris* and *O. maya* and is presented in more detail in the succeeding text.

2.4.1 *Sepia officinalis*

There have been many studies on the cuttlefish digestive system enzymes, their biochemical localization, their characterization and the effect of diet quality on both digestive gland cells and digestive physiology. The structure and function of the digestive system of cuttlefish are directly related to diet (Mangold and Bidder, 1989). Many enzymatic activities have been allocated in the cuttlefish digestive system, and the main activities were found to be nonspecific proteolytic enzymes, mostly alkaline and acid phosphatases, trypsin, chymotrypsin and cathepsins. Acid phosphatase and cathepsins are found in lysosomes and have an important role in intracellular digestion (Boucaud-Camou et al., 1985; Perrin et al., 2004). Trypsin and chymotrypsin are associated with extracellular proteolytic activity in adults, but the activity of cathepsins reflects total acid proteolytic activity in juveniles. Perrin (2004) traced the digestive system organs in which these main enzymes are present, finding that trypsin activity is restricted mainly to the digestive gland and chymotrypsin activity to the posterior salivary gland. Alkaline and acid phosphatase activities occur both in the digestive gland and in the digestive gland appendages, while cathepsin B and E activities are found only in the digestive gland (Le Bihan et al., 2004; Perrin et al., 2004).

Among the enzymes acting extracellularly, acid phosphatase activity varies little but diet has an effect on chymotrypsin and trypsin secretion:

Chymotrypsin activity is higher from 30 min to 1 h after feeding, while trypsin activity increases after 4–8 h (Perrin, 2004). It is not yet known whether alkaline phosphatase and cathepsins act only intracellularly or also extracellularly in adult cuttlefish.

Extracellular enzyme activity depends on diet quality (Perrin et al., 2004). Frozen prey items, for example, induce a decrease in extracellular enzyme activity (trypsin and chymotrypsin) and an increase in intracellular enzyme activities (acid phosphatase), even when frozen prey is previously enriched with polyunsaturated fatty acids (PUFAs). These results also indicate that there is an adjustment of the secretion depending on the enzymes required (extracellular enzymes).The effects of heavy metals on the digestive physiology of cuttlefish have been thoroughly examined by Le Bihan et al. (2004), who found that the digestive gland is the main organ of detoxification. Cathepsins were the intracellular lysosomal proteases implicated in lysosomal digestion and detoxification.

2.4.2 Octopus maya

This species readily accepts both fresh and frozen sources of protein, especially if the source is crustaceans. As for other cephalopods, external digestion in *O. maya* is acidic with maximum enzyme activity in the gastric juice at pH 6 fluctuating mainly between 5.2 and 6 during crab digestion. Cathepsins B, D, L and H are present in the gastric juice, confirming that an acidic medium is characteristic of this species, too.

This acidic medium has several consequences for the formulation of diets. For example, any cooking involved in preparing feed results in carbonylation of proteins, which eliminates the active sites for cysteine proteases, thus affecting protein digestibility (Santé-Lhoutellier et al., 2008). This probably explains why feeds for octopus based on fish or other marine meals in recent years have shown poor digestibility and therefore disappointing growth rates (Table 1.1; Rosas et al., 2013). It is now evident that clams (fresh or processed) and diets with fish flour are not well digested by octopus, resulting in negative growth rates. New experiments conducted with this information in mind have obtained higher growth rates for *O. maya* early juveniles when either crustaceans (mainly fresh, frozen or lyophilized crab bound with gelatin), squids or a mix of both is used as feed (Figure 1.2).

Recent studies have shown that a diet elaborated with freeze-dried crab and squid meat and a mix of vitamins and minerals all bound gelatin produced higher hatchling growth rates than those obtained with freeze-dried crab or squid diets, the former also yielding 100% survival rates (Martínez et al., 2014). Previous results of a feed made with native protein from

Table 1.1 Effects of different types of feed on the growth of Octopus maya

	Octopus initial weight (g WW)	Temperature (°C)	AGR (g d^{-1})	DGC (% d^{-1})	References
Crustaceans					
Artemia salina adults	0.13	25	0.009	4.8	Baeza-Rojano et al. (2013)
Crab (*Callinectes sapidus*)	316.4±9.8	25	10.24	3.71	Aguila et al. (2007)
Crab (*C. sapidus*)	0.23	27	0.007	2.05	Briceño et al. (2010)
Crab (*C. sapidus*)	0.17	27	0.004	1.92	Briceño et al. (2010)
Crab (*C. sapidus*)	0.12	27	0.009	2.07	Briceño et al. (2010)
Crab (*C. sapidus*)	472	28	24	2.7	Domingues et al. (2007)
Crab (*C. sapidus*)	0.71	25	0.015	1.72	Rosas et al. (2008)
Crab (*C. sapidus*)	458	28	11.81	1.07	Rosas et al. (2011)
Crab (*C. sapidus*)	0.13	28	0.009	4.46	Rosas et al. (2013)
Different crustacean species	0.1	25	12.60	4.07	Van Heukelem (1977)
Freshwater gammarid (*Hyale azteca*)	0.11	25	0.009	5.4	Baeza-Rojano et al. (2013)
Marine gammarid (*Hyale media*)	0.13	25	0.02	8.4	Baeza-Rojano et al. (2013)
Crustaceans mix					
Crab (*C. sapidus*), 84%; fish heads, 16% (*Lutjanus* spp.)	542	28	17.8	2.4	Domingues et al. (2012)

Continued

Table 1.1 Effects of different types of feed on the growth of Octopus maya—cont'd

	Octopus initial weight (g WW)	Temperature (°C)	AGR (g d^{-1})	DGC (% d^{-1})	References
Crab (*C. sapidus*), 84%; fish heads, 16% (*Lutjanus* spp.)	493	26	11.5	1.8	Domingues et al. (2012)
Crab (*C. sapidus*), 84%; fish heads, 16% (*Lutjanus* spp.)	321	24	10.3	2.2	Domingues et al. (2012)
Crab paste bound with gelatin *ad libitum*	0.50	27	0.17	10	Quintana et al. (2011)
Crab paste bound with gelatin ration: 30% WW	0.50	27	0.16	9.7	Quintana et al. (2011)
Crab paste bound with gelatin ration: 10% WW	0.40	27	0.12	9.7	Quintana et al. (2011)
Crab (*C. sapidus*) fresh meat bound with gelatin	0.61	28	0.012	1.43	Rosas et al. (2013)
Crab fresh meat bound with gelatin	0.73	25	0.02	1.9	Rosas et al. (2008)
Lyophilized crab (*C. sapidus*) meat bound with gelatin	0.77	28	0.014	1.55	Rosas et al. (2013)
Lyophilized crab (*C. sapidus*) meat bound with gelatin	0.68	28	0.011	1.71	Rosas et al. (2013)
Lyophilized crab (*C. sapidus*) and lyophilized squid (*Dosidicus gigas*) bound with gelatin	0.76	28	0.0095	1.12	Rosas et al. (2013)

Table 1.1 Effects of different types of feed on the growth of Octopus maya—cont'd

	Octopus initial weight (g WW)	Temperature (°C)	AGR (g d^{-1})	DGC (% d^{-1})	References
Squid					
CPSP70 lyophilized squid (*D. gigas*) bound with gelatin	0.75	28	0.002	0.26	Rosas et al. (2013)
Lyophilized squid (*D. gigas*) bound with gelatin	0.57	28	0.007	0.7	Rosas et al. (2013)
Clam					
Clam (Lamellibranchia) meal bound with gelatin	0.60	28	−0.025	−5.2	Rosas et al. (2013)
Clam lyophilized (Lamellibranchia) bound with gelatin	0.65	28	−0.02	−3.63	Rosas et al. (2013)
Pelletized diets					
CPSP70 lyophilized crab (*C. sapidus*) bound with gelatin	0.73	28	−0.0025	−0.35	Rosas et al. (2013)
CPSP70 bound with gelatin	0.69	28	−0.013	−2.16	Rosas et al. (2013)
Chilean prime fish meal bound with gelatin	0.63	28	−0.009	−1.88	Rosas et al. (2013)
Pelletized diet with fish hydrolysed[a] CPSP: 0%	316.4 ± 9.8	25	−0.69	−0.37	Aguila et al. (2007)
Pelletized diet with fish hydrolysed[a] CPSP: 5%	316.4 ± 9.8	25	0.45	0.40	Aguila et al. (2007)
Pelletized diet with fish hydrolysed[a] CPSP: 10%	316.4 ± 9.8	25	0.76	0.66	Aguila et al. (2007)

Continued

Table 1.1 Effects of different types of feed on the growth of Octopus maya—cont'd

	Octopus initial weight (g WW)	Temperature (°C)	AGR (g d^{-1})	DGC (% d^{-1})	References
Pelletized diet with fish hydrolysed[a] CPSP: 15%	316.4±9.8	25	1.31	0.86	Aguila et al. (2007)
Pelletized diet with fish hydrolysed[a] CPSP: 20%	316.4±9.8	25	1.1	0.70	Aguila et al. (2007)
Pelletized diet with 40% protein level[b]	446	28	−1.76	−0.43	Rosas et al. (2010)
Pelletized diet with different protein level[b] 60%	458	28	0.37	0.08	Rosas et al. (2011)
Pelletized diet with 8% CPSP level[c]	0.14	28	−0.001	−2.17	Rosas et al. (2013)
Pelletized diet with 15% CPSP level[c]	0.14	28	−0.001	−1.81	Rosas et al. (2013)
Pelletized diet[c]	0.13	28	−0.002	−2.49	Rosas et al. (2013)
Pelletized shrimp improved diet[d]	486	28	−1.3	−0.30	Domingues et al. (2007)
Processed crab meat					
Crab (fresh) bound with alginate	0.71	25	−0.02	−2.5	Rosas et al. (2008)
Crab (cooked) meat (*C. sapidus*) bound with gelatin	0.94	28	−0.05	−11.7	Rosas et al. (2013)

[a]Menhaden fish meal, CPSP70 (Sopropeche, France), fish hydrolyzate (BIOCP), profine, squid meal, milk serum, cornstarch, soybean lecithin, vitamin mix (Dibaq), vitamin C (Roche), filler, alginate–Na.
[b]Menhaden fish meal, CPSP70 (Sopropeche, France), fish hydrolyzate (BIOCP), profine, squid meal, dried whey, soybean lecithin, fish oil, wheat starch, vitamin mix (Dibaq), vitamin C (Roche), filler, alginate–Na.
[c]Chilean prime fish meal, CPSP70 (Sopropeche, France), clam meal (Lamellibranchia), lyophilized squid (*D. gigas*), fish oil, soybean lecithin, vitamin mix + Stay-C (DSM-NP), mineral mix (DSM-NP).
[d]Shrimp diet (Purina, 51%), CPSP90 (Sopropeche, France), fresh squid (*Lolliguncula brevis*), soybean lecithin, Rovimix Stay-C (Roche), mineral and vitamin mix (Dibaq).
Absolute growth rate (AGR, g d^{-1}) and daily growth coefficient (DGC, % d^{-1}) of octopus with different types of feeds. For groups' description, see text. WW, wet weight.

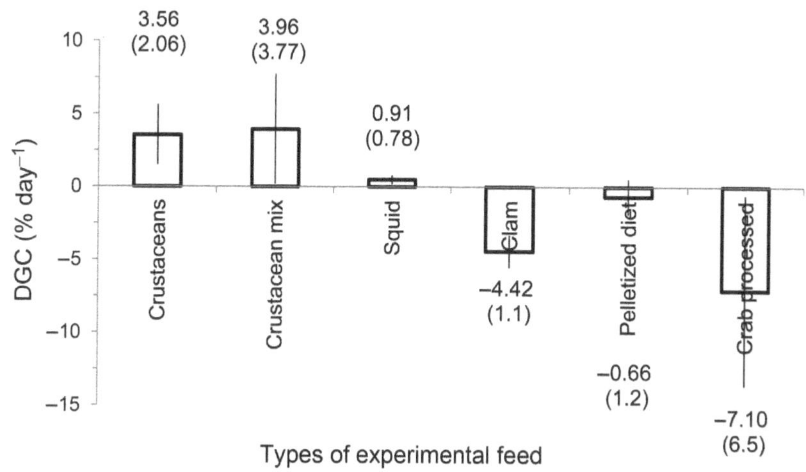

Figure 1.2 *Octopus maya.* Daily growth coefficients (mean ± SD) for octopus in response to different types of experimental feed. SD in parenthesis and from Table 1.1.

crustacean and squid-bound gelatin also produced higher growth and almost double survival rates than those from animals fed with raw crab meat (Rosas et al., 2008). This diet was tested in a semicommercial production system created to produce *O. maya* of 100–120 g. In this system, juveniles of 100 mg WW were seeded at a density of 25 individuals m^{-2} and fed with a ration of 30–50% WW d^{-1} (Quintana et al., 2011; Uriarte et al., 2011), yielding survival rates of 50% at harvest (see Sections 3.4.3 and 3.5.3).

2.4.3 *Octopus vulgaris*

Recent studies have demonstrated that maternal diet can influence embryonic development by reducing the initial weight of eggs when a crab or squid diet is used (Márquez et al., 2013), suggesting that the nutritional content of the diet can modulate the physiology of brooding females to produce eggs with certain biological characteristics.

Nutrition of paralarvae has been a major bottleneck for *O. vulgaris* aquaculture. The lack of specific diets to feed paralarvae has prevented the production of juveniles to seed subsequent culture systems. At present, juvenile survival is very limited, suggesting that the diets designed to date have marked nutritional and/or functional deficiencies (Iglesias et al., 2007). A recent study investigated the prey ingested by *O. vulgaris* paralarvae in the wild and found 20 different prey species in the stomach contents of paralarvae of which 17 were crustaceans and three fish (Roura et al., 2012), providing essential information to understand the nutrition of paralarvae.

Over the last decade or so, there have been several studies on the nutrition of *O. vulgaris* juveniles and subadults in an attempt to obtain a practical diet for the fattening phase of octopus culture in cages in Galicia, Spain. Experiments have been performed to determine the nutrient composition when designing commercial feed for octopuses (Domingues et al., 2010; García García and Aguado Giménez, 2002) and a balanced and elaborated diet was fed to *O. vulgaris* fattening in cages (Cerezo Valverde et al., 2008, 2012a). García-Garrido et al. (2010, 2011a) found that lipids are important nutrients for cephalopods, mainly as a source of energy. Among them, PUFAs and particularly eicosapentaenoic acid (EPA) were identified as the main source of energy of fasting octopus, suggesting that they should be included when an artificial diet is designed. These findings were confirmed when agglutinated moist diets were used to feed *O. vulgaris* juveniles in commercial cages (Estefanell et al., 2011). However, there are no balanced feeds performing as well as the fisheries discards currently used as feed for *O. vulgaris* grow-out in cages (see Section 3.5.4). Nevertheless, major efforts are still under way to design a formulated feed for the *O. vulgaris* aquaculture industry (Cerezo Valverde et al., 2012a).

Although to date there are many studies where subadults and adults of *O. vulgaris* were fed with different types of prey and diets (Cerezo Valverde et al., 2008, 2012b; Domingues et al., 2010; Estefanell et al., 2011; García García and Aguado Giménez, 2002; García-Garrido et al., 2010, 2011a; Miliou et al., 2005), only two of these studies address the lipid classes or amino acid content of different marine species and feed, focusing on the likely nutritional requirements of *O. vulgaris* (Cerezo Valverde et al., 2012a,c). Diets with crustacean and bivalve protein complemented with arginine or leucine were proposed as improved feeds for cephalopods (Cerezo Valverde et al., 2012c). It was also recognized that, although lipids in general are important for cephalopods, none of the 33 species and feeds analysed satisfied *O. vulgaris* requirements (Cerezo Valverde et al., 2012a). Monroig et al. (2012a) had shown that PUFAs (20:4n-3, 20:3n-6, 20:5n-3 and 20:4n-6) are essential for *O. vulgaris*, which explains in part why marine crustaceans and some molluscs with low-lipid content but high proportions of PUFA have been used with success as food for this species (Boucher-Rodoni et al., 1987; Domingues et al., 2010; García-Garrido et al., 2011a; Uriarte et al., 2011). Also, of all the feeds tested so far, only squid *L. vulgaris* fulfil the nitrogen requirements of *O. vulgaris* (Katsanevakis et al., 2005; Miliou et al., 2005).

Absolute growth rate ($g\,d^{-1}$) and daily growth coefficient (DGC, $\%\,d^{-1}$) of each feed type are shown grouped in Table 1.2, using DGC as an index for the diet with the best results on growth at the experimental level for *O. vulgaris*.

Table 1.2 Effects of different types of feed on the growth of Octopus vulgaris

	Octopus initial weight (g WW)	Temperature (°C)	AGR (g d^{-1})	DGC (% d^{-1})	References
Crustaceans					
Crab (several species)	500	17.5	15[a]	–	Aguado Gimenez and García García (2002)
Crab (several species)	2000	17.5	20[a]	–	Aguado Gimenez and García García (2002)
Crab	620–632	17	10.3–13.0	2.01–2.35	García García and Cerezo Valverde (2004)
Crab (*Carcinus mediterranus*)	819	17–21	16.2	1.61	Cerezo Valverde et al. (2008)
Maja crispata, 100%	614 ± 89	18	20.1	2.66	Prato et al. (2010)
Procambarus clarkii	1400	21	40.0	1.7	Domingues et al. (2010)
P. clarkii	620	21	7.6	1.1	Domingues et al. (2010)
Crustaceans mix					
Crab (*Carcinus mediterraneus*), 80%; *Boops boops*, 15%; and mussels *Mytilus galloprovincialis*, 5%	614 ± 89	18	10.81	1.56	Prato et al. (2010)
Maja crispata, 50%; *Diplodus vulgaris*, 50%	614 ± 89	18	19.32	2.07	Prato et al. (2010)

Continued

Table 1.2 Effects of different types of feed on the growth of Octopus vulgaris—cont'd

	Octopus initial weight (g WW)	Temperature (°C)	AGR (g d^{-1})	DGC (% d^{-1})	References
Shrimp (*Palaemonetes varians*) and gelatin	468.1	19	2.9	0.5	Quintana et al. (2008)
Fish					
B. boops (dietary group II)	614±89	18	18.44	2.04	Prato et al. (2010)
B. boops (fresh)	864	22	11.6	1.9	Estefanell et al. (2013)
B. boops (fresh)	889	22	16.4	1.5	Estefanell et al. (2013)
B. boops (moist diet)	882	22	11.0	1.1	Estefanell et al. (2013)
B. boops (moist diet)	918	22	17.1	1.5	Estefanell et al. (2013)
B. boops	500	17.5	7[a]	–	Aguado Gimenez and García García (2002)
B. boops	2000	17.5	10[a]	–	Aguado Gimenez and García García (2002)
B. boops	378–418	17	6.5–7.3	0.8–0.9	García García and Aguado García (2002)
Merluccius gayi	563	21	15.5	1.9	Domingues et al. (2010)
Merluccius gayi	1500	21	22.5	1.2	Domingues et al. (2010)

Table 1.2 Effects of different types of feed on the growth of Octopus vulgaris—cont'd

	Octopus initial weight (g WW)	Temperature (°C)	AGR (g d^{-1})	DGC (% d^{-1})	References
Sardina pilchardus	316–346	17	5.3–9.4	0.8–1.3	García García and Aguado García (2002)
Fish mix					
B. boops, 50%; prawn (*Hymenopenaeus muelleri*), 10%; alginate, 30%; calcium, 10%	810	17–21	6.26	0.71	Cerezo Valverde et al. (2008)
B. boops, 40%; water, 40%; prawn (*Hymenopenaeus muelleri*),10%; and gelatin, 10%	852	17–21	1.87	0.22	Cerezo Valverde et al. (2008)
[a]LL sodium alginate 0	842	17–20	0.53	0.06	Cerezo Valverde et al. (2013)
LL sodium alginate 5[a]	902	17–20	1.96	0.21	Cerezo Valverde et al. (2013)
LL sodium alginate 20[a]	859	17–20	4.93	0.53	Cerezo Valverde et al. (2013)
HL sodium alginate 0[b]	700	17–20	2.08	0.28	Cerezo Valverde et al. (2013)
HL sodium alginate 5[b]	685	17–20	3.61	0.33	Cerezo Valverde et al. (2013)
HL sodium alginate 20[b]	722	17–20	4.14	0.52	Cerezo Valverde et al. (2013)

Continued

Table 1.2 Effects of different types of feed on the growth of Octopus vulgaris—cont'd

	Octopus initial weight (g WW)	Temperature (°C)	AGR (g d^{-1})	DGC (% d^{-1})	References
Squid					
Doryteuthis gahi (fresh frozen)	713	15	12.06	1.36	García-Garrido et al. (2011a)
D. gahi (fresh frozen)	359.9	20	13.3	2.48	García-Garrido et al. (2011b)
D. gahi (fresh frozen)	710.8	20	18.3	1.92	García-Garrido et al. (2011b)
D. gahi	542	21	15.8	1.9	Domingues et al. (2010)
D. gahi (fresh)	486.3	19	9.5	1.4	Quintana et al. (2008)
D. gahi (paste and gelatin)	505.9	19	7.3	1.13	Quintana et al. (2008)
Squid mix					
Squid (*D. gahi*), 75%; CPSP, 10%; gelatin 15%	578.9	20	4.93	0.74	García-Garrido et al. (2011b)
Squid paste, 25%; CPSP90, 10%; fish flour, 55%; gelatin, 10%	692	15	0.68	0.09	García-Garrido et al. (2011a)
Squid (*D. gahi*), 50%; CPSP, 40%; gelatin, 10%	268.9	20	1.13	0.39	García-Garrido et al. (2011b)

Table 1.2 Effects of different types of feed on the growth of Octopus vulgaris—cont'd

	Octopus initial weight (g WW)	Temperature (°C)	AGR ($g\ d^{-1}$)	DGC ($\%\ d^{-1}$)	References
Squid (*D. gahi*), 65%; CPSP, 20%; gelatin, 15%	590.1	20	1.27	0.21	García-Garrido et al. (2011b)
Squid paste, 30%; CPSP90, 10%; fish flour, 50%; alginate, 10%	717	15	−3.3	−0.49	García-Garrido et al. (2011a)
Squid (*D. gahi*), 70%; CPSP, 20%; gelatin, 10%	318.1	20	−0.99	−0.32	García-Garrido et al. (2011b)
Mussel					
M. galloprovincialis	614±89	18	7.57	0.93	Prato et al. (2010)
Formulated diets					
Fish meal[c]	580±123	18	3.3	0.54	Querol et al. (2012)
Krill[c]	580±123	18	3.1	0.58	Querol et al. (2012)
Fish meal and krill[c]	580±123	18	3.8	0.62	Querol et al. (2012)

[a]Fish paste (*B. boops*), 50%; prawn paste (*Hymenopenaeus muelleri*), 10%; alginate + calcium, 27%; low-lipid (LL), 7.8–9.1 g/kg.
[b]Fish paste (*B. boops*), 50%; prawn paste (*Hymenopenaeus muelleri*), 10%; alginate + calcium, 27%; high-lipid (HL), 83.4–84 g/kg.
[c]Extruded diets bound with gelatin and mixed with maltodextrin and egg yolk powder.
Absolute growth rate (AGR, $g\ d^{-1}$) and daily growth coefficient (DGC, $\%\ d^{-1}$) of octopus with different types of feeds. For groups' description, see text. WW, wet weight.

These data were used to identify a relationship between groups of feed and their effect on *O. vulgaris* growth (Figure 1.3). Feeds composed of crustaceans, single or mixed (mainly crabs), fish (mainly *Boops boops*) and squid (*Doryteuthis gahi*) resulted in higher octopus growth rates than with fish or squid mixed diets (Figure 1.3). It is interesting to note that only once were mussels used as a

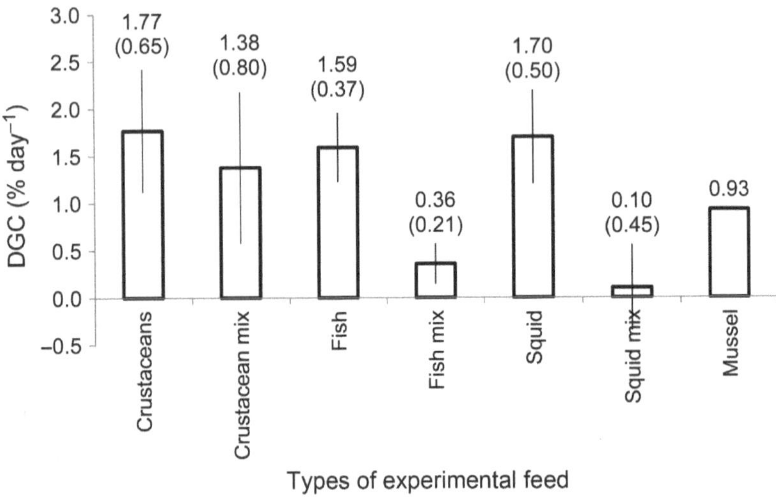

Figure 1.3 *Octopus vulgaris*. Daily growth coefficients (mean ± SD) for octopus in response to different types of experimental feed. SD in parenthesis and from Table 1.2.

potential diet for *O. vulgaris* (Prato et al., 2010). Although crabs yielded the highest growth rates for *O. vulgaris*, both *B. boops* and *D. gahi* resulted in growth rates high enough (>1.5 d^{-1}) that, under some circumstances perhaps related to freshness, these ingredients could be used to formulate feeds for culture (Estefanell et al., 2012a; García-Garrido et al., 2011a,b; Prato et al., 2010).

3. CULTURE REQUIREMENTS AT DIFFERENT LIFE PHASES

3.1. Broodstock

Cephalopod broodstocks for culture can be obtained directly from wild populations but the selection of individuals is a key factor: they should be healthy, with undamaged skin, arms and tentacles. Capture methods should be the most appropriate to minimize stress and maintain the physical integrity of the animals. The use of pots (Brazil, Japan, Portugal and Spain) or lines with crabs as bait without hooks is recommended for octopus capture. Size and weight should also be considered as key factors for the selection of potentially mature individuals. This section addresses the actual status of broodstock management of the four biological models under culture conditions.

3.1.1 Sepioteuthis lessoniana

This loliginid squid is an Indo-West Pacific species, widely distributed from Japan to northern Australia and New Zealand and from the East African coast to Hawaii. Common in coastal environments on sea grass beds, coral reefs and sandy bottoms, it occurs from surface waters to about 100 m depth throughout its area of distribution. This species is of medium to large size, with mantle length (ML) usually around 200–300 mm and weight up to 2 kg (Jereb et al., 2010), and is considered to be one of the most adaptable cephalopods for laboratory culture (Hanlon, 1990).

Set nets, small purse nets, squid traps and jigging are the main fishing gear to capture young and adult *S. lessoniana* for broodstock. Set nets are particularly suitable because living squid can be maintained in good condition in the net before landing. Captured broodstock may be either kept at sea or transported to a laboratory facility (with extreme caution to avoid possible skin damage). Broodstock individuals have been maintained in floating net cages in the sea (Ahmad and Usman, 1997; Saso, 1979), where squids spawn and provide the seed for culture experiments. Experimental studies undertaken to evaluate the usefulness of keeping broodstock and growing out juveniles in floating cages yield good results using 4 m^3 net cages (Wada and Kobayashi, 1995) and 5 m^3 net cages with masking net to protect them from sun damage (Sugita, 2012).

Egg capsules are attached in clusters in the substrate in spawning grounds located in shallow coastal areas in Japan and Southeast Asian countries, so they are relatively easy to collect (Choe and Ohshima, 1961; Ohshima and Choe, 1961; Segawa, 1987; Tsuchiya, 1982). Broodstock can also be obtained from captive cultured individuals, but the viability of laboratory-spawned eggs may be lower than that of eggs from wild-caught individuals (Lee et al., 1994). Walsh et al. (2002) also had a problem with egg fertility and viability but observed positive traits when culturing squid through seven consecutive generations: The number of eggs laid increased in general with subsequent generations, adult size remained similar throughout the experiment (in contrast with Lee et al. (1994) who observed a decrease in maximum adult size), and subsequent generations produced healthy and disease-resistant offspring.

Appropriate spawning substrates must be provided for squid kept in tanks or cages, such as artificial reefs and sea grass (Lee et al., 1994; Walsh et al., 2002). The eggs are surrounded by white gelatinous fingerlike capsules of about 6 cm in length and 1.3 cm in width and containing up to 13 eggs each. After copulation, mature females easily spawn eggs in captivity several times in one spawning season. Wada and Kobayashi (1995) observed a female reared with a male in a sea pen (4 m^3) spawn 11 × at intervals of 1–9 d, laying a total of 1540 egg capsules containing 7780 eggs. Under culture conditions, first mating was observed as early as 130–160 d after hatching (DAH) and spawning at 153 DAH at 23–25 °C (Hanlon, 1990).

Sex identification of adult *S. lessoniana* is simple because of their obvious sexual dimorphism, with males showing a specific colour pattern and the conspicuous hectocotylus at the tip of the left fourth arm. Females and immature squid have iridophores (round green spots) on the dorsal mantle, while mature males have relatively larger iridophores with horizontally elongated shapes (Ikeda, 1933). This species is relatively easy to maintain in captivity, given the availability of ample good quality sea water, tanks or cages of sufficient size, and good live feed, although frozen food of sufficient size is readily accepted by older individuals.

3.1.2 *Sepia officinalis*

This common European cuttlefish has a broad distribution from the North Atlantic, throughout the English Channel, and south into the Mediterranean Sea to the coast of West Africa. A demersal, neritic species occurring predominantly on sandy to muddy bottoms from the coastline to about 200 m depth, it has a maximum size of 450 mm ML and weight up to 4 kg in temperate waters but around 300 mm ML and 2 kg in subtropical seas (Reid et al., 2005).

This species produces large eggs (20–30 mm in length), attains sexual maturity at very different sizes and weights (Hanlon, 1990; Sykes et al., 2006a) and has a maximum estimated potential fecundity of up to 8000 eggs in nature (Laptikhovsky et al., 2003). Females mate repeatedly (Hanlon et al., 1999) and display intermittent or chronic spawning (Boletzky, 1987), depending on captive conditions. It is not advisable to separate females from males after copulation since this will not promote a reproduction resembling wild conditions and might have an indirect influence on egg quantity and quality (Figure 1.4). According to Boletzky (1983), larger females will lay bigger eggs, although this does not agree with recent observations by Sykes et al. (2013a). No parental care of eggs has been observed in cuttlefish, and senescent females, which allocate all their energy reserves to the eggs, die after their last spawning.

As with *S. lessoniana*, there are two possible broodstock sources for breeding cuttlefish: captive cultured individuals or direct from the wild (Sykes et al., 2006b). Koueta et al. (2002) and Perrin (2004) obtained smaller eggs when these were produced by a captive broodstock, except when enriched feeds were used. When establishing broodstock, all phases of the life cycle require great care because reproduction in captivity is still one of the bottlenecks in cuttlefish culture (Sykes et al., 2006b).

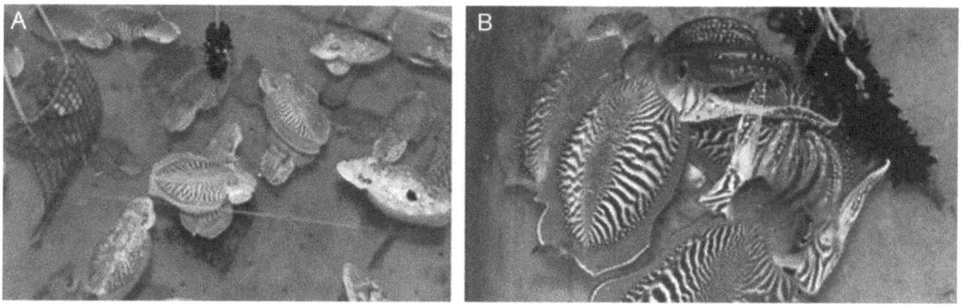

Figure 1.4 *Sepia officinalis.* (A) Cuttlefish breeders laying eggs inside 9000 L round tanks of large bottom areas and slow water movement. (B) Females laying eggs that are attached to a common cluster inside the tank. Original images (not to scale).

Closed systems can be used for experimental rearing and so can open systems if the quality of water is excellent. Semiopen systems in which 80% of water is renewed each day might also be a good alternative (Koueta and Boucaud-Camou, 1999).

Conditions to accommodate breeders have progressed from the use of 250 L tanks (Correia et al., 2005, Sykes et al., 2006a) up to 9000 L, and recommendations suggest that increased bottom areas should be used (Domingues and Márquez, 2010; Sykes et al., 2013a; Figure 1.5). Low disturbance areas should be elected to settle tanks, irrespective of their indoor or outdoor location on the husbandry facility. Tank design should allow correct water circulation inside the tank in order to ensure that no anoxic areas are potentially created. Additionally, there must be no sharp objects or rough surfaces inside the tank since they may cause skin damage. Round fibreglass tanks should be used and must comprise enough airlifts fixed on the tank walls and air stones deployed in the middle of the tank. These will ensure slow water movement and draining through the outlet piping located in the centre of the tank, reducing turbulence and improving conditions for egg laying and maintenance.

Broodstock must be kept under low light intensities (200 lux or less), and photoperiod should replicate that occurring during spawning in the wild for each particular geographic area. Ideal photoperiod will correspond to a combination of 12–14 hs light versus 12–10 hs dark, at a mean temperature of 23–25 °C. The combined effects of photoperiod and feeding can be used to increase survival and growth (Koueta and Boucaud-Camou, 2003; Perrin et al., 2004). When using outdoor tanks, the use of water-repellent masking nets is strongly advised to prevent excessively bright conditions and sunburn and decreases in pH and salinity caused by rainfall.

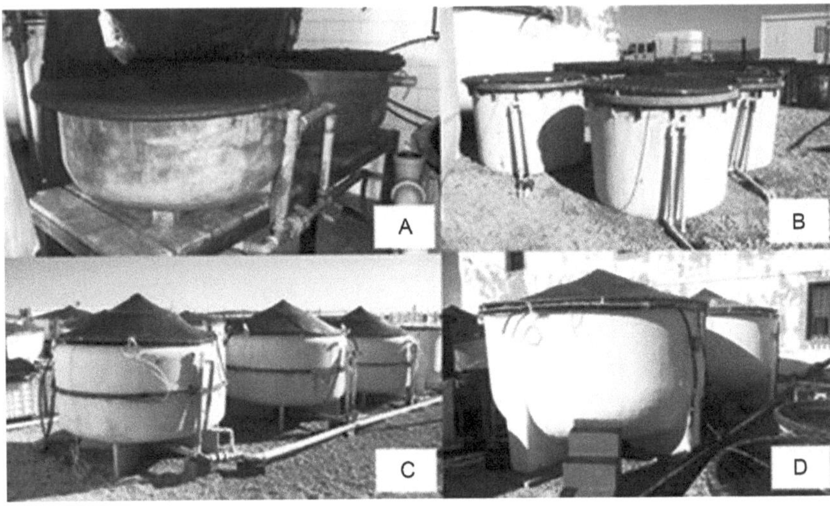

Figure 1.5 Cylindrical tanks of increasing volumes used for cuttlefish reproduction at the Ramalhete field station of the Centre for Marine Sciences, University of the Algarve, Portugal. (A) 250 L, (B) 750 L, (C) 3000 L and (D) 9000 L. Image not to scale. Original image.

Sex ratio and stocking densities have significant effects on the fecundity and fertility of this species (Boal and Golden, 1999; Forsythe et al., 2002). Forsythe et al. (1991) suggested a male/female sex ratio of 1:3 to limit the likelihood of male aggression and aggressive mating behaviour, and Forsythe et al. (1994) suggested a density of two cuttlefish m^{-2}. However, fecundity results obtained under different culturing conditions (tank volumes, stocking densities, sex ratios, temperature and food; Table 1.3) indicate that the sex ratio should be maintained at 2:1 and stocking densities kept low when setting up a broodstock. For instance, a 9000 L tank should have 21 individuals, 14 ♂♂ and 7 ♀♀, which is a stocking density of four cuttlefish m^{-2}.

The onset of maturation and reproduction in cuttlefish determines a shift in the diet, from a predominant composition of crustaceans (crabs, prawns and shrimps) during the juvenile phase to a mixture of fish and crustaceans as individuals mature and reproduce (Boletzky, 1983; Nixon, 1985). This agrees with published information on the diet of other cephalopod species, such as *L. vulgaris* (Coelho et al., 1997), in which changes in diet are also attributed to the onset of sexual maturation. In spite of this, Sykes et al. (2006a, 2013a) had demonstrated that it is possible to base successful cuttlefish culture on a diet composed exclusively of frozen grass shrimp, *Palaemonetes varians*. Perrin (2004) and Perrin et al. (2004), however,

Table 1.3 Fecundity values of *Sepia officinalis* obtained for different stocking densities, tank volume, sex ratio, temperature and food

Fecundity (egg. ♀$^{-1}$)	Density (no. m^{-2})	Tank volume (L)	Sex ratio (M/F)	Temperature (°C)	Food	Reference
144	38	250*	1:1	27±3	*Carcinus maenas* (F)	Domingues et al. (2001b)
225	19	250*	3:1	15±4	*Palaemonetes varians* (L)	Domingues et al. (2002)
150	19	250*	1:1	≈17	*P. varians* (L)	Domingues et al. (2003)
411	19	250*	1:1		*P. varians* (F)	
834	16	250*	3:1	24.5±1.4	*P. varians*, *C. maenas* and fish (L)	Correia et al. (2005)
290	76	250*	3:1			
370	9	250*	1:2	17.1±1.7	*P. varians* (L)	Sykes et al. (2006a)
301	9	250*	3:1	23.4±1.4		
247	13	250*	1:2	15.2±3.0		
478	8	250*	1:1	21.1±2.6		
293	3	250*	2:1	24.2±1.7		
787	15	400**	1:2	19.5±1.1	*P. varians* (F)	Sykes et al. (2009)
1383	4	9000*	1:2	20.5±2.9	*P. varians* (F)	Sykes et al. (2012)
223	15	750*	1:1	19.0±2.2	*P. varians* (F)	
325	29	250*	1:1	21.2±3.4	*P. varians* (F)	

*- Round tank; ** - Rectangular tank; L - live; F – frozen; M – male; F - female.

reported lower survival and growth rates and smaller egg size when captive animals are fed with frozen prey, noting that they adjust their enzymatic activity to the feed and that frozen feed delays the development of the digestive system. This may have a direct impact on juvenile growth and also on egg size once the animals reach maturity and spawn.

3.1.3 *Octopus maya*

Octopus maya is endemic to the Western Central Atlantic in the Gulf of Mexico, off the Yucatán Peninsula, Mexico. It is a benthic, shallow-water species occurring in sea grass beds, empty shells and hard substrates down to 50 m depth. Maximum total length and weight are 130 cm and 5 kg, respectively. Females may lay from 1500 to 2000 large eggs of up to 17 mm long, which produce large benthic hatchlings around 6–7 mm ML (Roper et al., 1984).

Studies conducted to date have shown that this species can be maintained under laboratory conditions for several generations (Solis, 1998; Van Heukelem, 1976, 1977, 1983). At the Universidad Nacional Autónoma de México (UNAM) facilities in Yucatán, Mexico, there is an experimental pilot unit for the production and rearing of *O. maya*. Between 2006 and 2012, more than 280 clutches were spawned from wild females with mean live weight of 815 ± 16 g, yielding a total of 250,000 eggs, with a mean WW for newly hatched juveniles of 0.13 ± 0.001 g ($N=553$).

Females are collected from the Yucatán continental shelf in the Gulf of Mexico, where lies the most important octopus fishery in America (Salas et al., 2006). They are caught from small artisanal boats (7 m length) with lines (around 10 boat^{-1}) suspended from long bamboo poles, using crabs as bait but without hooks (Figure 1.6). During the process, crabs are hauled through the bottom (8–10 m depth) in an attempt to stimulate the behaviour of their predators. Eventually, thanks to their voracity, *O. maya* attack the bait and tense the lines, which are hauled into the boat where the octopus is killed and stored in ice for 4–5 h. Octopuses captured alive are placed in tanks onboard and maintained with continuous sea water exchange.

In general, *O. maya* females (>300 g) are already functionally mature (Arkhipkin, 1992) and fertilized, such that growing oocytes lie within the

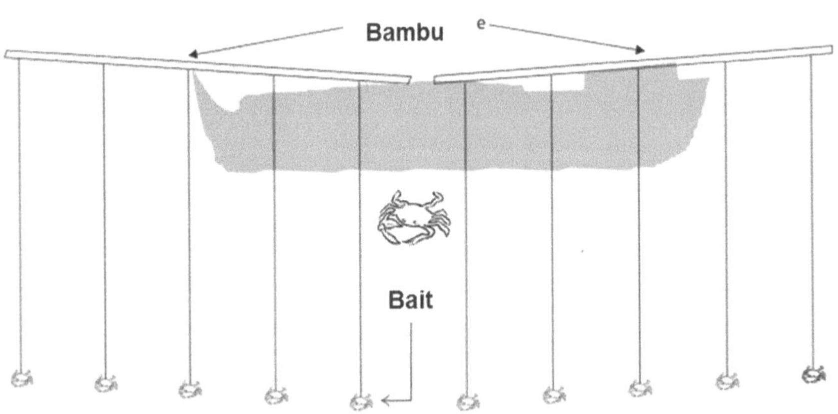

Figure 1.6 *Octopus maya* artisanal fisheries using crabs as bait without hook.

Figure 1.7 *Octopus maya*. Males and females during 10-day acclimation period at outdoor 6 m diameter ponds at the Universidad Nacional Autónoma de México (UNAM). Original image.

reproductive coelom. Nevertheless, to ensure that females and their stored sperm are of sufficient quality and quantity, they are maintained in outdoor ponds with males (sex ratio 1:1) for 10 d after capture and normally fed with blue crab (*Callinectes sapidus*) at a rate of 5% octopus WW d^{-1} (Figure 1.7).

After acclimation, *O. maya* females are conditioned for a 30-d period in 80 L dark tanks in a low light room (30 lux cm^{-2}), where sea water is maintained at 24 ± 1 °C, salinity > 32, pH 8, dissolved oxygen > 5 mg L^{-1}, ammonia < 1 mg L^{-1} and nitrite < 0.15 mg L^{-1}. During this period, females are fed with a mixed diet of crab (75%), mussel (*Mytilus* spp.) and/or fish heads (25%) at a rate of 5% octopus WW d^{-1}.

During the 2013 production season, the mean number of eggs per clutch was 800 ± 22 and only 15% of females produced unfertilized eggs. Wet weight of females laying eggs varies considerably (300–1300 g) and there is no apparent relationship between female live weight and fecundity. This weight variability might be related to the nutritional history of wild females, given the dependence of cephalopod gonad development on factors such as feeding activity and food availability.

In culture conditions, fertilized eggs hatched after 45–50 d at 24 ± 1 °C. At UNAM facilities, *O. maya* spawns are artificially incubated, facilitating spawn area management (Rosas et al., 2010) and releasing the holding tanks for accommodation of subsequent females ready for spawning.

3.1.4 *Octopus vulgaris*

Octopus vulgaris is a neritic species considered as cosmopolitan with unknown distributional limits in temperate and tropical seas. They are benthic and occur from the sublittoral to depths of 200 m, in a variety of habitats such as rocks, coral reefs and grass beds. Size is highly variable depending on the habitat, with a maximum total length of 130 cm and maximum weight up to 10 kg (common to 3 kg). This species has high fecundity (100,000–600,000 eggs), producing small hatchlings of about 2–3 mm ML (Mangold, 1983; Roper et al., 1984; Vidal et al., 2010).

The different teams working on *O. vulgaris* culture use different conditions for the spawners (Iglesias et al., 2007). The most obvious differences are the male/female ratios and the broad range of temperatures used (14–25 °C). However, the method of capture, transport conditions, food supply and light intensity are similar. There are no problems regarding the acclimation of spawners and obtaining viable eggs in captivity. The egg-laying process can be carried out by either creating a stock of spawners, mixing males and females (Iglesias et al., 2000) or individually separating already fertilized females captured at sea (Okumura et al., 2005; Villanueva, 1995). The females all mature and lay eggs in captivity, with successful hatching rates over 80%.

Comparing the systems used to maintain broodstock by different research teams (Iglesias et al., 2007) allows the establishment of a set of general recommendations for the reproduction of *O. vulgaris* that should enable more consistent production of viable eggs and paralarvae. Firstly, it is recommended that pots are used to capture broodstock octopuses. An appropriate feed is a mix of crustaceans and fish of low commercial value, with crustaceans comprising at least 30% of the diet. If octopuses are captured during the spawning period, it is sufficient to keep the females (>1 kg) until eggs are obtained. The rest of the year, males and females should be maintained at a ratio of 1:3. When the spawning has been completed, the female, together with her eggs, must be transferred to another tank, to avert disturbance of the brooding female by other spawners.

3.2. Embryonic phase

Embryonic incubation and maintenance does not represent a major problem for the production of paralarvae and juveniles in most species. However, egg care is one of the key factors to promote rearing success. Eggs can be obtained from broodstock, from the wild (Figure 1.8) or from *in vitro*

Figure 1.8 Eggs and egg masses. (A) *Sepioteuthis lessoniana* (original image), (B) *Sepia officinalis* (original image), (C) *Octopus maya* (original image) and (D) *Octopus vulgaris*. Scale bar 1 cm. *(D) Photograph courtesy of M.E. Garci.*

fertilization. Embryonic development should be monitored closely to evaluate developmental progress and to predict hatching time by following the different stages with an illustrated scale (e.g. Arnold, 1965; Naef, 1928) adapted to each species. Handling and maintenance of egg masses and brooding females should be done with care to avoid egg mass mortality. Husbandry conditions during embryonic development should be maintained ideal according to established requirements for each species; otherwise, the quality of the eggs will be compromised, which will inevitably be reflected in the quality of the hatchlings produced (i.e. normal, abnormal and/or premature). From a practical point of view in the laboratory, cephalopod species can be classified according to whether or not they require maternal care. The following section attempts to identify the main factors influencing embryonic development and methods to avoid death of the embryos and premature hatching.

3.2.1 Eggs with maternal care

Maternal care in cephalopods is only known to occur in incirrate octopods and in a few species of oceanic squid (Bush et al., 2012; Seibel et al., 2005).

The female will usually stop feeding and will take care of the eggs until they hatch. Maternal care in benthic octopods is well known and includes generally protecting the egg mass from potential predators, ventilation by flushing water through the eggs, cleaning the surface of the eggs and removing dead embryos. By means of mechanical stimulation, the female probably influences hatching, in such a way that it will occur more often during the preferred period of daily activity of the species. Usually, females live through to completion of embryonic development, but in some cases, depending on their condition, they might die before the eggs hatch, in which event embryo survival is compromised and usually the whole egg mass will perish.

To prevent egg losses, artificial incubation has been explored with the greatest success obtained with the large-egged *O. maya* (Avila-Poveda et al., 2009). Artificial incubation methods increase the possibilities for mass culture, promoting the exchange of material between laboratories and augmenting the collection of eggs from the wild. Further research is still required to develop artificial incubation methods for small-egged species yielding planktonic paralarvae.

3.2.2 Eggs without maternal care

Females of nautiluses, sepiids, cirrate octopods, loliginid squids and most oegopsid squids lay their eggs without maternal care, either in single encapsulated eggs (e.g. sepiids) (Figure 1.8B) or in capsules (Figure 1.8A) or in jelly masses containing several to thousands of eggs (as in many species of squid). Except for oegopsid squid that spawn pelagic eggs, most cephalopods attach eggs to a hard substrate, sea grass or macroalgae. Incubation procedures require appropriate egg density, aeration and water flow, which are crucial to promote adequate oxygenation (Vidal and Boletzky, 2014).

3.2.3 Main factors influencing embryonic development

During embryonic development, eggs require stable, optimal water quality. Eggs at different developmental stages show different degrees of vulnerability to stress, with tolerance limits intrinsically related to the normal habitat, season and geographic range (Kinne, 1971). The main factors influencing embryonic development are briefly reviewed later in the text.

Temperature. Incubation time in cephalopods is inversely proportional to egg size and also to temperature (Boletzky, 1987), which is the main factor influencing embryonic development. Temperature controls the rate of metabolic processes, such as the rate and efficiency of yolk utilization and consequently embryonic growth and size and weight at hatching (Boletzky, 1987; Vidal et al., 2002b, Villanueva et al., 2007). For each species, there is an optimal temperature range for embryonic development, so temperature can be used as a laboratory tool to manipulate the length of embryonic development and hatching time according to experimental requirements. When eggs are obtained from the broodstock or from *in vitro* fertilization, it is possible to accurately estimate the duration of embryonic development and thus the expected hatching period (Bouchaud and Daguzan, 1989; Villanueva et al., 2012). The hatching period is defined as the time difference between the hatching of the first and last individuals from a single egg mass and can last from several minutes to several days, depending on the species and temperature. This is useful information in order to organize laboratory procedures required to start and develop rearing experiments. Incubation temperature also strongly influences oxygen consumption by embryos, and as the ratio of oxygen supply to demand is increased at low temperatures (Woods, 1999), conversion efficiencies of yolk into tissue are improved, producing large hatchlings at low temperature, as for *S. officinalis* (Bouchaud and Daguzan, 1989). The reverse effect is observed at higher temperatures (Vidal et al., 2002b; Villanueva, 2000). Hence, there is a compromise between short and longer incubation periods, with intermediate periods and temperatures maximizing hatching rates (Sen, 2005a; Sykes et al., 2006a; Vidal and Boletzky, 2014).

Oxygen. Egg masses should be maintained under a well-oxygenated environment. Egg development is asynchronous either because they are not all laid on the same day or because they are exposed to different oxygen levels. Embryos located in the interior of a large egg mass will be exposed to reduced levels of oxygen, leading to slower development and low survival when compared to those located at the periphery (Steer and Moltschaniwskyj, 2007). Therefore, to provide ideal developmental conditions, oxygen levels should be maintained close to saturation and eggs without maternal care separated into small clusters suspended through an upwelling flow of water and/or exposed to high flow rates (Vidal and Boletzky, 2014). Oxygen becomes limiting in cephalopod eggs towards the end of development, when embryonic oxygen consumption and ammonia production are at their peak. The egg chorion, egg capsule and the presence of a jelly mass (depending on species) influence diffusion, a limitation that is known to be critical at least in large-egged species such as *S. officinalis* and *S. apama*. Thus, during late embryonic development, low oxygen diffusion may be a key factor leading to hatching (Cronin and Seymour, 2000;

Gutowska and Melzner, 2009). Late embryos increase the flow of water over the gills by intensifying both the frequency and amplitude of mantle contractions, presumably in response to declining oxygen (or increased carbon dioxide levels), and this seems to result in hatching. Thus, oxygen depletion should be minimized to prevent the death of embryos and premature hatching. In species with maternal care, such as *O. vulgaris*, oxygenation of the eggs is maintained by the female. In spite of this, it is estimated that the oxygen consumption of a medium-sized egg mass is approximately twice that of the oxygen uptake of the respective brooding female (Parra et al., 2000). Therefore, in effect, a brooding female and her egg mass require 3 × the amount of oxygen of a nonbrooding octopus.

Salinity. The influence of salinity on egg incubation and hatching of cephalopods has been considered only rarely. In one study, when eggs of *S. lessoniana* and *S. pharaonis* were incubated at different salinities over the range 12–44 psu (at 29 °C), hatching rates above 80% were recorded between 24 and 36 psu (Nabhitabhata et al., 2001). At salinities close to the upper limit of tolerance for these species, death of the embryos ensued before organogenesis and premature hatching was also observed, while low salinities led to abnormal development. There is also evidence to suggest that temperature has an important effect on salinity tolerance, as tolerance to low salinities seems to increase at higher temperatures (Nabhitabhata et al., 2001; see Table 1.1 of Palmegiano and D'Apote, 1983; Sen, 2005b).

pH. The aerobic performance of cephalopods is sustained by the highly pH-dependent affinity of their haemocyanin. As a consequence, changes in ambient pH are expected to have direct influences on their physiology (Pörtner and Zielinski, 1998), and this is especially true for embryos while their respiratory system is developing. To date, the influence of pH on the physiology and development of cephalopod embryos have not been quantified, but recent studies have shown that low pH has the potential to compromise the protective properties of the egg envelopes, leading to metal accumulation in embryos (Lacoue-Labarthe et al., 2011). Additionally, experiments undertaken with *L. vulgaris* intermediate stage eggs exposed to pHs between 7.1 and 9.1 showed that the normal development of eggs and hatching of viable paralarvae take place only within the pH range 7.8–8.4 (D'Aniello et al., 1989). This is therefore the pH range recommended for egg incubation.

Light conditions. In general, low light intensities (11–15 lux; Yang et al., 1983; 1.8–2.9 μE m^{-2} s^{-1}; Villanueva et al., 2007) should be maintained for most species, with a consistent photoperiod. Photoperiodicity has a key influence on the timing of hatching in cephalopod species without maternal care, as in the cuttlefish *S. officinalis* and the squid *L. vulgaris*. The transition from light to dark seems to function as a "zeitgeber" or synchronizer, stimulating hatching (Paulij et al., 1990b, 1991), hence the tendency for sunset and nocturnal hatching in octopus paralarvae (Villanueva and Norman, 2008). There is also evidence to suggest that continuous illumination might lengthen the embryonic development of *S. officinalis* and cause asynchronous hatching (Paulij et al., 1991) and also reduced statolith growth in both *L. vulgaris* and *Sepioteuthis australis* (Villanueva et al., 2007).

3.2.4 Mechanisms that promote normal hatching

Hatching time is a risky period for cephalopods, which is minimized by fast fracture of the chorionic membrane. Immediately before hatching, individuals use both chemical (protease enzymes produced by Hoyle's gland) and mechanical (mantle contractions and active use of arms and suckers) means to dissolve and open the chorionic membrane. The physiological or morphological mechanisms that promote the actual hatching process are unknown (Boletzky, 2003; D'Aniello et al., 1989).

In addition to embryonic rhythms, species-specific differences in the timing of hatching may be influenced by adult rhythms. During hatching, brooding octopus females sometimes forcibly expel water through the funnel over the eggs. This turbulence may act as a stimulus to instigate hatching (Villanueva and Norman, 2008). Mechanical stimulation provided by brooding female octopuses to their egg mass may differ between nocturnal and diurnal species, making maternal activity a possible behavioural factor in the hatching process of incirrate octopods. However, some physical, chemical and behavioural factors can instigate premature hatching.

3.2.5 Mechanisms that promote premature hatching and recommendations for prevention

A natural tranquillizer described in the perivitelline fluid of advanced loliginid squid eggs prevents premature hatching (Marthy et al., 1976). However, premature hatching is a major source of mortality in cephalopods in laboratory conditions. All female incirrate octopuses guard their eggs throughout embryonic development, after which they die, but brooding females under stress, disease or advanced senescence may modify their brooding behaviour, resulting in the death of the embryos or premature hatching.

Prematurely hatched paralarvae are recognized externally by the presence of the outer yolk sac within the arms and usually reduced body size (Figure 1.9). Normally, yolk reserves are absorbed within a few hours or days after hatching (depending on species and temperature). However, in premature hatchlings, the inner yolk sac is large and occupies a relatively large volume inside the mantle cavity, reducing the total water volume entering the mantle and expelled through the funnel during the jet propulsion cycle characteristic of cephalopods, reducing jetting capacity. Thus, both the outer yolk sac and the inner large yolk sac reduce swimming performance and

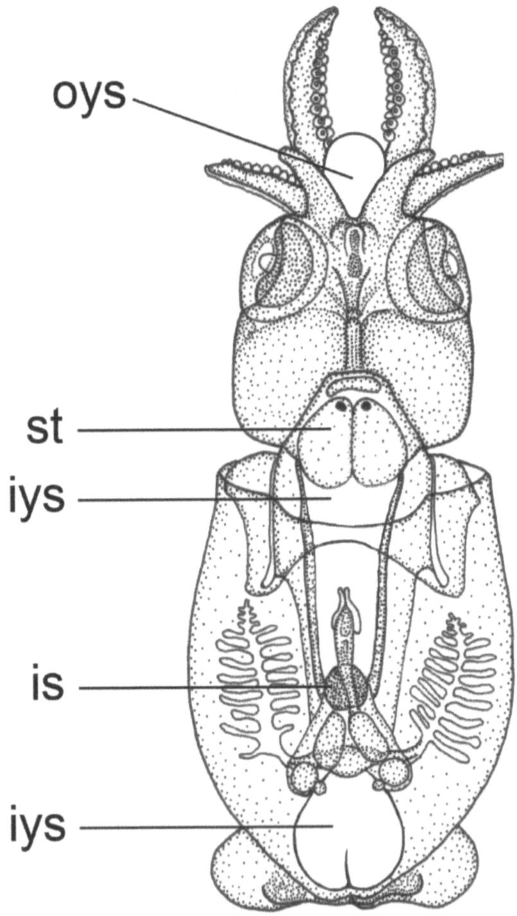

Figure 1.9 Prematurely hatched paralarva. oys, outer yolk sac; iys, inner yolk sac; st, statocyst; is, ink sac. *From Vidal and Boletzky (2014); adapted from Segawa et al. (1988).*

limit movements of newly hatched individuals. Inadequate absorption or loss of the outer yolk sac in premature hatchlings results also in poor post-hatching survival and growth, exacerbated by the fact that these hatchlings are not fully developed. Abrasion of the arms, tentacles, mantle and fins can be produced by friction against the bottom and walls of the rearing tanks (Vidal et al., 2002a) due to poor swimming and manoeuvring abilities. Resulting ulceration of the skin is a probable source of infection and death (see Sections 2.2.7 and 2.3.1).

Abrupt changes in the light conditions (from light to dark) or external mechanical stimulation of advanced embryos (from strong water fluxes or incorrect handling by the experimenter) and abrupt temperature rises are all sources of stress, promoting premature hatching and mortality. In general terms, rises in temperature should not be higher than 1 °C h^{-1} for tropical and subtropical coastal species. In contrast, gradual drops in temperature within the species temperature range retard hatching and can be used to adjust hatching time according to rearing needs. Therefore, when late stage eggs or an egg mass requires manipulation or moving, care should be taken to avoid mechanical impacts and abrupt light and temperature changes. Eggs should be always submerged in water and dry periods should be avoided. Handling of embryos should be limited to early developmental stages as far as possible. To avoid premature hatching in octopods, the egg mass should be handled before the second embryonic inversion typical of incirrate octopods (Boletzky and Fioroni, 1990), when embryos are positioned with the end of the mantle near the egg stalk. After the second embryonic inversion, the embryos are positioned with the mantle pointed at the end opposite the stalk, ready for hatching (Figure 1.10), and are thus sensitive to any stimulus that might trigger premature hatching.

3.3. Planktonic paralarval phase

Paralarval rearing is at present one of the main bottlenecks for cephalopod culture. The term paralarva defines a cephalopod that is planktonic after hatching and has a habitat different from older conspecific individual (Young and Harman, 1988). Newly hatched planktonic paralarvae are delicate and have relatively short arms and limited swimming ability (Villanueva et al., 1995). During the first days after hatching, paralarvae use a combination of endogenous (yolk) and exogenous (prey) food sources (Boletzky, 1975; Vidal et al., 2002b). Both the rate and efficiency of yolk absorption are temperature-dependent and the time required to deplete yolk

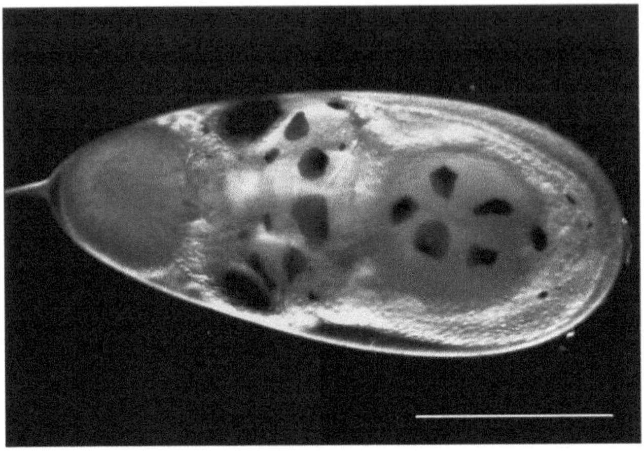

Figure 1.10 *Octopus vulgaris*. Late stage embryo after the second embryonic inversion, in position for hatching. Scale bar 1 mm. *Photograph courtesy of M. Nande and J. Iglesias.*

reserves decreases exponentially with increasing temperature. Feeding rates increase as the inner yolk sac is being absorbed and more space becomes available to accommodate meals for digestion (Vidal et al., 2002b).

The whole planktonic phase has been covered for 21 cephalopod species, but only at the experimental level. Eight sepiid cephalopods with a relatively brief planktonic phase have been reared: *Sepiella inermis* (Choe, 1966a; Nabhitabhata, 1997), *S. japonica* (Zheng et al., 2010), *Euprymna berryi* (Choe, 1966a), *E. hyllebergi* (Nabhitabhata et al., 2005a), *E. scolopes* (Hanlon et al., 1997), *E. tasmanica* (Moltschaniwskyj and Johnston, 2006), *Idiosepius paradoxus* (Natsukari, 1970) and *I. pygmaeus* (Nabhitabhata et al., 2005b). These sepioid species are relatively large at hatching and their planktonic phases range from a few hours to 1 month.

Six species of loliginid squid have been reared to the juvenile phase: *Doryteuthis opalescens* (Chen et al., 1996; Vidal et al., 2002a; Yang et al., 1986), *D. pealeii* (Hanlon et al., 1987), *Heterololigo bleekeri* (Ikeda et al., 2005), *Loligo forbesii* (Hanlon et al., 1989), *L. reynaudii* (Vidal et al., 2005) and *L. vulgaris* (Turk et al., 1986; Villanueva, 2000). Their planktonic phases differ, depending on the hatchling size of the species and temperature, but none lasts longer than 2 months. *Sepioteuthis lessoniana* (Choe, 1966a; Lee et al., 1994; Nabhitabhata, 1996; Segawa, 1990; Sugimoto and Ikeda, 2012) and *S. sepioidea* (LaRoe, 1971) hatch at a large size and can thus be considered juveniles.

Five species of octopus with a planktonic phase (merobenthic octopuses) have been reared to settlement: *Amphioctopus aegina* (Promboon et al., 2011),

Enteroctopus dofleini (Okubo, 1979, 1980; Snyder, 1986a,b), *O. joubini* (Forsythe and Toll, 1991), *O. vulgaris* (Carrasco et al., 2006; De Wolf et al., 2011; Iglesias et al., 2004; Itami et al., 1963; Villanueva, 1995) and *Robsonella fontaniana* (Uriarte et al., 2010). Their planktonic phase ranges from 3 weeks to 6 months, depending also on species and temperature.

There are still many questions to be answered in relation to paralarval rearing, particularly regarding their feeding, nutrition and husbandry requirements. Most information has been obtained from just a few species, and the most intense research during recent years has been on *O. vulgaris*, which is the focus of the next section.

3.3.1 *Octopus vulgaris*

No standard rearing technique for this species during the planktonic phase is yet available and present knowledge is based on experimental rearings. The transfer of paralarvae to their rearing tanks must be carried out with extreme care in order to avoid stress or physical damage, with monitoring carried out carefully, either by individual counting or by volumetric estimation. Small tank volumes (8–50 l) have been used during experimental rearing to concentrate natural prey that are difficult to obtain in large quantities, such as decapod crab zoeae (Carrasco et al., 2006; Itami et al., 1963; Villanueva, 1995). When using *Artemia* as food, larger tanks have been employed. Recently, Sánchez et al. (2013) compared survival and growth of paralarvae (Figure 1.11) in tanks of 100, 500 and 1000 l, obtaining better results in larger tanks. Usually, black tanks have been used but white tanks (Carrasco et al., 2006) or tanks with a white bottom and black walls (De Wolf et al., 2011) have also obtained settled individuals.

Both natural and continuous light photoperiods, too, have resulted in obtaining settled juveniles. Light intensities used have ranged from 60 (De Wolf et al., 2011) to 900 lux (Villanueva, 1995) at the water surface. Octopod paralarvae (Figure 1.11) show strong positive phototaxis, particularly during the first days after hatching, so light can be used as a tool to attract individuals to near the surface or midwater, away from the contact with tank walls and bottom to decrease the possibility of skin lesions, a source of mortality in the paralarvae (Vidal et al., 2002a). However, care should be taken to create a nearly homogeneous distribution of paralarvae and their prey to promote a maximum likelihood for successful interactions between them (Vidal et al., 2002a).

Nutrition seems to be the main problem to solve when rearing these paralarvae. Poor paralarval survival has been attributed to the lack of suitable

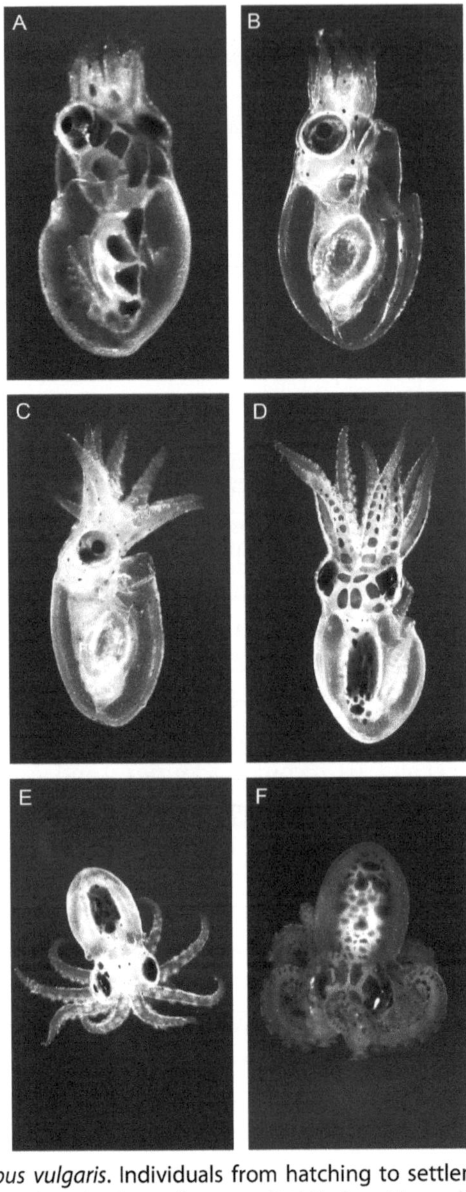

Figure 1.11 *Octopus vulgaris*. Individuals from hatching to settlement obtained from rearing experiments described in Villanueva (1995). Age (days, d) and mantle length (ML) of the individuals measured fresh are (A) 0 d, 2.0 mm ML; (B) 20 d, 3.0 mm ML; (C) 30 d, 4.3 mm ML; (D) 42 d, 5.9 ML; (E) 50 d, 6.6 mm ML; (F) 60 d, 8.5 mm ML. Octopuses from this experiment reared at 21 °C settled between 47 and 54 d. Individuals were photographed under anaesthesia (2% ethanol) causing chromatophore contraction in some cases. Images not to scale. *Photos by J. Lecomte (Observatoire Océanologique de Banyuls, CNRS); modified from Villanueva et al. (1995).*

prey, in both quality and quantity, and to their particular feeding requirements (Iglesias et al., 2007). High mortality rates occur particularly during the transition from yolk absorption to external feeding, due mainly to the high metabolic demands of paralarvae (Parra et al., 2000), their inability to withstand even short periods of starvation (Vidal et al., 2006) and the difficulties of offering prey in adequate quantity and quality.

To date, the best survival and growth results have been obtained when using live crustacean zoeae as single or complementary prey, but it is very difficult to obtain them in large quantities (Villanueva, 1994). *Artemia* is readily available but seems to have an unbalanced composition as prey for octopus paralarvae, particularly with regard to lipids (Fuentes et al., 2011; Monroig et al., 2012a,b; Navarro and Villanueva, 2000, 2003; Viciano et al., 2011), amino acids (Villanueva et al., 2004) and some elements such as copper (Villanueva and Bustamante, 2006). However, apparently, *Artemia* is not limiting in terms of vitamin A and E content (Villanueva et al., 2009). Recently, the lipid quality of *Artemia* has been improved using different microalgae as their food (Seixas et al., 2008, 2010a,b), and better results have been obtained when adding the microalga *Nannochloropsis* to the rearing tank as food for *Artemia* (Fuentes et al., 2011; Hamasaki and Takeuchi, 2000). Nevertheless, juvenile *Artemia* containing higher levels of PUFA was ineffective to improve rearing of *O. vulgaris* paralarvae (Seixas et al., 2010b). Experiments using inert food such as fish flakes (Kurihara et al., 2006; Okumura et al., 2005) or millicapsules (Villanueva et al., 2002) have not produced settled individuals and further research is necessary to design balanced diets for these paralarvae.

In relation to prey size, Iglesias et al. (2006) reported that larger *Artemia* (1.4 mm) was preferred to smaller *Artemia* (0.8 mm) at first feeding. Prey size requirement increases as paralarvae grow and notably when animals are near settlement, a delicate period associated with high peaks of mortality. At around 40–65 DAH, *O. vulgaris* paralarvae have between 17 and 20 suckers per arm, a mean ML of 6.5 mm and a mean DW of 20 mg (reared at 20–22 °C), and they change from their initial planktonic habit to a raptorial behaviour, fixing themselves to the walls and bottom of the tanks. Presettlement individuals (Figure 1.11E) tend to remain attached to the surfaces of the tank for most of the time and only swim to capture prey in the water column. This behaviour lasts for 2 weeks, before they acquire fully benthic behaviour. During the settlement period, several morphological changes took place, including positive allometric arm growth; the addition of new suckers, chromatophores, iridophores and leucophores; the development of skin sculptural components; and a horizontal pupillary response (Villanueva and Norman, 2008). At the same time, the Kölliker organs covering the body surface during planktonic life disappear, along with the lateral line system and the oral denticles of the beaks. This is a critical transitional period, which is poorly understood and requires focused research.

3.4. Juvenile phase

3.4.1 *Sepioteuthis lessoniana*

Hatchlings of this species are planktonic but hatch at a large size (4–6 mm ML and 44 mg WW; Hanlon, 1990) (Figure 1.12) compared to other loliginid squid and thus can be considered juveniles. Rearing from eggs has been successful in open sea water systems in temperate and tropical Asian countries (Ahmad and Usman, 1997; Choe, 1966a,b; Choe and Ohshima, 1961, 1963; Ikeda et al., 2003, 2009; Inoha and Sezoko, 1967; Nabhitabhata, 1978, 1996; Ohshima and Choe, 1961; Saso, 1979; Segawa, 1987, 1990, 1993; Sivalingam et al., 1993; Tsuchiya, 1982; Yu and Tung, 1989). Closed recirculating systems have also been used (Ikeda et al., 2003, 2009; Lee et al., 1994; Walsh et al., 2002) but these require costly upkeep and the addition of trace elements (to replace those lost by consumption, water filtration, and protein skimmers). Regarding tank design, several types have been used to rear *S. lessoniana* hatchlings and

Figure 1.12 *Sepioteuthis lessoniana* hatchling. Scale bar 1 mm. Original image.

juveniles, including small, cylindrical aquaria (Ikeda et al., 2003; Segawa, 1993) and raceways (Lee et al., 1994; Walsh et al., 2002).

Hatchlings start feeding just after hatching and, during rearing, have been observed to feed actively on live prey, such as juvenile atherinid fish and small mysid shrimp (Ikeda et al., 2003; Lee et al., 1994; Walsh et al., 2002). Ikeda et al. (2003) successfully fed them with frozen shrimp within a week of hatching. They have not been observed feeding on small zooplankton such as copepods, as other loliginid hatchlings do, probably because *S. lessoniana* hatchlings are much larger than other loliginid squid and prefer prey of their own size or slightly larger (Segawa, 1993). Juveniles have also been fed with a frozen diet (Segawa, 1993).

Feeding activity has been observed to change during early growth. Hatchlings and juveniles up to 40 mm ML feed actively during daytime mainly on small crustaceans and, as they grow, large juveniles change to nocturnal feeding and switch to larger benthic crustaceans and fish, although they seem to prefer the latter (Segawa, 1987, 1993). In spite of this pattern, feeding activity may also vary depending on prey type and behaviour. Cannibalism has been observed at times when a poor diet was provided and when individuals of different sizes (up to 4 × larger) were attacking the same prey (Segawa, 1993).

Daily food consumption rates of *S. lessoniana* reared at 23.5–26 °C and 33–34.5 psu varied with diet (Figure 1.13A), increasing with size up to 1.0 g WW reaching up to 70% (Segawa, 1995). Consumption rates ranged from 12% to 32% for individuals fed with atherinid fish and up to 72% for individuals between 0.1 and 1.0 g WW fed with mysids. Hatchling food conversion rates ranged from 18.6% to 43.8% (Segawa, 1990). Daily growth rates seem to be dependent on the diet provided. Hatchlings grow faster if fed with atherinid fish (attaining daily growth rates of 12–12.9% WW d^{-1}) instead of mysids (5.5–11% WW d^{-1}), but average daily growth rates were similar (Figure 1.13B) at around 6.5% WW d^{-1} for 10 mm ML squid (Segawa, 1990). Results from a study on the effects of temperature (10–30 °C range) on the oxygen consumption rate of *S. lessoniana* juveniles ($\leq$4.0 g WW) indicated a direct relationship between squid metabolism and body weight, showing sensitivity of hatchlings and juveniles (Figure 1.14A) to environmental temperatures (Segawa, 1995).

In a study spanning three consecutive generations (Ikeda et al., 2003), it was noted that overall squid survival could perhaps have been improved by increasing hatchling stocking density (to >150 individuals in tanks of 0.02–0.05 m^3). However, another study, which spanned seven consecutive

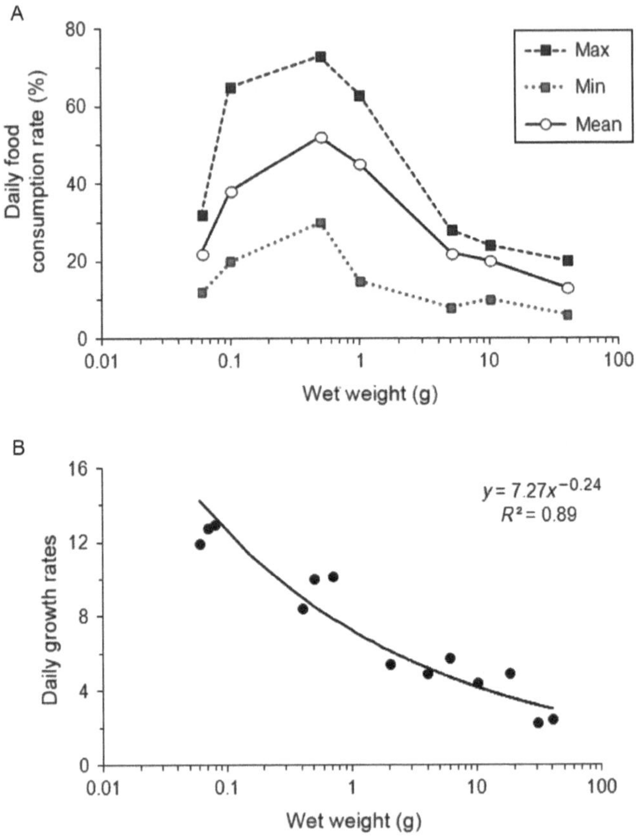

Figure 1.13 *Sepioteuthis lessoniana.* (A) Daily food consumption rates of juveniles; (B) daily growth rates of juveniles maintained between 23.5 and 26 °C from the first months of life. Note that daily growth rates decrease with increasing squid weight.

generations, observed survival rates for the transition from juvenile to grow-out ranging from 26% to 80%, even with much lower hatchling stocking densities (48 individuals in 3 m^3 or 16 individuals m^{-3}; Walsh et al., 2002).

3.4.2 *Sepia officinalis*

This species has pelagic–demersal hatchlings of 6–8 mm ML and 950 mg WW (Hanlon, 1990) and should be fed on live grass shrimp (*P. varians*) diet for the first 20–30 DAH and thereafter frozen grass shrimp (Sykes et al., 2006b). After this period, juveniles should have a fully developed digestive system and a mean WW of ~5 g (Sykes et al., 2006b). At this stage, they are ready to be transferred to tanks with increased bottom area, which can be a saline earth pond or an experimental fibre or plastic tank. If kept in tanks, the

Figure 1.14 Juveniles. (A) *Sepioteuthis lessoniana* (original image), (B) *Sepia officinalis* (original image), (C) *Octopus maya* (original image) and (D) *Octopus vulgaris*, 120-day-old (28 g) juvenile obtained in captivity with the utilization of PVC tubes and pebbles used as refuges during the settlement period (Iglesias et al., 2004). Images not to scale.

sea water systems and conditions used should be similar to those described for broodstock (see Section 3.1.2). Settings recommended for hatchlings and juveniles are the use of black tanks and low light intensities. At this point, new spawners should be selected from fast growers and deployed in outside tanks, while the other individuals should remain in the hatchery. For both hatchery and new spawner tanks, the number of individuals (up to 10 juveniles) and minimum bottom areas (1100 cm^2) should be taken into account (Sykes et al., 2003). The water height in the tanks should be low (less than 1 m) in order to favour prey encounters, thus reducing the energy allocated to catch prey (Figure 1.14B). Water height might be gradually increased as cuttlefish grow, providing increased water volume in the tanks (Forsythe et al., 1994).

If the individuals are to be cultured in ponds, abrupt changes in water temperature, salinity and pH must be avoided in order to reduce mortality rates associated with the transfer from the hatchery to the ponds. When individuals are kept in ponds, extensive culture should be used. In this case, live prey should be naturally present or deliberately introduced. Depending on the final size needed and the culture temperature, approximately 50–60 d are required to obtain individuals of 50 g at 25.0 ± 5.0 °C. Water inlet and outlet filters of appropriate size should be used to prevent both the entrance of predators and the escape of cuttlefish from the ponds. Predation by birds can be considerably reduced if a set of netting barriers is used to cover the tanks. The use of mechanical water oxygenators is recommended in order to keep adequate levels of DO (>80%), particularly during the night-time in the warmer months of the year.

A number of studies have addressed the densities to be used under different culturing conditions: closed, semiclosed or open water circulation. Forsythe et al. (1994) suggested a density of 20 cuttlefish m^{-2} in closed sea water systems during the juvenile phase but provided no information on the bottom area, an important parameter that is associated with culturing densities, as recognized by the ones from this paper authors. In a later study, Forsythe et al. (2002) suggested a density of 400 cuttlefish m^{-2} when using 1800 L circular tanks in a closed water system. These density values, coupled with culturing temperatures of around 25 °C, might be on the verge of impacting both growth and survival rates due to the increase of cuttlefish biomass in the tanks. Sykes et al. (2003) suggested culturing densities of 120 cuttlefish m^{-2} for 5 g juveniles and a minimum bottom area of around 1083 m^2 when using 10 L raceway tanks in a flow-through water system. These reference values might be used when culturing cuttlefish juveniles up to 25 g.

The effects of both density and bottom area on the growth of juvenile cuttlefish (Figure 1.14B) when using concrete raceway tanks in an open sea water system have been addressed by Domingues and Márquez (2010). Results from that investigation suggest that higher densities (33 individuals m^{-2}, mean weight of 9.5 g) and broader bottom areas might promote higher growth and survival rates and, furthermore, that irrespective of culture densities, higher survival and growth rates were obtained when using tanks with larger bottom areas. These results suggest that bottom area seems to be a more significant culturing parameter than density for cuttlefish. Thus, irrespective of the rearing/culture water system, density must decrease and tank bottom area increase with cuttlefish growth (Domingues and Márquez, 2010; Sykes et al., 2003). According to Forsythe et al. (1994), no substratum is necessary to obtain adequate growth and survival rates. In addition, this practice consistently facilitated cleaning, prevented disease and promoted good welfare conditions during cuttlefish rearing and culture (Sykes et al., 2012).

Culture density and bottom area are particularly important parameters when using earth ponds because high growth rates have been recorded on a pilot study under these conditions (Domingues, Sykes and Andrade, unpublished data). However, the rate of prey intake by cuttlefish can be reduced due to the high turbidity and the impossibility of cleaning ponds sufficiently, which might result in an excessive increase in nitrogenous compounds and a drop in DO. When coupled with increased carrying capacity, this situation could lead to mass mortalities and, eventually, a total loss of cuttlefish biomass.

As feed, live grass shrimp *P. varians* was used during the first 2 months after hatching and frozen *Carcinus maenas* for the remaining life cycle, with no significant effects on cuttlefish fecundity (Domingues et al., 2001b, Sykes et al., 2013a,b). However, *P. varians* was demonstrated to adequately replace *C. maenas* in the diet (Domingues et al., 2002) and a mixture of live *P. varians*, *C. maenas* and fish (Correia et al., 2005) delivered one of the highest individual fecundities ever recorded in the 250 L tanks (834 eggs $♀^{-1}$). The latter result might relate to temperature, cuttlefish density or sex ratio rather than to food quality, but this requires confirmation.

Information recently published by Sykes et al. (2012) provides the highest values for individual and overall fecundity (1383 and 16593 eggs, respectively) and acceptable individual fertility (72%) after feeding cuttlefish on live grass shrimp for the first 20–30 DAH and frozen grass shrimp from that period onwards. A hypothesis to explain this optimal result is that the use of frozen grass shrimp might influence cuttlefish fecundity by promoting lower energy expenditure associated with prey capture and feeding. However, tanks with increased bottom area were used, so factors other than diet might have affected fecundity and further research will be needed to clarify these findings.

3.4.3 *Octopus maya*

Like cuttlefish, *O. maya* hatches as a holobenthic animal with similar gross external morphological features as the adult. Nevertheless, it has been observed that hatchlings pass through a transition period before they acquire all the characteristics of juveniles (Moguel et al., 2010). Newly hatched individuals have a mean WW of 0.13 ± 0.03 g and a mean DW of 0.03 ± 0.01 g. To evaluate hatchling quality, fasting resistance tests are regularly used, where 25 hatchlings are isolated and fasted for 10 d at 24 ± 1 °C, 36 psu and DO > 5 mg L^{-1}. This test assumes that the hatchlings are of high quality when their survival rate is higher than 95%, since it means that their internal yolk reserves are sufficient to maintain the animals during this critical period (Moguel et al., 2010).

Juvenile growth is highly variable and influenced by biotic and abiotic factors such as temperature, hatchling size, food type and yolk reserves (André et al., 2008; Briceño et al., 2010; Leporati et al., 2007; Vidal et al., 2002b). Nevertheless, between 5% and 10% of hatchlings are born with an external yolk sac (depending on temperature stability), suggesting that *O. maya* passes through a postembryonic transition period during which the absorption of external and internal yolk supplies is completed and all the characteristics of juveniles are acquired. Histological, enzymatic and behavioural analyses concluded that this external yolk is a common characteristic of *O. maya* hatchlings (Moguel et al., 2010; cf. observations also for *O. joubini* and *Enteroctopus megalocyathus*; see Boletzky and Boletzky, 1969; Ortiz et al., 2006).

Octopus maya hatchlings at 1–4 DAH have the following characteristics: (1) necto-benthic behaviour; (2) no response to stimuli from prey; (3) a negative growth phase (cf. Vidal et al., 2002b); (4) arms proportionally shorter than the mantle; (5) reduction in total body lipid concentration, indicating that yolk lipids may be used as the main source of energy; (6) presence of residual inner yolk sac and immature digestive gland cells; and (7) digestive enzymatic activity directed to process internal yolk. At 5–10 DAH, food consumption is erratic while octopuses swim inside the tank, spending some time on the upper part of artificial grass (15 cm length; Figure 1.14C), which is offered as additional refuge during rearing (Rosas et al., 2008). Thus, during the first 10 DAH, morphometric studies reveal a phase where no net growth takes place, as observed also for loliginid squid (Vidal et al., 2002b, 2005).

Histological analysis reveals that digestive gland morphology changes with age (Martínez et al., 2011), from a simple tubular gland in 2 DAH octopuses to a tubulo-acinar and vacuolar structure with digestive cells characterized by vacuoles in 45 DAH octopuses. At 2–3 DAH, attacks are rarely seen in response to either visual or both visual and chemical stimuli from prey. At 4 DAH, octopuses respond to visual stimuli from crabs and palaemonids but do not display any preference in attacking either prey type. At 3–5 DAH, the digestive gland has no well-defined cellular and tubular structures, but a basement membrane delimits the undifferentiated tubules, demonstrating that the remaining yolk is used as the main source of energy in this transitional stage (Figure 1.15A). However, the development of the digestive system begins with the first feeding (even if some internal yolk

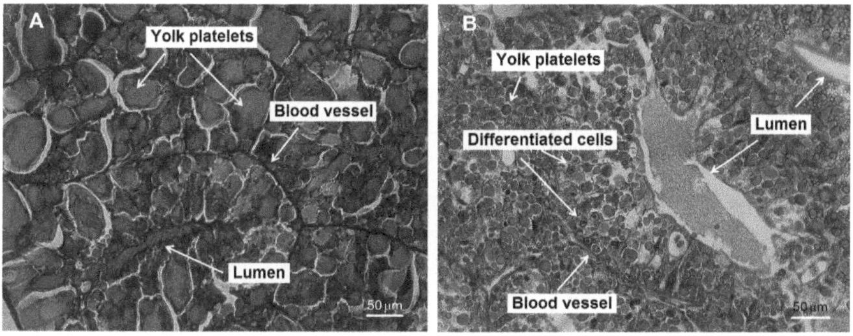

Figure 1.15 *Octopus maya*. Cytological ontogeny of the digestive gland. (A) Three days after hatching and (B) seven days after hatching. Scale bar 50 μm.

remains), and at 6–7 DAH, the subcellular organization pattern of the digestive gland becomes more evident and the lumen opens to accept chyme and begin absorption (Figure 1.15B), suggesting that the stimulation for cellular differentiation happens at this moment. In addition, digestive enzyme synthesis starts with the digestion and absorption of nutrients from the consumed food (Moguel et al., 2010). Digestive enzyme activity is erratic until 14 DAH, after which the activity begins to stabilize.

All of the above-mentioned characteristics of early *O. maya* juveniles have consequences for culture. Firstly, at 24 °C, immediately after hatching, only a half ration of feed should be given, considering that during at least the first 7–10 DAH, animals will consume the remaining yolk as their principal source of energy. Additionally, during yolk absorption, cannibalism is practically absent, allowing the maintenance of juveniles in high densities (150 octopus m^{-2}) without any mortality attributable to cannibalism. Once maturity of the digestive gland is complete, the voracity of juveniles will increase considerably, intensifying predatory behaviour and cannibalism if there is insufficient feed. From that stage, a density of 25–50 octopus m^{-2} is recommended.

3.4.4 Octopus vulgaris

Recently settled individuals (Figure 1.11F) show a preference for shade and exhibit reclusive behaviour, using refuges provided, such as gastropod shells, and capturing food from the bottom of the tank. At this time, a transition in diet is needed, and the concentration of *Artemia* offered should be gradually reduced. An inert diet composed of either mussel, crab muscle and sea urchin (Iglesias et al., 2004) or frozen mysidaceans (Carrasco et al., 2003,

2006) has produced some positive results in passing through the settlement period, but in both the studies reported, very high mortality rates were registered. During settlement, Iglesias et al. (2004) used 500 L square grey PVC tanks (1 × 1 × 0.5 m) and provided an artificial habitat made of gravel and pebble with small-angled PVC tubes as shelters (Figure 1.14D). However, these substrates and diets were only used on an experimental scale and focused research is required to understand the morphological, behavioural, physiological and nutritional requirements of this critical transitional period in the life cycle of *O. vulgaris*.

After the settlement period, which takes about 2 weeks, food is gradually changed to a thawed diet of 75% crab (*C. maenas*, *Macropipus depurator* and *Polybius henslowii*), 20% fish discard (*Micromesistius poutassou* or *Sardina pilchardus*) and 5% mussel (*Mytilus galloprovincialis*). The same 500 L tanks can be used but shelters consisting of larger PVC tubes should be supplied, and the habitat can be changed when cleaning the bottom of the tank is necessary. Under these feeding and habitat conditions and at ambient photoperiod and temperatures (17–19 °C), juveniles can reach 500–600 g after 6 months (Iglesias et al., 2004).

3.5. Grow-out

3.5.1 *Sepioteuthis lessoniana*

Intensive closed water systems can provide the high-quality water required for mass culture of this species but maintenance is costly. Lee et al. (1994) had grown-out squid in large raceways (6.1 m length, 2.4 m width, 0.9 m depth and 15,000 L) and reared early juveniles in a 6.5 m circular production tank equipped with a similar filtering system (1.8 m circular filtration tank) used for rearing early juveniles. Open water tank systems at shore-based facilities are an alternative method for culturing *S. lessoniana*, as long as good quality water is readily available. Floating net cages have also been used successfully for culturing this species during relatively short cultivation periods (Ahmad and Usman, 1997; Nabhitabhata, 1996; Nabhitabhata et al., 1984; Saso, 1979; Sugita, 2012). However, damage to squid by stormy weather was one of the main problems of the floating cages and Saso (1979) planked the sides of the cage to modify the influence of current and wave action. Experimental studies undertaken to evaluate the usefulness of keeping broodstock and grow-out juveniles in floating cages yielded good results using a 4 m^3 net cage (Wada and Kobayashi, 1995) and a 5 m^3 net cage with a shielding net to reduce light intensity (Sugita, 2012).

Adult *S. lessoniana* show crepuscular feeding activity, with variation according to changes in prey type and behaviour, although their main prey item seems to be fish (Segawa, 1993). Based on individual measurements, Segawa (1990) estimated daily food consumption rates for squid of >3.0 g WW to range from 5% to 36%, being higher when individuals were fed with squid hatchlings and lower with a mysid diet. Food conversion rates ranged from 15.8% to 43.8%, 18.6% to 35.6% and 25% to 34.1% for diets of atherinid fish, mysid and squid hatchlings, respectively. Daily growth rates decreased with increasing individual size for both ML and WW measurements, ranging from 1% to 3% WW d^{-1} in squids >50 mm ML and from 1.8% to 4.1% WW d^{-1} in squids weighing 30–50 g WW (Figure 1.13B) (Segawa, 1990). Instantaneous growth rates calculated for the entire life cycle have been observed to decline with subsequent generations (Lee et al., 1994).

Juveniles, subadults and adult squids grow fast when fed exclusively on live prey. Nevertheless, it has been observed in several studies that older individuals will eat frozen/dead food in captivity, which represents an economic alternative for mass culture (Ikeda et al., 2003, 2009). Walsh et al. (2002) pointed out that, in mass culture (Figure 1.16), there may be a trade-off between the high cost and labour-intensive provision of live food and the intensive behavioural training required to enable squid to feed successfully on freshly dead food. As such, food supply, including the exploitation of artificial diets, is the main bottleneck for the implementation of mass culture for *S. lessoniana.*

Figure 1.16 *Sepioteuthis lessoniana.* Grow-out in raceways (6.1 m length, 2.4 m width, 0.9 m depth and 15,000 L) at the National Resource Center for Cephalopods in Galveston, Texas, the United States. *Photo courtesy of J. Forsythe.*

Temperature is one of the most important environmental factors for grow-out in this species: higher temperatures result in shorter developmental times, higher activity, active feeding and higher growth rates within the range of suitable water temperature for the species (Ueta, 2000).

The lifespan of *S. lessoniana* in captivity has been found to range between 5.6 and 8.7 months (Lee et al., 1994; Walsh et al., 2002), although strongly affected by temperature, as indicated by comparing different studies (Lee et al., 1994; Nabhitabhata, 1995; Segawa, 1987; Walsh et al., 2002). It seems that laboratory culture may also lengthen the life cycle of this squid species (Lee et al., 1994). The maximum size attained in captivity was 360 mm ML and 2200 g in 184 d (Hanlon, 1990). Age at first mating was observed as early as 130–160 DAH and spawning at 153 DAH at 23–25 °C in the experiments by Walsh et al. (2002), over seven consecutive generations (but size at maturity decreased with subsequent generations). Marketable size can be attained about 4 months after hatching.

3.5.2 *Sepia officinalis*

Cuttlefish are usually conditioned in open sea water systems due to logistics, welfare and economic aspects. Nonetheless, as technology used in closed and semiclosed systems is becoming cheaper and aquaculture environmental concerns are scaled up, the use of closed sea water systems similar to those reported by Hanlon (1990) for cephalopods and recently by Martins et al. (2010) for fish is predicted. Recent investigations suggest an ideal tank volume of 9000 L (Sykes et al., 2013a), cylindrical in shape and made of fibreglass, with sufficient airlifts fixed on the tank walls and air stones arranged centrally to maintain steady water flow towards the outlet piping at the centre of the tank. Since cuttlefish are benthic organisms (Boletzky, 1983), culture tanks should have maximum base area (Domingues and Márquez, 2010; Sykes et al., 2013a). If closed and semiopen systems are used, UV sterilization will be necessary to avoid disease (see Section 2.2.7).

The tank environment should enable cuttlefish to camouflage themselves easily; tanks should be kept under low light intensities (<200 lux) and a normal photoperiod replicating natural geographic conditions during spawning in the wild. This may be achieved using a combination of natural or artificial light sources and tank colour. If placed outdoors, reduced light intensity can be afforded with water-repellent masking nets, which will also prevent pH and salinity decline due to rain fall. Temperature should be maintained at 23 °C with high levels of DO.

Currently, the grow-out of cuttlefish juveniles uses crustacean feed, mainly grass shrimp *P. varians* (Sykes et al., 2006a). Nonetheless, several different food items have been tested, either solely or as mixed diets (e.g. Almansa et al., 2006; DeRusha et al., 1989; Domingues et al., 2001a,b, 2002; Sykes et al., 2006a). Despite all this effort, it is not yet economically viable to produce cuttlefish in large numbers with a grow-out based on wild-collected food because the amount of feed biomass required is high and availability is too low. Attempts to feed cuttlefish on a prepared diet have failed to achieve suitable growth and survival rates (Castro, 1991; Castro and Lee, 1994; Castro et al., 1993; Domingues et al., 2005, 2008; Ferreira et al., 2010; Lee et al., 1991), so further research is necessary to produce a well-designed, inexpensive and storable artificial diet for *S. officinalis*.

Development time to attain marketable size will greatly depend on the product desired. For instance, in some Mediterranean countries, such as Portugal and Italy, undersized cuttlefish individuals (5–25 g) are prized and their commercial value is higher than that of animals weighing over 100 g. Time to market will also be dependent on culture conditions, especially temperature and food. *Sepia officinalis* is cultured extensively in Portugal, Italy and Tunisia, where human action is not involved at the feed level, with eggs caught and left in earthen ponds with naturally occurring prey, for animals to grow and be collected a few months later on reaching marketable size. According to Palmegiano and Sequi (1981), in this type of culture, 150–300 g of eggs will produce 800–1200 kg of cuttlefish with a mean WW of 40–80 g. The semi-intensive experiments performed in Italy in the 1980s obtained fast growth rates: 14.2 g in 60 d at 21–24 °C in ponds, 25 g in 40 d and 80 g in 100 d at 21–24 °C in concrete tanks (Palmegiano and Sequi, 1981, 1984) and 80 kg/Ha in 90 d at 21–24 °C in net cages in ponds (Sequi and Palmegiano, 1984).

Analysis of growth rates obtained during a set of trials using earthen ponds (Domingues, Sykes, Andrade, unpublished data) indicates that if sufficient food is available and other important conditions are met, cuttlefish will grow faster outdoors than indoors, which is in accordance with the findings of Domingues et al. (2006). Thus, it is possible to grow individuals to marketable size in 2–3 months or even less at water temperatures ranging from 23 to 25 °C (Figure 1.17). Market prices for *S. officinalis* are high and the short life cycle will support an easy short-term return on investment. In fact, the return on investment can be shortened to become more profitable if cuttlefish culture uses integrated aquaculture. The increase in nitrogenous compounds originating in the excretion products from cuttlefish metabolism will boost primary production, providing an increase in secondary production that will benefit other cultured species inside the ponds.

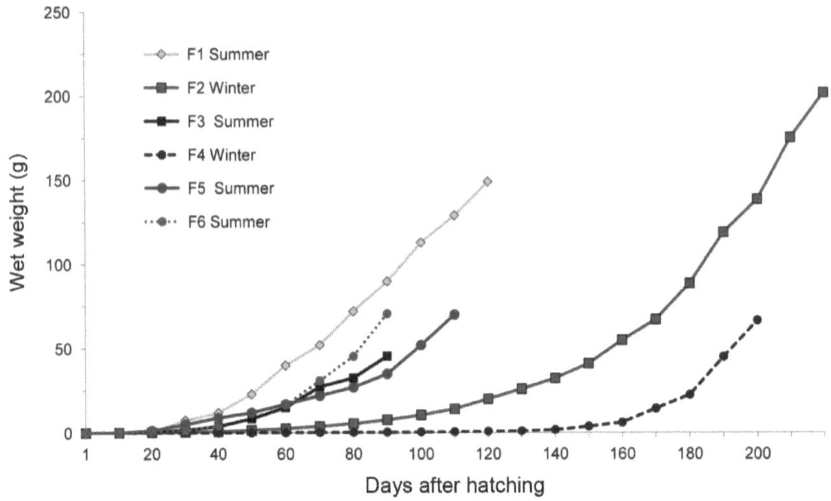

Figure 1.17 *Sepia officinalis*. Growth rates obtained during six consecutive generations (F1 to F6), fed with live *Paramysis nouveli* during the first 15 d after hatching and live or frozen *Palaemonetes varians* from this period onwards.

3.5.3 Octopus maya

Juveniles are seeded at a density of 50 individuals m^{-2} in prefattening dark (30 lux cm^{-2}) ponds of 7.5 m^2 at 25 ± 1 °C (Figure 1.18). Conch shells (*Melongena corona bispinosa* or *Strombus pugilis*) are used as refuges because they have only one entrance, which reduces the vulnerability of animals under attack attempts by other octopuses.

Temperatures around 24 °C favour growth and the energetic balance of *O. maya*, probably due to several factors acting simultaneously: (1) a reduction of basal metabolism and consequent reduction in overall activity and (2) a reduction of digestion rates and a better assimilation of nutrients, including fatty acids, from food (Noyola et al., 2013a,b). It has been shown that high ingestion rate and efficient use of fatty acids (as well as other nutrient categories) occur at lower temperatures, beyond the automatous effect that temperature has on enzyme kinetics. It has also been observed that juveniles maintained at a mean temperature of 24.3 °C channel a high proportion of ingested energy to biomass production, suggesting that this is the optimum temperature for the growth of *O. maya* (Noyola et al., 2013b).

The prefattening production system has been designed to maintain water quality using a partially recirculating system in which sea water drains to a reservoir where it is filtered through a vertical filter (15 μm), a protein skimmer, filter bags (100 μm) and a rotational water filter. A sea water exchange

Figure 1.18 Prefattening 7.5 m^2 tanks designed to maintain *Octopus maya* juveniles during the first 60 culture days. Conch shells are offered as refuges for the octopuses.

of 3% d^{-1} is supplied to compensate for water evaporation and to maintain salinity between 36 and 40 psu. Octopuses are weighed and separated into groups of similar size at 30 DAH in an attempt to avoid cannibalism, helping to maintain survival rates around 60–70%. In general, due to their voracity, *O. maya* juveniles (as other cephalopods) are highly sensitive to lack of food (see Vidal et al., 2006). Cannibalism is enhanced under culture conditions when the quality or quantity of food does not satisfy their basic requirements (Aguila et al., 2007; Domingues et al., 2007; Rosas et al., 2011). For that reason, *O. maya* juveniles are fed with a daily ration of 90–100% of their WW, divided into three portions. When this ration is not offered, the incidence of cannibalism increases and survival can be lower than 5% per tank, producing few but larger octopuses.

After the prefattening period (60 d), octopuses are placed in 6 m diameter outdoor ponds coupled to a system of recirculating sea water and protected with masking net to reduce the intensity of the ambient light (Figure 1.19A). Airlifts are used for aeration to avoid problems with bubbles, which arise if direct aeration is supplied. In ponds where aeration is supplied only with air stones, air bubbles can be trapped by the interbrachial membrane, producing serious injuries to the octopuses. When airlifts are used, this problem is eliminated and, at the same time, a better water circulation can be obtained (Figure 1.19B).

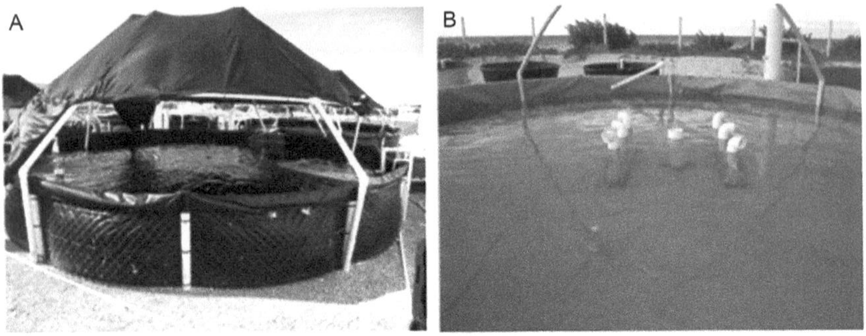

Figure 1.19 Outside ponds used for *Octopus maya* culture. (A) Black liner; (B) grey liner. Note the airlift system placed either laterally (A) or at the centre of the pond (B).

Animals of approximately 1.5–3 g WW are placed in outside ponds at a density of 28 individuals m^{-2} and maintained in these ponds for another 60 d until harvest. During this period, octopuses are weighed every 30 d and separated according to weight to avoid cannibalism. In the Yucatán Peninsula where *O. maya* is cultured, the temperature in the ponds fluctuates annually between 22 and 30 °C and the best growth results have been obtained between 24 and 26 °C (Domingues et al., 2012).

During the prefattening period in the outside ponds, octopuses are fed with a paste made with a combination of fresh squid and crustacean meat, vitamins and minerals, which is placed into bivalve shells for support (Figure 1.20A) and stored frozen.

Three refuges per individual are provided, reducing interaction, and consequently cannibalism, among juveniles. Survival in the outdoor culture is usually 50% at harvesting, when animals reach 120 g WW. Typical growth rates for the first 120 DAH are around 6.8% WW d^{-1}, when octopuses are cultivated at a density of 28 individuals m^{-2} in outdoor ponds at 28 ± 2 °C (Figure 1.21).

At present, the wives of fishermen belonging to the *Moluscos del Mayab* cooperative are cultivating *O. maya* juveniles in a semipilot-scale facility in Yucatán, Mexico (Figure 1.20B). The cooperative operates the entire grow-out system successfully, demonstrating that culturing this species can be used to improve the quality of life of economically constrained coastal communities.

Octopus maya production is currently channelled to gourmet markets where chefs have been testing the product. In the local gourmet market, the price of cultured *O. maya* juveniles reaches values around € 27 kg^{-1}, with production costs of € 18.75 kg^{-1} in land-based tanks on a semipilot

Figure 1.20 (A) Bivalve shells with paste to feed *Octopus maya* juveniles. (B) The shells with feed are prepared by the fishermen's wives of Sisal, Yucatán, Mexico.

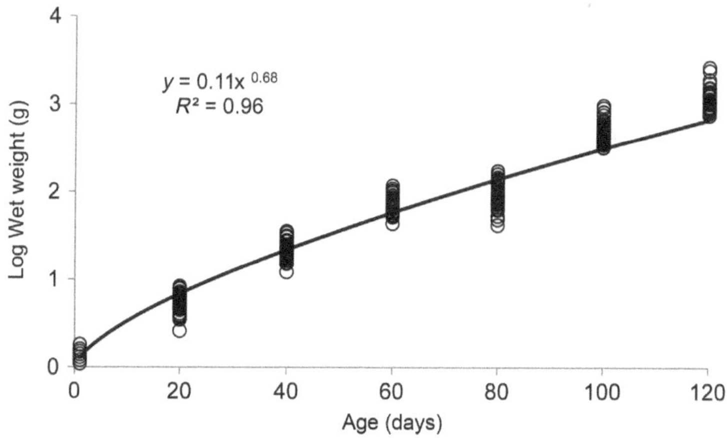

Figure 1.21 *Octopus maya*. Growth curve when cultivated at a density of 28 individuals m^{-2} in outdoor ponds at 28 ± 2 °C ($N = 300$).

experimental level. Although this small-scale production cannot be translated to industrial levels, the figures are a reasonable enough estimate to demonstrate that industrial production of *O. maya* in land-based tanks could be profitable.

3.5.4 *Octopus vulgaris*

Industrial production of *O. vulgaris* subadults in Galicia (NW Spain) began in 1996 after information on growth rates was published by Iglesias et al. (1997) and Rama-Villar et al. (1997). In 1997, 12 t of octopus was produced in Galicia, increasing to 15 t in 2000. At present, however, only one company is in the market, with annual production at 8–10 t. The ongrowing production system consists of small metallic sea cages (3–8 m^{-3}; Figures 1.22 and 1.23) suspended in the traditional mussel rafts of the Galician Rías (Figure 1.22). Fishermen grow *O. vulgaris* subadults of 0.8–1.0 kg up to 2.5–3.0 kg in a period of 3–4 months at temperatures of 12–19 °C. Initial density is 10 kg m^{-3}, yielding 40 kg m^{-3} after the 3–4 month period (Rey-Méndez et al., 2003). Survival is higher in winter than in summer, and under optimal environmental conditions (autumn), it is possible to attain 80% survival.

Depending on water temperature range and variation, from one to two fattening cycles may be carried out per year in the Mediterranean and up to three in the open sea on the Atlantic Coast. However, Aguado Giménez and García García (2002) had pointed out that mortality can increase considerably in the Mediterranean area if temperatures rise above 22 °C. The Galician octopus grow-out model was tested by Rodríguez et al. (2006)

Figure 1.22 Facilities for ongrowing *Octopus vulgaris* used in NW Spain, consisting of 8 m^3 metallic sea cages suspended from traditional mussel rafts in the Galician Rías, Spain. Original image.

Figure 1.23 Detail of the octopus refuge arrangement inside a 3-m^3 sea cage suspended in the Galician Rías, Spain. *Photograph courtesy of J. Sánchez and J. Iglesias.*

on the northern coast of Spain and by Estefanell et al. (2012b) in the Canary Islands, comparing the performance of benthic cages versus floating cages. Both studies obtained similar growth and survival results.

The diet used by fishermen during grow-out is composed of fisheries discards, mainly crustaceans and fish, and occasionally mussels (Chapela et al., 2006). García García and Cerezo Valverde (2006) demonstrated that the best combination of food is 50% *Carcinus* spp. and 50% bogue (*Boops boops*). Estefanell et al. (2012a) also found the best growth results using *B. boops* discarded from fish farms (40%) combined with blue crab (60%). Recent research on the elaboration of octopus artificial feeds has been reported by Cerezo Valverde et al. (2008), Quintana et al. (2008) and Estefanell et al. (2012a), who developed formulated feeds in a form acceptable to octopus, obtaining good growth rates (0.7–1.5% WW d^{-1}). These feeds use a paste containing fish, crustaceans or molluscs mixed with different binders (alginates or gelatins), producing a moist feed (>70% water) with a texture suitable for manipulation and ingestion by octopus. Gelatine was found to be more suitable than alginates for elaborating feeds because of its better acceptability and digestibility.

Recent economic analysis of profitability in the Mediterranean (García García and García García, 2011; García García et al., 2009) found that production costs for land-based tanks (€ 8.97 kg^{-1}) are noticeably higher than for open sea cages (€ 6.61 kg^{-1}), a difference attributable to the energy cost required for pumping water and the necessary maintenance of land-based facilities. To decrease production costs, it was suggested that octopus fattening could be integrated as a complement to fish farming, such as gilthead bream, sea bass or tuna, particularly during winter when low temperatures cause a decrease in the feeding rates of these species, which is perhaps the most important economic variable in the fish growing process.

According to the octopus-producing companies consulted, obtaining subadults and their feed is the main problem impacting the profitability of the ongrowing process, as it is also dependent on the large annual catch fluctuations of the fishery. Consequently, securing massive production of subadults from hatcheries and the provision of a commercial feed rendering high performance in terms of growth, conversion and survival rates are the future aims for the *O. vulgaris* fattening industry.

4. WELFARE AND ETHICS WHEN WORKING WITH LIVING CEPHALOPODS IN THE LABORATORY

It has recently become generally accepted institutionally that cephalopods, and indeed other invertebrates, should be accorded the respect and treatment required for vertebrate animals in assessing appropriate standards and protocols for keeping them, performing experiments on them and killing them. The topics of welfare and ethics concerning cephalopods have received much recent attention because it is only relatively recently that they have been considered overtly (e.g. Mather and Anderson, 2007). Legislative changes, particularly in Europe, have focussed attention on these important issues and the research community has been quick to recognize and respond to the necessity of compiling appropriate guidelines. With reference to some of these guidelines (particularly the extensive review by Moltschaniwskyj et al., 2007), this section summarizes the main points.

The species chosen for this chapter are the best known and understood, so using these species satisfies the ethical requirement for *refinement* (i.e. using species for which the details of handling and feeding are well known and controllable; so the chances of animals suffering or dying through lack of information about how to care for them should be minimal). The other two of the three "R's" (see, e.g. Moltschaniwskyj et al., 2007) are *replacement* (of animals for experimentation; e.g. by substituting appropriate cell culture preparations or computer models where feasible) and *reduction* (paring back the absolute number of individual animals required for a given experiment).

A basic ethical tenet for considering animal welfare is "do as you would be done by"—to treat animals with respect and ensure that they are not exposed to conditions that would cause pain, distress or suffering if the animal were a human in similar circumstances. This rule of thumb is useful because the inability of most animals to communicate their state of being to humans can make it extremely difficult to gauge the extent of suffering that they might be experiencing.

Appropriate ethics therefore require that the level of possible suffering during experiments is considered and alternative ways to reduce suffering are sought. For example, daily intramuscular injections of a drug are very traumatic for the animals concerned, but trauma can easily be avoided if the animals are first anaesthetized (e.g. by immersion in sea water containing a suitable anaesthetic, such as magnesium chloride or ethanol; Gleadall, 2013a; Gleadall et al., 1993).

The term "welfare" encompasses the entire range of human contact with animals, including the manner of their capture and transportation, housing, feeding, well-being, their use in experiments, and finally humane killing at the end of terminal experiments or in preparation for use as food (Andrews et al., 2013; Buchanan et al., 2013; Crook and Walters, 2011; Harvey-Clark, 2011; Leary et al., 2013; Mather, 2011; Mather and Anderson, 2007; Moltschaniwskyj et al., 2007). Recommended capture methods include pots for octopuses, jigs for squid and hand nets for small cephalopods and sepiolids (Moltschaniwskyj et al., 2007). Transportation and suitable accommodation are dealt with in other parts of this chapter. This section considers the remaining aspects of welfare relevant to culturing cephalopods. For further breadth and depth, see the references just mentioned.

It is also necessary to consider factors of the life of cephalopods that are less obvious to us because of the requirement for living in sea water. Thus, welfare includes ensuring that aquaria are clean and contain sea water in good condition: Small deviations from normal temperature, pH, salinity, light intensity and dissolved organic and inorganic substances can cause stress; and large deviations can result in suffering and death. Ink released during startle responses must be efficiently removed by the system; otherwise, it can be inhaled and will fatally interfere with the respiratory functioning of the gills.

Checking on captive animals regularly is also important, to ensure that they are not sick or injured: for example, checking for skin lesions, which can indicate a number of different problems that might not be observed directly, such as aggressive behaviour, physical manifestation of stress or disease through gnawing their own arms or desperate attempts to escape the confines of the aquarium where an animal may repeatedly dash itself against the aquarium wall. Since they are soft-bodied creatures (basically shellfish without a shell), keeping cephalopods in featureless aquaria in large numbers is in itself stressful, so provision of individual refuges (such as suitably sized pots or pieces of nontoxic plastic tubing, well in excess of the number of animals present) is a minimum requirement for the comfort and security of octopuses; and patches of real or artificial sea grass can provide a refuge zone for squid or cuttlefish.

Provision of ample suitable food is very important not only to maintain health but also to discourage cannibalism. Feeding should not be excessive, however, as food that is not eaten will quickly affect water quality. Strategies must be in place for dealing with egg masses laid in captivity, as these can be another source of pollution if they do not hatch; and if they are to remain, the system must be able to retain young animals in the aquarium rather than submitting them to suffering and unintentional death within filtration units, and unintentional death within filtration units and equipment such as protein skimmers.

Abnormal behaviour (e.g. stereotyped behaviour, such as continuous swimming back and forth) may indicate stress at merely being in the captured condition, so it is important to find ways to overcome this stress through consideration of appropriate forms of enrichment (such as providing interesting objects that animals can play with, novel ways for the animals to obtain food or even just presenting food items at different times of day; see, e.g. Anderson and Blustein, 2006; Anderson and Wood, 2001).

Finally, the death of cephalopods is another important consideration. Frequent inspection of cephalopod aquaria (at least twice a day) is crucial to ensure that dead animals are quickly recognized and removed before they can pollute the system, as this can quickly result in the death of all system occupants. Cephalopod flesh decomposes extremely quickly and is a potential health hazard, so carcasses should be frozen and stored until appropriate hygienic disposal, preferably by incineration. Animals destined for market should be free of disease and swiftly and humanely killed. Rapid freezing is one practical method (e.g. Badiani et al., 2013), and another is humane killing by destroying the brain with a cut between the eyes that also severs the cephalic artery and major nerves to the body. Recognizing that suffering and stress also result in inferior quality of seafood products, a procedure now routine in many companies specializing in the supply of high-quality fresh squid in Japan is a brain-destroying coup de grâce by swift insertion of a spike into the mouth, up through the beaks and buccal mass into the head (Masami Abe, personal communication). This procedure also leaves the animals outwardly intact, maintaining a high market value. Further processing should follow standard Hazard Analysis and Critical Control Point (HACCP) guidelines (e.g. FAO, 2009).

5. CURRENT LIMITATIONS

As can be understood from the preceding account, there has been significant progress in cephalopod culture methods and it is now possible to maintain many species for relatively long periods in laboratories and aquarium facilities. However, the more rigorous demands required for aquaculture mean that for many species, there are still some major obstacles to overcome before it becomes possible to sustain them throughout multiple life cycles to a level of production that is industrially stable and economically viable. This section emphasizes the major obstacles defining the current limitations on developing cephalopod aquaculture into an industrial-scale enterprise.

Basic culture system design for cephalopod aquaculture has progressed from rectangular–cubic to circular–cylindrical designs, populated with simple pipe-based housing units for octopuses arranged more or less at random but yet to take full advantage of the available three-dimensional water volume. Feed delivery in large aquaria is generally by hand and random, with a few animals tending to dominate feed reception, resulting in variable growth rates and consequently a broad size range at the end of each production cycle.

Water quality requirements raise the risks for large-scale (particularly land-based) aquaculture in the absence of an open sea water system that is stable in composition over the long term, and of high quality, since all cephalopods are highly vulnerable to natural and man-made pollutants and to sudden changes in parameters such as temperature, salinity, pH and concentrations of various solutes. It is therefore necessary to budget for sufficient monitoring, particularly regular measurement of pH and the content of nitrogenous compounds as the cultured biomass increases, taking care that the capacity of the system is not exceeded. Seeding aquaculture systems requires that temperatures are kept close to those of the source population (Boyle, 1991), which demands careful budgeting of the energy requirements for cooling and heating if local and source temperatures are significantly different.

One of the greatest bottlenecks to culture is the production of sufficient quantities of quality offspring. This is directly related to the control of reproduction that remains as a cornerstone for the development of cephalopod aquaculture. The quality and quantity of eggs and hatchlings produced will be reliant on the environmental influences during the maturation process and during embryonic development. It has been shown that egg viability is reduced after several culture generations due to inbreeding (e.g. Walsh et al., 2002); thus, control over reproduction is necessary to achieve mass production. The paralarval phase and the settlement period found in a number of small-egged species have proven to be critical bottlenecks: thus far, all species with a paralarval phase can only be raised experimentally in numbers too small to be commercially viable.

Nutrition is a problem common to culturing most species and many phases and stages in the life cycle because little is known about (1) what cephalopods at different life stages eat in the wild, (2) their nutritional physiology and (3) how and when to present them with appropriate food items in an enclosed system. Advances have been made in artificial feed formulations but there are still no available routine methods for rearing and providing suitable living prey items to paralarvae on a commercial basis and for relating this impact on profitability as there is yet no completely reliable artificial feed available for any cephalopod. The main limitations here are that cheap carbohydrate sources of feed cannot be used for cephalopods, which require high-protein diets fortified with certain lipids. There is ample evidence for different quantitative (as well as qualitative) requirements at different life stages, particularly during the posthatching period when the presence of yolk can delay the start of predatory responses to food items. Consequently, incomplete information about the timing of feed introduction and the appropriate amounts to introduce is therefore another limiting aspect for several species.

The extraordinarily high intelligence of cephalopods has led to the recognition of the importance of enrichment (e.g. Wood and Wood, 1999), such as play and other forms of stimulation that might provide distraction from negative behavioural traits such as cannibalism (Budelmann, 2010). The absence of opportunities for enrichment in the captive environment, particularly for octopuses, may detract from optimum health and therefore from the optimum growth required for commercial aquaculture.

In terms of possible future limitations, the value of polyculture (to overcome limitations of yield per unit water volume) at the commercial level has yet to be investigated. Algae have been used as part of in-series filtering during water reprocessing in closed circulations (see, e.g. Yang et al., 1989), but polyculture in terms of attempts to raise scavengers on waste and uneaten food from octopus culture systems is largely unexplored (cf. Al-Hafedh et al., 2012; Martínez-Porchas et al., 2010).

The ethical dilemma of culturing such intelligent animals as cephalopods as a food source has yet to be explored as a possible limitation. There is the potential for groups in some countries to shun the aquaculturing of octopuses, in particular, in ways that parallel taboos against the use of certain vertebrate animals as a food source for human consumers.

6. CONCLUSIONS AND RESEARCH PRIORITIES

While there has been significant progress in cephalopod culture methods, particularly during the last three decades, much remains to be done. This section explores the research priorities critical to advancing efforts to culture cephalopods according to the current limitations identified in the previous section and presents the main conclusions of this chapter.

There is no commercial-scale culture of cephalopods but some species, such as the four biological models discussed here, may represent the bulk of mass-cultured cephalopods in the near future. These species were selected for their very attractive aquaculture characteristics, such as high rates of growth and food conversion, which for aquaculture translates into short culture cycles, high ratios of production to biomass, high cost-effectiveness and well-established markets. The scientific objectives of laboratories actively working with live cephalopods around the world determine the target species selected for experiments, which usually correspond to local coastal species easily adapted to water systems in captivity.

From current knowledge, it is known that water quality is of critical importance and should be closely monitored according to each species and developmental stage. Water quality should be maintained during housing and transportation to ensure optimal survival and to avoid diseases. Studies on the behaviour and neurophysiology of cephalopods have underscored their highly developed nervous system, the most advanced among invertebrates. Indeed, their highly developed sensory capacity also stands as a clear indication of their capability to suffer pain. Accordingly, in recent years, it has become generally accepted that cephalopods should be maintained under approved ethical and welfare conditions similar to those required for fishes in relation to maintenance, anaesthesia and killing humanely with suitable techniques (Fiorito et al., 2014; Moltschaniwskyj et al., 2007).

One of the main bottlenecks for the culture of cephalopods and that should be ranked as a high-priority research topic is the control of reproduction. Research on methods to optimize broodstock efficiency, including accelerating or decelerating sexual maturity and inducing or delaying spawning, will help advance efforts to culture these animals. Genetic selection or manipulation of cephalopod species for aquaculture suitability is largely unexplored (cf. Benzie, 2009; Gjedrem, 2010; Guo, 2009; Nguyen et al., 2009). For instance, the hormone progesterone is thought to regulate vitellogenesis in *Octopus*, as its application to immature ovary initiates the process (Di Cristo et al., 2008). As a steroid, it is likely that progesterone is bound by a nuclear receptor and acts as a transcription factor to regulate gene expression after ligand binding (Bain et al., 2007). Thus, analysis of the nuclear receptor of progesterone in *Octopus* ovary could be used to identify the set of genes indirectly targeted by progesterone during vitellogenesis. The use of genetic improvement in broodstock selection to maintain genetic diversity may also help to sustain cultured populations through multiple generations. Current efforts to achieve this goal initially succeed, but after several generations, offspring are typically smaller and less healthy, and without outcrossing to wild-caught individuals, the population eventually dwindles (Sykes et al., 2006b; Walsh et al., 2002). This problem of reduced viability of eggs over several generations has yet to be resolved, especially in comparison with the success obtained from culture over several generations for some aquacultured fishes and crustaceans.

Further research into the aspects of embryonic development will also be important in improving cephalopod husbandry. Defining parameters to describe egg, embryo and hatchling quality is necessary in order to correctly evaluate the health of the population and the culturing methods being used and may provide insights into variability between experiments and individual variation between siblings, thereby optimizing culture methodologies. Developing reliable methods of artificial incubation for small-egged species such as *O. vulgaris* is necessary to take full advantage of their high fecundity. Recent studies on the aptitude for embryonic learning in the cuttlefish *S. officinalis* have demonstrated the impact of different environmental stimuli on embryos (Darmaillacq et al., 2006, 2008; Guibe et al., 2012). These poorly known embryo capabilities have potential for use in conditioning the prey preferences of hatchlings for paralarval and juvenile rearing, an as yet unexplored research topic.

Life stage and phase transitions are points where populations maintained in artificial settings are prone to collapse. Developing techniques to facilitate transitions of the embryo to hatching, paralarva to settlement, juvenile to subadult and subadult to adult will be important advances in the effort to culture cephalopods. At present, all 21 species for which paralarval development has been described can be raised only at the experimental level: none has yet been raised on a commercially viable basis. Another critical transition is the settlement period in *Octopus* species with a paralarval phase. Little attention has been directed towards understanding the high mortality rates and the major morphological changes that are associated with this transition. The development of culturing systems designed specifically to accommodate paralarvae through their planktonic phase with higher survival rates will be required to optimize the transition from paralarvae to juvenile and to make culturing of these species possible. Presently, the directly developing *O. maya* is the most advanced example of attempts at commercially sound cephalopod aquaculture, but even for this species, it has been necessary to rely on targeting a specialized gourmet-level market and cannot yet compete directly with the common wild-captured product.

Another topic that should be categorized as top priority for research is nutrition at all phases of the life cycle. The lack of a formulated feed for cephalopods limits industrial production due to their feeding habits and consumption of large quantities of feed particularly in the early stages of development. Paralarval nutritional requirements are still poorly understood, and providing appropriate diets (artificial, live prey or mixed) remains an important challenge. Research into the feeding behaviour of paralarvae and on the diet of wild paralarvae (Roura et al., 2012) will provide insights into their requirements in artificial settings. Continued efforts to understand the requirements for the juvenile and grow-out phases will also contribute greatly to efforts to enhance cephalopod culture techniques. The basic lipid and protein requirements at each stage are still incompletely understood (cf. also Glencross, 2009); and the effects of different feeds on feed-to-body mass conversion on the gut microbiota have yet to be explored (cf. Askarian et al., 2013).

Due to environmental concerns and with fisheries resources in decline, the use of high quantities of protein and high-quality lipids obtained mostly from marine resources to feed this group of voracious carnivores is a serious and costly problem. Thus, the challenge will be to obtain a sustainable, environmentally friendly artificial feed independent from fisheries products and formulated from vegetarian sources, as has been achieved recently for some marine carnivorous fishes (Watson et al., 2013). To help accomplish this task, cephalopod digestive physiology should be investigated in depth, since still there are considerable gaps in our knowledge of the digestion and absorption mechanisms of cephalopods. This information combined with genome sequencing of distinct regions of the digestive tract would make possible to identify enzymes in the gut during feeding under different feeding regimes and at different life stages to advance optimization of dietary requirements and formulated feeds. Availability of subadults and their feed are the main constraints influencing the profitability of the *O. vulgaris* ongrowing process. Thus, mass production of subadults from hatcheries and the provision of a commercial feed promoting high growth and survival rates are future priorities for the octopus fattening industry.

The remarkable learning activity of cephalopods also has yet to be explored in terms of improving feeding methods, while adequate culturing procedures to limit stress are essential to reduce pathologies.

The effects of parasites and disease remain poorly understood but they can greatly influence growth, reproduction and survival. Treatments for cephalopod diseases and pathogens are rare and may be a key tool in maintaining and culturing cephalopod populations. The study of infectious diseases during the culture of marine species has led to important advances in recognition and diagnosis of the causative agents and a better understanding of infection processes. A better understanding of the cephalopod immune system and its interaction with the confined culture environment is also necessary. Several studies have shown that the hemocytes of cephalopods have the capacity to engulf and neutralize foreign cells and are involved in radical production, inflammatory processes and regeneration of wounds (Beuerlein et al., 2002; Castellanos-Martínez and Gestal, 2013; Collins et al., 2012; Féral, 1988; Ford, 1992; Rodríguez-Domínguez et al., 2006). Although considered as central in the invertebrate immunity response (Soudant et al., 2013), there have been few descriptions of cephalopod hemocytes and their efficiency to prevent infections. As immune humoral components, the agglutinating activity of lectins in cephalopod plasma has been described (Fisher and DiNuzzo, 1991; Rögener et al., 1985), but their exact effects on pathogens remain to be determined.

Finally, the continued identification of new potential species for commercial production or for research models will continue to be an important research topic. This ongoing effort not only may yield insight into the care and rearing of existing species but also may result in finding species that are particularly well suited to aquaculture and experimentation. In addition, the maintenance of deep-sea, cold water and oceanic cephalopods in the laboratory would represent a considerable achievement towards understanding their biology and life cycle, providing new cephalopod species as models for research to many new areas of study, including toxicology, biomimetics and symbioses. Surely, new methods will be required to transport and rear species from these very unusual environments, an opportunity to meet new challenges in culturing.

ACKNOWLEDGMENTS

We would like to thank P. Bustamante, J. Estefanell, J. Forsythe, M. Garci, M. Nande and J. Sánchez for providing original photos for this chapter. EAGV was supported by the Brazilian National Research Council (CNPq-Pro. 485653/2012-5). RV was funded by the research project CALOCEAN-2 (AGL2012-39077) from the Ministry of Economy and Competitiveness of Spain (MINECO). IGG thanks Masami Abe (aquaculture expert and Executive Chairman of Gurumeito Ltd., Ishinomaki, Japan) for useful discussions concerning procedures and practices in the Japanese seafood industry and Hotland Co., Ltd., Ishinomaki, for help with expenses. IGG is supported by the Japan Science and Technology Agency Grant No. J130000263.

REFERENCES

Aguado Giménez, F., García García, B., 2002. Growth and food intake models in *Octopus vulgaris* Cuvier (1797): influence of body weight, temperature, sex and diet. Aquacult. Int. 10, 361–377.

Aguila, J., Cuzon, G., Pascual, C., Domingues, P.M., Gaxiola, G., Sánchez, A., Maldonado, T., Rosas, C., 2007. The effects of fish hydrolysate (CPSP) level on *Octopus maya* (Voss and Solis) diet: digestive enzyme activity, blood metabolites, and energy balance. Aquaculture 273, 641–655.

Ahmad, T., Usman, 1997. Bigfin squid culture: the Indonesian experience. Phuket Mar. Biol. Cent. Spec. Publ. 17, 285–287.

Al-Hafedh, Y.S., Alam, A., Buschmann, A.H., Fitzsimmons, K.M., 2012. Experiments on an integrated aquaculture system (seaweeds and marine fish) on the Red Sea coast of Saudi Arabia: efficiency comparison of two local seaweed species for nutrient biofiltration and production. Rev. Aquacult. 4, 21–31.

Almansa, E., Domingues, P., Sykes, A., Tejera, N., Lorenzo, A., Andrade, J.P., 2006. The effects of feeding with shrimp or fish fry on growth and mantle lipid composition of juvenile and adult cuttlefish (*Sepia officinalis*). Aquaculture 256, 403–413.

Anderson, R.C., Blustein, D.H., 2006. Smart octopus? The Festivus 38, 7–9.

Anderson, R.C., Wood, J.B., 2001. Enrichment for giant Pacific octopuses: happy as a clam? J. Appl. Anim. Welf. Sci. 4 (2), 157–168.

André, J., Pecl, G.T., Semmens, J.M., Grist, E.P.M., 2008. Early life-history processes in benthic octopus: relationships between temperature, feeding, food conversion, and growth in juvenile *Octopus pallidus*. J. Exp. Mar. Biol. Ecol. 354, 81–92.

Andrews, P.L.R., Darmaillacq, A.S., Dennison, N., Gleadall, I.G., Hawkins, P., Messenger, J.B., Osorio, D., Smith, V.J., Smith, J.A., 2013. The identification and

management of pain, suffering and distress in cephalopods, including anaesthesia, analgesia and humane killing. J. Exp. Mar. Biol. Ecol. 443, 46–64.
Appleford, P., Lucas, J.S., Southgate, P.C., 2003. General principles. In: Lucas J.S., Southgate P.C. (Eds.), Aquaculture. Farming Aquatic Animals and Plants. Wiley-Blackwell, West Sussex, UK, pp. 18–51.
Arkhipkin, A.I., 1992. Reproductive system structure, development and function in cephalopods with a new general scale for maturity stages. J. Northwest Atl. Fish. Sci. 12, 63–74.
Arnold, J.M., 1965. Normal embryonic stages of the squid, *Loligo pealii* (Leuseur). Biol. Bull. 128, 24–32.
Askarian, F., Sperstad, S., Merrifield, D.L., Ray, A.K., Ringø, E., 2013. The effect of different feeding regimes on enzyme activities of gut microbiota in Atlantic cod (*Gadus morhua* L.). Aquac. Res. 44 (5), 841–846.
Avila-Poveda, O.H., Colin-Flores, R.F., Rosas, C., 2009. Gonad development during the early life of *Octopus maya* (Mollusca: Cephalopoda). Biol. Bull. 216, 94–102.
Badiani, A., Bonaldo, A., Testi, S., Rotolo, M., Serratore, P., Giulini, G., Pagliuca, G., Gatta, P.P., 2013. Good handling practices of the catch: the effect of early icing on the freshness quality of cuttlefish (*Sepia officinalis* L.). Food Control 32 (1), 327–333.
Baeza-Rojano, E., Domingues, P., Guerra-García, J.M., Capella, S., Noreña-Barroso, E., Caamal-Monsreal, C., Rosas, C., 2012. Marine gammarids (Crustacea: Amphipoda): a new live prey to culture *Octopus maya* hatchlings. Aquac. Res. 44 (10), 1602–1612.
Bain, D.L., Heneghan, A.F., Connaghan-Jones, K.D., Miura, M.T., 2007. Nuclear receptor structure: implications for function. Annu. Rev. Physiol. 69, 201–220.
Benzie, J.A.H., 2009. Use and exchange of genetic resources of penaeid shrimps for food and aquaculture. Rev. Aquacult. 1 (3–4), 232–250.
Berger, E., 2010. Aquaculture of *Octopus* species: present status, problems and perspectives. Plym. Stud. Sci. 4 (1), 384–399.
Beuerlein, K., Löhr, S., Westermann, B., Ruth, P., Schipp, R., 2002. Components of the cellular defense and detoxification system of the common cuttlefish *Sepia officinalis* (Mollusca, Cephalopoda). Tissue Cell 34 (6), 390–396.
Boal, J.G., Golden, D.K., 1999. Distance chemoreception in the common cuttlefish, Sepiaofficinalis (Mollusca, Cephalopoda). J. Exp. Mar. Biol. Ecol. 235, 307–317.
Boletzky, S.V., 1974. Élevage de céphalopodes en aquarium. Vie Milieu 24 (2), 309–340.
Boletzky, S.V., 1975. A contribution to the study of yolk absorption in the Cephalopoda. Z. Morphol. Tiere 80, 229–246.
Boletzky, S.V., 1983. Sepia officinalis. In: Boyle P.R. (Ed.), Cephalopod Life Cycles, Volume I—Species Accounts. Academic Press, London, pp. 31–52.
Boletzky, S.V., 1987. Embryonic phase. In: Boyle P.R. (Ed.), Cephalopod Life Cycles, Volume II—Comparative Reviews. Academic Press, London, pp. 5–32.
Boletzky, S.V., 2003. Biology of early life stages in cephalopod molluscs. Adv. Mar. Biol. 44, 143–203.
Boletzky, S.V., 2004. A brief survey of cephalopod culture techniques. In: Öztürk B., Salman A. (Eds.), 1st National Malacology Congress, Izmir—Turkey. In: Turk. J. Aquat. Life, , vol. 2, pp. 229–240.
Boletzky, S.V., Boletzky, M.V.V., 1969. First results in rearing *Octopus joubini* Robson 1929. Verhandl. Naturf. Ges. Basel 80 (1), 56–61.
Boletzky, S.V., Fioroni, P., 1990. Embryonic inversions in incirrate octopods: the state of an enigma. J. Ceph. Biol. 1 (2), 37–57.
Boletzky, S.V., Hanlon, R.T., 1983. A review of the laboratory maintenance, rearing and culture of cephalopod molluscs. Mem. Natl. Mus. Victoria 44, 147–187.
Boucaud-Camou, E., Boucher-Rodoni, R., 1983. Feeding and digestion in cephalopods. In: Saleuddin A.S.M., Wilbur K.M. (Eds.), The Mollusca, Volume 5: Physiology, Part 2. Academic Press, New York, pp. 149–187.

Boucaud-Camou, E., Yim, M., Tresgot, A., 1985. Feeding and digestion of young *Sepia officinalis* L. (Mollusca: Cephalopoda) during post-hatching development. Vie Milieu 35 (3/4), 263–266.

Bouchaud, O., Daguzan, J., 1989. Study on the development of the egg of *Sepia officinalis* L. (Cephalopoda: Sepiidae) under experimental conditions. Haliotis 19, 189–200.

Boucher-Rodoni, R., Mangold, K., 1985. Ammonia excretion during feeding and starvation in *Octopus vulgaris*. Mar. Biol. 86, 193–197.

Boucher-Rodoni, R., Boucaud-Camou, E., Mangold, K., 1987. Feeding and digestion. In: Boyle P.R. (Ed.), Cephalopod Life Cycles, Volume II—Comparative Reviews. Academic Press, London, pp. 85–108.

Bower, J.R., Sakurai, Y., Yamamoto, J., Ishii, H., 1999. Transport of the ommastrephid squid *Todarodes pacificus* under cold-water anesthesia. Aquaculture 170, 127–130.

Boyle, P.R., 1991. The UFAW Handbook on the Care and Management of Cephalopods in the Laboratory. Universities Federation for Animal Welfare, Potters Bar, UK.

Briceño, F., Mascaró, M., Rosas, C., 2010. GLMM-based modelling of growth in juvenile *Octopus maya* siblings: does growth depend on initial size? ICES J. Mar. Sci. 67, 1509–1516.

Buchanan, K., de Perera, T.B., Carere, C., Carter, T., Hailey, A., Hubrecht, R., Jennings, D., Metcalfe, N., Pitcher, T., Péron, F., Sneddon, L., Sherwin, C., Talling, J., Thomas, R., Thompson, M., 2013. Guidelines for the use of animals: guidelines for the treatment of animals in behavioural research and teaching. Anim. Behav. 85 (1), 285–297.

Budelmann, B.U., 1995. The cephalopod nervous system: what evolutions has made of the molluscan design. In: Breidbach O., Kutsuch W. (Eds.), The Nervous System of Invertebrates: An Eolutionary and Comparative Approach. Birkhauser Verlag, Basel, pp. 115–138.

Budelmann, B.U., 2010. Cephalopoda. In: Hubrecht R., Kirkwood J. (Eds.), The UFAW Handbook on the Care and Management of Laboratory and Other Research Animals, eighth ed. Willey-Blackwell, Oxford, pp. 787–793.

Budelmann, B.U., Bleckmann, H., 1988. A lateral line analogue in cephalopods: water waves generate microphonic potentials, in the epidermal head lines of *Sepia* and *Lolliguncula*. J. Comp. Physiol. A. 164, 1–5.

Budelmann, B.U., Schipp, R., Boletzky, S.V., 1997. Cephalopoda. In: Harrison F.W., Kohn A. (Eds.), Microscopic Anatomy of Invertebrates, Volume 6A, Mollusca II. Wiley-Liss, New York, pp. 119–414.

Bush, S.L., Hoving, H.J.T., Huffard, C.L., Robison, B.H., Zeidberg, L.D., 2012. Brooding and sperm storage by the deep-sea squid *Bathyteuthis berryi* (Cephalopoda: Decapodiformes). J. Mar. Biol. Assoc. UK 92 (7), 1629–1636.

Carrasco, J.F., Rodríguez, C., Rodríguez, M., 2003. Cultivo intensivo de paralarvas de pulpo (*Octopus vulgaris*, Cuvier 1797) utilizando como base de la alimentación zoeas vivas de crustáceos. In: IX Congreso Nacional de Acuicultura, pp. 191–194. Edited by Consejería de Agricultura y Pesca de la Junta de Andalucía (S. 1.400-2005), Cádiz, Spain.

Carrasco, J.F., Arronte, J.C., Rodríguez, C., 2006. Paralarval rearing of the common octopus, *Octopus vulgaris* (Cuvier). Aquac. Res. 37, 1601–1605.

Castellanos-Martínez, S., Gestal, C., 2013. Pathogens and immune response of cephalopods. J. Exp. Mar. Biol. Ecol. 447, 14–22.

Castro, B.G., 1991. Can *Sepia officinalis* L. be reared on artificial food? Mar. Behav. Physiol. 19, 35–38.

Castro, B.G., Lee, P.G., 1994. The effects of semi-purified diets on growth and condition of *Sepia officinalis* L. (Mollusca: Cephalopoda). Comp. Biochem. Physiol. A 109 (4), 1007–1016.

Castro, B.G., Dimarco, F.P., Derusha, R.H., Lee, P.G., 1993. The effects of surimi and pelleted diets on the laboratory survival, growth, and feeding rate of the cuttlefish *Sepia officialis* L. J. Exp. Mar. Biol. Ecol. 170, 241–252.

Caverivière, A., Domain, F., Diallo, A., 1999. Observations on the influence of temperature on the length of embryonic development in *Octopus vulgaris* (Senegal). Aquat. Living Resour. 12 (2), 151–154.

Cerezo Valverde, J., García García, B., 2004. Influence of body weight and temperature on post-prandial oxygen consumption of common octopus (*Octopus vulgaris*). Aquaculture 233, 599–613.

Cerezo Valverde, J., García García, B., 2005. Suitable dissolved oxygen levels for common octopus (*Octopus vulgaris* Cuvier, 1797) at different weights and temperatures: analysis of respiratory behavior. Aquaculture 244, 303–314.

Cerezo Valverde, J., Hernández, M.D., Aguado-Giménez, F., Garcia Garcia, B., 2008. Growth, feed efficiency and condition of common octopus (*Octopus vulgaris*) fed on two formulated moist diets. Aquaculture 275, 266–273.

Cerezo Valverde, J., Hernández, M.D., García-Garrido, S., Rodríguez, C., Estefanell, J., Gairín, J.I., Rodríguez, C.J., Tomás, A., García García, B., 2012a. Lipid classes from marine species and meals intended for cephalopod feeding. Aquacult. Int. 20, 71–89.

Cerezo Valverde, J., Hernández, M.D., Aguado-Giménez, F., Morillo-Velarde, P.S., García García, B., 2012b. Performance of formulated diets with different level of lipids and glutamate supplementation in *Octopus vulgaris*. Aquac. Res. 44 (12), 1952–1964.

Cerezo Valverde, J., Martínez-Llorens, S., Vidal, A.T., Jover, M., Rodríguez, C., Estefanell, J., Gairín, J.I., Domingues, P.M., Rodríguez, C.J., García García, B., 2012c. Amino acids composition and protein quality evaluation of marine species and meals for feed formulations in cephalopods. Aquacult. Int. 21 (2), 413–433.

Cerezo Valverde, J., Tomás, A., Martínez-Llorens, S., Jover, M., García, S., Rodríguez, C., Estefanell, J., Gairín, I., Domingues, P.M., Rodríguez, C.J., García García, B., 2013. Amino acids composition and protein quality evaluation of marine species and meals for feed formulations in cephalopods. Aquac. Int. 21, 413–433.

Chapela, A., Gonzalez, A.F., Dawe, E.G., Rocha, F.J., Guerra, A., 2006. Growth of common octopus (*Octopus vulgaris*) in cages suspended from rafts. Sci. Mar. 70 (1), 121–129.

Chen, D.S., Dykhuizen, G.V., Hodge, J., Gilly, W.F., 1996. Ontogeny of copepod predation in juvenile squid (*Loligo opalescens*). Biol. Bull. 190, 69–81.

Chiao, C.-C., Wickiser, J.K., Allen, J.J., Genter, B., Hanlon, R.T., 2011. Hyperspectral imaging of cuttlefish camouflage indicates good color match in the eyes of fish predators. Proc. Natl. Acad. Sci. U. S. A. 108, 9148–9153.

Choe, S., 1966a. On the Eggs, Rearing, Habits of the Fry, and Growth of Some Cephalopoda. Bull. Mar. Sci. 16 (2), 330–348.

Choe, S., 1966b. On the growth, feeding rates and the efficiency of food conversion for cuttlefishes and squids. Korean J. Zool. 9 (2), 12–20.

Choe, S., Ohshima, Y., 1961. On the embryonal development and the growth of the squid, *Sepioteuthis lessoniana* Lesson. Venus 21 (2), 462–476.

Choe, S., Ohshima, Y., 1963. Rearing of cuttlefishes and squids. Nature 197, 307.

Coelho, M., Domingues, P., Balguerias, E., Fernandez, M., Andrade, J.P., 1997. A comparative study of the diet of *Loligo vulgaris* (Lamarck, 1799) (Mollusca: Cephalopoda) from the south coast of Portugal and the Saharan Bank (Central-East Atlantic). Fish. Res. 29 (3), 245–255.

Collins, A.J., Schleicher, T.R., Rader, B.A., Nyholm, S.V., 2012. Understanding the role of host hemocytes in a squid/*Vibrio* symbiosis using transcriptomics and proteomics. Front. Immunol. 3, 91.

Correia, M., Domingues, P.M., Sykes, A., Andrade, J.P., 2005. Effects of culture density on growth and broodstock management of the cuttlefish, *Sepia officinalis* (Linnaeus, 1758). Aquaculture 245 (1–4), 163–173.

Cronin, E.R., Seymour, R.S., 2000. Respiration of the eggs of the giant cuttlefish *Sepia apama*. Mar. Biol. 136, 863–870.

Crook, R.J., Walters, E.T., 2011. Nociceptive behavior and physiology of molluscs: animal welfare implications. Inst. Lab. Anim. Res. J. 52 (2), 185–195.

Cruz, H., Lizama, G., Arena, L., Méndez, M.J., Pascual, C., 2008. Cuantificación y aislamiento de bacterias en piel y hemolinfa de la población del pulpo rojo *Octopus maya*. In: IV Congreso Regional de Biotecnología y Bioingeniería del Sureste". Mérida, Mexico.

D'Aniello, A., D'Onofrio, G., Pischetola, M., Denucé, J.M., 1989. Effect of pH, salinity and Ca^{2+}, Mg^{2+}, K^+ and SO_4^{2+} ions on hatching and viability of *Loligo vulgaris* embryo. Comp. Biochem. Physiol. A 94 (3), 477–481.

Darmaillacq, A.S., Chichery, R., Shashar, N., Dickel, L., 2006. Early familiarization overrides innate prey preference in newly hatched *Sepia officinalis* cuttlefish. Anim. Behav. 71, 511–514.

Darmaillacq, A.S., Lesimple, C., Dickel, L., 2008. Embryonic visual learning in the cuttlefish, *Sepia officinalis*. Anim. Behav. 76, 131–134.

Delgado, M., Gairín, J.I., Carbó, R., Aguilera, C., 2011. Growth of *Octopus vulgaris* (Cuvier, 1797) in tanks in the Ebro Delta (NE Spain): effects of temperature, salinity and culture density. Sci. Mar. 75 (1), 53–59.

DeRusha, R.H., Forsythe, J.W., Dimarco, F.P., Hanlon, R.T., 1989. Alternative diets for maintaining and rearing cephalopods in captivity. Lab. Anim. Sci. 39, 306–312.

De Wolf, T., Lenzi, S., Lenzi, F., 2011. Paralarval rearing of *Octopus vulgaris* (Cuvier) in Tuscany, Italy. Aquac. Res. 42, 1406–1414.

Di Cristo, C., Paolucci, M., Di Cosmo, A., 2008. Progesterone affects vitellogenesis in *Octopus vulgaris*. Open Zool. J. 1, 29–36.

Dobell, C., 1925. The life-history and chromosome cycle of *Aggregata eberthi* [Protozoa: Sporozoa: Coccidia]. Parasitology 17 (1), 1–136.

Domingues, P., 1999. Development of alternative diets for the mass culture of the European cuttlefish *Sepia officinalis* (Ph.D. thesis). University of the Algarve, Portugal.

Domingues, P., Márquez, L., 2010. Effects of culture density and bottom area on growth and survival of the cuttlefish *Sepia officinalis* (Linnaeus, 1758). Aquacult. Int. 18, 361–369.

Domingues, P.M., Kingston, T., Sykes, A., Andrade, J.P., 2001a. Growth of young cuttlefish, *Sepia officinalis* (Linnaeus 1758) at the upper end of the biological distribution temperature range. Aquac. Res. 32, 923–930.

Domingues, P.M., Sykes, A., Andrade, J.P., 2001b. The use of *Artemia* sp. or mysids as food source for hatchlings of the cuttlefish (*Sepia officinalis* L.); effects on growth and survival throughout the life cycle. Aquacult. Int. 9, 319–331.

Domingues, P.M., Sykes, A., Andrade, J.P., 2002. The effects of temperature in the life cycle of two consecutive generations of the cuttlefish *Sepia officinalis* (Linnaeus, 1758), cultured in the Algarve (South Portugal). Aquacult. Int. 10, 207–220.

Domingues, P., Sykes, A., Sommerfield, A., Andrade, J.P., 2003. Effects of feeding live or frozen prey on growth, survival and the life cycle of the cuttlefish, *Sepia officinalis* (Linnaeus, 1758). Aquacult. Int. 11, 397–410.

Domingues, P., Sykes, A., Sommerfield, A., Almansa, E., Lorenzo, A., Andrade, J.P., 2004. Growth and survival of cuttlefish, (*Sepia officinalis*) of different ages fed crustaceans and fish. Effects of frozen and live prey. Aquaculture 229, 239–254.

Domingues, P.M., Dimarco, F.P., Andrade, J.P., Lee, P.G., 2005. Effect of artificial diets on growth, survival and condition of adult cuttlefish, *Sepia officinalis* Linnaeus, 1758. Aquacult. Int. 13, 423–440.

Domingues, P.M., Bettencourt, V., Guerra, A., 2006. Growth of *Sepia officinalis* in captivity and in nature. Vie Milieu 56 (2), 109–120.

Domingues, P.M., López, N., Muñoz, J.A., Maldonado, T., Gaxiola, G., Rosas, C., 2007. Effects of a dry pelleted diet on growth and survival of the Yucatan octopus, *Octopus maya*. Aquacult. Nutr. 13 (4), 273–280.

Domingues, P., Ferreira, A., Marquez, L., Andrade, J.P., López, N., Rosas, C., 2008. Growth, absorption and assimilation efficiency by mature cuttlefish (*Sepia officinalis*) fed with alternative and artificial diets. Aquacult. Int. 16, 215–229.

Domingues, P., Garcia, S., Hachero-Cruzado, I., Lopez, N., Rosas, C., 2010. The use of alternative prey (crayfish, *Procambarus clarki*, and hake, *Merlucius gayi*) to culture *Octopus vulgaris* (Cuvier 1797). Aquacult. Int. 18, 487–499.

Domingues, P., López, N., Rosas, C., 2012. Preliminary trials on the use of large outdoor tanks for the ongrowing of *Octopus maya* juveniles. Aquac. Res. 43, 26–31.

Elston, R., Wood, G.S.L., 1983. Pathogenesis of vibriosis in cultured juvenile red abalone, *Haliotis rufescens* Swainson. J. Fish Dis. 6 (2), 111–128.

Establier, R., Pascual, E., 1983. Efecto del cadmio y el cobre sobre el desarrollo de los huevos de *Sepia officinalis* Linneo. Invest. Pesq. 47 (1), 143–150.

Estefanell, J., Socorro, J., Tuya, F., Izquierdo, M., Roo, J., 2011. Growth, protein retention and biochemical composition in *Octopus vulgaris* fed on different diets based on crustaceans and aquaculture by-products. Aquaculture 322–323, 91–98.

Estefanell, J., Roo, J., Guirao, R., Afonso, J.M., Fernández-Palacios, H., Izquierdo, M., Socorro, J., 2012a. Efficient utilization of dietary lipids in *Octopus vulgaris* (Cuvier 1797) fed fresh and agglutinated moist diets based on aquaculture by-products and low price trash species. Aquac. Res. 44 (1), 93–105.

Estefanell, J., Roo, J., Guirao, R., Izquierdo, M., Socorro, J., 2012b. Benthic cages *versus* floating cages in *Octopus vulgaris*: Biological performance and biochemical composition feeding on *Boops boops* discarded from fish farms. Aquacult. Eng. 49, 46–52.

Estefanell, J., Roo, J., Guirao, R., Afonso, J.M., Fernández-Palacios, H., Izquierdo, M., Socorro, J., 2013. Efficient utilization of dietary lipids in *Octopus vulgaris* (Cuvier 1797) fed fresh and agglutinatedmoist diets based on aquaculture by-products and low price trash species. Aquac. Res. 44, 93–105.

Fadhlaoui-Zid, K., Knittweis, L., Aurelle, D., Nafkha, C., Ezzeddine, S., Fiorentino, F., Ghmati, H., Ceriola, L., Jarboui, O., Maltagliati, F., 2012. Genetic structure of *Octopus vulgaris* (Cephalopoda, Octopodidae) in the central Mediterranean Sea inferred from the mitochondrial COIII gene. C. R. Biol. 335, 625–636.

FAO, 2009. Food Quality and Safety Systems: A Training Manual on Food Hygiene and the Hazard Analysis and Critical Control Point (HACCP) System. FAO, Rome, Italy.

FAO, 2011. FishStatJ: Universal Software for Fishery Statistical Time Series. FAO Fisheries and Aquaculture Department, Statistics and Information Service, Rome, Italy.

Farías, A., Uriarte, I., Hernández, J., Pino, S., Pascual, C., Caamal, C., Domíngues, P., Rosas, C., 2009. How size relates to oxygen consumption, ammonia excretion, and ingestion rates in cold (*Enteroctopus megalocyathus*) and tropical (*Octopus maya*) octopus species. Mar. Biol. 156, 1547–1558.

Farto, R., Armada, S.P., Montes, M., Guisande, J.A., Pérez, M.J., Nieto, T.P., 2003. *Vibrio lentus* associated with diseased wild octopus (*Octopus vulgaris*). J. Invertebr. Pathol. 83 (2), 149–156.

Féral, J.-P., 1988. Wound healing after arm amputation in *Sepia officinalis* (Cephalopoda: Sepioidea). J. Invertebr. Pathol. 52, 380–388.

Ferreira, A., Marquez, L., Almansa, E., Andrade, J.P., Lorenzo, A., Domingues, P.M., 2010. The use of alternative diets to culture juvenile cuttlefish, *Sepia officinalis*: effects on growth and lipid composition. Aquacult. Nutr. 16 (3), 262–275.

Fiorito, G., Scotto, P., 1992. Observational learning in *Octopus vulgaris*. Science 256, 545–547.

Fiorito, G., Affuso, A., Anderson, D.B., Basil, J., Bonnaud, L., Botta, G., Cole, A., D'Angelo, L., De Girolamo, P., Dennison, N., Dickel, L., Di Cosmo, A., Di Cristo,

C., Gestal, C., Fonseca, R., Grasso, F., Kristiansen, T., Kuba, M., Maffucci, F., Manciocco, A., Mark, F.C., Melillo, D., Osorio, D., Palumbo, A., Perkins, K., Ponte, G., Raspa, M., Shashar, N., Smith, J., Smith, D., Sykes, A., Villanueva, R., Tublitz, N., Zullo, L., Andrews, P., 2014. Cephalopods in neuroscience: regulations, research and the 3Rs. Invert. Neurosci. 14 (1), 13–36. http://dx.doi.org/10.1007/s10158-013-0165-x.

Fisher, W.S., DiNuzzo, A.R., 1991. Agglutination of bacteria and erythrocytes by serum from six species of marine molluscs. J. Invertebr. Pathol. 57 (3), 380–394.

Ford, L.A., 1992. Host defense mechanisms of cephalopods. Annu. Rev. Fish Dis. 2, 25–41.

Ford, L.A., Alexander, S.K., Cooper, K.M., Hanlon, R.T., 1986. Bacterial populations of normal and ulcerated mantle tissue of the squid, *Lolliguncula brevis*. J. Invertebr. Pathol. 48 (1), 13–26.

Forsythe, J.W., Hanlon, R.T., 1988. Effect of temperature on laboratory growth, reproduction and life span of *Octopus bimaculoides*. Mar. Biol. 98, 369–379.

Forsythe, J.W., Toll, R.B., 1991. Clarification of the Western Atlantic ocean pygmy octopus complex: the identity and life history of *Octopus joubini* (Cephalopoda: Octopodinae). Bull. Mar. Sci. 49 (1–2), 88–97.

Forsythe, J.W., Van Heukelem, W.F., 1987. Cephalopod growth. In: Boyle P.R. (Ed.), Cephalopod Life Cycles, Volume II—Comparative Reviews. Academic Press, London, pp. 135–155.

Forsythe, J.W., Hanlon, R.T., DeRusha, R., 1991. Pilot large-scale culture of *Sepia* in biomedical research. In: Boucaud-Camou E. (Ed.), La seiche—The Cuttlefish. Actes du Premier Symposium International sur la Seiche. Centre de Publications de l'Université de Caen, Caen, France, pp. 313–323.

Forsythe, J.W., DeRusha, R.H., Hanlon, R.T., 1994. Growth, reproduction and life span of *Sepia officinalis* (Cephalopoda, Mollusca) cultured through seven consecutive generations. J. Zool. 233 (2), 175–192.

Forsythe, J.W., Walsh, L.S., Turk, P.E., Lee, P.G., 2001. Impact of temperature on juvenile growth and age at first egg-laying of the Pacific reef squid *Sepioteuthis lessoniana* reared in captivity. Mar. Biol. 138, 103–112.

Forsythe, J., Lee, P., Walsh, L., Clark, T., 2002. The effects of crowding on growth of the European cuttlefish, *Sepia officinalis* Linnaeus, 1758 reared at two temperatures. J. Exp. Mar. Biol. Ecol. 269 (2), 173–185.

Fox, J.D., Capadona, J.R., Marasco, P.D., Rowan, S.J., 2013. Bioinspired water-enhanced mechanical gradient nanocomposite films that mimic the architecture and properties of the squid beak. J. Am. Chem. Soc. 135, 5167–5174.

Fuentes, L., Sánchez, F.J., Lago, M.J., Iglesias, J., Pazos, G., Linares, F., 2011. Growth and survival of *Octopus vulgaris* (Cuvier 1797) paralarvae fed on three *Artemia*-based diets complemented with frozen fish flakes, crushed zooplankton and marine microalgae. Sci. Mar. 75 (4), 771–777.

Gamboa, M.M., 2011. Evaluación microbiológica y diseño de un PCR para la detección de *Vibrio* sp. en el pulpo Octopus maya (Ph.D. thesis). Instituto Tecnológico de Mérida, Mexico.

García García, B., Aguado Giménez, F., 2002. Influence of diet on ongrowing and nutrient utilization in the common octopus (*Octopus vulgaris*). Aquaculture 211, 171–182.

García García, B., Cerezo Valverde, J., 2004. Influencia del número de días de ayuno a la semana sobre el crecimiento, el índice de conversión y la supervivencia en el pulpo de roca (*Octopus vulgaris* Cuvier, 1797). Revista AquaTIC 21, 34–41.

García García, B., Cerezo Valverde, J., 2006. Optimal proportions of crabs and fish in diet for common octopus (*Octopus vulgaris*) ongrowing. Aquaculture 253, 502–511.

García García, J., García García, B., 2011. Econometric model of viability/profitability of octopus (*Octopus vulgaris*) ongrowing in sea cages. Aquacult. Int. 19, 1177–1191.

García García, B., Cerezo Valverde, J., Aguado-Giménez, F., García García, J., Hernández, M.D., 2009. Growth and mortality of common octopus *Octopus vulgaris* reared at different stocking densities in Mediterranean offshore cages. Aquac. Res. 40 (10), 1202–1212.

García-Garrido, S., Hachero-Cruzado, I., Garrido, D., Rosas, C., Domingues, P., 2010. Lipid composition of the mantle and digestive gland of *Octopus vulgaris* juveniles (Cuvier, 1797) exposed to prolonged starvation. Aquacult. Int. 18, 1223–1241.

García-Garrido, S., Domingues, P., Navarro, J.C., Hachero-Cruzado, I., Garrido, D., Rosas, C., 2011a. Growth, partial energy balance, mantle and digestive gland lipid composition of Octopus vulgaris (Cuvier, 1797) fed with two artificial diets. Aquacult. Nutr. 17, e174–e187.

García-Garrido, S., Hachero-Cruzado, I., Domingues, P., López, N., Rosas, C., 2011b. Effects of fish hydrolysate (CPSP®) on growth and digestive gland lipid composition of *Octopus vulgaris* (Cuvier, 1797) juveniles. Aquacult. Nutr. 17, e825–e839.

Gestal, C., 2000. Epidemiología y patología de las coccidiosis en cefalópodos (Ph.D. thesis). Universidad de Vigo, Spain.

Gestal, C., Abollo, E., Pascual, S., 2002a. Observations on associated histopathology with *Aggregata octopiana* infection (Protista: Apicomplexa) in *Octopus vulgaris*. Dis. Aquat. Org. 50, 45–49.

Gestal, C., Páez de la Cadena, M., Pascual, S., 2002b. Malabsorption syndrome observed in the common octopus *Octopus vulgaris* infected with *Aggregata octopiana* (Protista: Apicomplexa). Dis. Aquat. Org. 51, 61–65.

Gestal, C., Guerra, A., Pascual, S., 2007. *Aggregata octopiana* (Protista: Apicomplexa): a dangerous pathogen during commercial *Octopus vulgaris* ongrowing. ICES J. Mar. Sci. 64, 1743–1748.

Gestal, C., Pascual, S., Hochberg, F.G., 2010. *Aggregata bathytherma* sp. nov. (Apicomplexa: Aggregatidae), a new coccidian parasite associated with a deep-sea hydrothermal vent octopus. Dis. Aquat. Org. 91 (3), 237–242.

Gilbert, D.L., Adelman Jr., W.J., Arnold, J.M., 1990. Squid as Experimental Animals. Plenum Press, New York, NY.

Gjedrem, T., 2010. The first family-based breeding program in aquaculture. Rev. Aquacult. 2 (1), 2–15.

Gleadall, I.G., 2013a. The effects of prospective anaesthetic substances on cephalopods: summary of original data and a brief review of studies over the last two decades. J. Exp. Mar. Biol. Ecol. 447, 23–30.

Gleadall, I.G., 2013b. Low dosage of magnesium sulphate as a long-term sedative during transport of firefly squid, Watasenia scintillans. J. Exp. Mar. Biol. Ecol. 447, 138–139.

Gleadall, I.G., Ohtsu, K., Gleadall, E., Tsukahara, Y., 1993. Screening-pigment migration in the octopus retina includes control by dopaminergic efferents. J. Exp. Biol. 185, 1–16.

Glencross, B.D., 2009. Exploring the nutritional demand for essential fatty acids by aquaculture species. Rev. Aquacult. 1 (2), 71–124.

Globefish, 2012. Globefish highlights. FAO/Globefish (3/2012), p. 58, 15/08/2012.

González, A.F., Pascual, S., Gestal, C., Abollo, E., Guerra, A., 2003. What makes a cephalopod a suitable host for parasite? The case of Galician waters. Fish. Res. 60 (1), 177–183.

Gore, S.R., Harms, C.A., Kukanich, B., Forsythe, J., Lewbart, G.A., Papich, M.G., 2005. Enrofloxacin pharmacokinetics in the European cuttlefish, *Sepia officinalis*, after a single i. v. injection and bath administration. J. Vet. Pharmacol. Ther. 28, 433–439.

Grigoriou, P., Richardson, C.A., 2004. Aspects of the growth of cultured cuttlefish *Sepia officinalis* (Linnaeus 1758). Aquac. Res. 35 (12), 1141–1148.

Grigoriou, P., Richardson, C.A., 2008. The effect of ration size, temperature and body weight on specific dynamic action of the common cuttlefish *Sepia officinalis*. Mar. Biol. 154, 1085–1095.

Grigoriou, P., Richardson, C.A., 2009. Effect of body mass, temperature and food deprivation on oxygen consumption rate of common cuttlefish *Sepia officinalis*. Mar. Biol. 156, 2473–2481.
Grimpe, G., 1928. Pflege, Behandlung und Zucht der Cephalopoden für zoologische und physiologische Zwecke. In: Handbuch der Biologischen Arbeitsmethoden. Urban und Schwarzenberg, Berlin, pp. 331–402, 5, Abteilung, Teil IX.
Guibe, M., Poirel, N., Houdé, O., Dickel, L., 2012. Food imprinting and visual generalization in embryos and newly hatched cuttlefish, *Sepia officinalis*. Anim. Behav. 84, 213–217.
Guo, X., 2009. Use and exchange of genetic resources in molluscan aquaculture. Rev. Aquacult. 1 (3–4), 251–259.
Gutowska, M.A., Melzner, F., 2009. Abiotic conditions in cephalopod (*Sepia officinalis*) eggs: embryonic development at low pH and high pCO_2. Mar. Biol. 156, 515–519.
Hamasaki, K., Takeuchi, T., 2000. Effects of the addition of *Nannochloropsis* to the rearing water on survival and growth of planktonic larvae in *Octopus vulgaris*. Saibai Giken 28 (1), 13–16.
Hanlon, R.T., 1987. Mariculture. In: Boyle P.R. (Ed.), Cephalopod Life Cycles, Volume II—Comparative Reviews. Academic Press, London, pp. 291–305.
Hanlon, R.T., 1990. Maintenance, rearing, and culture of teuthoid and sepioid squids. In: Gilbert D.L., Adelman W.J., Arnold J.M. (Eds.), Squid as Experimental Animals. Plenum Press, New York, pp. 35–62.
Hanlon, R.T., Forsythe, J.W., 1985. Advances in the laboratory culture of octopuses for biomedical research. Lab. Anim. Sci. 35 (1), 33–40.
Hanlon, R.T., Forsythe, J.W., 1990. Diseases of Mollusca: Cephalopoda. Diseases caused by microorganisms. In: Kinne O. (Ed.), In: Diseases of Marine Animals: Introduction, Cephalopoda, Annelida, Crustacea, Chaetognatha, Echinodermata, Urochordata, vol. III. Biologische Anstalt Helgoland, Hamburg, Germany, pp. 23–46.
Hanlon, R.T., Forsythe, J.W., Cooper, K.M., 1984. Fatal penetrating skin ulcers in laboratory-reared octopus. J. Invertebr. Pathol. 44, 67–83.
Hanlon, R.T., Turk, P.E., Lee, P.G., Yang, W.T., 1987. Laboratory rearing of the squid *Loligo pealei* to the juvenile stage: growth comparisons with fishery data. Fish B-NOAA 85 (1), 163–167.
Hanlon, R.T., Yang, W.T., Turk, P.E., Lee, P.G., Hixon, R.F., 1989. Laboratory culture and estimated life span of the Eastern Atlantic squid, *Loligo forbesi* Steenstrup, 1856 (Mollusca: Cephalopoda). Aquat. Fish. Manage. 20, 15–34.
Hanlon, R.T., Claes, M.F., Ashcraft, S.E., Dunlap, P.V., 1997. Laboratory culture of the sepiolid squid *Euprymna scolopes*: a model system for bacteria-animal symbiosis. Biol. Bull. 192, 364–374.
Hanlon, R.T., Ament, S.A., Gabr, H., 1999. Behavioral aspects of sperm competition in cuttlefish, *Sepia officinalis* (Sepioidea: Cephalopoda). Mar. Biol. 134, 719–728.
Harms, C.A., Lewbart, G.A., McAlarney, R., Christian, L.S., Geissler, K., Lemons, C., 2006. Surgical excision of mycotic (*Cladosporium* sp.) granulomas from the mantle of a cuttlefish (*Sepia officinalis*). J. Zoo Wildl. Med. 37 (4), 524–530.
Harvey-Clark, C., 2011. IACUC challenges in invertebrate research. Inst. Lab. Anim. Res. J. 52 (2), 213–220.
Hendrix Jr., J.P., Hulet, W.H., Greenberg, M.J., 1981. Salinity tolerance and the responses to hypoosmotic stress of the bay squid *Lolliguncula brevis*, a euryhaline cephalopod mollusc. Comp. Biochem. Physiol. A 69 (4), 641–648.
Hilborn, R., 2012. The evolution of quantitative marine fisheries management 1985–2010. Nat. Resour. Model. 25, 122–144.
Hochberg, F.G., 1990. Diseases of Mollusca: Cephalopoda. Diseases caused by protistans and metazoans. In: Kinne O. (Ed.), In: Diseases of Marine Animals. Introduction, Cephalopoda, Annelida, Crustacea, Chaetognatha, Echinodermata, Urochordata, vol. III. Biologische Anstalt Helgoland, Hamburg, Germany, pp. 47–227.

Hulet, W.H., Villoch, M.R., Hixon, R.F., Hanlon, R.T., 1979. Fin damage in captured and reared squids. Lab. Anim. Sci. 29 (4), 528–533.

Ibáñez, C.M., Keyl, F., 2010. Cannibalism in cephalopods. Rev. Fish Biol. Fisher. 20 (1), 123–136.

Iglesias, J., Sánchez, F.J., Otero, J.J., 1997. Primeras experiencias sobre el cultivo integral del pulpo (*Octopus vulgaris*) en el Instituto Español de Oceanografía. In: Costa J., Abellán E., García García B., Ortega A., Zamora S. (Eds.), Actas del VI Congreso Nacional de Acuicultura. Cartagena, Spain, pp. 221–226.

Iglesias, J., Sánchez, F.J., Otero, J.J., Moxica, C., 2000. Culture of octopus (*Octopus vulgaris*, Cuvier): present knowledge, problems and perspectives. Cah. Opt. Médit. 47, 313–321.

Iglesias, J., Otero, J.J., Moxica, C., Fuentes, L., Sánchez, F.J., 2004. The completed life cycle of the octopus (*Octopus vulgaris*, Cuvier) under culture conditions: paralarval rearing using *Artemia* and zoeae, and first data on juvenile growth up to 8 months of age. Aquacult. Int. 12, 481–487.

Iglesias, J., Fuentes, L., Sánchez, J., Otero, J.J., Moxica, C., Lago, M.J., 2006. First feeding of *Octopus vulgaris* Cuvier, 1797 paralarvae using *Artemia*: effect of prey size, prey density and feeding frequency. Aquaculture 261, 817–822.

Iglesias, J., Sánchez, F.J., Bersano, J.F.G., Carrasco, J.F., Dhont, J., Fuentes, L., Linares, F., Muñoz, J.L., Okumura, S., Roo, J., van der Meeren, T., Vidal, E.A.G., Villanueva, R., 2007. Rearing of *Octopus vulgaris* paralarvae: present status, bottlenecks and trends. Aquaculture 266, 1–15.

Iglesias, J., Fuentes, L., Villanueva, R., (Eds.), 2014. Cephalopod Culture. Springer, Netherlands. 494 p. ISBN 978-94-017-8647-8.

Ikeda, H., 1933. Sex-correlated marking in *Sepioteuthis lessoniana* Férussac. Venus 3, 324–329.

Ikeda, Y., Sakurazawa, I., Sakurai, Y., Matsumoto, G., 2003. Initial trials of squid rearing, maintenance and culture at the Brain Science Institute of RIKEN. Suisan Zoshoku 51 (4), 391–400.

Ikeda, Y., Ueta, Y., Sakurazawa, I., Matsumoto, G., 2004. Transport of the oval squid *Sepioteuthis lessoniana* Férussac, 1831 in Lesson 1830–1831 (Cephalopoda: Loliginidae) for up to 24 h and subsequent transfer to an aquarium. Fisheries Sci. 70, 21–27.

Ikeda, Y., Sakurazawa, I., Ito, K., Sakurai, Y., Matsumoto, G., 2005. Rearing of squid hatchlings, *Heterololigo bleekeri* (Keferstein 1866) up to 2 months in a closed seawater system. Aquac. Res. 36, 409–412.

Ikeda, Y., Ueta, Y., Anderson, F., Matsumoto, G., 2009. Reproduction and life span of the oval squid *Sepioteuthis lessoniana* (Cephalopoda: Loliginidae): comparison between laboratory-cultured and wild-caught squid. Mar. Biodivers. Rec. 2, e50.

Inoha, M., Sezoko, M., 1967. The experiment on the culture of Aori-ika (white squid) *Sepioteuthis lessoniana*: 1967 Progress report. Ryukyu (Okinawa) Fish Res Lab, pp. 139–145.

Itami, K., Izawa, Y., Maeda, S., Nakai, K., 1963. Notes on the laboratory culture of the octopus larvae. Bull. Jpn. Soc. Sci. Fish. 29 (6), 514–520.

Jereb, P., Vecchione, M., Roper, C.F.E., 2010. Family loliginidae. In: Jereb P., Roper C.F. E. (Eds.), Cephalopods of the World. An Annotated and Illustrated Catalogue of Species Known to Date. In: Myopsid and Oegopsid Squids, Vol. 2. FAO, Rome, pp. 38–117, FAO Species Catalogue for Fishery Purposes, No. 4.

Katsanevakis, S., Stephanopoulou, S., Miliou, H., Moraitou-Apostolopoulou, M., Verriopoulos, G., 2005. Oxygen consumption and ammonia excretion of *Octopus vulgaris* (Cephalopoda) in relation to body mass and temperature. Mar. Biol. 146, 725–732.

Kinne, O., 1971. Salinity: 3. Animals: 1. Invertebrates. In: Kinne O. (Ed.), Marine ecology: A Comprehensive, Integrated Treatise on Life in Oceans and Coastal Waters, Volume 1—Environmental Factors, Part 2. Wiley-Interscience, Chichester, pp. 821–995.

Koueta, N., Boucaud-Camou, E., 1999. Food intake and growth in reared early juvenile cuttlefish *Sepia officinalis* L. (Mollusca Cephalopoda). J. Exp. Mar. Biol. Ecol. 240, 93–109.

Koueta, N., Boucaud-Camou, E., 2003. Combined effects of photoperiod and feeding frequency on survival and growth of juvenile cuttlefish *Sepia officinalis* L. in experimental rearing. J. Exp. Mar. Biol. Ecol. 296, 215–226.

Koueta, N., Castro, B.G., Boucaud-Camou, E., 2000. Biochemical indices for instantaneous growth estimation in young cephalopod *Sepia officinalis* L. ICES J. Mar. Sci. 57, 1–7.

Koueta, N., Boucaud-Camou, E., Noel, B., 2002. Effect of enriched natural diet on survival and growth of juvenile cuttlefish *Sepia officinalis* L. Aquaculture 203, 293–310.

Koueta, N., Andrade, J.P., Boletzky, S.V., 2006. The cuttlefish *Sepia officinalis*: a working model in cephalopod research. Vie Milieu 56 (2), 56–201.

Kreuzer, R., 1984. Cephalopods: handling, processing and products: FAO Fisheries Technical Paper 254. FAO, Rome, Italy.

Kurihara, A., Okumura, S., Iwamoto, A., Takeuchi, T., 2006. Feeding pacific sandeel enhances DHA level in common octopus paralarvae. Aquacult. Sci. 54 (4), 413–420.

Lacoue-Labarthe, T., Réveillac, E., Oberhänsli, F., Teyssié, J.L., Jeffree, R., Gattuso, J.P., 2011. Effects of ocean acidification on trace element accumulation in the early-life stages of the squid *Loligo vulgaris*. Aquat. Toxicol. 105, 166–176.

Laptikhovsky, V., Salman, A., Önsoy, B., Katagan, T., 2003. Fecundity of the common cuttlefish, *Sepia officinalis* L. (Cephalopoda, Sepiidae): a new look at an old problem. Sci. Mar. 67 (3), 279–284.

LaRoe, E.T., 1971. The culture and maintenance of the loliginid squids *Sepioteuthis sepioidea* and *Doryteuthis plei*. Mar. Biol. 9, 9–25.

Laschi, C., Cianchetti, M., Mazzolai, B., Margheri, L., Follador, M., Dario, P., 2012. Soft robotic arm inspired by the octopus. Adv. Robotics 26, 709–727.

Leary, S., Underwood, W., Anthony, R., Cartner, S., Corey, D., Grandin, T., Greenacre, C., Gwaltney-Brant, S., McCrackin, M.A., Meyer, R., Miller, D., Shearer, J., Yanong, R., 2013. AVMA Guidelines for the Euthanasia of Animals: 2013 Edition. American Veterinary Medical Association, Schaumburg, IL.

Le Bihan, E., Perrin, A., Koueta, N., 2004. Development of a bioassay from isolated digestive gland cells of the cuttlefish *Sepia officinalis* L. (Mollusca Cephalopoda): effect of Cu, Zn and Ag on enzyme activities and cell viability. J. Exp. Mar. Biol. Ecol. 309, 47–66.

Lee, P.G., 1994. Nutrition of cephalopods: fueling the system. Mar. Freshw. Behav. Physiol. 25 (1–3), 35–51.

Lee, P.G., Forsythe, J.W., Dimarco, F.P., DeRusha, R.H., Hanlon, R.T., 1991. Initial palatability and growth trials on pelleted diets for cephalopods. Bull. Mar. Sci. 49, 362–372.

Lee, P.G., Turk, P.E., Yang, W.T., Hanlon, R.T., 1994. Biological characteristics and biomedical applications of the squid *Sepioteuthis lessoniana* cultured through multiple generations. Biol. Bull. 186, 328–341.

Leibovitz, L., Meyers, T.R., Elston, R., Chanley, P., 1977. Necrotic exfoliative dermatitis of captive squid (*Loligo pealei*). J. Invertebr. Pathol. 30, 369–376.

Leporati, S.C., Pecl, G.T., Semmens, J.M., 2007. Cephalopod hatchling growth: the effects of initial size and seasonal temperatures. Mar. Biol. 151, 1375–1383.

Mangold, K., 1983. Octopus vulgaris. In: Boyle P.R. (Ed.), Cephalopod Life Cycles, Volume I—Species Accounts. Academic Press, London, pp. 335–364.

Mangold, K., Bidder, A., 1989. L'appareil digestif et la digestion. In: Grassé P.P., Mangold K. (Eds.), Traité de zoologie: Anatomie, systematique, biologie. Fascicucle IV: Céphalopodes. Masson, Paris, pp. 321–373.

Mangold, K., Boletzky, S.V., 1973. New data on reproductive biology and growth of *Octopus vulgaris*. Mar. Biol. 19, 7–12.

Márquez, L., Quintana, D., Lorenzo, A., Almansa, E., 2013. Biometrical relationships in developing eggs and neonates of *Octopus vulgaris* in relation to parental diet. Helgol. Mar. Res. 67, 461–470.

Marthy, H.J., Hauser, R., School, A., 1976. Natural tranquilliser in cephalopod eggs. Nature 261, 496–497.

Martínez, R., López-Ripoll, E., Avila-Poveda, O.H., Santos-Ricalde, R., Mascaró, M., Rosas, C., 2011. Cytological ontogeny of the digestive gland in post-hatching *Octopus maya*, and cytological background of digestion in juveniles. Aquat. Biol. 11, 249–261.

Martínez, R., Santos, R., Mascaró, M., Canseco, L., Caamal-Monsreal, C., Rosas, C., 2012. Digestive dynamics during chyme formation of *Octopus maya* (Mollusca, Cephalopoda). Aquac. Res. 43, 1119–1126.

Martínez, R., Gallardo, P., Pascual, C., Navarro, J.C., Sánchez, A., Caamal-Monsreal, C., Rosas, C., 2014. Growth, survival and physiological condition of Octopus maya when fed a successful formulated diet. Aquaculture 426–427, 310–317.

Martínez-Porchas, M., Martínez-Córdova, L.R., Porchas-Cornejo, M.A., López-Elías, J.A., 2010. Shrimp polyculture: a potentially profitable, sustainable, but uncommon aquacultural practice. Rev. Aquacult. 2 (2), 73–85.

Martins, C.I.M., Eding, E.H., Verdegem, M.C.J., Heinsbroek, L.T.N., Schneider, O., Blancheton, J.P., Roque d'Orbcastel, E., Verreth, J.A.J., 2010. New developments in recirculating aquaculture systems in Europe: a perspective on environmental sustainability. Aquacult. Eng. 43 (3), 83–93.

Mather, J.A., 2011. Philosophical background of attitudes toward and treatment of invertebrates. Inst. Lab. Anim. Res. J. 52 (2), 205–212.

Mather, J.A., Anderson, R.C., 2007. Ethics and invertebrates: a cephalopod perspective. Dis. Aquat. Org. 75, 119–129.

Melzner, F., Bock, C., Pörtner, H.O., 2007. Allometry of thermal limitation in the cephalopod *Sepia officinalis*. Comp. Biochem. Physiol. A 146, 149–154.

Messenger, J.B., Young, J.Z., 1999. The radular apparatus of cephalopods. Philos. Trans. R. Soc. B 354, 161–182.

Miliou, H., Fintikaki, M., Kountouris, T., Verriopoulos, G., 2005. Combined effects of temperature and body weight on growth and protein utilization of the common octopus, *Octopus vulgaris*. Aquaculture 249, 245–256.

Miserez, A., Schneberk, T., Sun, C., Zok, F.W., Waite, J.H., 2008. The transition from stiff to compliant materials in squid beaks. Science 319, 1816–1819.

Mladineo, I., Bočina, I., 2007. Extraintestinal gamogony of *Aggregata octopiana* in the reared common octopus (*Octopus vulgaris*) (Cephalopoda: Octopodidae). J. Invertebr. Pathol. 96 (3), 261–264.

Moe, M.A., 1993. The Marine Aquarium Reference: Systems and Invertebrates. Green Turtle Publications, Plantation, FL.

Moe, M.A., 2009. The Marine Aquarium Handbook: Beginner to Breeder, third ed. T.F.H. Publications, Neptune City, NJ.

Moguel, C., Mascaró, M., Avila-Poveda, O.H., Caamal-Monsreal, C., Sanchez, A., Pascual, C., Rosas, C., 2010. Morphological, physiological and behavioral changes during post-hatching development of *Octopus maya* (Mollusca: Cephalopoda) with special focus on the digestive system. Aquat. Biol. 9, 35–48.

Moltschaniwskyj, N., Johnston, D., 2006. Evidence that lipid can be digested by the dumpling squid *Euprymna tasmanica*, but is not stored in the digestive gland. Mar. Biol. 149 (3), 565–572.

Moltschaniwskyj, N.A., Hall, K., Lipinski, M.R., Marian, J.E.A.R., Nishiguchi, M., Sakai, M., Shulman, D.J., Sinclair, B., Sinn, D.L., Staudinger, M., Van Gelderen, R., Villanueva, R., Warnke, K., 2007. Ethical and welfare considerations when using cephalopods as experimental animals. Rev. Fish Biol. Fisher. 17, 455–476.

Monroig, O., Navarro, J.C., Dick, J.R., Alemany, F., Tocher, D.R., 2012a. Identification of a Δ5-like fatty acyl desaturase from the cephalopod *Octopus vulgaris* (Cuvier 1797) involved in the biosynthesis of essential fatty acids. Mar. Biotechnol. 14, 411–422.

Monroig, O., Guinot, D., Hontoria, F., Tocher, D.R., Navarro, J.C., 2012b. Biosynthesis of essential fatty acids in *Octopus vulgaris* (Cuvier, 1797): molecular cloning, functional characterisation and tissue distribution of a fatty acyl elongase. Aquaculture 360–361, 45–53.

Moreira, A.A., Tomás, A.R.G., Hilsdorf, A.W.S., 2011. Evidence for genetic differentiation of *Octopus vulgaris* (Mollusca, Cephalopoda) fishery populations from the southern coast of Brazil as revealed by microsatellites. J. Exp. Mar. Biol. Ecol. 407, 34–40.

Nabhitabhata, J., 1978. Rearing experiment on economic cephalopod-I: long-finned squid (*Sepioteuthis lessoniana* Lesson): Rayong Brackishwater fisheries station technical paper 1978. Brackishwater Fisheries Division, Department of Fisheries, Thailand.

Nabhitabhata, J., 1995. Mass culture of cephalopods in Thailand. World Aquacult. Mag 26, 25–29.

Nabhitabhata, J., 1996. Life cycle of cultured big fin squid, *Sepioteuthis lessoniana* Lesson. Phuket Mar. Biol. Cent. Spec. Publ. 16, 83–95.

Nabhitabhata, J., 1997. Life cycle of three cultured generations of spineless cuttlefish, *Sepiella inermis* (Férussac & D´Orbigny, 1848). Phuket Mar. Biol. Cent. Spec. Publ. 17 (1), 289–298.

Nabhitabhata, J., Kbinrum, S., Kumpang, P., 1984. On floating net cage culture of big finned squid, *Sepioteuthis lessoniana* Lesson, at different stocking densities: Rayong Brackishwater fisheries station technical paper 7/1985. Brackishwater Fisheries Division, Department of Fisheries, Thailand.

Nabhitabhata, J., Asawangkune, P., Amornjaruchit, S., Promboon, P., 2001. Tolerance of eggs and hatchlings of neritic cephalopods to salinity changes. Phuket Mar. Biol. Cent. Spec. Publ. 25 (1), 91–99.

Nabhitabhata, J., Nilaphat, P., Promboon, P., Jaroongpattananon, C., 2005a. Life cycle of cultures Bobtail squid, *Euprymna hyllebergi* Nateewathana, 1997. Phuket Mar. Biol. Cent. Res. Bull. 66, 351–365.

Nabhitabhata, J., Nilaphat, P., Promboon, P., Jaroongpattananon, C., Nilaphat, G., Reunreng, A., 2005b. Performance of simple large-scale cephalopod culture system in Thailand. Phuket Mar. Biol. Cent. Res. Bull. 66, 337–350.

Naef, A., 1928. Cephalopoda, embryology. In: Naef A. (Ed.), Fauna and Flora of the Bay of Naples, Volume II, Part I (Monograph 35), pp.1–461, Translated by the Smithsonian Institution Libraries, Washington, D.C.

Natsukari, Y., 1970. Egg-laying behavior, embryonic development and hatched larva of the pygmy cuttlefish, *Idiosepius pygmaeus paradoxus* Ortmann. Bull. Fac. Fish, Nagasaki Univ. 30, 15–29.

Navarro, J.C., Villanueva, R., 2000. Lipid and fatty acid composition of early stages of cephalopods: an approach to their lipid requirements. Aquaculture 183, 161–177.

Navarro, J.C., Villanueva, R., 2003. The fatty acid composition of *Octopus vulgaris* paralarvae reared with live and inert food: deviation from their natural fatty acid profile. Aquaculture 219, 613–631.

Nguyen, T.T.T., Davy, F.B., Rimmer, M.A., De Silva, S.S., 2009. Use and exchange of genetic resources of emerging species for aquaculture and other purposes. Rev. Aquacult. 1, 260–274.

Nixon, M., 1985. Capture of prey, diet and feeding of *Sepia officinalis* and *Octopus vulgaris* (Mollusca: Cephalopoda) from hatchling to adult. Vie Milieu 35 (3/4), 255–261.

Noyola, J., Caamal-Monsreal, C., Díaz, F., Re, D., Sánchez, A., Rosas, C., 2013a. Thermopreference, tolerance and metabolic rate of early stages juvenile *Octopus maya* acclimated to different temperatures. J. Therm. Biol. 38, 14–19.

Noyola, J., Mascaró, M., Caamal-Monsreal, C., Noreña-Barroso, E., Díaz, F., Re, D., Sánchez, A., Rosas, C., 2013b. Effect of temperature on energetic balance and fatty acid composition of early juveniles of *Octopus maya*. J. Exp. Mar. Biol. Ecol. 445, 156–165.

Nyholm, S.V., McFall-Ngai, M., 2004. The winnowing: establishing the squid-vibrio symbiosis. Nat. Rev. Microbiol. 2, 632–642.

O'Dor, R.K., Mangold, K., Boucher-Rodoni, R., Wellls, M.J., Wells, J., 1984. Nutrient absorption, storage and remobilization in *Octopus vulgaris*. Mar. Behav. Physiol. 11 (3), 239–258.

O'Dor, R.K., Wells, M.J., 1987. Energy and nutrient flow. In: Boyle P.R. (Ed.), Cephalopod Life Cycles, Volume II—Comparative Reviews. Academic Press, London, pp. 109–133.

Oestmann, D.J., Scimeca, J.M., Forsythe, J., Hanlon, R., Lee, P., 1997. Special considerations for keeping cephalopods in laboratory facilities. J. Am. Assoc. Lab. Anim. Sci. 36 (2), 89–93.

Ohshima, Y., Choe, S., 1961. On the rearing of young cuttlefish and squid. Bull. Jpn. Soc. Sci. Fish. 27 (11), 979–986.

Okubo, S., 1979. Spawning and rearing of hatchlings of the giant Pacific octopus. Anim. Nat. 9 (3), 2–6.

Okubo, S., 1980. Culture of the paralarvae and juveniles of *Octopus dofleini* in an aquarium for one year and two months. Shima Marineland Quarterly 25, 4–5.

Okumura, S., Kurihara, A., Iwamoto, A., Takeuchi, T., 2005. Improved survival and growth in *Octopus vulgaris* paralarvae by feeding large type *Artemia* and Pacific sandeel, *Ammodytes personatus*. Aquaculture 244, 147–157.

Ortiz, N., Ré, M.E., Márquez, F., 2006. First description of eggs, hatchlings and hatchling behaviour of *Enteroctopus megalocyathus* (Cephalopoda: Octopodidae). J. Plankton Res. 28 (10), 881–890.

Overstreet, R.M., Hochberg, F.G., 1975. Digenetic trematodes in cephalopods. J. Mar. Biol. Assoc. UK 55 (4), 893–910.

Palmegiano, G.B., D'Apote, M.P., 1983. Combined effects of temperature and salinity on cuttlefish (*Sepia officinalis* L.) hatching. Aquaculture 35, 259–264.

Palmegiano, G.B., Sequi, R., 1981. On the possibility of the introduction of cuttlefish (*Sepia officinalis* L.) in a management strategy for the coastal lagoons. In: Symposium International sur les Lagunes Cotieres, France, pp. 1–9.

Palmegiano, G.B., Sequi, R., 1984. Allevamento estensivo di *Sepia officinalis* L. Agric Ric 41, 49–56.

Parra, G., Villanueva, R., Yúfera, M., 2000. Respiration rates in late eggs and early hatchlings of the common octopus, *Octopus vulgaris*. J. Mar. Biol. Assoc. UK 80, 557–558.

Pascual, S., Gestal, C., Estévez, J.M., Rodríguez, H., Soto, M., Abollo, E., Arias, C., 1996. Parasites in commercially-exploited cephalopods (Mollusca, Cephalopoda) in Spain: an updated perspective. Aquaculture 142, 1–10.

Paulij, W.P., Zurburg, W., Denucé, J.M., van Hannen, E.J., 1990a. The effect of copper on the embryonic development and hatching of *Sepia officinalis* L. Arch. Environ. Contam. Toxicol. 19, 797–801.

Paulij, W.P., Herman, P.M.J., van Hannen, E.J., Denucé, J.M., 1990b. The impact of photoperiodicity on hatching of *Loligo vulgaris* and *Loligo forbesi*. J. Mar. Biol. Assoc. UK 70, 597–610.

Paulij, W.P., Herman, P.M.J., Roozen, M.E.F., Denucé, J.M., 1991. The influence of photoperiodicity on hatching of *Sepia officinalis*. J. Mar. Biol. Assoc. UK 71, 665–678.

Perrin, A., 2004. Etude expérimentale des capacités digestives chez la seiche Sepia officinalis L. (Mollusque, Céphalopode): Impact de l'alimentation, indice de condition nutritionnelle et formulation d'un aliment artificiel (Ph.D. thesis). Université de Caen, France.

Perrin, A., Le Bihan, E., Koueta, N., 2004. Experimental study of enriched frozen diet on digestive enzymes and growth of juvenile cuttlefish *Sepia officinalis* L. (Mollusca Cephalopoda). J. Exp. Mar. Biol. Ecol. 311, 267–285.

Petza, D., Katsanevakis, S., Verriopoulos, G., 2006. Experimental evaluation of the energy balance in *Octopus vulgaris*, fed *ad libitum* on a high-lipid diet. Mar. Biol. 148, 827–832.

Pierce, G.J., Portela, J., 2014. Fisheries production and market demand. In: Iglesias J., Fuentes L., Villanueva R. (Eds.), Cephalopod culture. Springer Netherlands, London, pp. 41–58. ISBN: 978-94-017-8647-8.

Pörtner, H.O., Zielinski, S., 1998. Environmental constraints and the physiology of performance in squids. S. Afr. J. Mar. Sci. 20, 207–221.

Prato, E., Portacci, G., Biandolino, F., 2010. Effect of diet on growth performance, feed efficiency and nutritional composition of Octopus vulgaris. Aquaculture 309, 203–211.

Promboon, P., Nabhitabhata, J., Duengdee, T., 2011. Life cycle of marbled octopus, *Amphioctopus aegina* (Gray) (Cephalopoda: Octopodidae) reared in the laboratory. Sci. Mar. 75 (4), 811–821.

Putnam, H.M., Edmunds, P.J., 2011. The physiological response of reef corals to diel fluctuations in seawater temperature. J. Exp. Mar. Biol. Ecol. 396 (2), 216–223.

Querol, P., Morillo-Velarde, P.S., Cerezo Valverde, J., Martinez Llorens, S., Moñino, A.V., Jover, M.T.A., 2012. Inclusion of fish and krill meal in extruded diets for *Octopus vulgaris* (Cuvier,1797): assessment of acceptance. Aquac. Res. http://dx.doi.org/10.1111/are.12093.

Quintana, D., Domingues, P., García, S., 2008. Effect of two artificial wet diets agglutinated with gelatin on feed and growth performance of common octopus (*Octopus vulgaris*) subadults. Aquaculture 280, 161–164.

Quintana, D., Rosas, C., Moreno-Villegas, E., 2011. Relationship between nutritional and rearing parameters of *Octopus maya* juveniles fed with different rations of crab paste. Aquacult. Nutr. 17, e379–e388.

Rama-Villar, A., Faya-Angueira, V., Moxica, C., Rey-Méndez, M., 1997. Engorde de pulpo (*Octopus vulgaris*) en batea. In: Costa J., Abellán E., García García B., Ortega A., Zamora S. (Eds.), Actas del VI Congreso Nacional de Acuicultura. Cartagena, Spain, pp. 245–250.

Reid, A., Jereb, P., Roper, C.F.E., 2005. Family sepiidae. In: Jereb P., Roper C.F.E. (Eds.), Cephalopods of the world. An annotated and illustrated catalogue of species known to date. In: Chambered Nautiluses and Sepioids (Nautilidae, Sepiidae, Sepiolidae, Sepiadariidae, Idiosepiidae and Spirulidae), Vol. 1. FAO, Rome, Italy, pp. 57–152, FAO Species Catalogue for Fishery Purposes, No. 4, Vol. 1.

Reimschuessel, R., Stoskopf, M.K., Bennett, R.O., 1990. Myocarditis in the common cuttlefish (*Sepia officinalis*). J. Comp. Pathol. 102 (3), 291–297.

Rey-Méndez, M., Tuñon, E., Luaces-Canosa, M., 2003. Estudio de los efectos del peso inicial y el sexo sobre el comportamiento, la mortalidad y el crecimiento del pulpo (*Octopus vulgaris*, Cuvier 1797) en cultivo industrial. In: IX Congreso Nacional de Acuicultura, Cádiz, Spain, pp. 276–277.

Rodhouse, P.G.K., Pierce, G.J., Nichols, O.C., Sauer, W.H.H., Arkhipkin, A.I., Laptikhovsky, V.V., Lipinski, M.L., Ramos, J., Gras, M., Kidokoro, H., Sadayasu, K., Pereira, J., Lefkaditou, E., Pita, C., Gasalla, M., Haimovici, M., Sakai, M., Downey, N., 2014. Environmental effects on cephalopod population dynamics: implications for management of fisheries. Adv. Mar. Biol. 67, 99–234.

Rodríguez, C., Carrasco, J.F., Arronte, J.C., Rodríguez, M., 2006. Common octopus (*Octopus vulgaris* Cuvier, 1797) juvenile ongrowing in floating cages. Aquaculture 254, 293–300.

Rodríguez-Canul, R., Pérez-Campos, R.A.A., Pérez-Vega, J.A., Pascual, C., Guillen Hernández, S., Arjona-Burgos, I.J., 2012. The red octopus *Octopus maya* as putative paratenic host for wssv: evaluation in experimental and natural conditions.

In: Abstract Book of the 2012 Cephalopod International Advisory Council Symposium, Florianópolis, Brazil, p. 231.

Rodríguez-Domínguez, H., Soto-Búa, M., Iglesias-Blanco, M., Crespo-González, C., Arias-Fernández, J., García-Estévez, J., 2006. Preliminary study on the phagocytic ability of Octopus vulgaris Cuvier, 1797 (Mollusca: Cephalopoda) haemocytes in vitro. Aquaculture 254, 563–570.

Rögener, W., Renwrantz, L., Uhlenbruck, G., 1985. Isolation and characterization of a lectin from the hemolymph of the cephalopod *Octopus vulgaris* (Lam.) Inhibited by α-D-lactose and N-acetyl-lactosamine. Dev. Comp. Immunol. 9 (4), 605–616.

Rohrbach, B., Schmidtberg, H., 2006. Sepia arms and tentacles: model systems for studying the regeneration of brachial appendages. Vie Milieu 56, 175–190.

Roper, C.F.E., Sweeney, M.J., Nauen, C.E., 1984. FAO Species Catalogue. 3. Cephalopods of the World. An Annotated and Illustrated Catalogue of Species of Interest to Fisheries. In: FAO Fisheries Synopsis No. 125, vol. 3. FAO, Rome, Italy.

Rosas, C., Tut, J., Baeza, J., Sánchez, A., Sosa, V., Pascual, C., Arena, L., Domingues, P., Cuzon, G., 2008. Effect of type of binder on growth, digestibility, and energetic balance of *Octopus maya*. Aquaculture 275, 291–297.

Rosas, C., Caamal, C., Cazares, R., 2010. Incubation Process for *Octopus* and Incubator. World Intelectual Property Organization (WIPO), Mexico, pp. 1–6, PCT/MX2009/000100.

Rosas, C., Sánchez, A., Pascual, C., Aguila, J., Maldonado, T., Domingues, P., 2011. Effects of two dietary protein levels on energy balance and digestive capacity of *Octopus maya*. Aquacult. Int. 19, 165–180.

Rosas, C., Valero, A., Caamal-Monsreal, C., Uriarte, I., Farias, A., Gallardo, P., Sánchez, A., Domingues, P., 2013. Effects of dietary protein sources on growth, survival and digestive capacity of *Octopus maya* juveniles (Mollusca: Cephalopoda). Aquac. Res. 44 (7), 1029–1044.

Roura, A., González, A.F., Redd, K., Guerra, A., 2012. Molecular prey identification in wild *Octopus vulgaris* paralarvae. Mar. Biol. 159, 1335–1345.

Salas, S., Mexicano-Cíntora, G., Cabrera, M., 2006. Hacia donde van las pesquerías de Yucatán? Tendencias, retos y perspectivas. Centro de Investigación y Estudios Avanzados, Departamento de Recursos del Mar, Unidad Mérida, Mérica, Yucatán, Mexico.

Sánchez, F.J., Fuentes, L., Otero, J.J., Lago, M.J., Linares, F., Pazos, G., Iglesias, J., 2013. Effect of tank volume on the growth and survival of reared *Octopus vulgaris* paralarvae. Aquac. Res. 44, 667–670.

Sangster, C.R., Smolowitz, R.M., 2003. Description of *Vibrio alginolyticus* infection in cultured *Sepia officinalis*, *Sepia apama*, and *Sepia pharaonis*. Biol. Bull. 205, 233–234.

Santé-Lhoutellier, V., Astruc, T., Marinova, P., Greve, E., Gatellier, P., 2008. Effect of meat cooking on physicochemical state and in vitro digestibility of myofibrillar proteins. J. Agric. Food Chem. 56 (4), 1488–1494.

Saso, N., 1979. On the culture of Aori-ika, *Sepioteuthis lessoniana*. In: Japan Association of Fisheries Cooperatives (Ed.), Proceedings of the 25th Japan National Fisheries Activity Convention, Tokyo, Japan, p. 204.

Saville, A., 1987. Comparisons between cephalopods and fish of those aspects of the biology related to stock management. In: Boyle P.R. (Ed.), Cephalopod Life Cycles, Volume II—Comparative Reviews. Academic Press, London, pp. 277–290.

Scimeca, J.M., 2012. Cephalopods. In: Lewbart G.A. (Ed.), Invertebrate Medicine, second ed. Wiley-Blackwell, Ames, IA, pp. 113–125.

Segawa, S., 1987. Life history of the oval squid, *Sepioteuthis lessoniana* in Kominato and adjacent waters, central Honshu, Japan. J. Tokyo Univ. Fish. 74, 67–105.

Segawa, S., 1990. Food consumption, food conversion and growth rates of the oval squid *Sepioteuthis lessoniana* by laboratory experiments. Nippon Suisan Gakk 56 (2), 217–222.

Segawa, S., 1993. Field and laboratory feeding studies of the Japanese oval squid, *Sepioteuthis lessoniana*. In: Okutani T., O'Dor R.K., Kubodera T. (Eds.), Recent Advances in Cephalopod Fisheries Biology. Tokai University Press, Tokyo, Japan.

Segawa, S., 1995. Effect of temperature on oxygen consumption of juvenile oval squid *Sepioteuthis lessoniana*. Fisheries Sci. 61 (5), 743–746.

Segawa, S., Hanlon, R.T., 1988. Oxygen consumption and ammonia excretion rates in *Octopus maya*, *Loligo forbesi* and *Lolliguncula brevis* (Mollusca: Cephalopoda). Mar. Behav. Physiol. 13, 389–400.

Segawa, S., Yang, W.T., Marthy, H.-J., Hanlon, R.T., 1988. Illustrated embryonic stages of the eastern Atlantic squid Loligo forbesi. Veliger 30 (3), 230–243.

Seibel, B.A., Robison, B.H., Haddock, S.H.D., 2005. Post-spawning egg care by a squid. Nature 438, 929.

Seixas, P., Rey-Méndez, M., Valente, L.M.P., Otero, A., 2008. Producing juvenile *Artemia* as prey for *Octopus vulgaris* paralarvae with different microalgal species of controlled biochemical composition. Aquaculture 283, 83–91.

Seixas, P., Otero, A., Valente, L.M.P., Dias, J., Rey-Méndez, M., 2010a. Growth and fatty acid composition of *Octopus vulgaris* paralarvae fed with enriched *Artemia* or co-fed with an inert diet. Aquacult. Int. 18, 1121–1135.

Seixas, P., Rey-Méndez, M., Valente, L.M.P., Otero, A., 2010b. High DHA content in *Artemia* is ineffective to improve *Octopus vulgaris* paralarvae rearing. Aquaculture 300, 156–162.

Sen, H., 2005a. Temperature tolerance of loliginid squid (*Loligo vulgaris* Lamarck, 1798) eggs in controlled conditions. Turk. J. Fish. Aquat. Sci. 5, 53–56.

Sen, H., 2005b. Incubation of european squid (*Loligo vulgaris* Lamarck, 1798) eggs at different salinities. Aquac. Res. 36, 876–881.

Sequi, R., Palmegiano, G.B., 1984. Some alternatives for the use of cuttlefish (*Sepia officinalis* L.) resources in coastal lagoons [Italy]. In: General Fisheries Council for the Mediterranean (Ed.), Food and Agriculture Organization of the United Nations (FAO), Management of coastal lagoon fisheries. FAO, Rome, Italy.

Shashar, N., Hanlon, R.T., Petz, A. de M., 1998. Polarization vision helps detect transparent prey. Nature 393, 222–223.

Sherrill, J., Spelman, L.H., Reidel, C.L., Montali, R.J., 2000. Common cuttlefish (*Sepia officinalis*) mortality at the national zoological park: implications for clinical management. J. Zoo Wildl. Med. 31 (4), 523–531.

Shvetsova, V.D., 1977. Cephalopods as a Probable Object of Cultivation. Review Information. In: Fishery Using of Resources of the World Ocean, Issue 5, Central Scientific-Research Institute of Information and Techno-economic investigations of Fishery of Ministry of Fisheries of USSR, Moscow, pp. 1–46 (In Russian).

Sivalingam, D., Ramadoss, K., Gandhi, A.D., Enose, M., 1993. Hatchery rearing of the squid, *Sepioteuthis lessoniana* and the cuttlefish, *Sepia pharaonis*. Mar. Fish. Infor. Serv. 122, 12–14.

Snyder, S., 1986a. Laboratory culture of *Octopus dofleini* from hatching to settlement. Am. Malacol. Bull. 4, 241.

Snyder, S., 1986b. Successful rearing of *Octopus dofleini* from hatchling to settlement. In: American Association of Zoological Parks and Aquariums (Ed.), Proceedings of the 1986 Annual Conference. American Association of Zoological Parks and Aquariums, Wheeling, WV, pp. 781–783.

Solis, M., 1998. Aspectos biológicos del pulpo: *Octopus maya* Voss y Solis. Instituto Nacional de la Pesca, Mexico, pp. 1–38, Contrib Invest Pesq.

Soudant, P., Chu, F.-L.E., Volety, A., 2013. Host-parasite interactions: Marine bivalve molluscs and protozoan parasites, *Perkinsus* species. J. Invertebr. Pathol. 114 (2), 196–216.

Steer, M.A., Moltschaniwskyj, N.A., 2007. The effects of egg position, egg mass size, substrate and biofouling on embryo mortality in the squid *Sepiotheutis australis*. Rev. Fish Biol. Fisher. 17, 173–182.

Sugimoto, C., Ikeda, Y., 2012. Ontogeny of schooling behavior in the oval squid *Sepioteuthis lessoniana*. Fisheries Sci. 78, 287–294.

Sugita, A., 2012. Aori-ika, *Sepioteuthis lessoniana*. Aquacult. Mag. 49 (4), 68–71.

Sykes, A.V., Domingues, P.M., Loyd, M., Sommerfield, A., Andrade, J.P., 2003. The influence of culture density and enriched environments on the first stage culture of young cuttlefish, *Sepia officinalis* (Linnaeus, 1758). Aquacult. Int. 11, 531–544.

Sykes, A.V., Domingues, P.M., Andrade, J.P., 2006a. Effects of using live grass shrimp (*Palaemonetes varians*) as the only source of food for the culture of cuttlefish, *Sepia officinalis* (Linnaeus, 1758). Aquacult. Int. 14, 551–568.

Sykes, A.V., Domingues, P.M., Correia, M., Andrade, J.P., 2006b. Cuttlefish culture - state of the art and future trends. Vie Milieu 56 (2), 129–137.

Sykes, A.V., Almansa, E., Lorenzo, A., Andrade, J.P., 2009. Lipid characterization of both wild and cultured eggs of cuttlefish (*Sepia officinalis* L.) throughout the embryonic development. Aquacult. Nutr. 15 (1), 38–53.

Sykes, A.V., Baptista, F.D., Gonçalves, R.A., Andrade, J.P., 2012. Directive 2010/63/EU on animal welfare: a review on the existing scientific knowledge and implications in cephalopod aquaculture research. Rev. Aquacult. 4, 146–162.

Sykes, A.V., Pereira, D., Rodríguez, C., Lorenzo, A., Andrade, J.P., 2013a. Effects of increased tank bottom areas on cuttlefish (*Sepia officinalis*, L.) reproduction performance. Aquac. Res. 44 (7), 1017–1028.

Sykes, A.V., Gonçalves, R.A., Andrade, J.P., 2013b. Early weaning of cuttlefish (*Sepia officinalis*, L.) with frozen grass shrimp (*Palaemonetes varians*) from the first day after hatching. Aquac. Res. 44 (12), 1815–1823.

Tsuchiya, M., 1982. On the culture of the squid *Sepioteuthis lessoniana* Lesson (Loliginidae), from hatching to the whole life in Iriomote Island, Okinawa. Bull. Inst. Oceanic Res. Dev., Tokai Univ. Notes 4, 49–70.

Turk, P.E., Hanlon, R.T., Bradford, L.A., Yang, W.T., 1986. Aspects of feeding, growth and survival of the European squid *Loligo vulgaris* Lamarck, 1799, reared through the early growth stages. Vie Milieu 36 (1), 9–13.

Ueta, Y., 2000. Fishery biological studies of the oval squid, *Sepioteuthis lessoniana* around Tokushima Prefecture. Bull. Tokushima Pref. Fish. Exp. Stn. 1, 1–80.

Uriarte, I., Hernández, J., Dörner, J., Paschke, K., Farías, A., Crovetto, E., Rosas, C., 2010. Rearing and growth of the octopus *Robsonella fontaniana* (Cephalopoda: Octopodidae) from planktonic hatchlings to benthic juveniles. Biol. Bull. 218, 200–210.

Uriarte, I., Iglesias, J., Domingues, P., Rosas, C., Viana, M.T., Navarro, J.C., Seixas, P., Vidal, E., Ausburger, A., Pereda, S., Godoy, F., Paschke, K., Farías, A., Olivares, A., Zuñiga, O., 2011. Current status and bottle neck of octopod aquaculture: the case of American species. J. World Aquacult. Soc. 42 (6), 735–752.

Van Heukelem, W.F., 1976. Growth, bioenergetics and life-span of *Octopus* cyanea and Octopus maya (Ph.D. thesis). University of Hawaii, Manoa.

Van Heukelem, W.F., 1977. Laboratory maintenance, breeding, rearing, and biomedical research potential of the Yucatan octopus (*Octopus maya*). Lab. Anim. Sci. 27 (5), 852–859.

Van Heukelem, W.F., 1983. Octopus maya. In: Boyle P.R. (Ed.), Cephalopod Life Cycles, Volume I—Species Accounts. Academic Press, London, pp. 311–323.

Vaz-Pires, P., Seixas, P., Barbosa, A., 2004. Aquaculture potential of the common octopus (*Octopus vulgaris* Cuvier, 1797): a review. Aquaculture 238, 221–238.

Viciano, E., Iglesias, J., Lago, M.J., Sánchez, F.J., Otero, J.J., Navarro, J.C., 2011. Fatty acid composition of polar and neutral lipid fractions of *Octopus vulgaris* Cuvier, 1797 paralarvae reared with enriched on-grown *Artemia*. Aquac. Res. 42, 704–709.

Vidal, E.A.G., Boletzky, S.V., 2014. *Loligo vulgaris* and *Doryteuthis opalescens*. In: Iglesias, J., Fuentes, L., Villanueva, R. (Eds.), Cephalopod Culture. Springer Netherlands, London, pp. 271–313.

Vidal, E.A.G., DiMarco, F.P., Wormuth, J.H., Lee, P.G., 2002a. Optimizing rearing conditions of hatchling loliginid squid. Mar. Biol. 140, 117–127.

Vidal, E.A.G., DiMarco, F.P., Wormuth, J.H., Lee, P.G., 2002b. Influence of temperature and food availability on survival, growth and yolk utilization in hatchling squid. Bull. Mar. Sci. 71 (2), 915–931.

Vidal, E.A.G., Roberts, M.J., Martins, R.S., 2005. Yolk utilization, metabolism and growth in reared *Loligo vulgaris reynaudii* paralarvae. Aquat. Living Resour. 18, 385–393.

Vidal, E.A.G., DiMarco, P., Lee, P., 2006. Effects of starvation and recovery on the survival, growth and RNA/DNA ratio in loliginid squid paralarvae. Aquaculture 260, 94–105.

Vidal, E.A.G., Fuentes, L., Silva, L.B., 2010. Defining *Octopus vulgaris* populations: a comparative study of the morphology and chromatophore pattern of paralarvae from North-eastern and Southwestern Atlantic. Fish. Res. 106, 199–208.

Villanueva, R., 1994. Decapod crab zoeae as food for rearing cephalopod paralarvae. Aquaculture 128, 143–152.

Villanueva, R., 1995. Experimental rearing and growth of planktonic *Octopus vulgaris* from hatching to settlement. Can. J. Fish. Aquat. Sci. 52, 2639–2650.

Villanueva, R., 2000. Effect of temperature on statolith growth of the European squid *Loligo vulgaris* during early life. Mar. Biol. 136, 449–460.

Villanueva, R., Bustamante, P., 2006. Composition in essential and non-essential elements of early stages of cephalopods and dietary effects on the elemental profiles of *Octopus vulgaris* paralarvae. Aquaculture 261, 225–240.

Villanueva, R., Norman, M.D., 2008. Biology of the planktonic stages of benthic octopuses. Oceanogr. Mar. Biol. Annu. Rev. 46, 105–202.

Villanueva, R., Nozais, C., Boletzky, S.V., 1995. The planktonic life of octopuses. Nature 377, 107.

Villanueva, R., Koueta, N., Riba, J., Boucaud-Camou, E., 2002. Growth and proteolytic activity of *Octopus vulgaris* paralarvae with different food rations during first feeding, using *Artemia* nauplii and compound diets. Aquaculture 205, 269–286.

Villanueva, R., Riba, J., Ruíz-Capillas, C., González, A.V., Baeta, M., 2004. Amino acid composition of early stages of cephalopods and effect of amino acid dietary treatments on *Octopus vulgaris* paralarvae. Aquaculture 242, 455–478.

Villanueva, R., Moltschaniwskyj, N.A., Bozzano, A., 2007. Abiotic influences on embryo growth: statoliths as experimental tools in the squid early life history. Rev. Fish Biol. Fisher. 17, 101–110.

Villanueva, R., Escudero, J.M., Deulofeu, R., Bozzano, A., Casoliva, C., 2009. Vitamin A and E content in early stages of cephalopods and their dietary effects in *Octopus vulgaris* paralarvae. Aquaculture 286, 277–282.

Villanueva, R., Staaf, D.J., Argüelles, J., Bozzano, A., Camarillo-Coop, S., Nigmatullin, C.M., Petroni, G., Quintana, D., Sakai, M., Sakurai, Y., Salinas-Zavala, C.A., De Silva-Dávila, R., Tafur, R., Yamashiro, C., Vidal, E.A.G., 2012. A laboratory guide to *in vitro* fertilization of oceanic squids. Aquaculture 342–343, 125–133.

Wada, Y., Kobayashi, T., 1995. On an iteroparity of the oval squid *Sepioteuthis lessoniana*. Bull. Jpn. Soc. Sci. Fish. 61 (2), 151–158.

Walsh, L.S., Turk, P.E., Forsythe, J.W., Lee, P.G., 2002. Mariculture of the loliginid squid *Sepioteuthis lessoniana* through seven successive generations. Aquaculture 212, 245–262.

Watson, A.M., Barrows, F.T., Place, A.R., 2013. Taurine supplementation of plant derived protein and n-3 fatty acids are critical for optimal growth and development of cobia, *Rachycentron canadum*. Lipids 48, 899–913.

Weiss, R.F., 1970. The solubility of nitrogen, oxygen and argon in water and seawater. Deep Sea Res. Oceanogr. Abstr. 17 (4), 721–735.

Wells, M.J., O'Dor, R.K., Mangold, K., Wells, J., 1983. Diurnal changes in activity and metabolic rate in *Octopus vulgaris*. Mar. Behav. Physiol. 9, 275–287.

Wells, M.J., Hanlon, R.T., Lee, P.G., Dimarco, F.P., 1988. Respiratory and cardiac performance in *Lolliguncula brevis* (Cephalopoda, Myopsida): the effects of activity, temperature and hypoxia. J. Exp. Biol. 138, 17–36.

Wood, J.B., Wood, D.A., 1999. Enrichment for an advanced invertebrate. Shape Enrich. 8 (3), 1–5.

Woods, H.A., 1999. Egg-mass size and cell size: effects of temperature on oxygen distribution. Am. Zool. 39, 244–252.

Worm, B., Hilborn, R., Baum, J.K., Branch, T.A., Collie, J.S., Costello, C., Fogarty, M.J., Fulton, E.A., Hutchings, J.A., Jennings, S., Jensen, O.P., Lotze, H.K., Mace, P.M., McClanahan, T.R., Palumbi, S.R., Parma, A.M., Rikard, D., Rosenberg, A.A., Zeller, D., Minto, C., 2009. Rebuilding global fisheries. Science 325, 578–585.

Yang, W.T., Hanlon, R.T., Krejci, M.E., Hixon, R.F., Hulet, W.H., 1983. Laboratory rearing of *Loligo opalescens*, the market squid of California. Aquaculture 31 (1), 77–88.

Yang, W.T., Hixon, R.F., Turk, P.E., Krejci, M.E., Hulet, W.H., Hanlon, R.T., 1986. Growth, behavior, and sexual maturation of the market squid, *Loligo opalescens*, cultured through the life cycle. Fish B-NOAA 84 (4), 771–798.

Yang, W.T., Hanlon, R.T., Lee, P.G., Turk, P.E., 1989. Design and function of closed seawater systems for culturing loliginid squids. Aquacult. Eng. 8, 47–65.

Young, R.E., Harman, R.F., 1988. "Larva", "paralarva" and "subadult" in cephalopod terminology. Malacologia 29 (1), 201–207.

Yu, T.C., Tung, T.Y., 1989. Studies on the culture of *Sepioteuthis lessoniana*. Bull. Taiwan Fish. Res. Inst. 46, 203–206.

Zheng, X., Lin, X., Wang, Z., Yu, R., Tian, C., Li, Z., 2010. Studies on life cycle of cultured spineless cuttlefish *Sepiella japonica*. Trans. Oceanol. Limnol. 3, 24–28.

CHAPTER TWO

Environmental Effects on Cephalopod Population Dynamics: Implications for Management of Fisheries

Paul G.K. Rodhouse[*,1], **Graham J. Pierce**[†,‡], **Owen C. Nichols**[§,¶], **Warwick H.H. Sauer**[||], **Alexander I. Arkhipkin**[#], **Vladimir V. Laptikhovsky**[**], **Marek R. Lipiński**[||], **Jorge E. Ramos**[††], **Michaël Gras**[‡‡,§§], **Hideaki Kidokoro**[¶¶], **Kazuhiro Sadayasu**[||||], **João Pereira**[##], **Evgenia Lefkaditou**[***], **Cristina Pita**[†,‡], **Maria Gasalla**[†††], **Manuel Haimovici**[‡‡‡], **Mitsuo Sakai**[§§§], **Nicola Downey**[||]

[*]British Antarctic Survey, Cambridge, United Kingdom
[†]Oceanlab, University of Aberdeen, Newburgh, Aberdeenshire, United Kingdom
[‡]CESAM & Departamento de Biologia, Universidade de Aveiro, Aveiro, Portugal
[§]School for Marine Science and Technology, University of Massachusetts – Dartmouth, Fairhaven, Massachusetts, USA
[¶]Center for Coastal Studies, Provincetown, Massachusetts, USA
[||]Department of Ichthyology and Fisheries Science, Rhodes University, Grahamstown, South Africa
[#]Falkland Islands Fisheries Department, Stanley, Falkland Islands
[**]CEFAS, Lowestoft, Suffolk, United Kingdom
[††]Institute for Marine and Antarctic Studies, Marine Research Laboratories Taroona, Nubeena Crescent, Taroona, Tasmania, Australia
[‡‡]Université de Caen Basse-Normandie, Institut de Biologie Fondamentale et Appliquée Department, UMR BOREA: Biologie des ORganismes et des Ecosystèmes Aquatiques, Esplanade de la paix, CS 14032, Caen, France
[§§]BOREA, UMR CNRS7208, IRD207, UPMC, MNHN, UCBN, Caen, France
[¶¶]Japan Sea National Fisheries Research, Institute, Fisheries Research Agency, Suido-cho, Niigata, Japan
[||||]Marine Fisheries Research and Development Center, Fisheries Research Agency, Yokohama, Kanagawa, Japan
[##]Instituto de Investigação das Pescas e do Mar (IPIMAR), Lisboa, Portugal
[***]Helenic Centre for Marine Research, Aghios Kosmas, Hellinikon, Athens, Greece
[†††]Fisheries Ecosystems Laboratory, Oceanographic Institute, University of São Paulo, São Paulo, Brazil
[‡‡‡]Institute of Oceanography, Federal University of Rio Grande, CEP, Rio Grande, Brazil
[§§§]National Research Institute of Far Seas Fisheries, Shizuoka, Japan
[1]Corresponding author: e-mail address: pgkr@bas.ac.uk

Contents

Advances in Marine Biology, Volume 67
ISSN 0065-2881
http://dx.doi.org/10.1016/B978-0-12-800287-2.00002-0

Abstract

Cephalopods are a relatively small class of molluscs (~800 species), but they support some large industrial scale fisheries and numerous small-scale, local, artisanal fisheries. For several decades, landings of cephalopods globally have grown against a background of total finfish landings levelling off and then declining. There is now evidence that in recent years, growth in cephalopod landings has declined. The commercially exploited cephalopod species are fast-growing, short-lived ecological opportunists. Annual variability in abundance is strongly influenced by environmental variability, but the underlying causes of the links between environment and population dynamics are poorly understood. Stock assessment models have recently been developed that incorporate environmental processes that drive variability in recruitment, distribution and migration patterns. These models can be expected to improve as more, and better, data are obtained on environmental effects and as techniques for stock identification improve. A key element of future progress will be improved understanding of trophic dynamics at all phases in the cephalopod life cycle. In the meantime, there is no routine stock assessment in many targeted fisheries or in the numerous by-catch fisheries for cephalopods. There is a particular need for a precautionary approach in these cases. Assessment in many fisheries is complicated because cephalopods are ecological opportunists and stocks appear to have benefited from the reduction of key predator by overexploitation. Because of the complexities involved, ecosystem-based fisheries management integrating social, economic and ecological considerations is desirable for cephalopod fisheries. An ecological approach to management is routine in many fisheries, but to be effective, good scientific understanding of the relationships between the environment, trophic dynamics and population dynamics is essential. Fisheries and the ecosystems they depend on can only be managed by regulating the activities of the fishing industry, and this requires understanding the dynamics of the stocks they exploit.

Keywords: Cephalopods, Population dynamics, Environment, Fluctuations, Stock assessment, Forecasting, Management, Governance

1. INTRODUCTION

There around 800 species of cephalopods living today. Fishery exploitation is mainly confined to coastal species of squid, cuttlefish and octopus and those oceanic squids whose migration routes regularly bring them within range of commercial fleets (see Fries, 2010).

Like most cephalopods, the exploited species typically live only 1 or 2 years, living fast and dying young. Their short life cycles, high metabolic rates and fast growth are associated with high plasticity in life history characteristics and marked sensitivity to environmental variation, reflected in large year-to-year fluctuations in population abundance. Cephalopod population dynamics are surprisingly poorly understood.

Empirical relationships between distribution or abundance and environmental conditions are widely documented (see Pierce et al., 2008 for a review), and some of these empirical relationships appear to be sufficiently predictable to be used for fishery forecasting (see Otero et al., 2008; Sobrino et al., 2002; Waluda et al., 2001a). However, caution is necessary before recommending such approaches. As Solow (2002) observed, relationships between time series have a habit of unravelling when longer time series become available. In short, there is no substitute for understanding the underlying mechanisms, in relation both to population dynamics and to how environmental variation effects change in dynamics.

Some authors (e.g. Pauly, 1998) have drawn attention to parallels with small pelagic fish, but many researchers working on cephalopods highlight the difference between fish and cephalopods (e.g. Boyle and Knobloch, 1983). It is, however, difficult to draw clear conclusions since cephalopods display a complex mixture of *r*- and *k*-selected traits, the balance varying between species (see Caddy, 1996). Some cephalopods produce relatively small numbers of eggs, for example, around 2500 eggs in *Eledone cirrhosa* (Regueira et al., 2013) and up to 8000 in *Sepia officinalis* (Laptikhovsky et al., 2003). In other species, tens or hundreds of thousands of eggs are produced: up to 74,000 eggs in *Loligo vulgaris* (Laptikhovsky, 2000), up to around 550,000 in *Octopus vulgaris* (Cuccu et al., 2013) and up to 800,000 in *Illex coindetii* (Laptikhovsky and Nigmatullin, 1999). Some species brood the eggs until hatching (e.g. *O. vulgaris* and *Gonatus onyx*) (Mangold and Boletzky, 1973; Seibel et al., 2000). All cephalopods lack true larval stages, but some have planktonic paralarvae, while in others, the hatchlings both are extremely similar in form to the adults and live in the same habitats.

Cephalopods can occupy similar trophic niches to fish, all commercial species being active predators, and they are also important prey of higher trophic levels, their significance accentuated by the high production to biomass ratio—see the series of reviews by Clarke (1996), Croxall and Prince (1996), Klages (1996) and Smale (1996), as well as many other papers by the late Malcolm Clarke. Recently, ecological modelling work has highlighted the fact that cephalopods can be keystone species (e.g. Gasalla et al., 2010).

1.1. Fisheries

The importance of cephalopods as fishery resources has risen dramatically since 1950. World cephalopod landings rose from around 500,000 t annually to a peak of over 4 million t in 2007. The most recent annual total (for 2010) is around 3.5 million t, an apparent decrease that is evident in trends from several regions (FAO, 2011). There are three main types of cephalopod fisheries: large-scale directed fisheries (e.g. jig fishing for ommastrephid squid), by-catch fisheries (e.g. a substantial proportion of landings of loliginid squids arise as by-catch from demersal trawling) and small-scale directed fisheries that use a range of gears to catch squid, cuttlefish and octopus. Cephalopods vary in their importance as fishery resources in different global regions and remain less important in the northeast Atlantic than in many other regions (see Caddy and Rodhouse, 1998; Hunsicker et al., 2010). Globally, the most important fisheries have been those for ommastrephid squid; while some are relatively stable, others (e.g. for *Dosidicus gigas* on the Pacific coast of the Americas) seem to be characterized by boom and bust dynamics.

Many cephalopod fisheries are regulated; again, Europe is something of an exception, with no routine assessment and no management specified by the Common Fisheries Policy, although local and regional regulatory systems are in place for some artisanal fisheries. Although biological characteristics such as the short life cycle and fast and variable growth rate, plus the difficulty of obtaining accurate age estimates, mean that some approaches to stock assessment (e.g. age-based methods) are unsuitable, a range of approaches, including in-season depletion methods, boat-based surveys and even production models, have been successfully applied (see Pierce and Guerra, 1994) and at least some cephalopod fisheries are routinely (and well) managed.

Particular issues arising in relation to assessment and management include stock identification, variability in abundance (and how to predict it) and prevention of damage to spawning areas. Globally, the comparatively fluid state of cephalopod taxonomy, as new molecular studies challenge (or sometimes support) traditional taxonomic units, and relatively slow progress of genetic stock identification studies also create challenges, compounded in many regions by a systemic failure to record fishery landings to species level.

We have already highlighted the environmental sensitivity of cephalopods and the fact that many if not all species show wide fluctuations in abundance and that this is most evident in the oceanic squid. For the demersal and benthic species that attach their eggs to the seabed or structures thereon, protection of spawning areas and eggs is critical. The use of fixed gear in spawning areas can be problematic in both squids and cuttlefish, with substantial losses of eggs when the gear is hauled, as seen for cuttlefish that lay eggs on cuttlefish traps in the English Channel and loliginid squids that lay their eggs on gill nets off western Portugal.

Management also presents particular challenges. Essentially, the main biological issue is one of escapement. In short-lived species, especially those with nonoverlapping generations, there is no buffer against recruitment failure (Caddy, 1983). In practice, cephalopods usually show nonsynchronous spawning and recruitment, which may help protect against total loss. The flip side of this coin is that cephalopod stocks are generally seen as resilient, rapidly bouncing back after overexploitation. However, it is also possible that the large natural fluctuations have obscured collapses caused by overfishing, even (or perhaps especially) in the ommastrephids.

There are also clear technical, social and economic challenges. Where cephalopods are taken as a by-catch of multispecies fisheries for demersal fish, regulating fishing mortality is difficult. In the case of the large-scale directed fisheries and indeed the small-scale directed fisheries, contingency plans are needed for low abundance years. Small-scale fisheries are probably more adaptable, since they routinely exploit a range of target species—but the social cost of failure is high due to the dependence of many small coastal communities on cephalopod fishing.

These challenges are increasingly relevant, not least because global fishery data (FAO, 2011) suggest that cephalopod fishery landings peaked in 2007 and overexploitation of cephalopod stocks may already be taking place.

1.2. Future challenges and the rationale for a new review

While nowadays cephalopods are routinely fished in coastal waters of most regions of the world, our knowledge of many aspects of their taxonomy, biology and ecology remains limited. The need to manage those cephalopod fisheries that are presently unregulated is becoming increasingly apparent; it is also evident that the toolbox of assessment methods and management measures—and even governance systems, traditionally used to ensure sustainability, needs to be updated to accommodate cephalopods.

An additional driver, which is already changing the way we manage fisheries, certainly in the European Union (EU), is the adoption of the so-called ecosystem approach, enshrined in the EU's revised Common Fishery Policy and supported by a range of other marine-related legislations including the (EU) Marine Strategy Framework Directive. The new paradigm includes the following:

1. Evaluation of effects of fishing on nontarget species and the wider ecosystem
2. Explicit consideration of social and economic consequences of possible management actions and accounting for implementation issues
3. Placing fisheries in the broader content of integrated marine management, for example, recognizing the impacts of multiple stressors on marine ecosystems, evaluating their effects on ecosystem function and ecosystem services and assessing the status of marine ecosystems through the Integrated Ecosystem Assessment, definition of Good Environmental Status and development of monitoring and management systems to deliver action

Such considerations are central to the new Science Plan (due in 2014) of the International Council for the Exploration of the Sea (covering the North Atlantic and its fisheries) but to a greater or lesser extent are also achieving global recognition.

Finally, we must consider the background of global climate change and the ever-rising human population of the world. Well-managed fisheries represent just one component of food security. Fisheries must currently provide both protein for direct human consumption and the base of much of the world's aquafeed, although the latter challenge falls beyond the scope of this chapter. While still a small component of global fisheries production, cephalopods are likely to be increasingly targeted unless overexploited finfish stocks are allowed to recover. Even in the relatively well-managed fisheries of the eastern North Atlantic, the collapse of traditional stocks like hake or cod is leading fishermen to target squid.

The warming and acidification of the oceans, falling salinity due to melting ice, shoaling of the oxygen minimum layer and changes in current systems are expected to have profound effects on marine ecosystems. As environmentally hypersensitive species, cephalopods may be seen as sentinels of future change. As fast-growing molluscs with calcareous statoliths and a high demand for oxygen, the effects of acidification and ocean warming may be significant, as a range of studies are already beginning to suggest (e.g. Rosa and Seibel, 2008).

This chapter arises from a workshop held at the 2012 CIAC (Cephalopod International Advisory Council) conference on *population dynamics, environmental effects, stock assessment and management.* The purpose of the workshop and review was to synthesize the state of the art, identify knowledge gaps and look forwards to the future of cephalopod fisheries, taking into account the many and demanding challenges that lie ahead.

2. POPULATION DYNAMICS

Currently, population dynamics mainly uses large-scale field observations and laboratory data, often from rodents (Turchin, 2003) or insects (e.g. *Drosophila* and *Tribolium*; Mueller and Joshi, 2000) or from fisheries stock assessment research (Quinn and Deriso, 1999). Spatial aspects of population dynamics are rarely considered in fisheries science (Quinn and Deriso, 1999).

Turchin (2003) defined a population as a group of individuals of the same species living together in an area of sufficient size to permit normal dispersal and migration behaviour and in which population changes are largely determined by births and deaths. This definition stems mainly from experience with terrestrial animals and does not provide much information on how to differentiate between populations. "Living together" in the context of mobile marine animals might imply a high probability of reproducing together and being together for important large-scale events (e.g. feeding and migration). Population dynamics is thus "the study of how and why population numbers change in time and space, documenting empirical patterns of population change and attempting to determine the mechanisms explaining the observed patterns" (Turchin, 2003), including consideration of population numbers and structure, population stability, temporal change, spatial change and demographic and genetic effects.

2.1. Population dynamics theory

Population dynamics research requires comparable and standardized data collection over many subsequent generations and long time series, which are not always available in a fisheries context.

A central point in population regulation is the realized per capita rate of population change, $r_t = \ln(N_t/N_{t-1})$, where $\ln(N_t)$ is the natural logarithm of population density at time t. Change is inversely related to population density and/or time-lagged density, but this function may be complex and nonlinear. Analogous to chemical reactions, population growth can be viewed as a zero-order (exponential), first-order (logistic) or second-order (cyclic, e.g. Lotka–Volterra-type equations) process (see Quinn and Deriso, 1999; Turchin, 2003).

Stability and oscillations in nature (a cornerstone of population dynamics science, representing two sides of the same coin) are invariably linked to trophic interactions: specialist predation is considered to be the most frequent cause of second-order oscillations in natural populations, with the second being food availability. Nevertheless, within this basic framework, there is no universal mechanism underlying population cycles.

Common questions in population dynamics (quoted from Mueller and Joshi, 2000) include the following:

- Are generations discrete or overlapping? If the latter, are cohorts segregated in space?
- What kinds of interactions exist among life stages? Which life stages are likely to be the triggers of density-dependent regulatory mechanisms? Often, the trigger stage is the primary consumer of resources.
- Which life stages are the likely targets of density-dependent regulatory mechanisms? If the target is the first juvenile stage, does the regulatory mechanism act primarily through fecundity or mortality?
- How do the trigger and target map onto the ontogeny, especially in the context of whether cohorts are spatially segregated? What are the time delays between the triggering of a regulatory mechanism and its effect

on the target and between the effect on the target and its final effect on the triggering life stage?

- If fecundity and mortality are density-independent, what is the magnitude of each?
- What is the census life stage? If this stage is not the trigger life stage, how does it map onto the ontogeny, relative to the trigger life stage, and the first juvenile stage to which recruitment is governed through fecundity?

Models and empirical studies of *Drosophila* populations suggest that the relative levels of food given to larval and adult stages are crucial for the ultimate stability of the populations. High levels of food for larvae and low levels of food for adults favour stability; the reverse situation leads to cycles and other departures from stable-point equilibriums. Model results show that time delays in density dependence destabilize populations when generations overlap (Mueller and Joshi, 2000).

Because of its commercial and applied significance, the study of fish population dynamics has developed a huge literature and a multitude of approaches. However, much of this research focuses on harvesting, and indeed fisheries management goals and policies' impact on the research approach. This is unfortunate, because trophic relationships, the dominant issue of theoretical ecology (see above, in third chapter of this section), are more often than not ignored. An exception is the multispecies approach, which is however generally regarded as too complicated and parameter-hungry for most practical applications (Quinn and Deriso, 1999). The recent consideration of trophic relationships in modelling of harvesting options (Overholtz et al., 2000, 2008; Tyrrell et al., 2008, 2011) is a step in the right direction although it also suffers from a weak link to theoretical ecology.

2.2. Recruitment

In fisheries science, the use of the term "recruitment" is often at odds with its usual meaning in ecology. While some authors propose purely biological definitions, for example, "an addition of new fish to the vulnerable population by growth from among smaller size categories" (Ricker, 1975; see also Boletzky, 2003; Quinn and Deriso, 1999), others acknowledge the reality that recruitment in fisheries is measured in a way that depends on gear selectivity: new recruits will be the smallest fish taken by a particular gear. Thus, Bloor et al. (2013a) defined recruitment as the renewal of harvestable stages in a population. O'Dor (1998a,b) noted that "from a fisheries perspective recruitment is quantitative, but from a population perspective it is also qualitative. All genes are not of equal value in all environments". Recruitment may in principle refer to the first or repeated appearance (i.e. at a moment in time linked to the value or characteristics of a given parameter) of a specific life stage, size, weight, age or maturity stage—or indeed when an animal with a specific gene enters the population.

A further theoretical challenge lies in the relationship, if any, between recruitment and spawner abundance. For some short-lived animals, stock–recruitment relationships have been found (e.g. shrimps; Ye, 2000). However, in schooling marine animals, due to density-dependent population regulation, the spawning biomass of a cohort is not necessarily proportional to the numbers recruited (Rochet, 2000).

In the cephalopod literature, there is an emphasis on strong links between recruitment and favourable oceanographic regimes (e.g. temperature, water masses and winds; see review by Pierce et al., 2008). Dawe and Warren (1993) and Dawe et al. (2000) found that *Illex illecebrosus* recruitment was positively related to negative values of the North Atlantic Oscillation index, high water temperatures off Newfoundland and a southward shift in the various water masses associated with the Gulf Stream. Models derived from such empirical relationships (see also Challier et al., 2005b; Garofalo et al., 2010; Nevárez-Martínez et al., 2010; Roberts and van den Berg, 2002; Waluda et al., 1999, 2001a) often have good predictive capability, at least in the short term—although Solow's (2002) warning about the transience of relationships between short time series should be heeded. Links between the recruitment and the trophic relationships are rarely addressed, although Moustahfid et al. (2009) included predation mortality in a surplus production model; see also the very general approach of Gaichas et al. (2010).

2.3. Defining populations: Concepts

The most comprehensive summaries of the population ecology of cephalopods are those by Boyle and Boletzky (1996) and Boyle and Rodhouse (2005). Saville (1987) and Lipinski et al. (1998a) discussed ecological differences between fish and cephalopods related to fisheries. These accounts offer general, descriptive reviews of questions, approaches and difficulties. Out of around 750–800 cephalopod species, of which some are not yet described, 59 have been researched relatively well and are therefore suitable candidates for the analysis of population dynamics (Table 2.1). An assessment of the current level of knowledge of trophic and environmental relationships, summarized by family, appears in Table 2.2.

Table 2.1 Species of cephalopods with sufficient data accumulated to be suitable for population dynamics analysis

		(1) Taxonomic issues		(2) Exploitation level		(3) Fishery	(4) Ecological information			
Family name	**Species name**	**Yes**	**No**	**FAO db**	**Nat. db**	**Regulated?**	**Pop. ID**	**Review**	**Envir.**	**Trophic**
Nautilidae (29%)	*Nautilus pompilius*		x	No/yes	A, Pa, Ph	No	Yes	Yes	Yes	Yes
	Nautilus macromphalus		x	No/no	A, Pa, Ph	No	Yes	Yes	Yes	Yes
Sepiidae (6%)	*Sepia apama*		x	No/yes	A	Partial	Yes	Yes	Yes	Yes
	Sepia australis		x	No/yes	SA	No	No	Yes	Yes	Yes
	Sepia elegans		x	No/yes	EU	Partial	No	No	No	No
	Sepia officinalis		x	Yes/yes	EU, M	Partial	Yes	Yes	Yes	Yes
	Sepia orbignyana		x	No/yes	EU	Partial	No	No	No	No
	Sepia esculenta		x	No/yes	J, Ch	No	No	No	No	Yes
	Sepia pharaonis	x		No/yes	A, J, Ch	Partial	No	No	No	No
	Sepiella inermis	x		No/yes	I, Th	No	No	No	No	No
Sepiolidae (5%)	*Sepietta oweniana*		x	No/yes	EU	Partial	No	No	No	Yes
	Rossia pacifica pacifica	x		No/yes	None[a]	No	No	No	No	No
	Heteroteuthis dispar	x		No/yes	None[a]	No	No	No	No	Yes

Continued

Table 2.1 Species of cephalopods with sufficient data accumulated to be suitable for population dynamics analysis—cont'd

		(1) Taxonomic issues		(2) Exploitation level		(3) Fishery	(4) Ecological information			
Family name	**Species name**	**Yes**	**No**	**FAO db**	**Nat. db**	**Regulated?**	**Pop. ID**	**Review**	**Envir.**	**Trophic**
Loliginidae (20%)	*Loligo vulgaris*		x	No/yes	EU	Partial	No	Yes	No	Yes
	Loligo forbesii		x	No/yes	EU	Partial	Yes	Yes	No	Yes
	Loligo reynaudii		x	No/yes	SA	Yes	Yes	Yes	Yes	Yes
	Alloteuthis media	x		No/yes	EU	Partial	No	No	No	No
	Alloteuthis subulata	x		No/yes	EU	No	No	No	No	No
	Doryteuthis plei	x		No/yes	B?	No	No	No	No	No
	Doryteuthis gahi		x	Yes/yes	FI, Ar	Yes	Yes	Yes	Yes	Yes
	Doryteuthis opalescens		x	No/yes	United States	Yes	Yes	Yes	Yes	Yes
	Doryteuthis pealeii	x		Yes/yes	United States	Yes	Yes	Yes	Yes	Yes
	Doryteuthis sanpaulensis		x	No/yes	B, Ar	No	No	Yes	No	Yes
	Heterololigo bleekeri		x	No/yes	J	Yes	Yes	Yes	Yes	Yes
	Lolliguncula brevis	x		No/no	United States	No	No	Yes	Yes	Yes
Loliginidae (20%)	*Sepioteuthis sepioidea*		x	No/yes	United States	No	No	Yes	No	No
	Sepioteuthis australis	x		No/yes	A	No	Yes	No	Yes	Yes
	Sepioteuthis lessoniana	x		No/yes	J, A, Th	Partial	Partial	No	Yes	Yes
	Uroteuthis edulis	x		No/yes	Ch, A, J	Partial	Partial	No	Yes	Yes
	Uroteuthis duvaucelii[b]	x		No/yes	I, Th, A	No	Partial	No	No	Yes

Gonatidae (10%)	*Berryteuthis magister*		x	No/yes	R	No	Yes	Yes	No	Yes
	Gonatus fabricii		x	No/yes	None[a]	No	No	No	No	Yes
Lycoteuthidae (17%)	*Lycoteuthis lorigera*		x	No/no	None[a]	No	No	No	No	No
Ommastrephidae (62%)	*Illex illecebrosus*		x	Yes/yes	C, United States	Yes	Partial	Yes	Yes	Yes
	Illex argentines		x	Yes/yes	FI, Ar	Yes	Partial	Yes	Yes	Yes
	Illex coindetii		x	No/yes	EU, M	No	No	Yes	No	Yes
	Ommastrephes bartramii[b]		x	No/yes	J, R, Ch	Partial	Partial	Yes	Yes	Yes
	Dosidicus gigas		x	Yes/yes	Me, United States, J	Partial	No	Yes	Yes	Yes
	Ornithoteuthis antillarum		x	No/yes	None[a]	No	No	No	No	Yes
	Sthenoteuthis oualaniensis	x		No/no	None[a]	No	No	Yes	No	Yes
	Sthenoteuthis pteropus[b]		x	No/yes	None[a]	No	No	Yes	No	Yes
	Todarodes sagittatus		x	No/yes	EU, No	Partial	No	No	No	Yes
	Todarodes pacificus		x	Yes/yes	J, R, Ch	Yes	Yes	Yes	Yes	Yes
	Nototodarus sloanii		x	Yes/yes	NZ	Yes	No	Yes	No	Yes
	Nototodarus gouldi		x	No/yes	A, NZ	Yes	No	No	No	Yes
	Todaropsis eblanae		x	No/yes	EU, SA	No	Yes	No	No	Yes
Thysanoteuthidae (100%)	*Thysanoteuthis rhombus*		x	No/yes	J	Yes	No	Yes	No	Yes

Continued

Table 2.1 Species of cephalopods with sufficient data accumulated to be suitable for population dynamics analysis—cont'd

		(1) Taxonomic issues		(2) Exploitation level		(3) Fishery	(4) Ecological information			
Family name	**Species name**	**Yes**	**No**	**FAO db**	**Nat. db**	**Regulated?**	**Pop. ID**	**Review**	**Envir.**	**Trophic**
Eledonidae (33%)	*Eledone cirrhosa*		x	No/yes	EU	Partial	No	Yes	Yes	Yes
	Eledone moschata		x	No/yes	EU	Partial	No	No	No	Yes
Enteroctopodidae (7%)	*Enteroctopus dofleini*	x		No/yes	United States	Yes	No	No	No	Yes
	Enteroctopus megalocyathus		x	No/yes	Ar, CE	No	No	No	No	Yes
Octopodidae (7%)	*Octopus vulgaris*	x		Yes/yes	World	Variable	Partial	Yes	Yes	Yes
	Octopus maya		x	No/yes	Me	Yes	No	Yes	Yes	Yes
	Octopus pallidus		x	No/no	A	No	No	No	No	Yes
	Octopus bimaculatus		x	No/no	United States	No	No	No	No	Yes
	Octopus cyanea		x	No/no	A	No	No	Yes	No	Yes
	Octopus tehuelchus		x	No/yes	Ar	No	No	No	No	Yes
	Octopus insularis		x	No/no	B	No	No	No	No	Yes
	Octopus mimus		x	No/no	CE	No	No	No	No	Yes

[a]No existing fisheries or by-catch or subsistence (small-scale); therefore, future modelling attempts must be based on research data.
[b]Spelt differently in the FAO aquatic species list (e.g. *Uroteuthis duvauceli*, *Ommastrephes bartrami* and *O. pteropus*).
Percentages in the first column refer to the number of species listed for each family as a percentage of the total number of species in that family. Results on availability of information about each topic are then given by species as follows: (1) The existence of issues with taxonomic status is indicated by yes/no (no = no issues). (2) Availability of information on exploitation level in (a) the Food and Agriculture Organization of the United Nations two databases (FAO db) (yes/no) and (b) national databases (Nat. db). Countries: A, Australia; B, Brazil; CE, Chile; Ch, China; Ar, Argentina; FI, Falkland Islands; M, Morocco; SA, South Africa. (3) Existence of fishery regulation (yes/no/partial). (4) Ecological information: the existence of studies is scored as yes/no. Pop. ID = population identity researched and recognized; Review = broad ecological review published; Envir. = environmental relationships studied; Trophic = trophic relationships studied. Main source of information: FAO, in particular Jereb and Roper (2005, 2010) and references therein, also FAO Aquatic Species Fact Sheets (www.fao.org/fishery/species/search/en) and FAO Aquatic Species Portal (http://termportal.fao.org/faoas/main/start.do).

Table 2.2 Level of basic knowledge of trophic and environmental relationships for all cephalopod families

Family name	(1) Number of species			(2) % of species well investigated			(3) Commercial value/potential	(4) Ecological role
	Common	Rare	Undetermined	Trophic	Environmental	Others		
Nautilidae	2	3	2	14	14	14	8	4
Spirulidae	1	0	0	0	0	100	1	2
Sepiidae	55	22	32	3	4	50[a]	10	10
Sepiolidae	37	10	16	5	3	24[a]	4	6
Sepiadariidae	3	4	0	29	0	29[a]	1	2
Idiosepiidae	1	0	7	12	12	12[a]	1	2
Loliginidae	47	0	0	36	26	47[a]	10	10
Australiteuthidae	0	1	0	0	0	0	1	2
Gonatidae	14	0	6	40	10	50	5	8
Octopoteuthidae	2	0	6	12	0	25	1	5
Pyroteuthidae	2	0	4	0	0	33	1	5
Ancistrocheiridae	1	0	0	0	0	100	1	3
Enoploteuthidae	11	0	32	1	1	2	5	8
Onychoteuthidae	9	0	6	53	0	60	7	10
Pholidoteuthidae	2	0	0	100	0	100	2	6
Lepidoteuthidae	1	0	0	100	0	100	2	4

Continued

Table 2.2 Level of basic knowledge of trophic and environmental relationships for all cephalopod families—cont'd

Family name	(1) Number of species			(2) % of species well investigated			(3) Commercial value/potential	(4) Ecological role
	Common	Rare	Undetermined	Trophic	Environmental	Others		
Chtenopterygidae	1	2	0	0	0	0	1	3
Batoteuthidae	0	0	1	0	0	0	1	2
Brachioteuthidae	3	?[b]	5[c]	0	0	0	1	6
Lycoteuthidae	1	3	2	0	0	17	3	7
Histioteuthidae	5	0	12	18	0	18	2	8
Bathyteuthidae	1	0	2	0	0	33	1	3
Psychroteuthidae	1	0	0	100	0	100	3	5
Neoteuthidae	1	0	3	25	0	25	1	5
Architeuthidae	1	0	0	100	0	100	1	7
Ommastrephidae	21	0	0	86	24	90	10	10
Thysanoteuthidae	1	0	0	100	100	100	10	7
Chiroteuthidae	1	7	8	0	0	0	1	2
Mastigoteuthidae	4	0	14	0	0	0	1	3
Joubiniteuthidae	0	1	0	0	0	0	1	1
Magnapinnidae	0	0	3	0	0	0	1	3
Cycloteuthidae	0	0	4	0	0	0	1	2

Promachoteuthidae	0	5	0	0	0	0	1	1
Cranchiidae	19	1	10	7	0	13	2	10
Vampyroteuthidae	1	0	0	100	0	100	1	3
Cirroteuthidae	0	0	3	0	0	33	1	3
Stauroteuthidae	0	0	2	0	0	0	1	1
Alloposidae	1	0	0	100	0	100	1	4
Tremoctopodidae	1	0	3	0	0	25	1	4
Argonautidae	2	0	2	0	0	50	1	6
Ocythoidae	1	0	0	0	0	0	1	3
Eledonidae[b]	6	0	0	33	33	33	7	10
Octopodidae[b]	24	0	95	8	8	20	10	10
Enteroctopodidae	4	0	25	3	3	3	7	10
Bathypolypodidae	2	0	4	0	0	17	2	6
Megaeledonidae	6	0	24	0	0	10	5	10
Amphitretidae	3	0	2	0	0	0	1	3

[a]Note: includes species investigated for trophic and environmental relationships.
[b]The number of undescribed species is high.
[c]Under revision.

(1) Number of species refers to the described species only, categorized according to their abundance (common, rare and undetermined). (2) The percentage of species well investigated is given in relation to knowledge of trophic relationships, environmental relationships and other studies. The last two columns indicate the (3) likely commercial value or potential and (4) the ecological importance of the species, in both cases on a scale of 1–10 where a score of 10 indicates the highest potential or importance. Based on Jereb and Roper (2005, 2010).

Boyle and Rodhouse (2005) stated that "The life cycle characteristics and ecology of the oceanic and mesopelagic cephalopod fauna, in particular, need to be established before current estimates for global cephalopod biomass can be reconciled with their biological productive capacity and that of the marine ecosystem in total. We must begin to understand whether the life cycle features established for the coastal species represent special cases, or the degree to which they may be generalised to the much greater oceanic and deepwater fauna". It might be expected that species exploited commercially on a large scale would be the best candidates for population dynamics analysis. However, of the three dominant species in the world fisheries, arguably only one (*Todarodes pacificus*) has been researched well enough to meet the standards described in the preceding text (see Table 2.1).

Previous reviews identified limitations to understanding of cephalopod population ecology. Boyle and Boletzky (1996) stated that "the study of cephalopod populations currently lacks the means to define populations adequately and to resolve basic systematic confusions", while Boyle and Rodhouse (2005) indicated that "no cohesive description of cephalopod population ecology is yet available. (...) Modelling of population ecology for fisheries purposes is confounded by a lack of consensus among workers as to the form of the growth model to be applied, and also by a lack of data to define populations". However, for a number of families and species (Tables 2.1 and 2.2), systematic problems have been largely resolved over the last 20 years, and a wealth of life cycle and ecological data have accumulated as a result of fisheries, fisheries research and biological projects.

Furthermore, Turchin's (2003) definition of the population requires only that the animals "live together in an area of sufficient size to permit normal dispersal and migration behaviour", which can be easily investigated (e.g. Augustyn et al., 1992, 1994; Sauer et al., 1992, 2000), and "in which population changes are largely determined by birth and death processes". The latter statement implies exclusion of situations in which exchange of animals with other areas is known but suggests that useful work could be done even in the absence of stock identification based on comprehensive genetic analysis. However, fine-tuned understanding of population identity and stock structure is possible only by combining multilevel ecological research, well-thought out molecular biology research and modelling. Such research is under way, setting standards (O'Dor, 1998a) for sustainable utilization of these resources.

In the past, there has been limited interest in cephalopods by theoretical ecologists. For example, the degree of overlap between consecutive generations, crucial in population dynamics analysis, was poorly understood even in the better-known families (e.g. Melo and Sauer, 1999, 2007). In practice, the only biological feature that results in nonoverlapping generations is strict semelparity, when an animal produces offspring and dies before the hatching of its progeny. In most cephalopods (especially squid, not octopods), the spawning period for an individual female can be quite prolonged; certainly, multiple modes in egg size distributions in the ovary suggest that eggs may be spawned in several batches (although it is not certain that all eggs present in the ovary are finally spawned). However, in colder regions, the embryonic phase may be relatively protracted, reducing the likelihood of overlap.

In addition, the lack of synchrony between spawning in different individuals can lead to protracted spawning seasons, evidenced by the existence of several microcohorts, and there may also be multiple spawning seasons, blurring the distinction between different generations (even if individuals do not overlap with their own progeny). This contrasts with the usual situation in fish and other iteroparous organisms, in which overlapping generations are also created by the occurrence of multiple discrete spawning events, often over a period of several years: parents thus coexist with their progeny over an extended period.

In their review, Boyle and Rodhouse (2005) discussed the problem of finding and researching separate populations of the same species, listing 31 species for which some information is available. They mentioned the use of molecular biology, morphometrics and parasite tags, as well as knowledge of "timing and location of breeding or the recruitment of young" (e.g. in *T. pacificus*, *O. vulgaris*, *Sthenoteuthis oualaniensis* and *I. illecebrosus*) and information on population structure, particularly emphasizing size and age.

Some authors (e.g. Yeatman and Benzie, 1993) have questioned the validity of separating populations using a morphological approach such as the one used by Nesis (1993), but see Vidal et al. (2010a). This may be even more relevant for cephalopods than other organisms (especially long-living fish), because of the importance of spatial considerations linked to survival of consecutive generations (Lipiński, 1998; Lipinski et al., 1998b; O'Dor, 1998a,b; Ranta et al., 1997). O'Dor (1998a) felt that "management of squid stocks according to the 'Precautionary Principle' requires defining individual microcohorts genetically, temporally and spatially". This requirement lies at the base of any rigorous ecological testing in the field and in the laboratory. What gives even more credence to this statement is the fact that, in cephalopods, there is a possibility to mistake different (time-, space- or temperature-wise) groups (broods) for biologically and/or genetically separate populations—see discussions in Boyle and Rodhouse (2005) and Forsythe (1993, 2004). This can arise as a result of intertwined generations with different biological characteristics, either alternating generations, as hypothesized by Mesnil (1977), or when individuals of the same cohort breed at two (or more) different times. This pattern is well documented in the genus *Sepia* (Boletzky, 1983; Hall et al., 2007; Le Goff et al., 1998), in which the only certain method to separate populations is by molecular biological methods, for example, Pérez-Losada et al. (2007). Therefore, simple indicators in isolation (especially length–frequency) are not good enough to diagnose separate populations of cephalopods. Also, while it may be easy to generate consecutive generations in a model (indeed, this is frequently done), it is difficult to identify them in the wild.

2.4. Defining populations: Examples

Yatsu et al. (1998), Nagasawa et al. (1998) and Chen (2010) have provided fisheries and biological evidence, collected over vast area during more than 20 years of exploitation, indicating that *Ommastrephes bartramii* from the northern Pacific comprises four groups. At least two of these groups are clearly distinct populations (eastern and western), as confirmed by Katugin (2002). Discriminating factors included hatching time and area, length–frequency data, maturities, paralarval occurrence, parasitic infestation, fleet operations and environmental factors. However, it is still not possible to distinguish different generations.

On the other hand, numerous studies on three exploited species of squids (*Nototodarus gouldi*, *N. sloanii* and *D. gigas*) revealed no clear differentiation into separate populations, because of the complexity of their distribution and biological characteristics (Jackson et al., 2005; Keyl et al., 2011; Masuda et al., 1998; Uozumi, 1998; Zavala et al., 2012). *D. gigas* has relatively recently expanded its distribution northwards, probably due to a combination of favourable environmental conditions and fishery impacts (Keyl et al., 2008) interacting with physiological mechanisms (e.g. related to oxygen debt; Rosa and Seibel, 2010).

T. pacificus is one of the best-studied squids in the world. There is evidence (Katugin, 2002; Kidokoro et al., 2010) that this huge resource comprises a distinct autumn-spawning population and less distinct non-autumn-spawning population that is dominated by the winter cohort. Both these populations migrate between spawning grounds in the south and feeding grounds in the north. The biology of the more diverse "non-autumn-spawning population" is especially interesting as it sheds light on how, when and where various splinter cohorts and microcohorts make up one large population, with complicated structure, variable life cycle parameters, long migration pathways and large fluctuations in abundance (Nakata, 1993; Song et al., 2012; Takayanagi, 1993). It is believed that it is possible to track consecutive generations of the winter population by research in the Tsugaru Strait between Honshu and Hokkaido during years of high yield (Takayanagi, 1993). Indexes of maturity may be used for this purpose. In two consecutive years, maturity indexes of the winter population were similar, which indicates stability during which generations may be identified and compared (Figure 2.1).

I. illecebrosus was intensely exploited in the northwest Atlantic in the 1970s, with a subsequent stock collapse, and has never regained its former numbers. Nevertheless, it is one of the best-researched ommastrephid squids

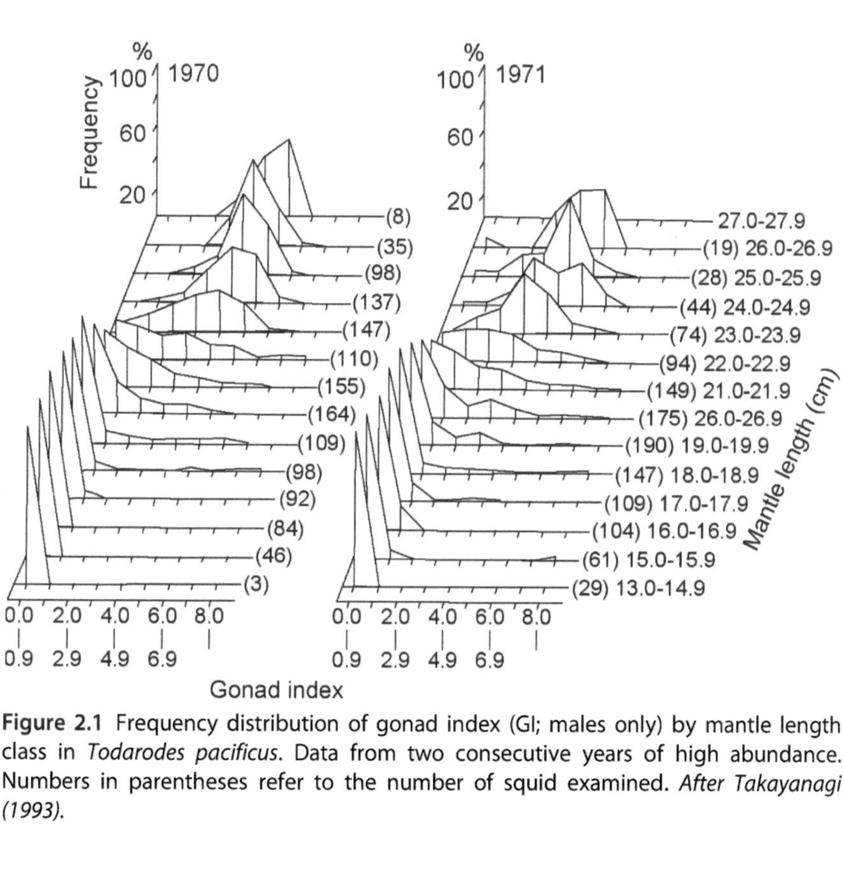

Figure 2.1 Frequency distribution of gonad index (GI; males only) by mantle length class in *Todarodes pacificus*. Data from two consecutive years of high abundance. Numbers in parentheses refer to the number of squid examined. *After Takayanagi (1993).*

in the world (O'Dor and Dawe, 1998). Dawe and Hendrickson (1998) and Hendrickson (2004) provided evidence that there is a single population of *I. illecebrosus* extending over a wide area (from Cape Hatteras to Newfoundland). This stock is however very variable and its structure is complicated, so it would be difficult to track consecutive generations in this species.

Illex argentinus is one of the three most abundant squids in the world. It is relatively well researched (Arkhipkin, 1993, 2000; Brunetti et al., 1998; Haimovici et al., 1998; Sacau et al., 2005; Uozumi and Shiba, 1993). It is distributed over a large area and has a complicated population structure with many and variable microcohorts. At present, two populations are recognized: winter spawning and summer spawning; differentiation criteria are temporal, spatial and biological (Sacau et al., 2005). However, such divisions may be transient, and recent work by Crespi-Abril and Baron (2012) and Crespi-Abril et al. (2013, 2014) suggests inshore spawning of *I. argentinus* over a wide area year round. This would most likely create one large panmictic metapopulation. Thus, as in *I. illecebrosus*, it would be difficult to track subsequent generations in *I. argentinus*.

Todaropsis eblanae is an ommastrephid that is bottom-dwelling and not so heavily exploited. Nevertheless, it is relatively well studied (Dillane et al., 2000, 2005; Hastie et al., 1994; Lordan et al., 1998; Rasero et al., 1996; Zumholz and Piatkowski, 2005). Based on the analysis of one minisatellite and four microsatellite loci, Dillane et al. (2005) concluded that there are at least three genetically isolated populations in the east Atlantic. Again as in *Illex*, there is no immediate prospect of tracking consecutive generations in this species.

Berryteuthis magister from the northern Pacific has been well studied, mainly by Russian scientists—see Jelizarov (1996) and Katugin (2002). It is a bottom-dwelling squid, abundant and targeted by a bottom trawl fishery; there are three subspecies and population structure is complicated. One subspecies, *B. magister magister*, has been the object of detailed ecological and genetic analysis and appears to comprise three populations, occurring in the Alaskan Gyre system, the western subarctic gyre and the Sea of Japan, respectively. Variability is clinal (Katugin, 2002). As in the *Illex* spp. and *T. eblanae*, there is no immediate prospect of tracking consecutive generations in this species.

Some squids from the family Loliginidae have also been intensively studied. Accounts of a few of the best known are given in the following text.

Doryteuthis gahi is an unusual loliginid, thriving in cool and relatively deep waters. It is heavily exploited and researched well in a fairly narrow area around the Falkland Islands (Agnew et al., 1998a,b; Hatfield, 1996; Hatfield and des Clers, 1998; Hill and Agnew, 2002; Patterson, 1988; Shaw et al., 2004). There is one population and at least two cohorts (based on seasonal appearances on the fishing grounds but with inferred separate spawning and recruitment). It is a good candidate for tracing consecutive generations. In fact, a model of optimal harvest strategy proposed by Hill and Agnew (2002) assumes a single generation each year, which undergoes continuous depletion due to natural mortality. The main equation of this model is $C_y/C_x = (1 + aT)e^{-MT}$, where C_y and C_x are catch weights at specific points in time, a is growth rate, $T = y - x$ (in weeks) and M is natural mortality. Figure 2.2 illustrates how this model works. Rates of immigration and emigration are also incorporated in the model.

Doryteuthis opalescens is heavily exploited off California (Fields, 1962; Recksiek and Frey, 1978; Zeidberg et al., 2006). Initial investigations concerning population structure were inconclusive (Ally and Keck, 1978; Christofferson et al., 1978; Kashiwada and Recksiek, 1978). Population structure in this species has also been studied by Jackson (1998), Vojkovich (1998), Jackson and Domeier (2003), Reiss et al. (2004), Macewicz et al. (2004), Maxwell et al. (2005), Brady (2008), Warner et al. (2009) and Dorval et al. (2013) and appears to be complex. A study of microsatellite loci by Reichow and Smith (1999, 2001) concluded that there is a single large, possibly panmictic, population. However, further analysis of local "cohorts" over consecutive spawning cycles would still be useful. Because of the complex population structure, despite the fact that population is apparently genetically uniform, tracing consecutive generations in this species may be difficult.

Loligo vulgaris is one of the most-studied loliginids, but its population structure is still imperfectly known. Most research has focused on particular regions within its distribution (e.g. Coelho et al., 1994; Guerra and Rocha, 1994; Krstulović Šifner and Vrgoč, 2004; Marques Moreno, 2012; Moreno et al., 1994, 2005; Vila et al., 2010). The only large-scale synthesis was that by Moreno et al. (2002), which, by using multivariate analysis of biological indexes, demonstrated significant differences between regions. Existing evidence suggests that these differences may be ascribed to large-scale environmental phenomena. Despite the large number of studies, population structure cannot be confidently described for this species.

Loligo forbesii, another large European loliginid, is probably better studied than *L. vulgaris*. Several studies on both species appeared in a special volume of Fisheries Research (Boyle and Pierce, 1994). Shaw et al. (1999)

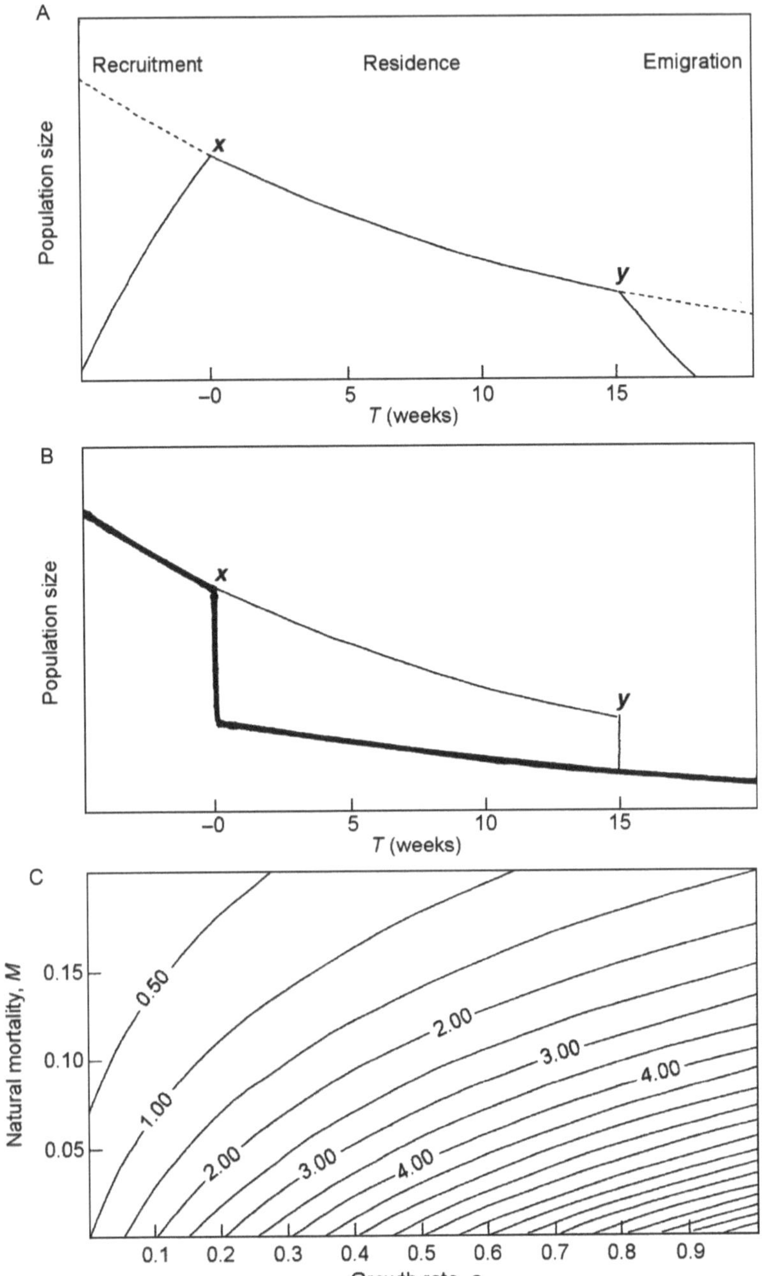

Figure 2.2 (A) A population consisting of a single generation and experiencing constant natural mortality (dotted line). The solid line shows the population vulnerable to fishing, where 100% of the population is resident (between points *x* and *y*).

demonstrated the possible existence of three populations: the Azores population (see also Brierley et al., 1993, who considered it might be a separate subspecies), a population inhabiting NE Atlantic offshore banks such as Rockall and Faroe, and the shelf population. Papers that discuss various aspects of its population structure include Holme (1974), Martins (1982), Collins et al. (1995, 1997, 1999), Pierce et al. (1998, 2005), Bellido et al. (2001), Pierce and Boyle (2003), Young et al. (2004), Challier et al. (2005a) and Chen et al. (2006); as in other squids, winter and summer breeders and varying numbers of different microcohorts have been documented. The identification of consecutive generations in this species is most likely to be successful in the isolated parts of the range (e.g. Azores).

Loligo reynaudii is one of the best-studied loliginids in the world. Population structure of this species was reported in a number of studies (Augustyn, 1989; Augustyn et al., 1992, 1993, 1994; Lipinski et al., 1998b; Martins et al., 2014; Olyott et al., 2006, 2007; Roberts, 2005; Roberts and Sauer, 1994; Roberts and van den Berg, 2002; Roel, 1998; Sauer, 1991, 1993; Sauer et al., 1997). However, population differentiation has been studied in detail only recently (Stonier, 2012; van der Vyver, 2013). There are three populations, the first located along the eastern part of the south coast of South Africa, the second from Agulhas Bank and the west coast of South Africa and the third off Angola. The change was clinal, differences increased with the geographic distance. The best candidate for the study of consecutive generations is the most isolated and distant site of the overall distribution, that is, the Angolan population. South African populations have a complicated structure and are subject to mixing of many microcohorts, taking into consideration the dynamic character of each spawning aggregation (Lipinski et al., 1998b; Sauer et al., 2000).

Sepioteuthis australis is relatively well researched; Pecl (2000), Jackson and Pecl (2003), Pecl et al. (2006) and Hibberd and Pecl (2007) all described a complicated population structure in the eastern Tasmanian population. However, Triantafillos and Adams (2001, 2005) detected two cryptic species

A single instantaneous fishing event may occur at either of these points and will result in the capture of a fixed proportion of the vulnerable population. *T* is time relative to point *x*. (B) A population experiencing a fixed fishing mortality at point *x* (thick line; catch there is C_x) or point *y* (thin line; catch there is C_y). The number of survivors at time 20 (escapement) is the same under either scenario. (C) Contour lines showing combinations of natural mortality and growth rates that produce various values of the ratio C_y/C_x (increasing in steps of one, left to right). The residence period (*T*) is 10 weeks for all lines. *From Hill and Agnew (2002).*

instead of the one previously described, as well as hybrid forms. A similar situation was described for the sister species, *S. lessoniana* (see Cheng et al., 2014). More data are required to explain population structure(s) in these species. Because of the "conveyor belt" of recruits found in an ideal spot for research (Great Oyster Bay), this species is an unlikely candidate to research consecutive generations. The related species, *S. lessoniana*, has been cultured in the laboratory for multiple generations (Lee et al., 1994; Walsh et al., 2002).

Several species of cuttlefish are relatively well studied (Jereb and Roper, 2005), but *S. officinalis* is perhaps the most representative example of this group; research on its population structure is summarized in Guerra (2006) and Pierce et al. (2010). Several authors have detected a multitude of fairly localized separate populations throughout the distribution range (Pérez-Losada et al., 1999, 2002, 2007; Turan and Yaglioglu, 2010; Wolfram et al., 2006). *S. officinalis* offers excellent prospects for investigating consecutive generations both in the field and in the laboratory. The same can be said about the largest cuttlefish in the world, the Australian *Sepia apama* (Hall et al., 2007).

Likewise, several species of octopus are well researched (Roper et al., 1984), but *O. vulgaris* is perhaps the best representative example of this group, despite the fact that there is a view that it may be a large complex of species (Mangold, personal communication). Its population structure was described by Mangold and Boletzky (1973), Hatanaka (1979), Smale and Buchan (1981), Mangold (1983), Sanchez and Obarti (1993), Oosthuizen and Smale (2003) and Robert et al. (2010). Population differentiation has been studied throughout the world, using both genetic and morphological approaches (e.g. Vidal et al., 2010a), most frequently in Europe and in South Africa (Cabranes et al., 2008; Greatorex et al., 2000; Maltagliati et al., 2002; Moreira et al., 2011; Murphy et al., 2002; Oosthuizen et al., 2004; Robert et al., 2010; Teske et al., 2007). A good summary is given by Pierce et al. (2010). All these authors detected a multitude of localized populations throughout the distribution range, similar to that of *S. officinalis*. As in *S. officinalis*, *O. vulgaris* offers excellent prospects for investigating subsequent generations both in the field and in the laboratory.

2.5. Population dynamics of cephalopods: Models

As reviewed in the preceding text, an understanding of the population structure of a species is a fundamental introduction to population dynamics and involves two steps: the description of the biological parameters (length–frequency, maturity, abundance, age, growth rate, recruitment, environmental relations, trophic relations, etc.) and molecular biological studies of intraspecific variability, to identify populations (e.g. Shaw, 2002; Triantafillos and Adams, 2001, 2005; Yeatman and Benzie, 1993). An effort to discriminate between consecutive generations is the next logical direction to follow.

The first step requires the choice of model(s) to describe growth and maturity. Existing models include the primitive linear three-stage model (Lipinski, 2001), which was followed by Keyl et al. (2011) and Zavala et al. (2012); the ontogenetic growth model for squids of Arkhipkin and Roa-Ureta (2005), followed by many authors, for example, Schwarz and Perez (2010, 2013); the bioenergetic models of Grist and Jackson (2004) and O'Dor et al. (2005), followed by André et al. (2009); and the physiological model of Moltschaniwskyj (1994, 2004), followed by many authors (e.g. Kuipers, 2012; Semmens et al., 2011). The maturity model of Macewicz et al. (2004) has been further developed by Dorval et al. (2013) into a good management tool.

With the basic data available, it is possible to devise a model that addresses the two main issues of population dynamics: *per capita* rate of population change and stability versus oscillations. Models can then be used to test possible explanations of the observed change. In theoretical ecology, more often than not, this explanation lies in trophic relationships (e.g. specialist predation is thought to be the most frequent cause of second-order oscillations). Two older reviews of the population dynamics of cephalopods (Caddy, 1983; Pauly, 1985) underlined the differences and similarities of cephalopod population biology compared with fish, utilizing both the traditional fisheries framework of stock assessment and resource management.

Recently, however, the most frequently pursued direction has been to focus on understanding external effects of environmental systems and variables. Does the environment govern cephalopod life cycles?

Given the apparent unsuitability of traditional approaches to stock assessment arising from the complexity of the squid life cycle and the sensitivity to extrinsic factors touched on in the preceding text, it can be argued (Pierce et al., 2008) that an understanding of the traditional population dynamics parameters (fecundity, mortality and growth) may be fruitless; stock–recruitment relationships are absent and much of the predictability in population dynamics may derive from knowledge of external effects, particularly the physical environment. In fact, there are several examples of models in the published literature, often investigating in detail the impact of temperature upon growth rate, mantle length at age, maturity and ultimately fecundity described by Forsythe (1993, 2004). There is an inference that higher temperatures may reduce life span, which in turn will result in oscillations of abundance linked only to change in a population structure, but not effected in the long series of subsequent generations (Pecl and Jackson, 2008); see Figure 2.3. Roberts (2005) presented a simple model whereby he calculated the relationship between maximum summer SST as a monthly average and biomass of squid (*L. reynaudii*) the following autumn (and/or annual catch). The linear relationship obtained (Figure 2.4) shows a clear problem for rational management of the resource: catch is more strongly correlated with SST the previous summer than with stock biomass. Also, Roberts' model suffers from intense data manipulation (all relationships are based upon pooling massive database and on averages) and simplistic treatment of changes in the population; the model does not consider population structure.

Reiss et al. (2004) constructed an age-based temperature-dependent model of squid (*D. opalescens*) growth and a simple population dynamics model based on the aforementioned to drive the population growth rates. The results of this model are presented in Figure 2.5. A surprising result was that growth rate was negatively related to temperature, contrary to the predictions by Forsythe (1993, 2004). Jackson and Domeier (2003) were first to detect this inverse relationship; they also detected a relationship between the intensity of upwelling and the size and age of squid, as might be expected. Although conceptual or quantitative proof is lacking, they

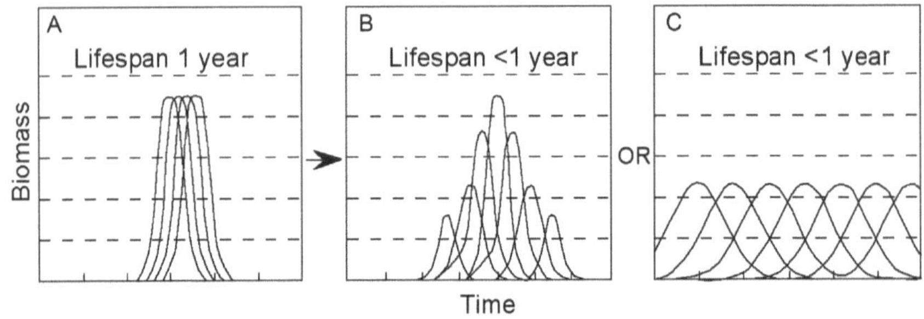

Figure 2.3 Diagrammatic representation of fluctuations in biomass of squid over 1-year period. (A) Aggregative spawning over an extended spawning season of up to several months resulting in successive waves of recruitment, however, a clear peak is present. (B) Breeding season is extended beyond a few months as the lifespan of squid becomes shorter, although seasonal peaks in biomass are still evident. (C) Uncoupling of seasonal and synchronous spawning cues resulting in aseasonal pulses of recruitment with no obvious dominant peak in biomass. *From Pecl and Jackson (2008); adapted from Boyle and von Boletzky (1996).*

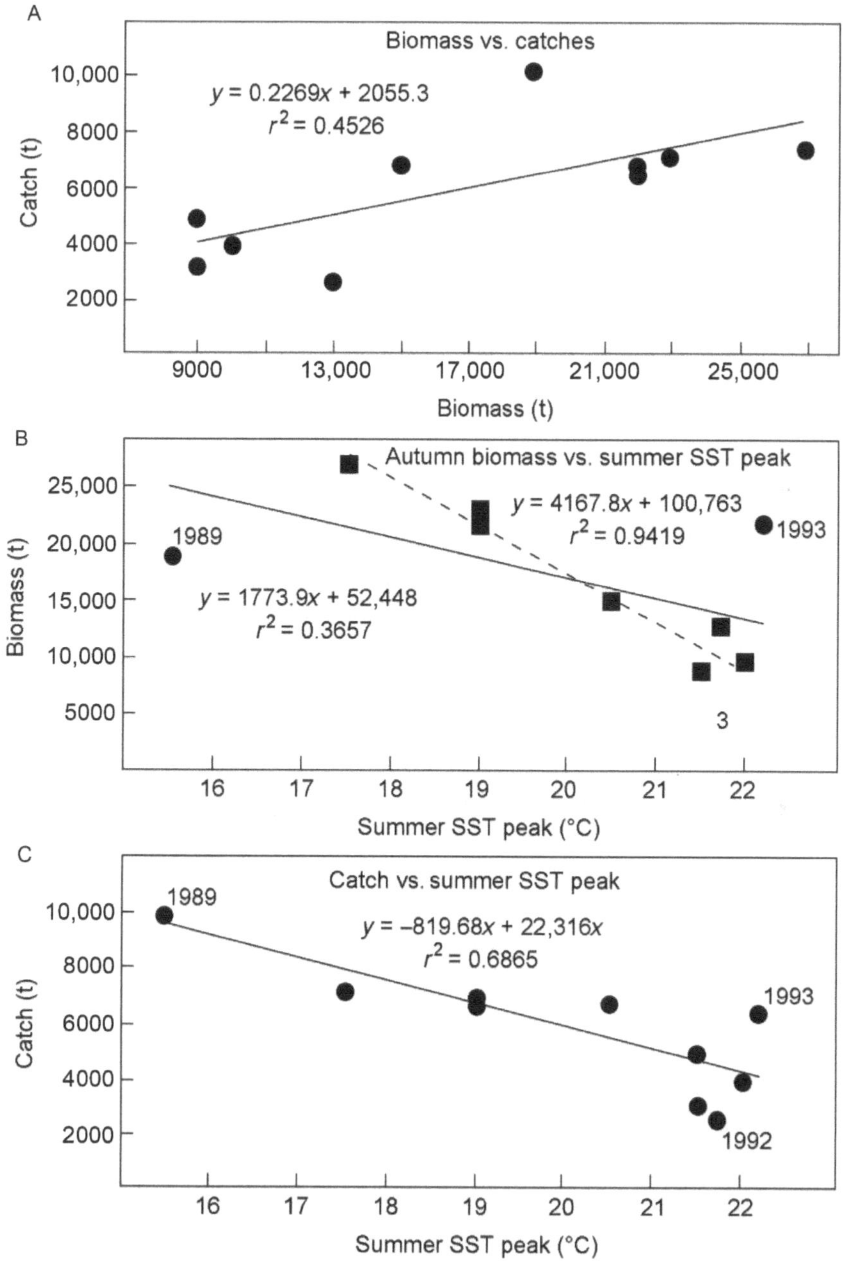

Figure 2.4 (A) Estimated *Loligo reynaudii* biomass versus total annual jig catch (1988–1997). (B) Biomass versus maximum monthly average SST (sea surface temperature). The linear fit is improved (dashed line) if the anomalous years 1989 and 1993 are excluded. (C) Total annual jig catch versus maximum monthly average SST. *From Roberts (2005).*

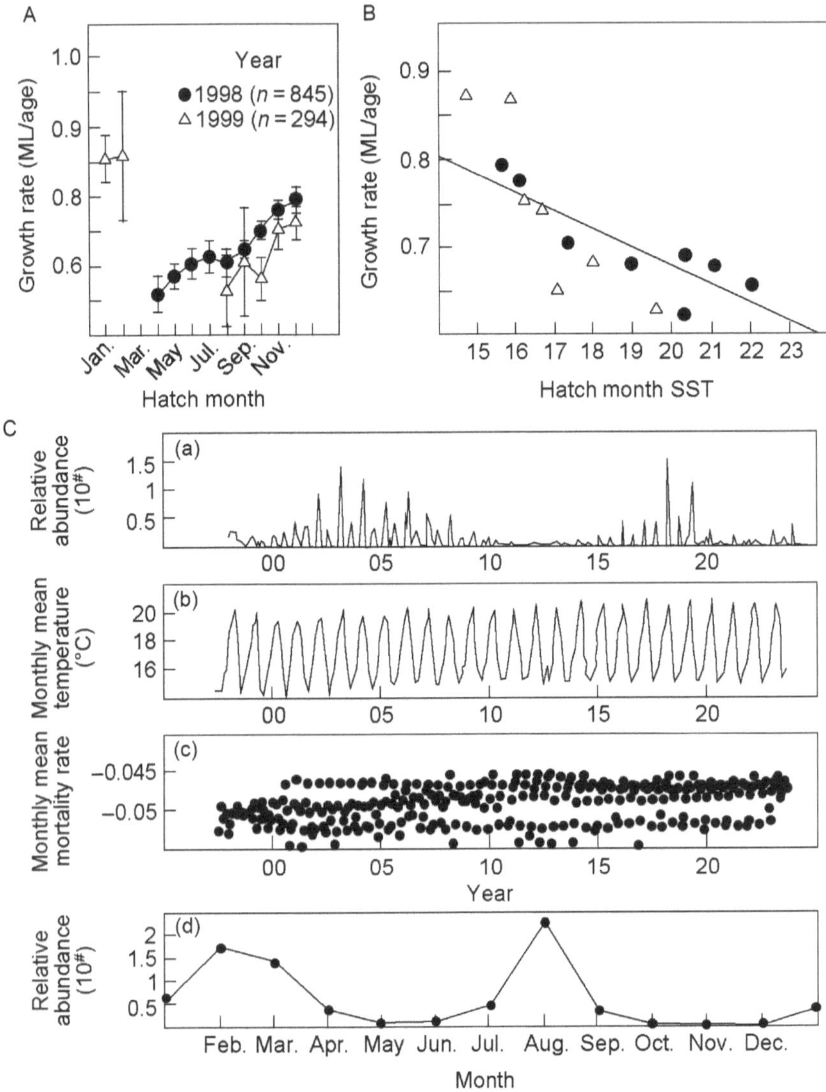

Figure 2.5 (A) Mean growth rate (±1 SE) of mature *Loligo opalescens* from the Southern California Bight commercial fishery in 1998 and 1999 plotted by month of hatch; (B) means of growth rates in relation to hatch-month SST (sea surface temperature) as calculated from the monthly mean temperature recorded at Scripps Pier (California, USA); (C) 25-year population simulations of using an age-based temperature-dependent growth model. (a) Time series of monthly population abundance; (b) monthly average temperature; (c) monthly mortality rates; and (d) seasonal pattern of recruitment. *From Reiss et al. (2004).*

propose that these relationships reflect a trade-off between physical environmental effects and food availability. Reiss et al. (2004) did not include trophic relationships or density-dependent processes in their model or indeed test their model against real data. However, they suggested that including food in the model would have affected the empirically derived growth relationship. They predicted that this inclusion would shift the period of maximum growth rate from winter to late spring, to coincide with low temperatures and high abundance of prey.

André et al. (2010) used a combination of individual-based bioenergetics and stage-structured population models to describe the capacity of cephalopods (represented by *Octopus pallidus*) to respond to climate change. Results of this model are given in Figure 2.6. This very useful model predicts possible consequences of climate change. The model assumed a linear increase in the mean annual temperature from 17.32 in 2005 to 19.43 °C in 2070. Results indicated that the response of the *O. pallidus* population to climate change would be nonlinear. Assuming the survivorship schedule remained constant, an increase in water temperature could lead to a shift from exponential population growth to exponential decline within a matter of years. Egg incubation period was predicted to fall (from 186 to 95 days), coupled

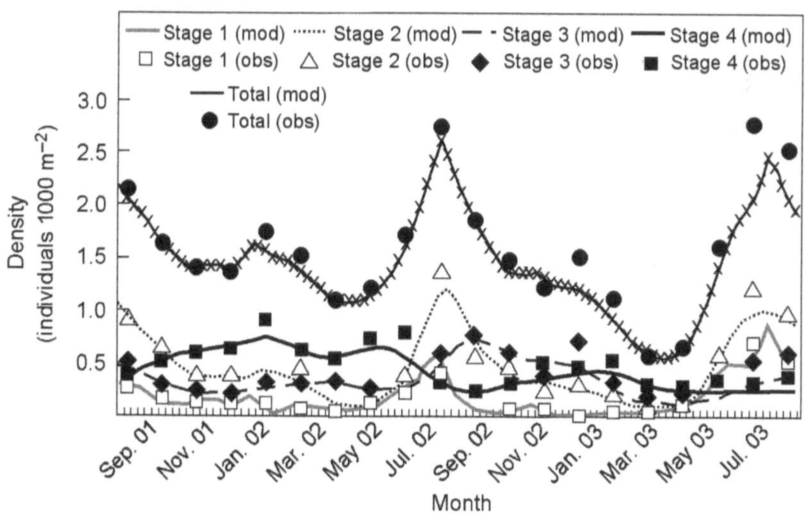

Figure 2.6 Model predictions concerning densities of *Octopus vulgaris* in Greece over time. Predictions start from an initial density vector $\mathbf{n}_1$ equal to the observed vector at that time. Predictions are compared with observed densities (real data). Lines represent model estimations and markers represent real data. *From Katsanevakis and Verriopoulos (2006).*

with reduced hatchling size (0.34 to 0.23 g), small weight at reproductive maturity (466.0 to 395.8 g) and a shorter generation time (12 to 9 months). One conclusion, therefore, is that successful adaptation to climate change may come at the cost of substantial change in population structure and dynamics, resulting in a potential decrease in generation time, streamlining of the life cycle, lower fecundity and possible loss of resilience to catastrophic events. Secondly, cephalopods may be bad climate indicators. However, it should be noted that, again, the authors did not include trophic relationships in their model. Instead, they speculated why the exponential growth is not observed in reality and ascribed this to environmental factors (such as extreme weather events and various environmental variations). The lack of exponential growth in the real population can however be related to trophic relationships, and this should be taken into consideration in future research. This is underlined by the fact that the change illustrated in the model can lead theoretically to decoupling of predator–prey relationships. The authors speculated what implications this may have for cephalopods and indeed for whole marine ecosystems.

The existence of numerous empirical models that link environmental variables with distribution, abundance and recruitment of several cephalopod species (e.g. Sobrino et al., 2002; Waluda et al., 1999, 2001a,b; Wang et al., 2003) led Pierce et al. (2008) to acknowledge the environment as a key factor in determining, leading and varying cephalopod life cycles and their population dynamics. However, they also recognized the importance of trophic relationships, specifically the role of prey availability (alongside environmental factors) in determining growth and mortality rates of early life stages. The same view (adding density-dependent effects) is underlined by Otero et al. (2008) who investigated abundance fluctuations of *O. vulgaris* and their possible causes. In addition, Vidal et al. (2006) provided empirical data to demonstrate the importance of prey availability for the survival and growth rates of squid paralarvae.

Katsanevakis and Verriopoulos (2006) constructed a simple model of *O. vulgaris* population dynamics in the eastern Mediterranean. The basis for this model was a monthly visual census (July 2001–September 2003), using scuba diving, of octopus abundance along 14 fixed transects within an area of 1600 m^2. The census was run monthly from July 2001 to September 2003. All octopuses sighted were assigned to one of four estimated weight classes (<50, 50–200, 200–500 and >500 g). To explain densities by weight class and to estimate life cycle parameters, a time-variant, weight class-based matrix population model was developed. Annual and semiannual density cycles were found, with the main peak of benthic settlement in summer and a secondary, irregular settlement during late autumn. On the basis of the model, spawning peaks, mortality, lifespan and growth rates for various stages were predicted, and the model achieved good prediction capability (Figure 2.6). However, modelling the complete life cycle would require information on fecundity as well as egg and paralarval densities, parameters that would be difficult to estimate for the population study because of the possible disturbance of the spawning process in octopuses' dens (although literature values of fecundity could be used, e.g. Mangold, 1983d) and because knowledge of hatching success and mortality of paralarval mortality in the plankton is lacking. Other aspects not covered by the model include trophic relationships and environmental influences.

Trophic relationships of cephalopods are extensively covered in the literature, primarily from a classical descriptive point of view (e.g. Amaratunga, 1983; Dawe and Brodziak, 1998; Jackson et al., 2007; Lipiński, 1987, 1992; Lipinski and David, 1990; Lipinski and Jackson, 1989; Lipinski et al., 1991, 1992; Lordan et al., 1998; Pierce et al., 1994; Rodhouse and Nigmatullin, 1996). As can be seen in Table 2.1, trophic relationships are the most often researched topic in the best-known species of cephalopods.

However, the use of these data in the generation of biological ideas and models is rare. Nevertheless, there has been a trend to use the wealth of basic field and laboratory data that are available for some form of ecological modelling. This modelling is based not only on stomach content analyses but also in bioenergetics research (which is mentioned but not reviewed here; see O'Dor and Wells, 1987; Wells and Clarke, 1996). In recent years, some of the first ecosystem models that explicitly examine the importance of squids have been produced, for example, Jackson et al. (2007), Gasalla et al. (2010) and Wangvoralak (2011).

Amaratunga (1983) in his early review of the role of cephalopods in the marine ecosystem presented conceptual models of cephalopod predation for various groups of cephalopods in the form of block diagrams. He mentioned briefly energy requirements, balance and change of generations, but he did not discuss the issue of overlapping generations. His block model of biomass change in a squid (*I. illecebrosus*) is shown in Figure 2.7A. He calculated the prey biomass taken by 1000 g of squid under various assumptions and linked growth rate to feeding rate (Figure 2.7B) following Jones (1976) and O'Dor et al. (1980). He also addressed mortality in a population using yield-per-recruit analysis (after Mohn, 1982). However, this very simplistic description of biomass change, driven by trophic relationships, relies on

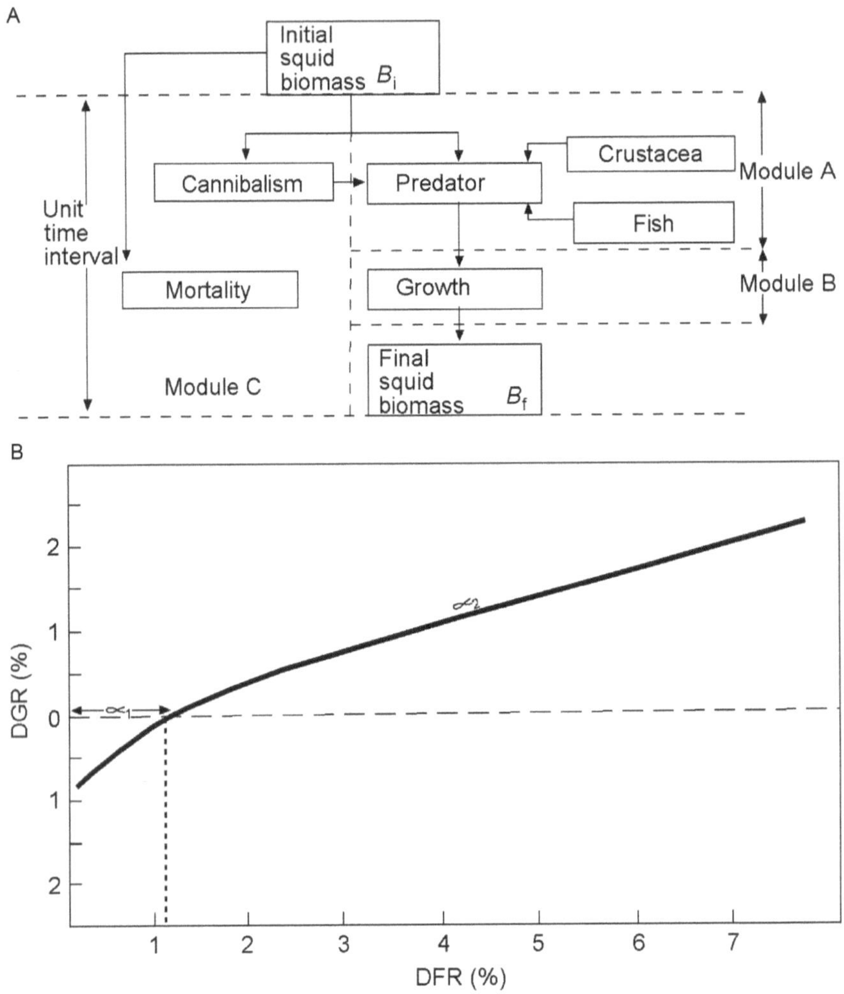

Figure 2.7 (A) Block diagram of the biomass change of *Illex illecebrosus*, affected by trophic relationships. Module A represents predation, module B represents growth. (B) Mean daily growth rate (DGR) plotted against mean daily feeding rate (DFR) for *Illex illecebrosus* maintained in the aquarium. *From Amaratunga (1983).*

sweeping assumptions about constancy of feeding rate and a linear relationship between the percentage of animals' feeding and time (month).

In a considerable improvement of this approach, Pierce and Santos (1996) modelled month-to-month changes in the population size and amount of different prey species removed, using data on fishery landings, size composition and diet of *L. forbesii* in Scottish waters, along with literature estimates for natural mortality and daily energy requirements.

In their review, Rodhouse and Nigmatullin (1996) not only provided a descriptive reflection of trophic relationships in cephalopods but also covered quantitative impacts on prey populations. Their summary of the life energetics of a squid, specifically the winter-spawning population of *I. argentinus* (the best available at the time), is given in Figure 2.8. In another review (concerning squid of the genus *Illex*), Dawe and Brodziak (1998) listed difficulties in incorporating trophic relationships into quantitative population dynamics analysis of cephalopods, as follows:

- If *Illex* recruitment is substantially influenced by environmental variation, then trophic interactions may be difficult to discern.
- If important trophic interactions occur primarily between the youngest stages of *Illex* and other species during the oceanic phase of the life cycle, then abundance data concerning older life stages may be inadequate to discern the cause of recruitment variability.
- If spatial aggregation and temporal aggregation of relative abundance data conceal the effects of local processes, then correlations based on aggregated data may be impossible to measure.
- If species that interact with *Illex* through competition for prey or through the sharing of predators are not considered, then important indirect trophic effects may be impossible to measure.

Their diagnosis stands firm to the present day. An example of Dawe and Brodziak's approach is given in Figure 2.9.

Jackson et al. (2007) provided a general analysis of the role of squid using the Atlantis model (Fulton et al., 2004). This is a holistic ecosystem model based on trophic interactions in many modules, including fisheries. An example of the use of this model to assess an impact of fisheries on squids in the Bass Strait is given in Figure 2.10.

Gasalla et al. (2010) included the squid *Doryteuthis plei* within an Ecopath model for which the mixed trophic impact and "keystoneness" were calculated for all component groups and/or species. The main finding was that *D. plei* had the third highest "keystoneness" as well as a high overall mixed trophic effect index. It appears that "squid on squid" effects are very important in these interactions. The interactions matrix (i.e. for mixed trophic impact) for *D. plei* is shown in Figure 2.11.

Gaichas et al. (2010) used a food web model to incorporate data on trophic relationships into stock assessment under an "ecosystem approach to fisheries" (EAF) perspective. They included squids without specifying species and compared resources with high fishing mortality (halibut, skate and walleye pollack) with those that are incidentally fished (squids), noting that

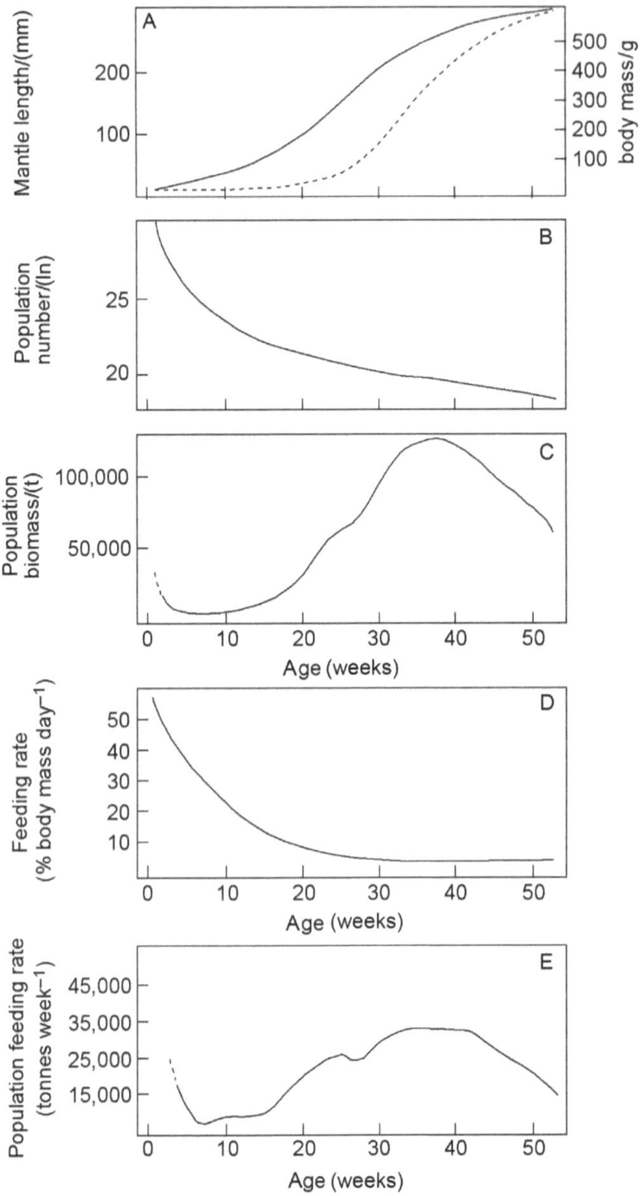

Figure 2.8 Lifetime energetics of an *Illex argentinus* cohort from the winter-spawning southern Patagonian Shelf population. *From Rodhouse and Nigmatullin (1996).*

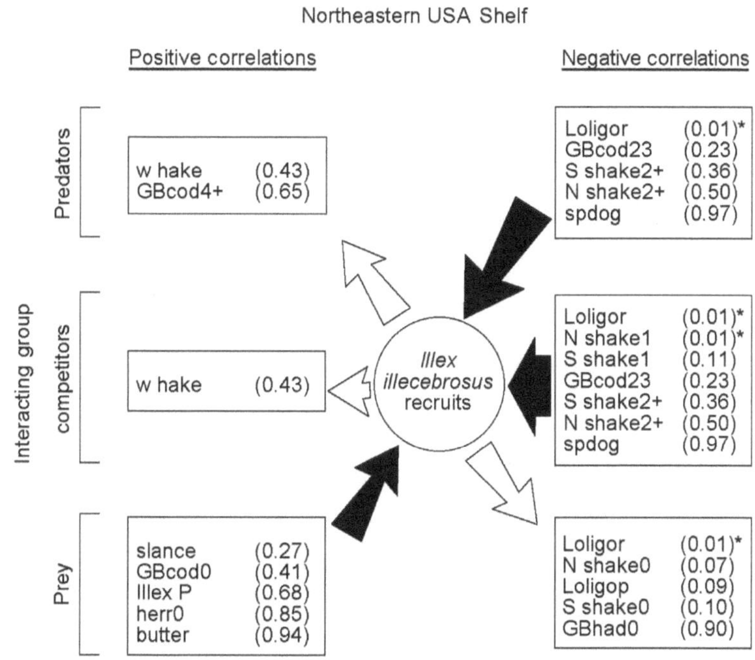

Figure 2.9 Schematic representation of the relative importance of three types of trophic interactions on *Illex illecebrosus* recruitment on the northeastern U.S. shelf, based on the occurrence of positive versus negative correlations with relevant fish stocks and age groups (*p*-values marked by * were judged to be statistically significant); thickness of the dark arrows represents relative importance of interactions which could affect I. illecebrosus recruitment. For definitions of interacting groups, see Dawe and Brodziak (1998) p. 131 (Table 7.1) from where this figure was reproduced.

natural predation in squids is much greater than fishing mortality and should therefore be considered in ecosystem modelling.

Moustahfid et al. (2009) concluded that incorporating trophic relationships (predation) into a surplus production model is feasible, providing a demonstration of an alternative to the present approach to management of *Doryteuthis pealeii*.

Roel (1998) identified several mechanisms that play a role in determining recruitment levels in chokka (*L. reynaudii*), for example, predation on the spawners and on eggs and cannibalism. She concluded that cannibalism is likely to be a density-dependent cause of mortality, while environmental events such as the frequency of westerly winds in winter and of upwelling events in summer appear to have a direct influence on the extent of spawning inshore (and are positively correlated with abundance).

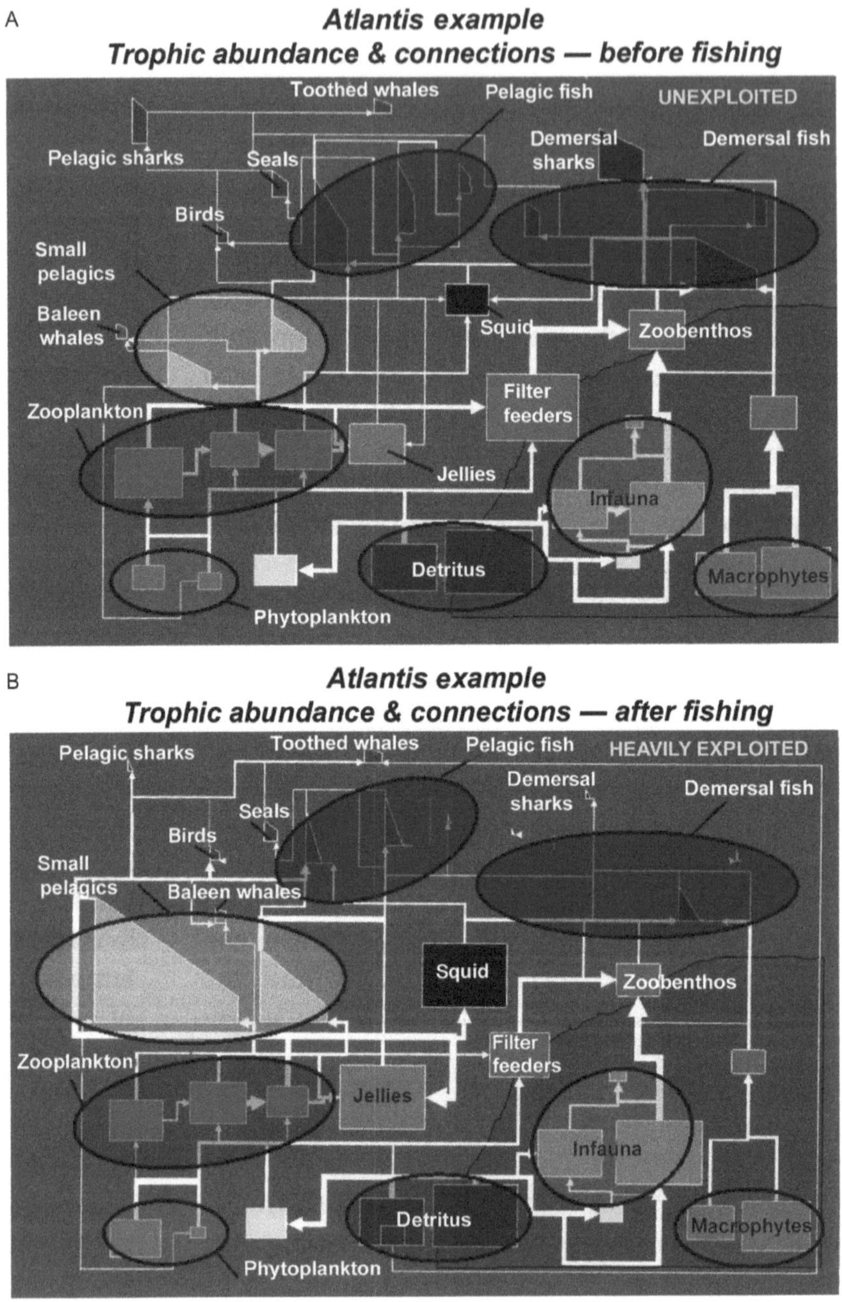

Figure 2.10 The framework of the "Atlantis" model. The framework represents a natural ecosystem using a nutrient-based biogeochemical model that is coupled in the biological/physical sense through differential equations. An assessment, using "Atlantis", of the potential effect of fishing pressure on trophic abundances and connections in the Bass Strait (Australia), by showing the food web equilibrium before fishing (a) and after fishing (b). *From Grist et al. (2007) in Jackson et al. (2007).*

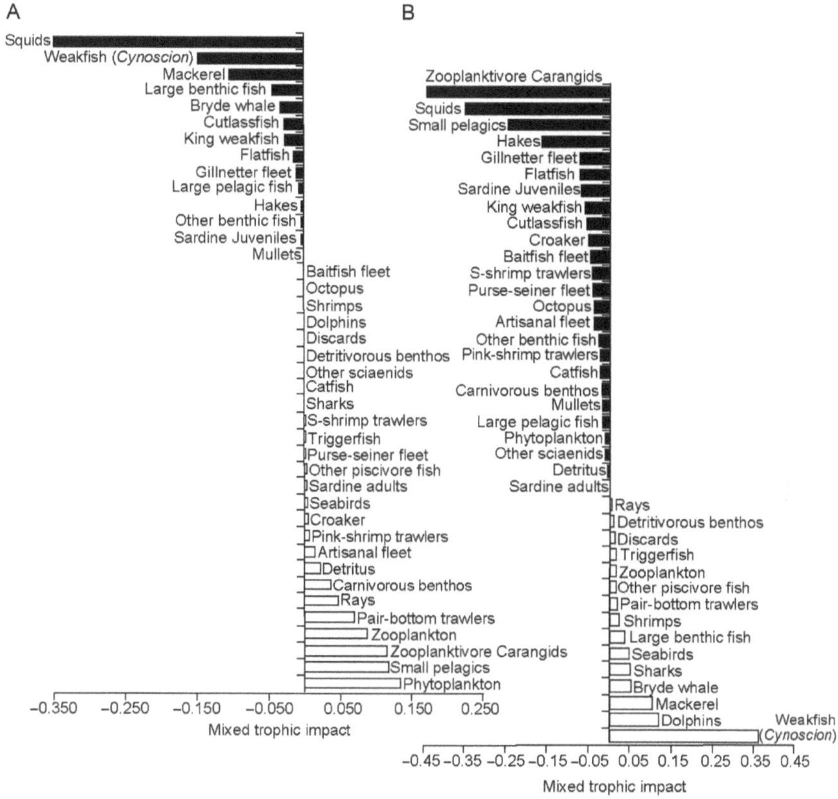

Figure 2.11 Trophic role of the *Loligo plei* in the South Brazilian Bight. The mixed-trophic impact matrix analysis was used. (A) Impacts of other groups upon squid. For example, weakfish, cutlassfish, whales, large pelagic fish and mackerel seem to negatively impact squid as predators or indirectly (top-down). Producers and plankton groups, small pelagic fish and carangids seem to impact squid positively via bottom-up process. (B) Squid as impacting species upon other groups or species. Negative impacts are seen for several prey species such as zooplanktivorous carangids and small pelagic fish. Positive impacts are seen among "predators" of squid and/or as indirect links. *From Gasalla et al. (2010).*

Fisheries biomass and various aspects of applied population dynamics of cephalopods are covered by a large literature base stretching over many years, probably starting with Sasaki's (1921, 1929) remarks about *T. pacificus* exploitation in Japan. Generally, neither environmental impacts nor trophic relationships are explicitly included in stock assessment models (e.g. Basson et al., 1996; Beddington et al., 1990; Khoufi et al., 2012; Lu et al., 2013; Morales-Bojorquez et al., 2001a, 2008, 2012; Nevárez-Martínez et al., 2006, 2010; Robert et al., 2010; Tomas and Petrere,

2012), except perhaps *T. pacificus* in Japan (see below). Relevant studies on specific cephalopod species include those of Mohamad Kasim (1985), Vidyasagar and Deshmukh (1992), Karnik et al. (2003) and Thomas and Kizhakudan (2011) for *Photololigo duvaucelii*; Sundaram and Khan (2009) for *Sepiella inermis*; Mohamad Kasim (1993) for *Sepia elliptica*; Arreguín-Sánchez et al. (2000) for *Octopus maya*; Alvarez Perez (2002) for *D. plei*; and Augustyn et al. (1993) and Roel (1998) for *L. reynaudii*. A notable exception to this rule is stock assessment for *T. pacificus*, where environmental effects are included (Kidokoro and Mori, 2004; see also Section 4).

There is certainly enough evidence that intrinsic elements and trophic relations are no less important than the environment in shaping cephalopod life cycles and their population dynamics and recruitment in particular. However, their incorporation into workable management strategies is a challenge. At present, empirical models of abundances (or catches) based on external (environmental) factors may appear to be better candidates for fisheries management tools than traditional stock assessment approaches, but our lack of knowledge about underlying mechanisms, rooted in ecological theory, is also a serious weakness.

2.6. Synthesis and the future

Stability, oscillations in abundance and occurrence and chaotic behaviour of populations have been studied only selectively in cephalopods. The linking of population dynamics to molecular biology is at an early stage, although its importance has been recognized.

The problem of overlapping generations has not been solved (neither is it satisfactorily solved in theoretical ecology either). The population biology of octopods offers excellent experimental opportunities in this regard because of their strict semelparity. The understanding of interaction between various life stages in studies on both fish and cephalopods has just only started receiving attention (see Bloor et al., 2013a), with the understanding that larval and paralarval phases are the most vulnerable, and account for the greatest mortality in each generation because of starvation and predation pressure. However, paralarvae and juveniles differ profoundly: newly hatched paralarvae must learn within a very short time period how to catch food (e.g. Chen et al., 1996) and are unable to withstand even short periods of starvation due to their high metabolism and short-lasting yolk reserves (Vidal et al., 2006); they are therefore very prone to death from starvation. Cephalopod juveniles are highly visible, occur together in large numbers and energetically

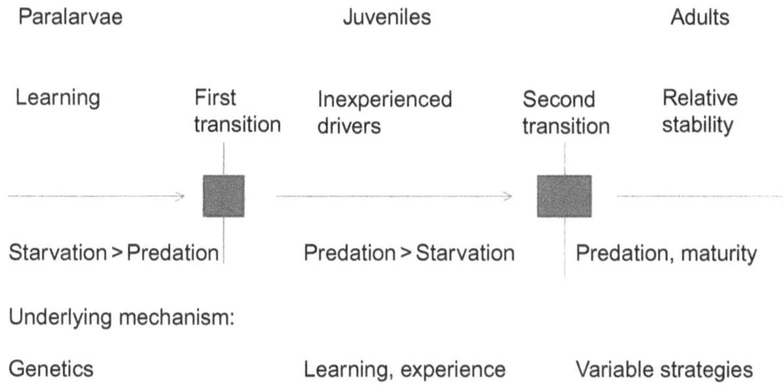

Figure 2.12 Schematic representation of a cephalopod life cycle. Mortality of paralarvae is mainly due to starvation; mortality of the juveniles is mainly due to predation. Transitions are morphological as well as physiological and behavioural.

offer greater energetic rewards for predators than do paralarvae, suggesting that their mortality from predation must be considerable (Figure 2.12).

Pierce et al. (2008) had reviewed the importance of environmental factors for the structure and dynamics of cephalopod life cycles (including population dynamics). Their assessment is corroborated by further research upon several species. This effort makes a strong connection with recent trends in general marine ecology. Indeed, Lehodey et al. (2010) and Buckley and Buckley (2010) had strongly advocated further the development of an approach linking ocean models and environmental effects to population dynamics of large marine predators. However, the cornerstone of the current population dynamics theory–trophic relationships remains to be adequately addressed in cephalopods. This is the most important area for future research and one that could result in an original contribution to theoretical ecology. One of the main inferences from theory is that specialist predation is a key driver of population dynamics of prey. However, predation upon and by cephalopods is a network of opportunistic links, governed by a different set of models (Turchin, 2003). This is untapped research territory in theoretical ecology. Another important area of research is the interplay of temporal and spatial considerations and issues of cephalopod survival (e.g. Challier et al., 2006b; Crespi-Abril and Baron, 2012; Crespi-Abril et al., 2013, 2014; Lipinski et al., 1998a,b; see also Olyott et al., 2006, 2007). Researchers wishing to pursue this line of research have plenty of data (see Table 2.1). Myers (2000) had written about the Ocean Biogeographic Information System (OBIS; http://www.iobis.org/), while Turchin (2003) mentioned the Global Population Dynamics Database (http://www3.imperial.ac.uk/cpb/databases/gpdd). There are also the database of the Food and Agriculture Organization of the United Nations (http://www.fao.org/fishery/en) and data from various large projects funded by the European Union. There are also large databases of the individual countries (or scores of countries under bilateral and multilateral agreements) at the forefront of exploitation and research of cephalopods (Japan, the United Kingdom and the United States).

3. CAUSES OF FLUCTUATIONS IN POPULATIONS

3.1. Physical and biological effects

3.1.1 Temperature effects on metabolism and survival

Body temperature is perhaps the most important ecophysiological variable affecting the performance of ectotherms. Performance functions including metabolism and growth rates steadily increase with temperature from the critical thermal minimum until achieving the thermal optimum with a further abrupt drop to zero at the critical thermal maximum (Anguiletta et al., 2002). Such an asymmetric function, in which performance is maximized at an intermediate temperature, is especially marked in cephalopods because of an incipient oxygen limitation of metabolism at the species-specific thermal maximum (Meltzner et al., 2007).

Cephalopod egg survival seems to be very stable within the thermal optimum (ca. 90–100%), then, either abruptly dropping to zero at thermal limits as in squids *D. gahi* and *D. opalescens* (Cinti et al., 2004; Zeidberg et al., 2011) or gradually decreasing because of a simultaneous increase of developmental anomalies, particularly in the last stages of development as in *L. reynaudi* and *T. pacificus* (Oosthuizen et al., 2002; Sakurai et al., 1996). Fluctuations in temperature have a negative impact on survival ofloliginid eggs, and upward fluctuations are comparatively more deleterious than downward fluctuations. Earlier stages of embryonic development are more sensitive to such temperature variability (Gowland et al., 2002; Oosthuizen et al., 2002; Segawa, 1995).

Within the optimum thermal range for reproduction, populations reproducing at lower temperatures (e.g. winter vs. summer), or in higher latitudes, produce eggs of larger size (Laptikhovsky, 2006). Because of this higher amount of yolk in the egg, squids and cuttlefish hatched at lower temperatures not only are larger but also have a proportionally larger yolk sac, so increase in hatching size is not coming at the cost of diminishing of yolk reserves (Bouchaud, 1991; Vidal et al., 2002; Villanueva et al., 2011). Taking into account the fact that lower temperature also invokes lower rates of metabolism, survival on yolk reserves is longer at lower temperatures (Vidal et al., 2005), and these hatchlings possibly have more time to find a suitable stable food source before their yolk reserves run out and they die (Bouchaud, 1991). Therefore, lower temperatures within the optimum species-specific range might generally be more favourable for recruitment survival and cohort strength. However, it was demonstrated under experimental conditions that temperature might have negative impact on hatchling survival within normal developmental temperature range (Vidal et al., 2002), so this supposition should be taken cautiously. Hatchlings of *S. officinalis* are known to vary in weight from 0.053 to at least 0.180 g (Bloor et al., 2013a), so differences in size-related survival rates might be quite important.

Temperature has a crucial impact on paralarval survival, growth rate, age of juvenile benthic settlement and timing of reproduction. Recruitment strength (hence commercial catch) of cephalopods is often related to environmental temperatures during the first months of life (Caballero-Alfonso et al., 2010; Pierce and Boyle, 2003; Vargas-Yáñez et al., 2009), but it is not clear if this is a direct temperature impact or whether temperature acts as an indicator of some other environmental changes influencing survival at early stages. At lower temperatures, growth rates are reduced, increasing size-dependent mortality by predation, while high temperatures increase growth rate, reducing size-dependent mortality but will tend to increase mortality due to starvation if food is scarce. These relationships are not straightforward due to numerous other factors, in turn affected by temperature, impacting recruitment strength.

Generally, once cohort strength is established, stock size depends mostly on growth rates and mortality. Squids grow faster at higher temperatures within the optimum range, and because of this, squids hatched at different times from the same population achieve different adult sizes. Beyond this range, at the thermal limits of the species survival, growth is slower.

At higher temperatures, cephalopods also mature faster, so their growth rates slow down at an earlier age. However, this earlier start of reproductive activity might not always impact on the population-specific duration of the life cycle, but rather, it will extend the duration of the adult phase (Arkhipkin and Laptikhovsky, 1994; Forsythe, 2004; Forsythe et al., 2001; Rodhouse and Hatfield, 1990). Extension of the adult phase might support a manyfold increase in fecundity and thus in recruitment strength, particularly in cuttlefish with their highly flexible reproductive strategies ranging from a single batch to multiple spawning within the same population

(Boletzky, 1987, 1988; Laptikhovsky et al., 2003). However, populations of some squids with wide temperature tolerance might exhibit either an annual or a half-annual life cycle (Arkhipkin et al., 2000) depending mostly on environmental temperatures, thus making this factor very important in determining the entire life cycle.

As oceans warm, marine species that live near their upper thermal limit are likely to undergo changes in distribution as they follow their optimum thermal window (Parmesan and Yohe, 2003; Pinsky et al., 2013; Sunday et al., 2012). Most coleoid cephalopods are short-living species (Boyle and Rodhouse, 2005) with annual or subannual life cycles. This has recently raised the interest of studying them as models to examine possible changes in the life histories of long-lived range-shifting species (Hoving et al., 2013).

3.1.2 Other physical environmental effects

The effect of salinity on embryonic development is generally similar to that of temperature: hatching rates are close to 90–100% within the normal environmental range quickly dropping to zero beyond its limits, where metabolic processes slow down and embryos develop malformations (Paulij et al., 1990). Generally, salinity fluctuations in the natural environment are less than as those of seasonal temperature. However, some short-term large-scale changes caused by torrential rains in coastal areas are possible, particularly in closed bights and bays used by many cephalopod species for spawning. Human-induced effects can also occur. It was shown that a desalination plant, which discharges concentrated brine into the vicinity of *S. apama's* breeding aggregation, could possibly be detrimental to the future survival of the population (Dupavillon and Gillanders, 2009). Survival changes during gradual changes in salinity are to those following sudden change. Outside the optimum range, the salinity tended to cause premature hatching and death of the embryos before organogenesis at high salinity and abnormal development at low salinity (Nabhitabhata et al., 2001).

Oxygen tension is important for egg development and survival. It is not usually a problem for octopod egg clusters that are brooded and aerated by the female, but in loliginid squids, the large size of egg masses is likely to restrict water flow causing hypoxic conditions and higher occurrence of developmental abnormalities. Egg mortality is higher in large egg masses, as well as in embryos located near the attachment point where oxygen is likely to become most depleted (Gowland et al., 2002; Murray, 1999; Steer and Moltschaniwskyj, 2007). Biofouling has been shown to have an interactive effect with egg position (thus oxygenation) on egg survival. In heavily fouled masses, egg survival on the distal end of a strand was lower, and on a proximal end, higher than in clean egg masses (Steer and Moltschaniwskyj, 2007).

Light and photoperiod might also have the impact on survival and growth of cephalopods, which are visual predators especially if their prey lacks photophores and therefore cannot be seen in darkness. Adult cuttlefish *S. officinalis* died of starvation in experiments in complete darkness even when enough prey was available, and mortality of juveniles was higher at the shortest light period (Koueta and Boucaud-Camou, 2003; Richard, 1975).

Changes in dissolved carbon dioxide content and related acidification of oceanic waters might have an intensive impact on cephalopod survival and evolution as in the geologic past, they caused numerous mass extinctions of ammonites and belemnites including final extinction (Arkhipkin and Laptikhovsky, 2012). Epipelagic squids (e.g. Ommastrephidae, Gonatidae and Loliginidae) are hypothesized to be most severely impacted by the interference of CO_2 with oxygen binding at the gills, because their metabolic rates are higher than other aquatic animals (Seibel, 2007; Seibel and Drazen, 2007). Rosa and Seibel (2008) subjected the squids (*D. gigas*) to elevated concentrations of CO_2 equivalent to those likely to be found in the oceans in 100 years due to anthropogenic emissions. They found that routine oxygen consumption rate was reduced under these conditions and their activity levels declined, presumably enough to have an effect on their feeding behaviour. Kaplan et al. (2013) reported on their experiments hatching Atlantic long-fin squid, *D. pealeii*, in both regular ocean water and acidified ocean water, mimicking the conditions likely to be seen in the oceans in 100 years. The squid eggs placed in the acidified water hatched later were smaller and their statoliths were smaller, more porous and less dense, and the small crystals that comprise the statolith were organized more irregularly than those in a normal squid.

3.1.3 Oceanographic effects: Currents/transport and upwelling

Paralarval dispersal plays an important role in the population dynamics and survival of a generation. On one hand, it permits a species to occupy new habitats and expand its range wherever and whenever it is possible while simultaneously diminishing intraspecific competition at early stages of the life cycle. On another hand, it invokes critically high mortality when paralarvae are transported into area with unfavourable environmental conditions.

A particular case is a situation when a substantial proportion of the generation might be carried away from the normal population range into remote areas where the environment is still favourable for growth and reproduction. Even if these squids survive and spawn, they are lost from the original stock, which thus declines. In some years, offshore feeding aggregations of immature *D. gahi* are transported by the Falkland Current >500–700 km from their usual feeding ground to the east of the Falkland Islands (51–52°S) to the high seas area (46–47°S) off the Patagonian Shelf. These squids do not migrate back to the Falkland Islands to spawn, which causes a decrease in parent spawning stock. Due to changes in the position and intensity the Falkland Current is thus responsible for major variability in the recruitment of this species to the fisheries (Arkhipkin et al., 2006).

Recruitment strength of the winter-spawning cohort of the squid *I. argentinus* is strongly influenced by retention/transport of eggs within the spawning grounds off Uruguay and adjacent waters. Recruitment is stronger when the suitable SST habitat around spawning and nursery grounds increases in area (Rodhouse et al., 2013; Waluda et al., 1999, 2001a), thus diminishing intraspecific competition and possibly predation because of wider dispersal. A similar cause of recruitment variability was recorded in another ommastrephid squid, *T. pacificus* (Sakurai et al., 2000, 2013). Since the late 1980s, the autumn- and winter-spawning areas of this species have expanded over the continental shelf and slope in the East China Sea, promoting an increase in abundance with respective expansion of the summer feeding grounds.

In upwelling areas, oceanographic events can be an important factor determining cephalopod distribution and abundance. For example, the short-fin squid *I. coindeti* appeared in Galician waters at very high level of abundance in 1987, when the upwelling season was a particularly favourable (Gonzalez and Guerra, 2013). Similarly, the abundance of the squid *L. reynaudi* in waters of the South Africa seems to be influenced by upwelling intensity (Sauer et al., 2013). For species spawning in upwelling areas, paralarval retention is a crucial factor of survival of a generation. In a common octopus, *O. vulgaris*, retention in coastal waters appears to be a key factor for the recruitment success off the Arguin Bank (Mauritania). Paralarvae have been shown to benefit from increased retention in spring due to enrichment and limited mixing and dispersion, whereas hatching in autumn is less beneficial to the recruitment because at the beginning of the upwelling season, there is less coastal retention and only weak wind-induced turbulence (Demarcq and Faure, 2013; Faure et al., 2000). Wind stress structure and related upwelling intensity were found to affect the early life phase of this species off Galicia and explain up to 85% of the total variance of the year-to-year variability of the adult catch (Otero et al., 2008).

The range of the Humboldt squid, *D. gigas*, has recently expanded and stock structure has changed. This is a species in which abundance is strongly driven by ENSO events around the Peruvian coastal upwelling system. Abundance is at a maximum when temperature anomalies are moderate, and decreases during intense warm and cold events, probably because of lower survival of early stages and adult spawning outside areas of optimal conditions (Rosa et al., 2013; Waluda and Rodhouse, 2006; Waluda et al., 2004).

Cold upwelling water might block distribution of warm water species in the area, and occasional intensive inflows of anoxic waters with sporadic occurrence of high H_2S concentrations might be deleterious for entire cephalopod groups. Over a huge area of intensive upwelling in Namibian waters, between 21°S and 27°S outbreaks of toxic H_2S gas are a seasonally recurrent feature (Emeis et al., 2004) poisoning near-bottom layers, whereas surface layers remain clean. Only those cephalopods with small pelagic eggs are able to reproduce there (Ommastrephidae, Enoploteuthidae and *Lycoteuthis diadema*), whereas to the north and to the south of this area, large-egged bottom dwellers laying eggs on the sea floor (Sepiida, Sepiolida and *L. vulgaris reynaudii*) are very common (Laptikhovsky et al., 2013).

3.2. Migrations

Migrations cause a significant source of population variability among cephalopods. They may be a reflection of population redistribution, when animals move either diurnally or throughout their life cycle during ontogenetic migrations. Diurnal and ontogenetic migrations may take place either horizontally or vertically or both. Diurnal migrations mainly reflect changes in feeding behaviour with alternating periods of feeding activity and resting at different times of the day. Ontogenetic migrations happen when the species range (or population area) is separated into spawning/nursery grounds and feeding grounds to take an advantage of favourable environmental conditions for the development of egg masses and juveniles and maximally exploit food resources, therefore releasing some competitive pressure between different ontogenetic stages (Nesis, 1985). Changes in environmental conditions at any ontogenetic stage may cause changes in natural mortality resulting in variability in population abundance. Various cephalopods are characterized by a wide spectrum of both diurnal and ontogenetic migrations (Boyle and Rodhouse, 2005; Nesis, 1985; Roper and Young, 1975).

The chambered *Nautilus* spp. do the shortest spatial migrations among all recent cephalopods. The animals exhibit complicated diurnal behaviour, with continuous movement and feeding at night between depths of 130 and 700 m and with some animals resting during the day at relatively shallow depths and others actively forage in the deep. There is no separation between nursery and feeding grounds, as juveniles are distributed within the same habitat as adults (Dunstan et al., 2011). Tagging studies have revealed some long-term movements of up to 150 km in 332 days by living nautiluses (Saunders and Spinosa, 1979).

Benthic species of octopods have limited migrations, generally moving no more than tens of kilometres during their whole life. The largest species, the Pacific giant octopus *Enteroctopus dofleini*, remains stationary or hiding for 94% of the time with maximum movement being 4.8 km in a 3-month period (Scheel and Bisson, 2012). In Hokkaido waters, giant octopuses make short distance offshore–inshore nondistant migrations twice a year. These movements probably coincide with active choice of optimum ambient temperature inshore, when they avoid water that is too cold (in winter) and too warm (in summer) (Rigby and Sakurai, 2004).

Cuttlefish (Sepiida) are characterized by separate feeding grounds, which are usually offshore, and the inshore spawning grounds. Their ontogenetic migrations are quite short (tens to low hundreds of kilometres) but could result in a strong seasonal variability in abundance especially in their localized spawning grounds. Dense breeding aggregations are formed by the Australian cuttlefish *S. apama* during the austral winter when mature animals converge in a highly localized area (60 ha) of rocky reef in the northern Spencer Gulf (South Australia) to breed. For the rest of the year, young and juvenile cuttlefish disperse and forage over a much wider area of the gulf (Hall and Hanlon, 2002). The European cuttlefish *S. officinalis* has extensive offshore feeding grounds around Hurd Deep (100–170 m depth) in the English Channel, where juveniles and subadults spend the autumn and winter months in deeper and warmer waters. In spring, maturing and mature adults move to their spawning grounds located along the French coasts of Normandy, with mature males arriving there about a month earlier than females. Mating and spawning occur in summer in shallow water where the warmest ambient conditions for egg development are found. In inshore waters, some adult cuttlefish might stay within a small spawning area for weeks, while others might travel some 20–35 km in 2–6 weeks during the spawning season (Bloor et al., 2013b). Mass postspawning mortality drastically reduces the biomass of cuttlefish in inshore waters until the next spawning migration the following year (Boletzky, 1983). A similar pattern of ontogenetic migration was also revealed for the abundant Pacific cuttlefish in Japanese waters. Mature adults of *Sepia esculenta* and *Sepiella japonica* migrate to shallow bays and inlets from their offshore feeding grounds in summer to spawn and lay their eggs in the warmest time of the year (May–July) (Natsukari and Tashiro, 1991).

Squids (Teuthida) have the longest migrations among cephalopods. Among them, neritic squids of the suborder Myopsida migrate shorter distances between their spawning and feeding grounds than some large nerito-oceanic and oceanic squids of the suborder Oegopsida families Ommastrephidae and Onychoteuthidae (Nesis, 1985).

The nearshore species *Sepioteuthis* (Loliginidae) have a similar lifestyle to cuttlefish, including relatively short migrations. Southern calamari *S. australis* moves extensively within their inshore spawning grounds (30–60 km, one squid even 600 km). However, no movement was observed between the two spawning grounds separated by 25–30 km only in the southeastern part of Tasmania, Australia (Pecl et al., 2006). It means that during breeding season, an individual squid might move significantly within a habitat favourable for reproduction, but do not cross unfavourable areas from one breeding site to another.

Shelf squids of *Loligo* (and *Doryteuthis*) have longer migrations (up to several hundred kilometres) between their well-defined inshore spawning/nursery grounds and offshore feeding grounds. They also move substantial distances during foraging within their feeding grounds. European loliginids *L. vulgaris* and *L. forbesi* do not aggregate into dense migrating schools neither in their feeding nor in their spawning grounds, laying their eggs at wide range of depths between 10 and 300 m (Guerra and Rocha, 1994; Pierce et al., 1994). Other loliginids such as the southeast Pacific population of *D. gahi*, northeast Pacific *D. opalescens* and South African *L. reynaudii* disperse on their offshore feeding grounds but form dense spawning aggregations in their localized inshore spawning grounds (Sauer et al., 1992; Villegas, 2001; Vojkovich, 1998). The location of spawning sites depends on physical oceanographic dynamics with squids actively choosing areas on the shelf with bottom temperatures and dissolved oxygen concentrations, which are at optimal levels for egg development (Roberts, 2005). On the contrary, the southwest Atlantic population of *D. gahi* does not aggregate in shallow-water spawning grounds, but forms dense feeding schools offshore. The time of offshore migrations of *D. gahi* (thus its availability to predators and fishery) depends on water temperatures in their inshore spawning/nursery grounds.

Warmer temperatures induced earlier emigration, and colder temperatures delayed those emigrations. The variability in the extent and locations of *D. gahi* offshore migrations on the Falkland Shelf is determined by the location of their preferable offshore feeding habitat, the so-called transient zone that is a mixture between the shelf waters and the subantarctic superficial waters (Arkhipkin et al., 2004).

As the juveniles move offshore, they start segregating by depth with females migrating deeper than males. The segregated feeding period lasts several months, with males predominating in shallower waters and females in deeper waters. After maturation, both sexes move to the inshore spawning grounds (kelp forests between 5 and 20 m depth) separately with males arriving earlier than females (Arkhipkin and Middleton, 2002a).

Nektonic oegopsid squids of the families Ommastrephidae and Onychoteuthidae are highly migratory animals (Boyle and Rodhouse, 2005). The spatial structure of their ranges is usually complex with spawning, nursery and feeding grounds often located in different ecosystems (O'Dor and Coelho, 1993). Large species such as *I. argentinus*, *I. illecebrosus*, *T. pacificus* and *D. gigas* migrate thousands of kilometres between their feeding and spawning grounds during their short (usually) annual life cycle (Arkhipkin, 1993; Froerman, 1986; Nigmatullin et al., 2001; O'Dor, 1983; Sakurai et al., 2002). These squids also transition vertically during diurnal migrations, rising to superficial epipelagic waters every night to feed and descending to deep mesopelagic and bathypelagic waters during daytime to rest (Hanlon and Messenger, 1996).

Nerito-oceanic squids *Illex* spp. use the waters of the polarward warm currents (Gulf Stream, Brazil Current) flowing along the continental slopes to transport their paralarvae and juveniles from the tropical/subtropical spawning grounds to temperate feeding grounds (O'Dor and Coelho, 1993). The abundance of their recruitment on the feeding grounds varies depending on the intensity, position of the stream and meandering of the current in a given year. Increased meandering of the Gulf Stream causes the enhanced shoreward transport of *I. illecebrosus* juveniles onto the shelf of Nova Scotia between 35°N and 45°N (Rowell et al., 1985). The intensity of the current determines how far downstream the juveniles of *I. illecebrosus* might be transported. In years of strong Gulf Stream intensity, the juveniles move as far as the Grand Banks to the east of Newfoundland (43–47°N); weakened Gulf Stream transport causes for recruitment to appear in Georges Bank and western part of the Nova Scotia shelf (35–42°N) (Dawe et al., 2000). Ambient temperature also determines the extent of feeding migrations in the southwestern counterpart—*I. argentinus*. During warm years, squids penetrate almost to the southern edge of the Patagonian Shelf (54°S), whereas in cold years, they migrate only down to 45–46°S (Arkhipkin, 2013). Water temperatures impacted not only the extent but also the migration routes of *T. pacificus* in the northwest Pacific. It has been shown that under a cool regime during the 1980s, migrations of *T. pacificus* were restricted to the Sea of Japan. When temperatures increased in 1989–1992, not only did biomass of the stock increased threefold, but also migration routes changed and included a new migration pattern in the Pacific side off Japan (Sakurai et al., 2002).

Several environmental factors like temperature regime shift and availability of preferred prey are believed to cause the dramatic change in the size of mature animals of the southeast Pacific Humboldt squid D. Gigas (Argüelles et al., 2008; Keyl et al., 2008, 2011). In the 1990s, most *D. gigas* populations consisted of small squids (<500 mm mantle length, ML) with the species range situated mainly in tropical and subtropical waters of the eastern Pacific between 30°N and 30°S. Since 2000, 2 years after the strong El Niño of 1997–1998, the average size of adult squids dramatically increased to 800–900 mm ML with simultaneous expansion of the species range to temperate and even subpolar waters of the Alaska 47°N in the north (Field et al., 2007; Zeidberg and Robison, 2007) and central Chile (43°S) in the south (Alarcón-Muñoz et al., 2008). Such an increase in adult squid size with simultaneous expansion of the species range has had important impacts on species abundance, diversity and community structure in the temperate and subpolar ecosystems of the eastern Pacific that are now under predation pressure of a large voracious predator that previously had been a less dominant member of the community.

3.3. Trophic ecology

Embryogenesis is probably the least vulnerable part of the cephalopod life cycle. Although egg masses of myopsid squids are laid on the bottom in large numbers and are easily available to potential predators, no major predation on eggs has been reported in spite of the fact that egg beds have been intensively explored by divers and ROVs in different oceans and at different latitudes. Some rare attempts of predation by benthic echinoderms and fish (Kato and Hardwick, 1975; Sauer and Smale, 1993) can probably be ignored and have little or no impact on survival of spawning products. However, eggs of some loliginids (primarily *D. opalescens*) collected in the wild might be infested with capitellid polychaete worms (Fields, 1962; Boletzky and Dohle, 1967; Vidal et al., 2002). These worms cause the deterioration of the external egg envelope and expose the chorion of the eggs causing premature hatching and subsequent high mortality of paralarvae (Vidal et al., 2002). Zeidberg et al. (2011) found an exception to this pattern and observed that the disturbances by the worms slightly increased hatching rate, but did not provide data on paralarval survival (Vidal and Boletzky, 2004).

Among oegopsid pelagic egg masses, only those of *Thysanoteuthis rhombus* are commonly observed, and among numerous descriptions and photographs, there is no sign on possible predatory impact or mass mortality. Benthic octopod egg masses are generally well hidden from large predators in sheltered places and protected from small predators scavengers by a defensive female.

Feeding conditions are of crucial importance for cephalopod population dynamics from the early life stages in spite they still have some endogenous yolk reserves during first days after hatching while already hunting prey (Boletzky, 2003). Food availability can induce growth plasticity in paralarvae in very short time periods (Vidal et al., 2006).

Females of a sepiolid, *Euprymna tasmanica*, maintained in captivity on a low ration produced smaller egg clutches, consisting of smaller eggs and exhibiting higher embryo mortality rates than females fed *ad libitum* (Steer et al., 2004). Loliginid squid hatchlings living in better foraging conditions and at lower temperatures utilize yolk more slowly and so conserve their reserves longer (Vidal et al., 2002).

Cannibalism is an important element in the life cycle of many squids. It has been reported for many cephalopods in genera *Illex*, *Octopus*, *Sepia*, *Dosidicus*, *Onychoteuthis*, *Todarodes*, *Ommastrephes* and *Loligo* (Ibáñez and Keyl, 2010 and references within). In nature, cannibalism increases when prey availability decreases and larger squids are more cannibalistic than smaller. In captivity, it takes about 3 days of starvation for *I. illecebrosus* squid to induce cannibalism. Selective removal of smaller animals not only decreases stock size but also gives a false impression of faster somatic growth when size–frequency data are analysed (Arkhipkin and Pérez, 1998; O'Dor and Dawe, 2013).

Generally, cannibalism is density-dependent and acts as a tool regulating population biomass within an optimum level. Years of high density of the schooling squid *I. illecebrosus* were associated with high rates of cannibalism in this species, though it is apparent that the shift to cannibalism does not merely reflect opportunity but is related to depletion of other suitable prey types (Dawe, 1988). Because of density dependence, cannibalism also increases in artificial aggregations of ommastrephid squids in light fields of a fishing boat (Zuev et al., 1985). In octopuses *Enteroctopus megalocyathus* and *O. vulgaris*, a higher frequency of cannibalism has been reported in areas and periods where this species is more abundant (Ibáñez and Chong, 2008; Oosthuizen and Smale, 2003).

Easiness with which cephalopods switch to cannibalism might help populations survive episodes of low food availability. However, cephalopods generally have broad feeding spectra that are very flexible in both prey species composition and prey size, so food availability likely is not a very common problem. During their sudden explosion in number, squids normally have detrimental impact on potential food sources (Alarcón-Muñoz et al., 2008; Laptikhovsky et al., 2013) during a range of years rather than die out of starvation.

Another factor impairing well-being of cephalopods and making them more susceptible to predators and both biotic and abiotic stressors—and thus influencing cephalopod population variability—is parasite load. Cephalopods are hosts to a diverse assemblage of parasites and symbionts including potentially pathogenic organisms such as viruses, bacteria, fungi, protozoans, nematodes, monogeneans, digeneans, cestodes, acanthocephalans, polychaetes, hirudineans, crustaceans, copepods and isopods (Hochberg, 1983). Quantifying the incidence of diseases in cephalopod populations and impact on its dynamics may be difficult because diseased and dead animals (especially as they are soft-bodied) are likely to be rapidly removed by predators or scavengers (Pierce et al., 2010). Some parasites, such as the copepod family Pennelidae, might have such negative impact on squid condition that they cause important losses in commercially important stocks (Pascual and Guerra, 2001; Pascual et al., 1998, 2005, 2007).

Cephalopods, particularly schooling squids, are important prey of many large marine predators, sometimes the most important prey for species including pilot whales and sperm whales (Clarke, 1996; Piatkowski et al., 2001; Santos et al., 2001; Smale, 1996). Thus, annual cephalopod abundance might have a strong impact on populations of predators. However, predators, though possibly being the most important factor regulating squid population dynamics (together with fisheries), should probably not be counted as a major factor responsible for the fluctuation of populations in commercially exploited cephalopods. In the virtual absence of whaling, cetacean populations are probably stable over annual and possibly decadal timescales. There has been no documented case of top-down control, when changes in predator abundance have been correlated with cephalopod prey abundance. However, when the predator is a cephalopod with highly fluctuating population biomass and distribution, top-down control has been observed. *I. argentinus* preying on *D. gahi* in Falkland Islands' waters have a negative impact on final cohort biomass. However, this inverse relation might be also (at least partially) explained by competition for the same food sources (Arkhipkin and Middleton, 2002b).

3.4. Fisheries

Fishing pressure is an important source of variability in any of the commercially fished stocks. Direct (targeted) fishing as well as indirect (by-catch) fishing may overexploit and therefore deplete the stocks as has been reported for finfish fisheries worldwide (Pauly et al., 1998). Alternatively, some stocks may also increase in abundance if main predators have been fished out or the competitive pressure has been relieved by fishing out the main competitor and therefore vacating the econiche (Boyle and Rodhouse, 2005; Caddy and Rodhouse, 1998).

Among cephalopods, the chambered *Nautilus* spp. are characterized by low growth rates, relatively slow maturation and low fecundity, and they are especially vulnerable to overfishing. The growing market demands for their ornamental shells have resulted in up to 80% declines in reported catch per unit effort (CPUE) from 1980 to the present time in the Philippines, where the fishery became unsustainable (Dunstan et al., 2010). The current status of *Nautilus* populations in various areas of the tropical Indo-West Pacific is being estimated in order to include them on the list of Appendix-II of the Convention on International Trade in Endangered Species of Wild Fauna and Flora (CITES) (De Angelis, 2012).

Other commercial species of octopods, cuttlefish and squids are short-living having high growth rates and high fecundity. Their populations have evolved to withstand substantial variations in abundance. Until the turn of the century, it seemed that only a few species of coleoid cephalopods had been locally overexploited, despite heavy fishing pressure on many stocks (Boyle and Rodhouse, 2005). More recently, annual world cephalopod catches reached 4.3 million t in 2007 but decreased to 3.6 million t in 2010, according to FAO statistics (FAO, 2011), having increased steadily from around 600,000 t in 1950 (Jereb and Roper, 2010).

Several collapses in squid fisheries (*I. illecebrosus* in the northwest Atlantic or *Todarodes sagittatus* in the northeast Atlantic) left unexplained what had been the main contributor of such a collapse—fishery or environmental conditions or probably both. The fishery for the short-fin squid *I. illecebrosus* developed quickly in the second half of the 1970s, peaking at the annual catch of >100,000 t in 1979. After only 5–6 years of intense fishing, it collapsed in 1983 to <1000 t per year and has never recovered to its earlier size. One of the possible reasons of such a collapse was suggested to be sequential fishing pressure on various seasonal cohorts of the squids (first the most abundant winter-spawning cohort and then on the spring spawning cohort) that impacted the general population structure (O'Dor and Coelho, 1993). Changes in position of the main flow of the Gulf Stream relative to the shelf edge that happened in 1980s also contributed to the decrease in juvenile transport of the winter-spawning cohort to their common feeding grounds in temperate waters of Canada (Dawe et al., 2000). Similarly, overfishing of the winter-spawning cohort of *T. pacificus* caused the southern shift of the commercial fleet to fish off the autumn-spawning cohort; this dramatically changed the population composition of the species (Nakata, 1993). The autumn-spawning cohort may have acted as a reservoir that buffered the effect of fisheries until it was itself exploited.

Periodic intrusions of large quantities of the arrow squid *T. sagittatus* into the Norwegian Sea have resulted in being targeted by trawling fleets, with a maximum annual reported catch in Norwegian waters of ~18,000 t in 1982–83. However, possibly partially due to extensive exploitation, partially to changes in the environment, the commercial aggregations disappeared from that area after 1988 (FIGIS, 2011). A similar phenomenon has been reported for *L. forbesi* in the Rockall Bank area, (the United Kingdom, northeast Atlantic, 54–58°N, 12–17°W) (Pierce et al., 1994).

Fishing might also be sex-selective, removing the more active sex from their spawning grounds. In southern calamari *S. australis*, the commercial fishery alters the population structure on the spawning beds by removing the large males first and leaving relatively large numbers of small males and females. Sex-ratio studies on the same spawning grounds over several years suggested that progressively longer closures allowed time for more males to accumulate on the spawning beds, therefore maintaining the natural sex ratio during the spawning season (Hibberd and Pecl, 2007). Size selectivity in a jig fishery has been also recorded for *I. argentinus* in the southwest Atlantic. It was found that sizes of both sexes caught by jigging were 7 mm ML (males) and 12 mm ML (females) larger on average than those caught by the trawlers over the same time. Moreover, fishing method may also affect the proportions of mature squids in the population, that is, maturity-selective. The artificial lit area created by jigging vessels at night might cause the differences in behavioural responses of immature and mature squids reacting to the lure. Proportions of mature squids caught by jiggers were greater than those caught by trawlers by 0.8% in males and 5.1% in females (Koronkiewicz, 1995). Fishing effort targeting inshore spawning aggregations of neritic species such as loliginid squids may cause population declines due to behavioural disruption or insufficient escapement of prespawning individuals (Hanlon, 1998; Iwata et al., 2010).

4. FORECASTING AND ASSESSMENT

Global fishery landings of cephalopods have increased over the past few decades, while fishery- or ecosystem-level assessments of many stocks have been undertaken infrequently or not at all (Anderson et al., 2011; Boyle and Rodhouse, 2005; Hunsicker et al., 2010; Payne et al., 2006). Rapid growth and a short lifespan render cephalopod fisheries difficult to assess and manage (Boyle and Rodhouse, 2005; Payne et al., 2006; Pierce and Guerra, 1994). Most cephalopod species targeted by fisheries are short-lived, usually with a 1-year life cycle and a semelparous life history strategy, with a single spawning soon followed by death (Boyle and Rodhouse, 2005). Individuals targeted in 1 year do not survive until the fishing season in the next year, meaning that even though the abundance in the current year may be relatively high, the stock size in the next year may decline greatly, due to high variability in abundance between generations. Boyle and Rodhouse (2005) noted that the effect of environmental variables on abundance of annual species at multiple scales is the main reason for difficulties in establishing reliable stock assessment and management procedures. While a greater understanding of environmental influences on recruitment is frequently recommended for improved management of squid fisheries (Agnew et al., 2005; Boyle and Rodhouse, 2005), knowledge of the relationships between squid distribution and environmental variables and the associated effects on squid availability to fisheries and assessment surveys also has the potential to improve fisheries management (Ish et al., 2004; Schön et al., 2002). An understanding of environmental effects on population dynamics and species distributions could be used to appropriately incorporate data into stock assessments to support cephalopod fisheries management. Environmental processes introduced into stock assessments generally include those that affect population dynamics in the form of recruitment, with only recent attention to spatially explicit effects on distribution and migration patterns (Keyl and Wolff, 2008). Extensive research has been conducted on environmental effects on fish and cephalopod distributions (Fréon et al., 2005; Pierce et al., 2008; references therein). Studies of environmental effects on spatial and temporal distribution of marine species can inform stock assessment by defining stock structure at larger scales (Link et al., 2011) and identifying effects on availability to fisheries or assessment surveys (Brill and Lutcavage, 2001; Fréon and Misund, 1999; references therein). Approaches that incorporate environmental factors are identified in the following sections.

4.1. Stock identification and structure

For the purposes of fisheries management and stock assessment, populations or segments thereof are assumed to be a single unit stock within spatial boundaries in which the components of production (e.g. recruitment and mortality) are considered spatially homogenous (Cadrin et al., 2013). Cephalopod stock identification is complicated by taxonomic/systematic confusion along with variable abundance and distribution due to life history traits and environmental factors (Boyle and Boletzky, 1996). For management and assessment purposes, stock boundaries are often delineated based on territorial boundaries or the range of fisheries or resource assessment surveys rather than the distributional range of a species (Cadrin et al., 2013), and cephalopods are no exception. For example, the long-fin inshore squid *D. pealeii* is distributed in continental shelf and slope waters of the northwest Atlantic Ocean from Newfoundland to the Gulf of Venezuela but is considered a single unit stock within a much smaller area from Cape Hatteras north to Georges Bank, which encompasses most of the fishery and area routinely surveyed (NEFSC, 2011; Roper et al., 1984; Shaw et al., 2010). Since the comprehensive review by Boyle and Boletzky (1996), many advances have been made in tagging and genetic techniques useful for cephalopod stock identification and understanding stock structure (Semmens et al., 2007; Shaw et al., 2010), and a greater understanding has been attained regarding environmental effects on distribution and abundance (e.g. Pierce et al., 2008 and this chapter). While the remainder of this section is devoted to cephalopod stock assessment methods, it is important to recognize the importance of stock identification and the associated research as a prerequisite to assessment.

4.2. Stock assessment

For the sustainable fishery of cephalopods, most of which are considered to be annual species, it is important to determine the appropriate level of fishing and the relationship between stock abundance and recruitment for each species (Boyle and Rodhouse, 2005; Caddy, 1983; Pierce and Guerra, 1994). Basically, stock assessment and forecasting for cephalopod species are methods widely used for fish stocks. However, estimating and forecasting cephalopod abundance using fishery-dependent data such as cohort analysis have had limited success because of their fast growth, maturing in less than 1 year and relatively short period of time for recruitment, although age-structured models are considered to be useful for forecasting stock size precisely (Boyle and Rodhouse, 2005; Caddy, 1983; Pierce and Guerra, 1994). Although some octopods live more than 3 or 4 years, it is considered to be difficult to apply age-structured models for octopods in the same manner as squids due to uncertainty in age estimation because their growth patterns and maturation process are highly variable and tend to be affected by environmental conditions and food availability. With relatively few exceptions to date (e.g. Royer et al., 2002), it has been most common to adapt simple population dynamics models for cephalopod stocks that do not require age composition data (Boyle and Rodhouse, 2005; Caddy, 1983; Pierce and Guerra, 1994). Generally, this requires assessment data (including catch statistics) and suitable models to forecast fisheries stocks, which are varied in the targets of what period should be forecast (e.g. stock size in the next week, next month and next year).

4.3. Assessment timescales/timing

Cephalopod stock assessment methods can be generally categorized based on the timing of their application relative to the fishing season: preseason, in-season and postseason (Boyle and Rodhouse, 2005; sensu Pierce and Guerra, 1994). Preseason assessment and forecasting usually use data obtained from experimental surveys on the individuals in the prerecruit stages. Preseason assessments are useful for forecasting stock size and proper management for cephalopod stocks of which life span is usually 1 year and all targets for fishing are composed of newly recruited individuals. In the *Illex* and *Doryteuthis* fisheries around the Falkland Islands, fishing effort is set based on the results of preseason assessments and previous experience of recruitment variability (Rodhouse, 2001). However, preseason assessments usually require special sampling gears to catch prerecruit stages, and it is difficult to obtain enough time to survey just before fishing seasons open. Preseason assessments are often conducted by using experimental methods like midwater trawl nets to observe the abundance of cephalopods in the juvenile stage (Brunetti and Ivanovic, 1992; Kawabata et al., 2006; Kidokoro et al., 2014) and plankton nets to be used for paralarval distribution surveys (Bower et al., 1999a; Goto, 2002; Murata, 1989; Yamamoto et al., 2007). Stock sizes are forecasted based on the suitable models (e.g. growth–survival model and correlation models) with the results of preseason assessment (e.g. juvenile abundance, paralarval and juvenile densities).

In-season assessment and forecasting methods generally use data on recruited individuals from commercial fisheries. In most cases, changes (usually declines) in stock abundance during a fishing season are monitored based on CPUE of commercial fisheries. Data are collected at high temporal resolution (daily) and abundance is modelled primarily with depletion models (see Section 4.4). In-season assessments include abundance estimates updated within a fishing season and compared to reference points to inform adaptive management measures such as fishery closures (Walters and Martell, 2004). Successful adaptive management systems depend upon the accumulation of experience regarding spatiotemporal patterns of abundance and impacts of regulatory measures (e.g. fishery openings) and are vulnerable to environmentally driven shifts in abundance or distribution (Walters and Martell, 2004). Most commercially important cephalopods (e.g. ommastrephid squids) are highly migratory species; therefore, in-season assessments are only used for a few special examples, such as the *Illex* fishery around the Falkland Islands (Arkhipkin et al., 2013; Basson et al., 1996; Rosenberg et al., 1990) and *T. pacificus* fisheries (Okutani, 1977). However, in-season assessments and knowledge about seasonal shifts in fishing grounds with migration patterns are useful for forecasting catches at individual fishing grounds, for example, in the Japanese *T. pacificus* fisheries (Kasahara, 1978; Kidokoro et al., 2010; Murata, 1989; Okutani, 1983). Postseason cephalopod stock assessments rely on data on former generations, generally using one or more of the methods described in the following section. It is also important to note that many cephalopod fisheries are assessed with multiple methods, conducted at some or all of the scales identified in the preceding text (e.g. Arkhipkin et al., 2013; Roa-Ureta and Arkhipkin, 2007).

4.4. Stock assessment methods

4.4.1 Surplus production models

The surplus production model, also referred to as a biomass dynamic model, is essentially the simplest stock assessment model. In theory, the surplus production model estimates the biomass of a resource for the year $y+1$ as the biomass of the year *y* plus the surplus production (an aggregated parameter of the recruitment and the growth) and minus the catch (Graham, 1935; Schaefer, 1954). However, the surplus production model presents a very simplified view of the population dynamic; age or spatial structure of the population is not considered. In addition, fitting a surplus production model is not easy, and the interpretation of the maximum sustainable yield can be difficult because the model assumes that the stock is at the equilibrium, a situation rarely encountered (Hilborn and Walters, 1992), and that has sometimes led to the overestimation of a stock. Finally, the surplus production model assumes that the recruitment is highly density-dependent, a situation rarely encountered in cephalopod species (Pierce and Guerra, 1994). Surplus production models have the advantage of quick application with very few data but give only mid- to long-term objectives in a stable environment, while cephalopod stocks are known to be highly variable resources and their abundance is generally linked to the environmental conditions (e.g. Pierce et al., 2008; Section 3).

Despite the aforementioned limitations, surplus production models have been adapted for cephalopod fisheries with some success. Some surplus production models have been developed to assess cephalopod stocks from the Saharan Bank, results of which indicated that octopus, squid and cuttlefish stocks in this area were overexploited (Bravo de Laguna, 1989; Sato and Hatanaka, 1983). More recently, a surplus production model with environmental effects was fitted to assess the *O. vulgaris* stock off Senegalese coast (Laurans et al., 2002). Results highlighted that the wind speed parallel to the shore that generates upwelling has a significant effect on the octopus abundance. A Fox surplus production model (Fox, 1970) was also fitted to assess the English Channel cuttlefish (*S. officinalis*) and the English Channel squid (*L. vulgaris* and *L. forbesii*, which are not distinguished by the fishermen) stocks in the framework of a bioeconomic modelling of the English Channel fishery (Ulrich et al., 2002). A surplus production model with quarterly time step was fitted, using fishery-dependent and fishery-independent abundance indexes, to assess *D. pealeii* (Cadrin and Hatfield, 1999). Recent assessments of *D. pealeii* have taken into account removals from the predation to define new reference points that are significantly different from those defined without accounting for predation (Moustahfid et al., 2009; NEFSC, 2011). Quetglas et al. (2013) studied combined effects of fishing and climate (using NAO effects) by fitting a surplus production model on several demersal species including *O. vulgaris*, *L. vulgaris* and *S. officinalis*.

4.4.2 Depletion method

The depletion method derived from Leslie and Davis (1939) and De Lury (1947) is the stock assessment methodology commonly used to assess cephalopod stocks, and it is considered as the least expensive methodology. In theory, this model estimates the consequences of the removal of individuals (natural or fishing mortality) on the population and determines the size of the population without fishing activity (Hilborn and Walters, 1992). The basic method (Leslie and Davis, 1939) uses the accumulated catch and the assumption of a closed population. Some variants use effort data and have been adapted to open populations with recruitment and natural mortality (Rosenberg et al., 1990). These methodologies are suitable to perform a real-time modelling of the data collected in a short period (in particular when exploitation does not exceed 1 year, they are favoured to assess cephalopod stocks; Boyle and Rodhouse, 2005). Its main limitation is the assumption of a population randomly distributed with a constant catchability, an important assumption rarely encountered in cephalopod stocks (Pierce and Guerra, 1994).

The depletion method has been applied to assess the Humboldt squid *D. gigas* stock in the Gulf of California (Ehrhardt et al., 1983). Morales-Bojórquez et al. (2001b) performed a stock assessment modelling three different fleets exploiting Humboldt squid in the same area and enabled the estimation of reference points. A modified standard Leslie–DeLury method (integrating a natural mortality parameter) was suggested to model the squid species exploited around the Falkland Islands with the proportional escapement as a management tool (Beddington et al., 1990; Rosenberg et al., 1990) and applied to *I. argentinus* (Basson et al., 1996). The modified standard Leslie–DeLury approach was then improved to assess the *D. gahi* stock by adding a migration parameter (Agnew et al., 1998a). This migration parameter was first introduced to improve the assessment of *D. pealeii* in the northwest Atlantic, which showed variations in abundance related to migration patterns (Brodziak and Rosenberg, 1993). Depletion estimates were then used to assess European squids *L. forbesii* and *L. vulgaris* (Royer et al., 2002; Young et al., 2004). A trial to assess the English Channel cuttlefish stock using the depletion method was carried out by Dunn (1999) using the UK fishery landings, which represents approximately a third of the total landings in the English Channel. Most recently, a state-space modelling framework was investigated for a DeLury depletion model applied to the *O. vulgaris* stock exploited off Morocco (Robert et al., 2010), and

Roa-Ureta (2012) extended the depletion approach to account for the several major in-season recruitment pulses occurring in the Falklands Islands *D. gahi* fishery.

4.4.3 Age-structured models

The virtual population analysis (VPA) is the most widespread method to assess long-lived finfish stocks in developed countries (Hilborn and Walters, 1992). In a VPA, the stock is considered to be composed of several annual cohorts. The number of individuals alive in each cohort is estimated by performing a back calculation from the last age class to the first one by adding the number of individuals lost to fishing and natural mortality during a year to the number of individuals at the end of the year to estimate the number of individuals at the beginning of the year. It is sometimes called cohort analysis because each cohort of the stock is analysed and followed separately from the other cohorts living at the same time in a given stock. In cephalopod stocks, as species have a short life cycle (generally 1 or 2 years), the VPA is generally implemented on a monthly basis and using microcohorts (Jouffre et al., 2002; Royer et al., 2002, 2006; Thiaw et al., 2011).

Ehrhardt et al. (1983) applied this methodology to assess the Humboldt squid *D. gigas* stock in the Gulf of California with microcohorts defined according to a polymodal decomposition of the length–frequency. VPA was also applied to the Senegalese *O. vulgaris* stock using age classes first defined using tagging methodology to estimate the growth rate (Domain et al., 2000) and then using landings split in 10 commercial categories in the fishery industry (Jouffre et al., 2002; Thiaw et al., 2011). VPA therefore enabled an estimation of the fishing closure effect on the exploitation of the stock (Jouffre et al., 2002), and the environmental and fishing effects on the cohorts were then established (Thiaw et al., 2011). VPA has also been used to assess stocks of the English Channel loliginid squids (Royer et al., 2002) and northwest Atlantic *D. pealeii* (Cadrin and Hatfield, 1999) indicating that exploitation levels of these resources were above the optimum at the time. In loliginid squids, length–frequency is not the most suitable tool to define monthly age classes but is used for practical reasons. Indeed, age determination based on statolith rings (Challier et al., 2005b, 2006b) is time-consuming and cannot be routinely implemented for a stock assessment trial. Further analysis suggested the introduction of interindividual growth variability into the cohort analysis (Challier et al., 2006a). VPA was also applied as a trial to the English Channel cuttlefish stock (Royer et al., 2006) to estimate fishing impact on the studied cohorts and interactions between fishing fleets involved in the cuttlefish exploitation in the English Channel (mainly trawlers and trappers). As in the loliginid squid case, estimating age classes using statolith rings is feasible until the age of 240 days (Bettencourt and Guerra, 2001) but time-consuming and was mainly used to explore the recruitment period (Challier et al., 2002, 2005a). However, at ages greater than 240 days, no tool is currently able to estimate monthly age of the cuttlefish and length–frequency is the only suitable methodology. An age-based cohort analysis was developed to assess *I. illecebrosus* (Hendrickson and Hart, 2006), primarily to estimate natural mortality, particularly during the spawning season when this mortality increases greatly.

4.4.4 Two-stage models

When the data are not accurate enough to implement a VPA but are accurate enough to distinguish a recruitment period and a fully recruited phase (Collie and Sissenwine, 1983), the two-stage model can be considered as a solution and can give consistent results with a VPA fitted with unsuitable age data (Mesnil, 2003). The model has also the advantage to be able to be fitted using several time series (Roel and Butterworth, 2000) and can be implemented using indifferently numbers or biomass. In theory, the population is modelled from the recruited stage when abundance is estimated by adding the recruitment strength to the fully recruited individuals. According to Collie and Sissenwine (1983) and Mesnil (2003), the population dynamics can be modelled using the following equation:

$$N_{t+1} = (N_t + R_t)e^{-M} - C_t e^{-M(1-\tau)}$$

where N_t is the population size in number of fully recruited animals at the beginning of the year t, R_t is the population size in number of recruits at the beginning of the year t, C_t is the catch in number during the year t, M is the instantaneous natural mortality rate assumed to be equal for the different stages and τ is the fraction of the year when the catch is assumed to occur as a pulse. An alternative version of the model exists in biomass and uses an aggregated instantaneous growth and natural mortality parameter (Gras et al., 2014).

The two-stage model is often used to assess invertebrate stocks (Cadrin et al., 1999; Collie and Kruse, 1998; Conser, 1991; Conser and Idoine, 1992; Zheng et al., 1997). In cephalopod stocks, the first trial was made on the chokka squid (*L. reynaudii*) stock exploited since the 1980s (Roel and Butterworth, 2000) off South Africa. It was then updated by adding process error using Bayesian methodology to the observation error (Glazer and Butterworth, 2006). The two-stage model was also fitted to assess the English Channel cuttlefish using four different abundance indexes coming from both fishery-dependant and fishery-independent data and enabled to highlight a correlation between the environmental conditions encountered during the early life stages and the recruitment strength that occurs at 1 year old (Gras et al., 2014). An application software was then developed to perform a routine assessment by the International Council for the Exploration of the Sea Working Group on Cephalopod Fisheries and Life History.

4.5. Forecasting methods and general/empirical models

Stock sizes can be forecasted based on the suitable models (e.g. growth–survival model and correlation models) with the results of preseason assessment conducted by using experimental methods like midwater trawl nets to observe the abundance of cephalopods in the juvenile phase (Brunetti and Ivanovic, 1992; Kawabata et al., 2006; Kidokoro et al., 2014) and plankton nets to be used for paralarval distribution surveys (Bower et al., 1999a; Goto, 2002; Murata, 1989; Yamamoto et al., 2007). For the Japanese *T. pacificus* fisheries, paralarval surveys have been conducted for over 40 years (Goto, 2002; Murata, 1989; Okutani and Watanabe, 1983). In the original plan, these surveys were conducted in order to forecast the stock size in the next year. However, the paralarval densities obtained in these surveys had weak power to predict the stock size in the next year while appearing highly related with the spawning stock size (escaped population number) in the former generation (Goto, 2002; Murata, 1989). These results meant that paralarval densities have almost the same power to forecast stock size in the next year as the forecasting methods based on spawner–recruitment relationships because survival rate in the prerecruit stages is highly variable. If cephalopod prerecruit abundance can be precisely estimated just before a fishing season opens, stock size can be estimated by the abundance of prerecruits using a simple model (e.g. proportional model). Surveys targeted for prerecruit abundance using midwater trawl nets have been conducted in the Japanese *T. pacificus* fisheries for a decade (Kidokoro et al., 2014). Midwater trawl surveys for the quantitative assessment of oceanic cephalopod populations are likely to provide serious underestimates of population density as well as biased size–frequency and species composition (Boyle and Rodhouse, 2005; Wormuth and Roper, 1983), resulting mainly from net avoidance. Therefore, the target of the midtrawl surveys for *T. pacificus* is limited in the individuals ranging mainly 3–10 cm ML size class, which are not the stages with strong swimming power (Kidokoro et al., 2014). However, net avoidance rate remains unclear, so stock size is not quantified using data obtained in these surveys. Instead, stock size is shown as an index (e.g. average individuals/tow) obtained in these surveys. The relationship between the results of preseason assessment and stock size of recruits may not be proportional because these preseason assessments usually contain large observation errors. In the example of *T. pacificus* assessments, this relationship is fitted with a linear model with large intercept (Kidokoro et al., 2013). Although there are some problems and difficulties in forecasting methods based on preseason assessments, these methods have the advantage that they do not require historical data, which can be a useful feature for newly assessed cephalopod stocks.

Data from in-season assessments and knowledge about seasonal shifts in fishing grounds with migration patterns can be used to forecast catches on individual fishing grounds. In *T. pacificus* fisheries, detailed migration patterns have been examined for a long time because of the need to forecast catch condition at each fishing ground (Kasahara, 1978; Kidokoro et al., 2010; Murata, 1989; Okutani, 1983). Fishing grounds of *T. pacificus* shift seasonally according to migration routes (Figure 2.13). Therefore, forecasting methods for the catches at each fishing ground have been examined by the relationship among catches in each fishing ground and relationship between

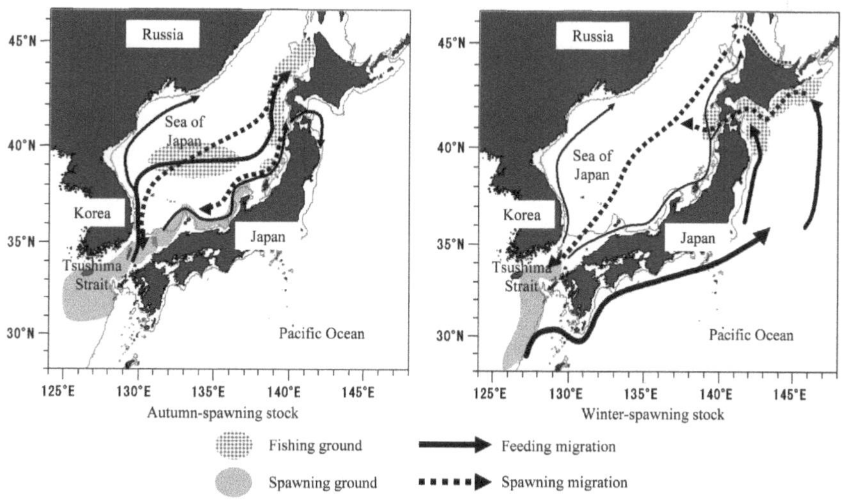

Figure 2.13 Schematic diagrams of the migration routes of *T. pacificus* autumn cohort and winter cohort. *Modified from Kidokoro et al. (2010).*

oceanographic conditions and CPUE (Kasahara, 1978). The site of fishing grounds of jigging fisheries and their oceanographic conditions are able to be monitored by the satellite images (Kiyofuji and Saitoh, 2004; Rodhouse et al., 2001), so it may be easy to understand the distribution area of squids, which is highly related with fishing grounds of squid jiggers. The ocean dynamics models (e.g. Regional Ocean Modeling System http://www.myroms.org/; Research Institute for Applied Mechanics Ocean Model http://dreams-i.riam.kyushu-u.ac.jp/vwp/) enable forecasts of oceanographic conditions at high resolution (mainly temperatures), which has become widely applied in fisheries. Based on these methods, shifts in fishing grounds can be forecasted in the next week or next month (Figure 2.14), with knowledge about the relationship of the distribution of cephalopods species to oceanographic conditions. Such forecasts of the distribution of fishing stocks (Figure 2.14) are considered to be useful for fishers to search fishing grounds at low cost.

Reliable spawner–recruitment relationships are quite important and useful for stock forecasting and management of cephalopod stocks. However, there is no clear relationship between spawning stock abundance and subsequent recruitment in cephalopod stocks (Basson et al., 1996; Pierce and Guerra, 1994; Uozumi, 1998). Annual variability in oceanographic conditions causes recruitment variability (e.g. Dawe et al., 2000; Waluda et al., 2001a), which leads spawner–recruitment relationships to be unreliable. In some cases, cephalopod spawner–recruitment relationships have been fitted into Ricker (1975) and Beverton and Holt (1957) models for forecasting recruitment in the next year (Agnew et al., 2000; Kidokoro, 2009). However, estimation parameters in these nonlinear models using spawner–recruitment data usually contain statistical problems that are mainly derived from observation errors in the explanatory variables (Walters and Martell, 2004). These problems lead to a tendency for stock size to be overestimated particularly in the case of low spawning stock size, which tends to mislead stock management strategies (Walters and Martell, 2004). State-space models that can be used to estimate both process and observation error may have promise for parameter estimation in stock dynamics models such as spawner–recruitment relationships (Bolker, 2008).

Empirical models based on oceanographic conditions data are often used to forecast recruitment strength for cephalopods, particularly for ommastrephid species. In the empirical models, recruitment strength is usually forecasted based on the oceanographic indexes that are often available as a long time series. Although these empirical models may be useful,

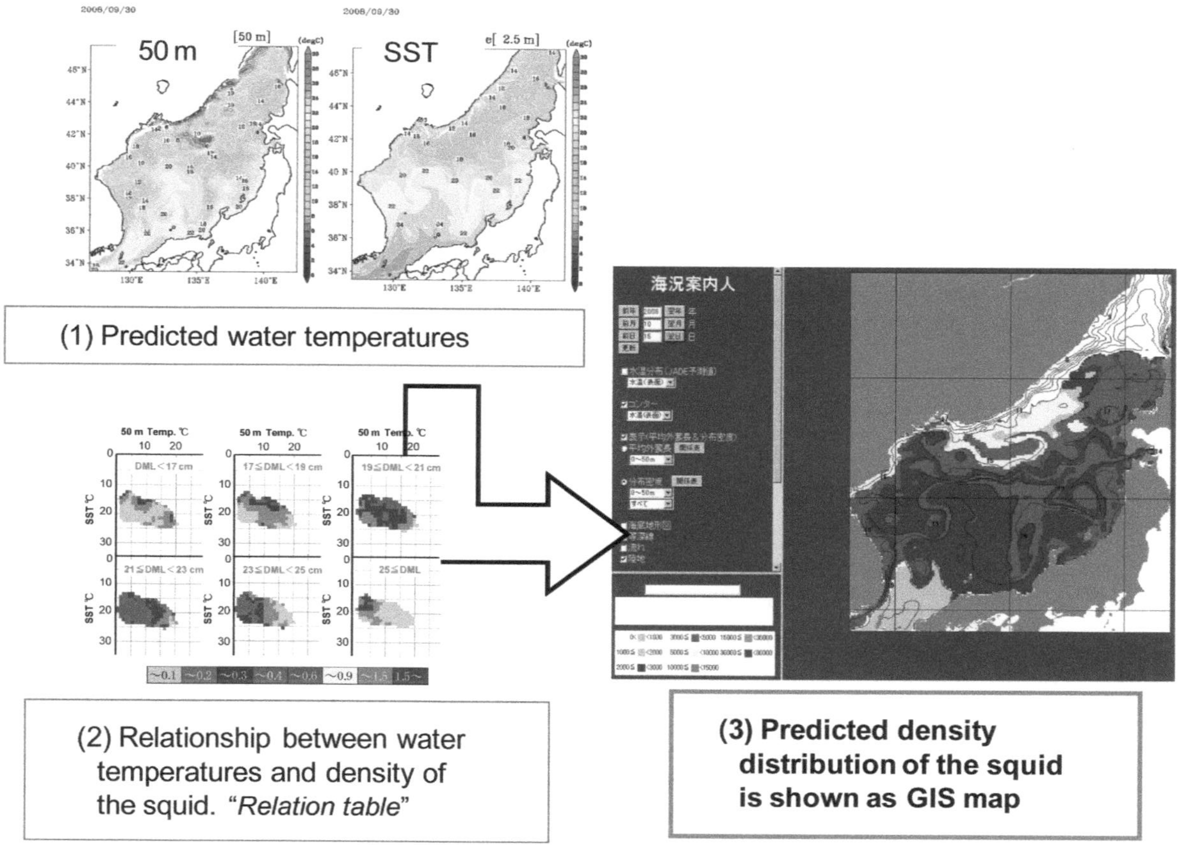

Figure 2.14 Forecasting system for distribution of *T. pacificus* in the Sea of Japan. In this system density distribution of *T. pacificus* (number of individuals/km^2) are water temperatures of sea surface and 50 m deep which are forecasted based on data assimilation ocean dynamic model (http://jade.dc.affrc.go.jp/jade/). Density distribution of *T. pacificus* (number of individuals/km^2) is able to be forecasted within the next month and can use freely through net work (http://jsnfri.fra.affrc.go.jp/shigen/kaikyo2/).

mechanisms explaining how these oceanographic conditions influence recruitment variability are necessary for reliable forecasting. Recruitment strength for *D. gigas* is highly correlated with ENSO events. Commercial catches of *D. gigas* off Peru tend to be small in the year of El Niño when primary productivity tends to be low (Waluda and Rodhouse, 2006). These features are useful to apply forecasting methods for recruitment strength. Distribution ranges (Field et al., 2007) and body size (Keyl et al., 2011) varied largely during the recent 20 years along with recruitment variability in *D. gigas*, and it is important to determine how oceanographic conditions affect such changes.

Oceanographic conditions around spawning grounds are often identified as a contributing factor in stock size fluctuations of *Illex* species (O'Dor, 1998b; Waluda et al., 2001a) and *T. pacificus* (Okutani and Watanabe, 1983; Sakurai et al., 2000). Recruit success of *I. illecebrosus* is considered to be related to larval transport by the Gulf Stream (Dawe et al., 2000, 2007), which can be an indicator for forecasting recruitment strength. Stock size fluctuations of *T. pacificus* are hypothesized to be influenced by favourable conditions for spawning grounds inferred by the ideal water temperature as estimated with rearing experiments (Sakurai et al., 1996). This hypothesis was verified by comparison with the relationship between stock size and variations in oceanographic conditions and inferred spawning grounds (Rosa et al., 2011; Sakurai et al., 2000) and may be particularly useful for forecasting stock size at decadal or interdecadal timescales.

In the Japanese stock management procedure for *T. pacificus*, annual total allowable catch (TAC) is set based on an allowable biological catch (ABC), which is calculated from the fishing mortality (F_{lim}) and forecasted stock abundance in a target year. In this procedure, a spawner–recruitment relationship is used to forecast stock size in the target year, which is composed of new recruitment only. In the spawner–recruitment relationship for *T. pacificus*, proportional models were applied to forecast recruits and estimate a biological reference point (F_{med}) while avoiding overestimation derived from biases pointed out by Walters and Martell (2004). In a proportional model in which a density-dependent effect is ignored, estimates of stock size are going to be relatively high (sometimes unreliable) but are considered to be safer using parameters estimated from proportional models than using those from density-dependent models (Hiramatsu, 2010).

Decadal or interdecadal changes in environmental conditions are assumed to influence stock status and spawner–recruitment relationships of *T. pacificus* (Kidokoro, 2009; Kidokoro et al., 2013; Sakurai et al.,

2000; Yamashita and Kaga, 2013). Therefore, the parameters used in spawner–recruitment relationships are estimated from data collected since 1990 following an apparent regime shift (Hare and Mantua, 2000), but when the current regime changes, such parameters should be revised accordingly. Unfortunately, the exact mechanism with which changing environmental conditions influence the stock size of *T. pacificus* remains unclear, and it is difficult to predict when regime shifts might occur. However, the results of investigations show that spawning grounds (Goto, 2002), migration routes (Kidokoro et al., 2010; Nakata, 1993) and body size (Takayanagi, 1993) all show changes that coincide with changing stock size. These changes are assumed to be closely connected with changing environmental conditions (e.g. regime shift). Therefore, forecasts of oceanographic conditions that may be favourable or unfavourable for stock size based on changes in ecological traits may be easier to observe relative to estimation of stock size itself. It remains necessary to understand the mechanism with which these ecological changes affect stock size in order to allow better forecasts of future trends in stock size. However, caution must be exercised when deriving empirical stock–environment relationships in the absence of long time series (Solow, 2002).

In their in-depth review of cephalopod stock assessment methods, Pierce and Guerra (1994) noted the promise of time-series models incorporating environmental information and multispecies models incorporating trophic dynamics. Georgakarakos et al. (2006) applied autoregressive integrated moving average techniques, artificial neural networks and Bayesian dynamic models incorporating environmental factors to forecast squid landings in Greek waters. Using a static mass-balance model, Gaichas et al. (2010) demonstrated the utility of food web-derived predation information to inform stock assessments of incidentally caught squids for which predation mortality exceeds fishing mortality.

4.6. Fishery-dependent assessment data

Stock assessment methods incorporate several types of data, including abundance indexes and biological data (e.g. age, length and maturity), which can be collected from fisheries (fishery-dependent) or fishery-independent methods such as surveys (Hilborn and Walters, 1992). Many assessed cephalopod fisheries incorporate data from multiple sources. For example, the northwest Atlantic *D. pealeii* and *I. illecebrosus* fisheries are assessed using fishery-independent resource assessment trawl survey data, supplemented by a host of fishery-dependent data, including standardized landings per unit effort (LPUE), size composition and discard data collected on board commercial fishing vessels (Johnson, 2011; NEFSC, 2011). Daily catch reports, fisheries observer data and electronic logbook data are all incorporated into assessments of *D. gahi* and *I. argentinus* off the Falkland Islands (Arkhipkin et al., 2013).

4.6.1 Landings/catch data

To fit a population dynamics or stock assessment model, a prerequisite is to collect suitable abundance indexes via fishery-independent surveys or by fishery-dependent catches. Fishery-independent surveys are generally designed to be standardized and data collected can be considered to be representative of trend in stock abundance (Hilborn and Walters, 1992), but fishery-independent data are generally available for a short period (or not at all in small-scale fisheries) and abundance indexes for the rest of the year are often derived from commercial fisheries (CPUE or LPUE). Effort data can be difficult to obtain for small-scale, diverse artisanal fisheries; data collected from sales at individual ports and auction sites can be used to generate CPUE indexes at a higher resolution than that possible with aggregated catch data (Lourenço and Pereira, 2006). In small-scale cephalopod fisheries, interviews can be a suitable method for collecting catch and effort data in the absence of the infrastructure noted in the preceding text (Young et al., 2006). However, in the absence of sufficient biological data collection, misidentification of similar species with different life histories or sex-specific catchability such as that observed for some octopus species can confound the use of CPUE to derive abundance indexes (Leporati et al., 2009). Even in some relatively well-sampled cephalopod fisheries, for example, those targeting California market squid (*D. opalescens*) and *D. pealeii*, management measures (catch limits) are set based on historical landings data in the absence of a functional assessment model or when confidence in assessment models is low (NEFSC, 2011; Zeidberg et al., 2006).

4.6.2 Abundance index standardization

A challenge when using fishery-dependent data (CPUE) to derive abundance indexes is that fishing fleets are generally heterogeneous and fishing performance may vary with time. CPUE should therefore be standardized considering that resource catchability can vary according to spatiotemporal variables (such as fishing area, years and seasons) as well as vessel class variables (vessel length or engine power). The abundance index standardization was developed by Allen and Punsly (1984) for the Inter-American Tropical

Tuna Commission. Hilborn and Walters (1992) made a more general presentation of the methodology using a Gaussian error generalized linear model (GLM; McCullagh and Nelder, 1989) to statistically model the log-transformed CPUE using various explaining variables mentioned in the preceding text. However, this methodology is difficult to use when the data set is composed by numerous null values as is the case in multispecies fisheries such as trawl fisheries. In this case, a Delta-GLM enables modelling of resource presence/absence using a binomial error GLM and, at the same time, modelling of resource abundance using a Gaussian error GLM (Acou et al., 2011; Fletcher et al., 2005; Gras et al., 2014; Le Pape et al., 2003, 2007; Rochette et al., 2010; Stefansson, 1996; Syrjala, 2000). The explaining variables used in the Delta-GLM can be similar to those mentioned previously.

The GLM methodology was used to standardize CPUE for various species of octopods, *O. vulgaris* (Diallo and Ortiz, 2002; Erzini et al., 2005; Tserpes and Peristeraki, 2002), *O. pallidus* (Leporati, 2008), *E. cirrhosa* and *E. moschata* (Belcari et al., 2002). This methodology was also used to standardize loliginid CPUE (Cao et al., 2011; Glazer and Butterworth, 2002, 2006; Roel and Butterworth, 2000; Roel et al., 2000; Royer et al., 2002; Tserpes and Peristeraki, 2002), and *Illex* CPUE (Chen et al., 2007). Finally, several studies performed the CPUE standardization for cuttlefish (Erzini et al., 2005; García-Rodriguez et al., 2006; Royer et al., 2006; Tserpes and Peristeraki, 2002). In the NW Pacific Ocean, Tian et al. (2009a) found that using various spatial scales to aggregate *O. bartramii* CPUE influences the CPUE standardization. Moreover, Tian et al. (2009b) studied the influence of three groups of variables, spatial (longitude and latitude), temporal (year and month) and environmental (sea surface temperature, sea surface salinity and sea level height), concluding that month was the most important variable influencing CPUE. Standardization using the delta-GLM methodology has been used to standardize the English Channel cuttlefish LPUE derived from French and UK bottom trawl fisheries prior to fitting a two-stage biomass model (Gras et al., 2014). Increasingly, generalized additive models (GAMs; Hastie and Tibshirani, 1986) are being used to complement or supplant GLMs for abundance index standardization (Venables and Dichmont, 2004), including applications to cephalopods (e.g. Tian et al., 2009b).

4.7. Fishery-independent data

Fishery-independent data used in cephalopod stock assessments can be collected in a variety of ways, including paralarval surveys, trawl surveys, jigging surveys and acoustic surveys.

4.7.1 Paralarval surveys

Paralarval surveys have been carried out to reveal spawning and nursery grounds for certain species in relation to annual variation in the marine environment, to understand population dynamics and to develop stock–recruitment models for commercially important species (Ichii et al., 2011). The most precise method of collecting cephalopods for estimating distribution and abundance is by sampling paralarvae in their early life stage, because conventional estimation techniques such as trawling for mature adults have biases due to avoidance and escapement from net openings (Bower et al., 1999b; Vecchione, 1987). The density of cephalopod paralarvae is relatively higher and more easily sampled than adults.

A large variety of sampling gear has been used for capturing cephalopod paralarvae. Among them, the gear used for collecting plankton is used in many cases, such as a ring net (Bower, 1996; Bower et al., 1999b; Goto, 2002), a Nansen net (Baron, 2003) and a bongo net (González et al., 2005; Jorgensen, 2007). A neuston net that collects at the surface (rectangular midwater trawl; Hatfield and Rodhouse, 1994; Vidal et al., 2010b) used for sampling larger paralarvae or juveniles and multilayer samplers such as a MOCNESS (Multiple Opening/Closing Net and Environmental Sensing System; Moreno et al., 2009) and BIOMOC (BIOlogical Multiple Opening/Closing; Diekmann et al., 2006) have also been applied to obtain paralarval samples.

There are three general categories of sampling gear deployment: horizontal tow, vertical tow and oblique tow. Horizontal tows are done when a vessel is moving at slow speed by attaching a weight to the net to ensure that the net is in a horizontal pattern. The length of the wire, vessel and winch speed are used to control the depth of the sampling gear. Vertical tows are usually used to study a particular layer and suitable to collect smaller paralarvae or eggs. Oblique tows combine features of both vertical tows and horizontal tows, which can sample from a desired depth layer between the surface and any depth (e.g. several metres off the bottom) while a vessel is running. It is done by slowly releasing the sampling gear from the surface to the given depth and then towing at that depth for a while before pulling the net towards the surface obliquely. A calibrated flowmeter is attached to the net mouth of sampling gear to measure the volume of water that passes through the net to quantify and standardize density. To confirm the accurate towing depth, a depth data logger or real-time depth sensor may be attached to the sampling gear.

Hatfield and Rodhouse (1994) used bongo net and RMT to determine the distribution and abundance of paralarvae of *D. gahi* around the Falkland

Islands, and the distribution on the coastal shelf was associated with water-column structure. Bower (1996) estimated ages and hatch dates of *O. bartramii* sampled by ring net near the Hawaiian Islands and inferred spawning sites from hatch dates by back calculating with physical data on the speed of ocean current. Bower et al. (1999b) sampled paralarvae of 58 cephalopod species using ring nets near the Hawaiian Islands and classified onshore and offshore distribution patterns of each species based on distance from the Island. Goto (2002) revealed that the extent and range of area of suitable habitat for spawning of *T. pacificus* was related to the adult stock size by analysing a long time series (1972–1999) of paralarval sampled by ring net.

González et al. (2005) found that the distribution and movement of *O. vulgaris* and *L. vulgaris* paralarvae collected in the western coast of the Iberian Peninsula followed the oceanographic circulation system associated with wind-driven seasonal upwelling. Vidal et al. (2010b) had shown that the patterns of distribution and density of *I. argentinus* paralarvae and small juveniles could be linked to oceanographic conditions (sharp pycnoclines) and high primary production in an upwelling area off southern Brazil.

It is critical to understand the biology and ecology of early life stages before routine quantitative sampling for paralarvae. The identification of paralarvae to species is essential. Morphological characteristics such as morphometrics and patterns of chromatophore distribution are usually useful for identifying species (Baron, 2003; Jorgensen, 2007). A molecular genetic method has also been employed for species-level identification of cephalopod paralarvae (Gilly et al., 2006; Wakabayashi et al., 2006).

It has also been reported that avoidance from the mouth of sampling gear in the survey of paralarvae may occur, as documented for adult and juvenile cephalopods (Collins et al., 2002; Haimovici et al., 2002). In order to collect more quantitative samples of paralarvae, larger openings of the net mouth and other modifications may be required, as well as an examination of the collection efficiency of each gear.

4.7.2 Trawl and jigging surveys

Assessing juvenile and adult stocks of cephalopods is a difficult task, and only a few assessments in the world have used fishery-independent data obtained by midwater or bottom trawl and jigging surveys. Nevertheless, long-term seasonal stratified random bottom trawl survey data have been used in the assessments of *I. illecebrosus* and *D. pealeii* in the continental shelf and slope waters of the northwest Atlantic Ocean (Hendrickson, 2004; NEFSC, 2011). Pierce et al. (1998) estimated distribution and abundance of

L. forbesii from length–frequency data collected during demersal trawling surveys in Scottish waters and denoted that abundance from the February survey was suitable for indicator of commercial catch rate in the autumn of the same year. Litz et al. (2011) used pelagic and midwater trawl nets to estimate the horizontal distribution and abundance of *D. gigas* in the northern California Current system and reveal the relationship between physical oceanographic features and spatial distribution of the squid and its predator or prey. As trawl surveys are usually used to sample juveniles or adults, net avoidance and catch efficiency of the used gear need to be considered. Cod-end mesh size of sampling gear may be an important factor in estimation of length–frequency and abundance; several studies have examined trawl cod-end selectivity for the target species (Fonseca et al., 2002; Hastie, 1996; Hendrickson, 2011; Ordines et al., 2006; Tosunoğlu et al., 2009).

The jigging method is a common way to catch squids and cuttlefish in a commercial fishery, but few studies using this method have been performed for the purpose of biomass estimation. One of the most important reasons for this is that colours and sizes of jigs and survey timing affect size selectivity and catch rate significantly because the jigging method employs a passive fishing gear exploiting the feeding behaviour of squids (Mercer and Bucy, 1983). Another problem with quantitative jigging surveys is the use of artificial lights to aggregate squid, so catch efficiency is affected by the spread and intensity of fishing light. Nonetheless, for the stock assessment of the Japanese common squid (*T. pacificus*), the results of a jigging survey (catch per time effort and number of jigging machines (CPUE)) and number of fishing machines) have been used for an abundance index (Kidokoro et al., 2013). Several studies of *D. gigas* in the Gulf of California have been conducted using jigging sampling for stock assessment (Nevárez-Martínez et al., 2000; Robinson et al., 2013). Nevárez-Martínez et al. (2000) estimated biomass of *D. gigas* by stratified random sampling and swept area by strata using jigging survey data, which covered a grid of stations.

4.7.3 Acoustic surveys

For several decades, significant advances have been made in acoustics as a direct stock assessment tool for marine species (Stanton, 2012). In particular, assessments of biomass using a quantitative echosounder have been widely used for various fish species (Koslow, 2009; Simmonds and MacLennan, 2005). The advantages of using acoustic methods over other traditional methods are the following: (1) nearly the whole vertical distribution can be obtained quickly; (2) horizontal extent is continuous along a survey line; and (3) data resolution is high, that is, less than a metre vertically and tens of metres horizontally. Although several early studies of cephalopod stocks using acoustic methods were conducted (Shibata and Flores, 1972; Starr and Thorne, 1998; Suzuki et al., 1974), these surveys tended to be qualitative rather than quantitative. Focusing on the more quantitative aspects, Goss et al. (2001) conducted two-frequency acoustic surveys for *D. gahi* around the Falkland Islands. These surveys revealed the potential of separation of squids from finfish using dual-frequency acoustics, but the accurate biomass of this species was not estimated because the information about target strength, which is an essential parameter for the estimation of stock assessment using echo integration method, was not well enough understood. More recently, quantitative surveys for estimating cephalopod abundance have been conducted using acoustic equipment after obtaining reliable target strength measurements (Benoit-Bird et al., 2008; Kang et al., 2005; Kawabata, 2005; Mukai et al., 2000; Soule et al., 2010). Distribution and density of *T. pacificus* were estimated by acoustic survey using a quantitative echosounder off the Sanriku Coast of Japan in the western North Pacific Ocean and compared with a catch rate index of commercial fishery (Kawabata, 2005). These results showed that both density estimations almost agreed and demonstrated that direct abundance estimation of this species using the acoustic method is possible. A sequence of acoustic studies for estimating the biomass in spawning schools of *L. reynaudii* have been conducted over 20 years off the southeast coast of South Africa (Augustyn et al., 1993; Lipinski and Soule, 2007; Soule et al., 2010). These surveys have combined acoustic and trawl methods to obtain the abundance of inshore spawning aggregations and deeper offshore dispersed aggregations, respectively. A combination of both methods enables an estimate of the total abundance of mature squids during the spawning season. Acoustic surveys for estimating the distribution and abundance of *D. gigas* and the lightfish (*Vinciguerria lucetia*), which is one of the important prey for *D. gigas*, conducted in the Humboldt Current system off Peru indicated that spatial and temporal distribution and abundance between both species were similar (Rosas-Luis et al., 2011).

One of the most important tasks facing an acoustic assessment is species classification and identification of acoustically detected targets. The identification of acoustic targets on the echogram is typically inferred from sampling results that include species composition, length–frequency and other fundamental biological parameters. Other supplementary information such as detected depths in the water column, geographical location, survey timing and knowledge of the species' habits is also essential and helped identify acoustic echo signals. For the acoustic approaches to noninvasive species identification, multifrequency acoustics have been used for not only fish–plankton separation (Kang et al., 2002; Swartzman, 1997) but also squid–fish separation (Goss et al., 2001). Recent advances in broadband acoustics, for example, dolphin mimetic sonar, may provide more accurate species identification and classification techniques (Imaizumi et al., 2008; Stanton et al., 2010), and broadband acoustic signal characteristics of live *D. pealeii* have already been estimated (Lee et al., 2012).

Extensions of traditional fisheries acoustics approaches are acoustic–optical platforms that combine traditional echosounders with cameras to allow simultaneous target detection, species identification, enumeration and target strength estimation (Miksis-Olds and Stokesbury, 2007; Sawada et al., 2004) and the application of high-frequency sonars, or "acoustic cameras", that can detect and identify individual squid in darkness and turbidity (Belcher et al., 2001; Iida et al., 2006). The acoustic–optical platform employed by Miksis-Olds and Stokesbury (2007) has been used to quantify the abundance of *D. pealeii* captured in shallow-water fish traps with some success, particularly when squids were a high percentage of total catch.

4.8. Way forwards for forecasting and assessment

It is difficult to forecast stock size for most cephalopod stocks, often rendering management unreliable. In many cases, it may be most effective to combine forecasting, monitoring and revising estimated stock size using the methods of preseason, in-season and postseason assessments. For example, in the case of *T. pacificus* stock management, an original TAC is set based on the stock size forecasted from postseason assessment data (e.g. spawner–recruitment relationship), and preseason assessments are conducted just before fishing season opens to forecast stock size again. If the results of preseason assessments are quite different from the forecasted stock size, it will be revised. After the fishing season opens, if real-time monitored data are quite different from the forecasted stock size used to set TAC, then management measures can be adapted accordingly. Indicator-based approaches such as the "traffic light" approach employed by Ceriola et al. (2007) may also provide a dynamic means of generating timely information for fishery managers.

Many recent developments have been made in stock assessment models that incorporate environmental processes, including those that affect population dynamics in the form of recruitment, and spatially explicit effects on distribution and migration patterns (Keyl and Wolff, 2008). As the body of research on environmental effects on cephalopod distribution and abundance grows, so too will the potential for application of such techniques. Similarly, as cephalopod stock identification techniques (e.g. tagging and genetics) improve (Semmens et al., 2007), the potential for application of the latest generations of spatially explicit stock assessment models (Cadrin and Secor, 2009; Goethel et al., 2011; Keyl and Wolff, 2008) will increase. Multispecies models or extensions of single-species models that incorporate trophic dynamics may improve assessments for some cephalopod species and are particularly timely with increasing emphasis on ecosystem-based approaches to fisheries management (e.g. Gaichas et al., 2010; Moustahfid et al., 2009).

Recent advances in data-poor stock assessment methods (Pilling et al., 2008; Starr et al., 2010) may have application for cephalopod species. Accounting for environmental effects on spatiotemporal distribution at multiple scales has implications for stock identification, survey design and interpretation of survey and landings data. For example, a revised understanding of stock structure based on an understanding of environmental effects on distribution has direct implications for assessment and management (e.g. *I. argentinus* off Patagonia; Crespi-Abril et al., 2013). At finer scales, a knowledge of habitat preference can provide data with which to design or interpret results of seasonal resource assessment trawl surveys such as those used to provide data to assess the *D. pealeii* stock in the western North Atlantic (Manderson et al., 2011). Environmental effects on cephalopod availability to surveys can be modelled as environmentally influenced variations in catchability (e.g. Fréon and Misund, 1999; Maunder and Watters, 2003), an important parameter in many stock assessment models that relates an index of abundance to population size (Arreguín-Sánchez, 1996; Hilborn and Walters, 1992; Wilberg et al., 2010). Adjustments can also be made to survey data before input to assessment models by incorporating data on environmental effects on distribution (e.g. Brodziak and Hendrickson, 1999). Environmentally induced distributional shifts render landings data difficult to interpret and create management challenges as new fisheries develop (e.g. *D. pealeii* off the northeast United States; Mills et al., 2013); this is particularly important for developing assessment and management strategies to address the effects of climate change (Link et al., 2011; Pinsky and Fogarty, 2012). Continued participation of fishing communities in data collection, stock assessment and management is critical for the sustainability of cephalopod fisheries (Arkhipkin et al., 2013; Johnson, 2011), particularly with respect to the performance of in-season monitoring and assessment (Walters and Martell, 2004). Advances in acoustic and optical survey techniques for benthic cephalopod egg masses (Young et al., 2011; Zeidberg et al., 2012) may have application as a fishery-independent index of future abundance of neritic species as well as a means of delineating important spawning habitat for protection.

5. MANAGEMENT AND GOVERNANCE

Globally, over half of all fish stocks are exploited. Of these, only 20% can be said to be moderately or underexploited, having the potential to expand (MRAG, 2010). Of those yielding less than their potential, 8% are depleted, 1% are recovering and 19% are overexploited (MRAG, 2010). According to MRAG (2010), the percentage of overexploited, depleted and recovering stocks has tripled since the 1970s. Worm et al. (2009) reported an 11% decline in total biomass across all ecosystems of regularly assessed stocks. Research survey data (targeted and nontargeted species) indicated a 32% decline in total biomass, a 56% decline in large demersal fish biomass (species $\geq$90 cm maximum length), 8% decline for medium-sized demersal fish (30–90 cm) and 1% decline for small demersal fish ($\leq$30 cm), whereas invertebrates biomass increased by 23% and pelagic species by 143% (Worm et al., 2009). As mentioned by Worm et al. (2009), these increases are likely due to decreases in the predator population.

The depletion of many finfish species throughout the world over the last few decades has led to an increase in the commercial importance of cephalopods (Chen et al., 2008; Pierce et al., 2010; Young et al., 2004). Cephalopods seem to be one of the remaining marine resources, in some areas, that still experience an increase in landings (Boyle and Rodhouse, 2005). This has led to both the exclusive target of cephalopods using a variety of gear types and the increased targeting of cephalopods by fisheries traditionally targeting finfish. For example, the Indian trawl fleet started targeting cephalopods along the west coast of India during certain seasons (Meiyappan et al., 2000), while the Tasmanian arrow squid, *N. gouldi*, fishery expanded rapidly between November 1999 and February 2000, requiring the immediate closure of state waters to the majority of large-scale automatic jig operators (Willcox et al., 2001). In the coastal waters of Africa, there has been an expansion of the foreign trawl fishery and a high international demand for octopus, resulting in overfishing in some instances (Sauer et al., 2011). However, the development of local pot fisheries and adequate management based on economic analysis are being developed in some cases (Oosthuizen, 2004; Raberinary and Peabody, 2011). Since 2003, industrial fishing using longlines of pots targeting *O. vulgaris* and *O. insularis* has developed in southern and northeastern Brazil (Barahona et al., 2010; VASCONCELLOS et al., in press), and large-scale international fisheries for jumbo flying squid (*D. gigas*), Argentine short-fin squid (*I. argentinus*) and neon flying squid (*O. bartramii*) have also developed beyond the 200 miles territorial limit with a concomitant lack of management.

5.1. General management challenges

The increase in exploitation of cephalopods has resulted in a number of management challenges, with managers faced with not only the unique aspects of cephalopod biology but a large selection of management strategies and tools available for teleost stocks, including EAF, integrated coastal zone management, marine protected areas (MPA), balanced harvesting and various input and output control tools. In deciding on appropriate management strategies and measures, a broad understanding of the current management frameworks available is essential.

Several management regimes exist for cephalopod fisheries, ranging from co-management regimes to rights-based management. A rights-based (access rights and withdrawal rights) approach is now broadly accepted as the most successful form of management for marine stocks, with a number of options applicable to cephalopods. As summarized by Charles (2009), access rights include both territorial use rights for fisheries (TURFs) where rights are assigned to individuals and/or groups to fish in certain locations, generally, although not necessarily, based on long-standing tradition, and limited entry access rights where governments issue a limited number of licences to fish. Withdrawal rights include quantitative input or effort rights. Effort-based use rights are where each fisher has the right to use a specified amount of gear or fish for a certain time period. In some instances, all fishers may have equal quantitative rights within a fishery or the rights may vary dependent on location, boat size or other criteria. Quantitative output rights or harvest quotas require the subdivision of TAC into quotas allocated to sectors of the fishery, individual fishers or communities. Harvest rights can also be allocated as trip limits, while quotas can be further subdivided into individual transferable quotas (ITQs) or individual nontransferable quotas (INTQs).

Co-management systems are also popular, particularly in developing countries. In co-management regimes, resource users, government agencies and, sometimes, other stakeholders are responsible for the management of a specific area or set of resources (Gutiérrez et al., 2011; IUCN, 1996). There are many advantages to involving fishers in the management of their activity. Co-management arrangements have the capacity to increase fishers' responsibility and accountability, decrease their scepticism towards management, increase fishers' likelihood of compliance with management policies and decisions and facilitate common understanding and establishing trust between fishers, government bodies and scientists (Coffey, 2005; Guidetti and Claudet, 2010; Jentoft and Kristoffersen, 1989; Jentoft and McCay, 1995; Mikalsen and Jentoft, 2001; Pita et al., 2010). However, true co-management is rare and, in practice, fishers often still have little or no say in management decisions (Mikalsen and Jentoft, 2008). Examples of both rights-based and comanaged cephalopod fisheries are given under Section 5.3, which addresses current management initiatives.

5.2. Limitations to management of cephalopod fisheries

While management measures for finfish can be said to be fairly well advanced, managers responsible for cephalopod fisheries face a number of significant challenges. Cochrane and Garcia (2009) provided a useful set of general guidelines for fishery managers, which are applicable to cephalopods, while Pierce and Guerra (1994) provided a list of the requirements for the successful assessment and management of cephalopod fisheries (see also Section 4). Certainly, a key challenge to successful management is the fact that all species are short-lived, often necessitating in-season assessment and real-time management of cephalopod populations (Pierce and Guerra, 1994). Further complicating management is the understanding of the role of cephalopods in exploited communities and ecosystems, particularly with respect to trophic interactions (Pierce and Guerra, 1994), and the fact that many species may have expanding numbers due to a decline of key predators and therefore thrive in disturbed environments.

Broadly, fisheries management can be divided into three general aims: sustaining the fish stock, sustaining the fishery and sustaining fishery-related employment (Pilling et al., 2008). While the determination of the distribution and boundaries of cephalopod stocks can be said to be a fundamental requirement for fisheries management, the identification of these is often not possible, and many species undertake fairly extensive migrations and have an expanding home range in years of good recruitment. A lack of species discrimination in official statistics, which often lump together species with very distinct distribution ranges and population dynamics (Cavaleiro, 2006), further complicates any attempt to determine stock boundaries. Complicating management is also the fact that many fisheries straddle political boundaries, requiring intergovernmental bodies (Pope, 2009), often not geared towards making rapid decisions around a short-lived resource (e.g. *D. gigas* in the eastern central Pacific, which is exploited in the Peru, Mexico and Chile Exclusive Economic Zone (EEZ).

Almost all management requires good long-term data on catch and effort and key biological information. This is a challenge to many cephalopod fisheries, particularly where cephalopods are caught as by-catch. The collection of appropriate fisheries statistical data by countries involved in cephalopod fisheries often lags that for finfish (Pierce and Guerra, 1994). Total nominal catches, fishing effort and CPUE and biological data are required for both simple and complex assessments of cephalopod stocks (Pierce et al., 2010).

As described in the previous section on assessment and forecasting, stock assessments can be categorized as preseason, in-season or postseason, with preseason assessments used to estimate stock biomass at the start of the fishing season, in-season assessment using incomplete CPUE data to adjust fishing level activity throughout the fishing season and postseason assessment using complete CPUE data sets to establish management goals for the following fishing season. For example, DeLury depletion methods of assessment have been the approach of choice in squid fisheries around the Falkland Islands (e.g. Agnew et al., 1998a; Beddington et al., 1990; Rosenberg et al., 1990), combined with limited entry access rights and quantitative output rights/harvest quotas. With modifications, this has been successfully used for a number of years. Production models (requiring only catch and effort data) have been applied with some success to cephalopod fisheries in the Saharan Bank (Bravo de Laguna, 1989), and modifications of this methodology, including additions for the influence of environmental factors, culminated in the implementation of a management plan in 2006 (Binet, 2012). Relatively accurate estimates of within-season stock size are essential for cephalopod fisheries managed by way of limiting catch (Young et al., 2004); however, the cost of such an undertaking is often prohibitive and not a realistic management option. However, as stated by Cochrane (2002), fisheries managed by way of effort require slightly lower precision as removal rates generated by fishing effort levels can be judged over a series of years rather than on a year-by-year basis.

As for all fishing resources, a comprehensive knowledge of cephalopod species life cycles, particularly with respect to the distribution of spawning sites and early life stages, can be essential when implementing management measures such as gear restrictions and spatial or temporal closures/restrictions. Gear restrictions and closed areas and/or seasons have been used to limit the harvest of specific life stages (e.g. the South African *L. reynaudii* fishery closed season during the peak spawning period), protect genetic reservoirs, protect habitat that is critical for the sustainability of a harvested resource, restrain excess fleet capacity and optimize the value of the catch, by limiting by-catch and protecting attributes of the ecosystem that are critical for its preservation (Cochrane, 2002).

Interactions with other fisheries operating in an area, as well as biological and fisheries information for mixed fisheries and/or by-catch species, further complicate management. Not only cephalopod fisheries using unselective fishing gear, as well as those fisheries where cephalopods form one component of a mixed fishery, need to consider operational interactions with other fisheries, but also management measures taken for one component of the fishery may conflict with measures taken for other components (Meiyappan et al., 1993). In the case of the western North Atlantic fishery for *D. pealeii*, by-catch of finfish in small-mesh trawls has led to management measures such as minimum mesh sizes and fishery closures when by-catch limits are exceeded (Mid-Atlantic Fishery Management Council, 2011).

A global problem in the management of fish stocks is the lack of information on the economic and social dimensions of fisheries with key information for managers often lacking. This is particularly important for both existing and new cephalopod fisheries, where stocks may go through a boom and bust period, and an understanding of the social and economic implications is essential when allocating access rights. For example, cephalopods are important fishery resources for several EU countries such as Spain, Portugal, Italy, Greece, France and the United Kingdom (Pierce et al., 2010). According to FAO (2004), the European market is the most important market in the world for this resource, and most imports go to the south of Europe. Spain and Italy, for instance, are major importers of cephalopods (Pierce et al., 2010). Small-scale, directed coastal fisheries for cephalopods have increased dramatically in Europe, particularly in southern European countries where cephalopods are traditionally consumed. The huge increase in landings over the last few decades is basically due to the replacement of traditional relatively inefficient fishing gears (e.g. clay pots and trammel nets) by more efficient gears, including modernized traditional gears and newly introduced gears such as plastic pots and fyke nets (Borges, 2001; Lefkaditou et al., 2002; Pierce et al., 2010; Young et al., 2006). Inshore local small-scale fishing fleets targeting squid, cuttlefish and octopus in Portugal, Spain, Italy and Greece are of considerable socioeconomic importance in terms of providing employment and income in coastal fishing communities (Pierce, 1999; Pierce et al., 2010; Shaw, 1994).

In Portugal, the economic importance of cephalopods is relatively high, and there is an increasing economic dependence of small-scale fisheries on cephalopods. In Italy, most landings of cephalopods come from bottom trawlers, but a substantial fraction still depends on small-scale and mixed fisheries. Both activities employ an important number of local fishers in many small coastal communities along the Italian coast. Cuttlefish (*S. officinalis*) is one of the most significant marine resources in Lower Normandy, France (Pierce et al., 2010). In the United Kingdom, the most important commercial cephalopod is the cuttlefish, and most catches of this species are from the English Channel. The cuttlefish fishery is based on a combination of trawling and artisanal fishing, the latter exclusively directed at cuttlefish. There is also increased targeting of squid (*L. forbesii)* by small-scale inshore trawlers in the Moray Firth (east coast of Scotland, the United Kingdom) (Young et al., 2006). For instance, the small inshore trawler fleet in Burghead (a small fishing community in the Moray Firth) can make up to 50% of their annual revenue targeting squids for 3–4 months of the year.

Therefore, the effect of management measures, such as effort or catch limitations and spatial or temporal closed seasons, on social and economic objectives needs to be considered. Also important is the effect of management measures on the opportunity window to fish short-lived and fast-growing species. Such measures can have large implications on catchability and therefore reduce profitability by reducing the available biomass production. Measures, often seen as used to secure the biological objectives of a fisheries management plan, must also secure the socioeconomic objectives. Pomeroy and Fitzsimmons (1998) mentioned that both social information and economic information can aid fishery managers by giving them a better understanding of how management measures will be received and affected by individuals involved.

Illegal, unreported and unregulated (IUU) fishing activities have become a global problem. They not only do have far-reaching economic and social impacts (DfID, 2007) but also are harmful to global fish stocks and undermine the effectiveness of management measures adopted regionally, nationally and internationally (Agnew et al., 2009; Berg and Davies, 2002; Schmidt, 2004). IUU has been estimated to account for annual catches of 11–26 million t of fish worth US $10–20 billion (Agnew et al., 2009). Addressing the global extent of illegal fishing, Agnew et al. (2009) calculated illegal and unreported cephalopod catch to be in the region of 25% of reported catch (lower and upper bounds of ~12.5% and ~38.5%, respectively). IUU is particularly difficult to control on the high seas as the only authority able to prosecute a vessel for illegal activities is the vessels own flag state. High seas squid fishing activity has resulted in a number of issues such as unreported catches from high seas waters, the use of illegal gear and the poaching of resources in adjacent EEZ waters (MRAG, 2005). In a review of the impacts of IUU, MRAG (2005) summarized IUU squid fishing activity on the high seas. The southwest Atlantic Ocean high seas fishery for *I. argentinus* has been estimated to catch 50–100,000 t per year. The operation of this fleet, composed of jigging vessels and trawlers from numerous countries, just outside the Argentine and Falkland Islands' EEZ has led to poaching within EEZ waters. Chinese vessels have been reported to use illegal driftnets to target neon flying squid *O. bartramii* in the Pacific Ocean. Off Peru and northern Chile, licensed Japanese and Korean vessels target jumbo flying squid *D. gigas*, mainly within the Peruvian EEZ. Combined, an annual catch of 45,000 t has been recorded. China, a recent entrant into this fishery, has been estimated to catch around 40,000 t from adjacent high seas waters alone. In total, annual high seas IUU squid landings have been estimated to have a value of US $108 million.

Most catches from small-scale octopus and squid fisheries in developing countries go unreported, in part because they are consumed locally and in part because keeping track of artisanal landings is difficult and expensive due to the large number of vessels involved and the geographically widespread nature of their activities.

Localized or regional IUU activity can have equally harmful results. Often, the lucrative nature of short-term gains overrides long-term interests in maintaining the sustainability of the resource (Hauck and Sweijd, 1999). IUU activity reduces the incentives to comply with rules, particularly for legal fishers (Schmidt, 2004). An example of an extreme case of IUU activity (poaching) would be the South African abalone fishery. Not only were management initiatives unsuccessful due to the high financial returns of illegal fishing, but also criminal syndicates developed, fuelling exploitation and providing the means to export the product (Hauck and Sweijd, 1999). The commercial South African abalone fishery has since been closed.

Therefore, apart from scientific, economic and social information, compliance with conservation-based measures is also essential to the proper management of fishery resources (Berg and Davies, 2002). Fishery monitoring, control and surveillance (MCS) contributes to good fisheries management by ensuring that appropriate controls are set, monitored and complied with (Berg and Davies, 2002). MCS involves both a preventative approach, encouraging voluntary compliance through understanding and support of management strategies by communities/fishers, and a deterrent/enforcement approach that ensures the compliance by fishers who resist the management regime to the detriment of both the fishery and the economic returns to fellow fishers (Flewwelling, 2001). No one MCS solution exists for all fisheries but instead should be developed for either a specific fishery or a group of interacting fisheries based upon, among other factors, available and cost-effective resources, the desired and expected level of compliance, the value of the fishery and the state of the stock (Flewwelling, 2001). Flewwelling (2001) stressed every MCS system requires regular assessment to determine success in achieving strategic targets in the most cost-effective and efficient manner and to ascertain its effectiveness on compliance over time.

5.3. Examples of current management of cephalopod fisheries worldwide

While not exhaustive, Tables 2.3 and 2.4 give a synthesis of current management initiatives for cephalopod fisheries, highlighting some of the differences in approaches. For example, cephalopod fisheries in Europe are excluded from quota regulations under the Common Fisheries Policy (CFP). Pierce et al. (2010) came to the conclusion that this is likely a good thing, since the management arrangements that have evolved in the various small-scale fisheries across the EU effectively operate under a precautionary principle, apparently successfully regulating exploitation, despite the unknown size of stock. Plus, most management arrangements in place are tailored at the local level; fishers participate in the management of their activity; and, in some cases, co-management arrangements are in place (Pierce et al., 2010). European management regimes for cephalopod fisheries are done through input and output controls. Input controls consist mostly of setting limits to the characteristics of the gear and the number of licences and output controls by limiting the size and weight of the specimens landed (Pierce et al., 2010). Southern European countries appear to be the ones that more actively manage their cephalopod fisheries, possibly a reflection of the

Table 2.3 Current management initiatives for small-scale cephalopod fisheries

Species	Location/country	Targeted/ by-catch	Gear type	Management regime	Management measures	References
Cuttlefish						
Sepia officinalis	Greece, Aegean and Ionian Seas		Trammel nets			Pierce et al. (2010)
Octopus						
Enteroctopus megalocyathus	Southern Chile, southeastern Pacific	Target and by-catch	Free divers using gaffs and by-catch in crab traps	Co-management	Four-month seasonal bans, minimum catch weight (1 kg), 3-year ban (2008–2011)	IInstituto de Fomento Pesquero (2010), Castilla (2010)
Enteroctopus megalocyathus	Patagonian Gulfs and Shelf, SW Atlantic	Targeted	Scuba diving and gaffs	No management		Ortiz et al. (2011)
Octopus insularis	Rio Grande do Norte State, northeastern Brazil, SW Atlantic	Targeted in summer, by-catch in winter	Free diving and compressor-aided diving (illegal) from small boats	Top-down management, no enforcement	Diving with air compressor forbidden, no enforcement of legislation	Vasconcellos et al. (in press)
Octopus insularis	Northeastern Brazil, SW Atlantic	Targeted	Walking on the reef flats in the low tide with gaffs	Top-down management, no enforcement		Vasconcellos et al. (in press)

Octopus insularis	Fernando de Noronha Islands, northeastern Brazil, SW Atlantic	Targeted	Free divers using gaffs	Co-management	Limited number of licences, minimum DML (80 mm)	Leite et al. (2008)
Octopus maya (and *O. vulgaris*)	Mexico: Yucatan Shelf	Targeted	Baited lines		Static 6.5-month closed season, minimum size (110 mm ML), prohibition of commercial diving and hooks	Diaz-De-León and Seijo (1992)
Octopus mimus	Northern Chile, southeastern Pacific	Targeted	Free divers using gaffs	Rights-based and co-management	Four-month closed season, minimum landing size (MLS: 1 kg), territorial use rights for fisheries (TURFs), exclusive fishing rights, extraexclusive fishing rights allocated to subsistence and small-scale artisanal communities	Rocha and Vega (2003)
Octopus tehuelchus	Nuevo and San Jose Northern Patagonian Gulfs, SW Atlantic	Targeted	Fishing with gaffs during low tide	No fisheries management but access restricted by MPA		Narvarte et al. (2007)

Continued

Table 2.3 Current management initiatives for small-scale cephalopod fisheries—cont'd

Species	Location/country	Targeted/ by-catch	Gear type	Management regime	Management measures	References
Octopus vulgaris	Asturias, Spain	Targeted	Traps	Rights-based and co-management	TURFs, exclusive fishing rights, territorial use rights, sea zoning, closed seasons, MLS	Fernández-Rueda and García-Flórez (2007)
Octopus vulgaris	Greece, North Aegean Sea (NE Mediterranean)		Plastic/PVC pots		<1500 pots/vessel, MLS (individual weight, >500 g), fishing ban: July–September, fishing depth > 10 m	Pierce et al. (2010)
Octopus vulgaris	Greece, North Aegean Sea (NE Mediterranean)	Targeted	Fyke nets		<1500 pots/vessel, individual weight >500 g, fishing ban: July–September, fishing depth > 10 m	Pierce et al. (2010)
Octopus vulgaris	Portugal	Targeted	Pots and traps		MLS (main measure, 750 g), technical measures regulating gear	Pereira (1999)

Squid						
Dosidicus gigas	Eastern Pacific Mexico EEZ and Gulf of California	Targeted	Hand jigging		Sonora (Mexico): effort control	Morales-Bojórquez et al. (2001b)
Loligo reynaudi	South Africa	Targeted	Hand-held jigs	Rights-based	Limited number of vessels, limits to number of crew depending on vessel size, 6-week annual closed season during peak spawning period	Sauer (1995)
Todarodes pacificus	Northwest Pacific off Japan and Sea of Japan	Targeted	Jigging machine and squid fishing light		Japan: Restricted by fishing areas, season and fishing light intensity. Allocation of TAC based on ABC.	Kiyofuji and Saitoh (2004)

DML: dorsal mantle length, ML: mantle length, MLS: minimum landing size, TAC: total allowable catch, ABC: available biological catch.

Table 2.4 Current management initiatives for industrial cephalopod fisheries

Species	Location/country	Targeted/ by-catch	Gear type	Management regime	Management measures	References
Cuttlefish						
Sepia officinalis	Lower Normandy, France	Targeted	Trawl and trap fish	Co-management	Closed season, limited licences, fishing seasons, MLS, technical measures (mesh size, number of pots and traps), control number of vessels operating (a limit of 180 trap vessels and 140 coastal trawlers), regulating trawling within the 3 mile limit (Normandy regional orders)	Pierce et al. (2010)
Sepia officinalis	Portugal	Targeted	Unbaited jigs		MLS, technical measures (type of mesh, mesh size)	Pierce et al. (2010)
Octopus						
Octopus insularis	Ceará state, northeastern Brazil, SW Atlantic	Targeted	Pots longlines	Co-management	15 boats under 15 m total length, 5000 pots per boat. Individual mean weight around 600 g	Vasconcellos et al. (in press)
Octopus vulgaris	Galicia, Spain	Targeted	Traps	Co-management	Technical measures (mesh size, number of vessels, number of pots/traps per vessel), time and area restrictions, limitation on the number of licences, MLS, closure period	Pierce et al. (2010), Bañón Diaz et al. (2006)

Octopus vulgaris	Southern Brazil, SW Atlantic	Targeted	Pots longlines		28 boats with up to 20000 pots each. Individual ML over 11 mm, around 1 kg	Barahona et al., 2010
Octopus vulgaris	Greece	Targeted	Traps and fyke nets		Technical measures (number of pots/traps per vessel), time restrictions, MLS, closure period	Pierce et al. (2010)
Octopus vulgaris	Gulf of Cadiz, Spain	Targeted	Pots and traps		Technical measures (number of pots/traps per vessel, length of line), area restrictions, MLS, closure period, ban on sport fishing for octopus	Pierce et al. (2010)
Squid						
Berryteuthis magister	Russia, northwest Pacific	Targeted	Bottom trawl		Technical measures (restricted areas and periods), TACC at about 45–55% of the total assessed biomass	Katugin et al. (2013)
Doryteuthis opalescens	California, United States	Targeted	Purse seine nets	Rights-based	Technical measures (closed or restricted areas), limitations to fishing effort and minimum sizes of mesh and species	California Department of Fish and Game (2007)
Doryteuthis pealeii	United States	Targeted	Trawl		ABC and DAH, fishery closure threshold, trip limits	Mid-Atlantic Fishery Management Council (2011)

Continued

Table 2.4 Current management initiatives for industrial cephalopod fisheries—cont'd

Species	Location/country	Targeted/ by-catch	Gear type	Management regime	Management measures	References
Illex argentinus	Southwest Atlantic in the EEZs of Argentina	Targeted	Jigging machine and squid fishing light		Jigging boat entry. Real-time control (until the cumulative catch reaching 40% escapement)	Agnew et al. (2005)
Illex argentinus	Southwest Atlantic FICZ of Falkland (Malvinas) Islands	Targeted, by-catch	Jigging machine and squid fishing light, bottom trawl catch		*Illex* licences. Real-time control (until the cumulative catch reaching 40% escapement). Restricted entry, closures	Arkhipkin et al. (2013)
Illex illecebrosus	United States	Targeted	Small-mesh trawl		TAC, ABC and DAH, depending on area	Mid-Atlantic Fishery Management Council (2011)
Loligo forbesi	Scotland, the United Kingdom	By-catch (whitefish directed trawl fishery)	Trawl	None	None	Hastie et al. (2009)
Doryteuthis gahi	Southwest Atlantic FICZ of Falkland (Malvinas) Islands	Targeted	Bottom trawl with small-mesh liner	Rights-based	Seasonal licences, two fishing seasons per year, real-time management (may lead to early closure of fishery if necessary), "Loligo box": grounds reserved for *D. gahi* fishing, i.e. finfish trawlers prohibited (coastal habitat is *de facto* MPA), individual transferable quotas	Hatfield and des Clers (1998), Arkhipkin et al. (2013)

Loligo vulgaris	Portugal				MLS	Pierce et al. (2010)
Nototodarus gouldi	Tasmania and Australia	Targeted	Automatic squid jigging gear, trawl	Rights-based	Limited entry licensing for all fisheries, TAE (Southern Squid Jig Fishery only)	Flood et al. (2012)
Nototodarus sloanii	New Zealand: southern islands only	Targeted	Trawl		TACC	Chilvers (2008)
Nototodarus sloanii and *N. gouldi*	Southern New Zealand	Targeted	Jigging machine and squid fishing light		TACC	Chilvers (2008)
Nototodarus sloanii and *N. gouldi*	Southern New Zealand	Targeted	Trawl		Squid TACC. Since 2001, sea lion exclusion devices	Chilvers (2008)
Uroteuthis duvauceli	India	By-catch but targeted seasonally	Trawl	None	None	Meiyappan et al. (2000)

ML, mantle length; MLS, minimum landing size; TAC, total allowable catch; TACC, total allowable commercial catch; ABC, available biological catch; DAH, domestic annual harvest; TAE, total allowable effort.

long history of the exploitation of these resources and of the local importance of the species. These countries are the ones with the greatest internal consumption of the resources and those in which the relative economic and/or social value of the fisheries is the greatest. For instance, in Portugal and Spain, the minimum landing weight for the common octopus is 750 g, and in Greece, it is 500 g (Pierce et al., 2010). In some parts of Spain (e.g. Galicia, northwest Spain and the Gulf of Cádiz, all in the Atlantic), the minimum landing weight for common octopus was increased to 1 kg in 2008 (Pierce et al., 2010). Examples of cephalopod fisheries management through co-management systems include the Asturian octopus artisanal fishery (Spain), cuttlefish in Lower Normandy (France) and the small-scale octopus fishery in northern Chile.

In contrast to the aforementioned European examples, cephalopod fisheries in Africa are often largely unmanaged, exceptions being a hand jig fishery for *L. reynaudi* in South Africa and some octopus fisheries. For *L. reynaudi*, management is by way of effort limitation. Effort has been capped at a level that secures the greatest catch in the long term without exposing the resource to the threat of reductions to levels at which recruitment success is compromised or catch rates become economically unviable (Anon, 2010), with limits on the number of fishers, a closed season and closed areas for fishing. Artisanal fishing for octopus is an important subsistence and economic activity practised by local coastal communities in the western Indian Ocean, particularly in Tanzania, Mozambique, Madagascar and Kenya (Guard, 2009; Otieno, 2011). Initially caught for local and inland consumption, export to European and Far Eastern markets has led to a rise in demand (Guard, 2009). An interesting community-based approach to managing octopus fisheries has recently been introduced in the south of Madagascar with pilot no-take zones introduced to demonstrate the tangible fisheries benefits of protecting one of the region's most economically important marine resources and to increase the involvement of local communities in marine resource management (Epps, 2007). The pilot no-take zone was successful and resulted in an increase of catches as well as an increase in the price paid by commercial buyers after the closure. This success increased the interest of the Andavadoaka community and adjacent fishing communities in Madagascar in developing additional no-take zones. A similar exercise has recently taken place in Rodrigues (Epps, 2007).

Tables 2.3 and 2.4 highlight the lack of routine assessment for cephalopods with few countries collecting detailed data on cephalopod fisheries, suggesting the introduction of a precautionary approach to management in many cases. In 1996, following a 1995 Technical Consultation on the Precautionary Approach to Capture Fisheries in Sweden, the FAO published a set of guidelines on the precautionary approach to capture fisheries and species introductions (FAO, 1996). These guidelines, as summarized by Punt (2006), include the following principles:

- A level of precaution commensurate to risk should be applied at all times to all fisheries.
- Potentially irreversible changes should be avoided (to maintain options for future generations).
- Undesirable outcomes should be anticipated and measures be taken to reduce their likelihood.
- Corrective measures should be applied immediately and be effective within an acceptable time.
- Precautionary limits should be placed on fishing capacity on highly uncertain resources.
- All fishing activities should be subject to prior authorization and periodic review.
- The burden of proof should be appropriately (realistically) placed.
- Standards of proof commensurate with the potential risk to the resource should be established.
- The approach should be formalized in a comprehensive legal and institutional framework.

Specific guidelines for applying the precautionary approach to artisanal fisheries and new or developing fisheries were also addressed (FAO, 1996). For artisanal fisheries, guidelines suggest limiting risks to resources and the environment with the use of closed areas, delegating certain management decisions to the community (co-management), limiting the influence of industrial fishing on resources harvested by the community and investigating the influence of social and economic factors on fishing pressure.

Precautionary principles that would apply to data-poor and unmanaged fisheries include (1) controlling access to the fishery; (2) conservatively capping fishing capacity and mortality rate by limiting effort or TAC until data analysis to justify increases can be carried out; (3) avoiding new investment in the fishery by temporarily licensing vessels from another fishery and allowing flexibility to phase out vessels if necessary; (4) closed areas to limit risks to the resource; (5) establishing precautionary, preliminary biological limit reference points; (6) encouraging responsible fishing through co-management or tenure of fishing rights; (7) encouraging the development of fisheries that are economically viable without long-term subsidies; (8) establishing a data collection and reporting system; (9) starting research programmes on the stock and fisheries; and (10) setting up experimental situations to generate information on the resources.

In the past, management approaches have been based on target reference points (TRPs), the specific values of indicators for catch, biomass and fishing mortality regarded as optimal (Caddy, 2004). However, it has been shown that once a TRP is overshot, the stock becomes vulnerable to overfishing (Caddy, 1998). Caddy and Mahon (1995) put forwards the concept of limit reference points (LRPs) (which has been recommended following the United Nations Conference on Environment and Development; Caddy, 2004) as one way of defining the limits to exploitation of a stock and so implementing a more precautionary approach. Integral to this approach is prenegotiating responses to unfavourable events and implementing these responses when a fishery approaches an LRP (Caddy, 1998). As noted by Caddy (1998), almost all RPs currently in use are based on the availability of age-structured data and on information on stock and recruitment accumulated over a significant period of time. Defining reference points for data-poor fisheries, such as the majority of cephalopod fisheries, could be somewhat difficult. Caddy (1998) however had proposed a number of ways LRP can be used in the management of data-poor fisheries. For example, it may be possible to set a single LRP that corresponds to serious but not catastrophic conditions and then pick a TRP based on estimates of variance and probability of overshoot. See Caddy (1998) for a full review on application in data-poor fisheries and Caddy (2004) for the potential application of RPs to invertebrate fisheries specifically.

However, in undeveloped or developing countries, in which many small-scale octopus fisheries occur, obtaining adequate data for fisheries management is only a small part of the problem. The lack of adequate institutions; management organizations; monitoring, control and surveillance (MCS) bodies; and enforcement prevents the efficient management of resources.

Factors contributing to the unsustainability and overexploitation of small-scale octopus, and other invertebrate, fisheries in Latin America have been explored by Narvarte et al. (2007). Unsustainability of these fisheries is explained as follows:

- The irregularity in recruitment of target species produces difficulties in long-term planning (complexity and lack of knowledge concerning biological/ecological processes).

- There are numerous incentives for entering the fishery, investments required are low, and there is a common belief held in the community that the fishery could mitigate all unemployment problems when reduction or collapse occurs in other economic activities.
- Violations of norms and regulations (evident in illegal captures of banned species, violation of established catch quotas, use of unpermitted methods, etc.) and the high prices obtained for the resources in closed seasons lead to irresponsible behaviour (e.g. clandestine harvesting) by some individuals.
- Lack of organization among fishers and lack of understanding of harvesting rights for the resources.
- Poverty and lack of satisfaction of basic living requirements in the coastal collector segment, especially related to those engaged in collecting octopus. One problem is the abandonment of fishing as an activity to pursue other nonfishing-related employment. When the trade is abandoned at an early age, it is difficult to recover abilities required for fishing.
- Conflicts with other activities (e.g. growth of tourism affects the natural habitat of the octopus).
- Lack of organization and a collective strategy by fishers in marketing their catches, which produces competition among them, thus lowering prices. This in turn puts pressure on increasing the catches to maintain incomes.
- Lack of initiatives for fishers to develop value-added products at the site of production. Although infrastructures exist for processing catches, catches are commercialized whole, at places far from their origin.
- Institutional difficulties to rapidly and efficiently respond to requests for information and technical assistance.

These points raised by Narvarte et al. (2007) also apply to other regions. For instance, cephalopod fisheries in Argentina, Brazil, Chile and Mexico have some data on landings, are subject to nominal rules and have some information on the biology of the resources, yet rules are seldom enforced and fishers usually violate the norms.

5.4. The use of Marine Protected Areas as a general conservation tool

MPAs may provide refugia for a variety of species including cephalopods, and they have occasionally been used as management tools (sometimes experimentally) for cephalopod fisheries (e.g. in Portugal and Spain).

In general, MPAs set up in areas of previous small-scale fisheries continue to allow small-scale exploitation of a variety of species including cephalopods, which is a special form of spatial planning. In Portugal legislation regulates cephalopod catches by recreational fishers within specific MPAs. The MPA "Parque Natural do Sudoeste Alentejano e Costa Vicentina" in the southwest coastal area of Portugal, for instance, limits octopus catches to two octopuses per fisher per day, while the number of entry licences is also restricted (legal diploma "Portaria" 115-A/2011). In northeastern Brazil, most small-scale octopus fisheries can function within MPAs, but legislation on MPAs is more strict and enforced more efficiently than fisheries legislation, which results in easier and more efficient implementation of catch limitations. In South Africa, no-take MPAs are used as a management tool for *L. reynaudi* (Sauer, 1995).

5.5. Recreational fishery data

Recreational fisheries for cephalopods do exist in countries with a traditional consumption of cephalopods, but they are often not well documented. As a long-standing recreational tradition, people hunt for octopus on the shore by employing dedicated gear to catch the animals from rocky tide pool areas or in shallow infratidal areas accessible during low tides. By luring them out of shelter with bait, octopuses become accessible to any spearing or hooking device, of which the "Bicheiro" has been described already in the nineteenth century by Baldaque da Silva (1891). Octopuses are also caught by recreational fishers from boats or other floating platforms by means of jiglike devices, which are essentially baited rods terminated by a number of hooks designed to penetrate the muscle of the attracted animal by means of a sudden jerking movement. Squids are caught with hand-held jigs in a number of countries. Cuttlefish are recreationally caught by means of snorkelling and spearing, throughout much of the distributional range of each species. Where legislation exists, it often sets an overall limit on daily allowable catches, regardless of species: for example, in Portugal, Ordinance n°14 of 23 January 2014, Article 12, n°1, states that recreational fishery catches cannot exceed the daily limit of 10kg per person for all species combined, excepting the possibility of one larger-than-the-limit specimen of fish or cephalopod; and n°2 further states that if fish or cephalopods are not included in the catches, the limit is lowered to 2 kg per person. Where legislation exists, it often sets an overall limit on numbers, regardless of species.

5.6. The way forwards: Balancing the many objectives of management

We suggest that in order to address the complexities discussed in this chapter, we should be exploring an ecosystem approach to cephalopod fisheries (EACF) (FAO, 2003) defined as "An ecosystem approach to fisheries strives to balance diverse societal objectives, by taking into account the knowledge and uncertainties about biotic, abiotic and human components of ecosystems and their interactions and applying an integrated approach to fisheries within ecologically meaningful boundaries". A set of general practical guidelines exists and a start is now being made in a number of countries in implementing EAF, including South Africa, Mauritius, Seychelles, Tanzania, Kenya, Madagascar and Uganda.

The goals of EAF are "to balance diverse societal objectives, by taking into account the knowledge and uncertainties about biotic, abiotic, and human components of ecosystems and their interactions and applying an integrated approach to fisheries within ecologically meaningful boundaries" (FAO, 2003). The approach thus intends to foster the use of existing management frameworks, improving their implementation and reinforcing their ecological relevance, and will contribute significantly to achieving sustainable development (Garcia, 2003).

However, implementation and effectiveness will undoubtedly benefit from reducing important uncertainties, and further research is needed for this purpose including better understanding of ecosystem structure and function and how fisheries affect them; integrating social, economic and ecological considerations into decision making; improving the management measures available to implement EAF; understanding the management process better; and improving monitoring and assessments (Garcia and Cochrane, 2005).

Fisheries production and yield are constrained by a number of factors that can be classified as biological, ecological and environmental, technological, social and cultural and economic considerations (Cochrane, 2002). Fisheries for cephalopods are often multispecific, which means that it is nearly impossible to manage them on a single-species basis. However, nearly all fisheries management to date is focused on biological reference points for single species. This means that measures taken to ensure the sustainable yield of a species completely ignore the remaining assemblage. Yet, it is a well-established fact that many fisheries will include destructive gear impacts on the sea floor, unwanted by-catch and eventually ecosystem effects such as changes in species richness and composition and relative species abundance.

We recognize the sometimes conflicting objectives and management aiming to maximize fisheries production often make it impossible to have a large abundance of both cephalopods and finfish (perhaps due mostly to predatory interactions). Management is often aimed at optimizing species assemblage abundance, either from a purely economical perspective (i.e. maximizing the abundance of the economically most-valuable species) or from a biodiversity (or species richness) perspective (i.e. maintaining the undisturbed species balance). More complex adaptive management systems are currently being explored, of which the responsive fisheries management system being developed through the EU-funded EcoFishMan project (www.ecofishman.com) is a good example.

Despite the increasing social and economic importance of cephalopod fisheries in Europe, very little information exists on the human dimensions of these fisheries. As identified by Pierce et al. (2010) on a major review of cephalopods fisheries in Europe, there is an urgent need for a detailed analysis of the economic and social importance of these fisheries locally as well as at the national levels, including bioeconomic studies of the fleets targeting cephalopods, and the evaluation of possible socioeconomic implications of alternative management strategies at the local level.

Small-scale and artisanal fisheries face different challenges. Guard (2003, 2009) had listed potential management actions that can be applied to artisanal octopus fisheries. They are as follows:

- Initiation of a stock assessment programme and continued catch monitoring
- Introduction of rotational or "pulsed" fishing regimes
- Collaborative agreements for restriction of fishing outside of spring tide periods
- Temporary reef closures and reduced fishing effort during brooding periods
- Introduction of size limits
- Collaborative licence scheme for octopus fishermen and the formation of community stewardship groups
- Introduction of recommended maximum sustainable yields and associated effort for each octopus fishery
- Dissemination of results and community awareness raising

Fishers input (through co-management) should always be utilized when developing management strategies. Fishers can contribute information regarding the stock structure (in terms of migrations patterns, spawning grounds and juvenile habitat), schooling behaviour, habitat preference, gear selectivity and effort changes as a result of new regulations (Johnson, 2011; Johnson and van Densen, 2007). Another positive outcome of co-management is the potential for cooperative research or research conducted in partnership with industry (Arkhipkin et al., 2013; Johnson, 2011; Johnson and van Densen, 2007).

Several approaches have promise for by-catch reduction in cephalopod fisheries, for example, finfish by-catch and the associated regulations in the United States. *D. pealeii* small-mesh trawl fishery has spurred experimentation with gear modifications to reduce by-catch (Bayse et al., 2014; Glass et al., 1999; Hendrickson, 2011) and research on environmental factors driving species co-occurrence (Lange and Waring, 1992; Manderson et al., 2011), as well as exploration of the potential utility of by-catch avoidance strategies (e.g. Bethoney et al., 2013).

There is a need for more effective monitoring, control and surveillance (MCS) while recognizing that there is no unique MCS solution for all fishery situations, nor inherently right or wrong approaches to the implementation of MCS systems (Berg and Davies, 2002). Large-scale cephalopod fisheries, especially international industrial squid fisheries, generally target straddling stocks or species occurring both within the exclusive economic zone and in areas beyond and adjacent to the zone. *I. argentinus*, *D. gigas*, *T. pacificus* and *O. bartramii* stocks are under proper management and regulation by each country or bilateral exploiting these resources and are well monitored, controlled and kept under surveillance. In contrast, however, those stocks caught on the high seas are generally under no control or management due to the lack of international regulation or no establishment of regional fisheries management organizations (RFMOs). Cooperation between RFMOs or multinational agreements are essential, especially for neon flying squid *O. bartramii*, jumbo flying squid *D. gigas*, Japanese common squid *T. pacificus* and Argentine short-fin squid *I. argentinus* fisheries operating in country EEZs and open seas. Small-scale cephalopod fisheries target species generally more local and nonmigratory and stocks are smaller. The lack of effective monitoring, control and surveillance and basic statistical information hampers effective management and should be addressed, including an understanding of migrant fishers. In many cases, policy/legislation is not adequate.

6. CONCLUSIONS

The understanding of cephalopod population dynamics is largely based on coastal and shelf species that are exploited by fisheries, but these may not represent the large populations of oceanic and deepwater species. The exploited species often have complex and variable cohort and microcohort structures within the population, which may vary annually in terms of life cycle parameters, migration and abundance. There is a risk that this complexity can obfuscate interpretation of population dynamics, especially when only short-term data sets are available. Population variability is linked to the environment, but trophic relationships, a dominant theme in theoretical ecology, are poorly understood over the whole life cycle in cephalopod populations and receive little attention in research on population variability. In most cases, neither environmental impacts nor trophic relationships are explicitly included in stock assessment models. Intrinsic elements and trophic relationships are probably no less important than environment in shaping cephalopod life cycles, population dynamics and recruitment, but their incorporation into workable management strategies is more difficult than incorporating environmental effects because of the mathematical and statistical challenges involved and high biological variability, which may in fact contain a strong environmental signal.

The effects of environmental parameters on cephalopod population variability will operate on different timescales. Fisheries management in the short term is concerned with interannual variability. In the long-term ecologists, fishery managers and policy makers are concerned with the effects of global climate change. Although much attention focuses on warming, others aspects of measured and predicted change include ocean acidification, changes in oxygen tension, salinity and macro- and mesoscale oceanography. All of these will drive changes to whole ecosystems, which will impact on particular species groups including cephalopods.

Cephalopod populations could be good indicators of short-term environmental variability, if we could only understand fully what they are telling us, but their response top short term variability means they are poor indicators of long-term change because the long-term signal is masked by the annual noise. Nevertheless, cephalopods could be long-term winners under global climate change because they are ecological opportunists and have plastic population dynamics. Also, because they are short-lived and can reproduce fast, they may be able to evolve more rapidly under high selection pressure relative to many fish competitors and predators.

A range of assessment methods is used in different cephalopod fisheries, depending on the characteristics of the species concerned, the particular fishery and the resources available to managers. Given the importance of forecasting in fisheries for short-lived species, there is a strong argument for combining appropriate stock assessment methods with environmental predictions based initially on empirical relationships between environmental variability and stock variability. In the longer term, as mechanisms explaining these relationships become better understood, this approach will become more robust. With current state of knowledge, the use of pre-fishing season assessments is strongly recommended. Empirical relationships between stock and environment, especially when they are based on short time series, should be used with caution as they may be revealed to have been illusory in the light of data collected over a longer period.

Given the current levels of uncertainty in even the best-managed fisheries, a precautionary approach is nearly always indicated. Ecosystem-based management of cephalopod fisheries should become the norm. Generally, this involves ensuring that harvesting does not decrease the size of any population below that which is required for stable recruitment over time, that ecological relationships are maintained between harvested populations and those populations that depend on them or are otherwise related to them and that the risk of causing irreversible changes to the ecosystem is prevented or minimized. The ecosystem approach to fisheries also implies consideration of the socioeconomic dimensions of fisheries. Cephalopod fisheries, in common with fisheries for other groups of organisms, will benefit in most cases from the introduction of participatory approaches to management, which engage all stakeholders from fishers to managers, scientists and policy makers. Where this approach has been adopted collaboration has increased understanding on all sides resulting in better regulations and better compliance and hence greater likelihood of long-term sustainability.

ACKNOWLEDGEMENTS

We warmly thank Erica Vidal and her committee who led the planning and organization of the Cephalopod International Advisory Council workshops and symposium at Florianópolis, Brazil, in November 2012. Rodrigo Martins (Federal University of Sao Paulo) captured the essence of the discussions during our workshop whether they were direct, meandering, terse or verbose. Several scientists contributed to the discussions of the workshop. In the group working on the causes of variability, they were Y. Sakurai, R. Martins, M. Sakai, S. Keller, Y. Kato, M. Haimovici, A. Tomas and T. Yamaguchi. In the group working on population dynamics, they were C. Augusto, C. Yamashiro, I. Sobrino, S. Arbuckle and S. Leporati.

Lead author(s) for each section, 1: G.J.P.; 2: M.R.L.; 3: A.I.A. and V.V.L.; 4: O.C.N.; 5: W.H.H.S.; 6: P.G.K.R.

REFERENCES

Acou, A., Rivot, E., van Gils, J.A., Legault, A., Ysnel, F., Feunteun, E., 2011. Habitat carrying capacity is reached for the European eel in a small coastal catchment: evidence and implications for managing eel stocks. Freshwater Biol. 56, 952–968.

Agnew, D.J., Baranowski, R., Beddington, J.R., des Clers, S., Nolan, C.P., 1998a. Approaches to assessing stocks of *Loligo gahi* around the Falkland Islands. Fish. Res. 35, 155–169.

Agnew, D., Nolan, C., Des Clers, S., 1998b. On the problem of identifying and assessing populations of Falkland Island squid *Loligo gahi*. S. Afr. J. Mar. Sci. 20, 59–66.

Agnew, D.J., Hill, S., Beddington, J.R., 2000. Predicting the recruitment strength of an annual squid stock: *Loligo gahi* around the Falkland Islands. Can. J. Fish. Aquat. Sci. 57, 2479–2487.

Agnew, D.J., Hill, S.L., Beddington, J.R., Purchase, L.V., Wakeford, R.C., 2005. Sustainability and management of southwest Atlantic squid fisheries. Bull. Mar. Sci. 76, 579–593.

Agnew, D.J., Pearce, J., Pramod, G., Peatman, T., Watson, R., 2009. Estimating the worldwide extent of illegal fishing. PLoS One 4, e4570. http://dx.doi.org/10.1371/journal.pone.0004570.

Alarcón-Muñoz, R., Cubillos, L., Gatica, C., 2008. Jumbo squid (*Dosidicus gigas*) biomass off central Chile: effects on Chilean hake (*Merluccius gayi*). CalCoFI Rep. 49, 157–166.

Allen, R., Punsly, R.G., 1984. Catch rates as indices of abundance in yellowfin tuna, *Thunnus albacares*, in the eastern Pacific Ocean. Bull. Inter-Am. Trop. Tuna Comm. 18, 31–379.

Ally, J.R.R., Keck, S.A., 1978. A biochemical-genetic population structure study of market squid, *Loligo opalescens*, along the California coast. Calif. Dept. Fish & Game Fish Bull. 169, 113–121.

Alvarez Perez, J.A., 2002. Biomass dynamics of the squid *Loligo plei* and the development of a small-scale seasonal fishery off southern Brazil. Bull. Mar. Sci. 71, 633–651.

Amaratunga, T., 1983. The role of cephalopods in the marine ecosystem. In: Caddy, J.F. (Ed.), Advances in Assessment of World Cephalopod Resources. In: Fisheries Technical Paper No. 231, FAO, Rome, pp. 379–415.

Anderson, S.C., Mills Flemming, J., Watson, R., Lotze, H.K., 2011. Rapid global expansion of invertebrate fisheries: trends, drivers, and ecosystem effects. PLoS One 6 (3), e14735.

André, J., Grist, E.P., Semmens, J.M., Pecl, G.T., Segawa, S., 2009. Effects of temperature on energetics and the growth pattern of benthic octopuses. Mar. Ecol. Prog. Ser. 374, 167–179.

André, J., Haddon, M., Pecl, G.T., 2010. Modelling climate-change-induced nonlinear thresholds in cephalopod population dynamics. Glob. Change Biol. 16, 2866–2875.

Anguiletta, M.J., Newiarowski, P.H., Navas, C.A., 2002. The evolution of thermal physiology in ectotherms. J. Therm. Biol. 27, 249–268.

Anon, 2010. Status of the South African Marine Fishery Resources 2010. Compiled by the Chief Directorate: Fisheries Research, Fisheries Branch, Department of Agriculture, Forestry and Fisheries, South Africa.

Argüelles, J., Tafur, R., Taipe, A., Villegas, P., Keyl, F., Dominguez, N., Salazar, M., 2008. Size increment of jumbo flying squid *Dosidicus gigas* mature females in Peruvian waters, 1989–2004. Prog. Oceanogr. 79, 308–312.

Arkhipkin, A.I., 1993. Age, growth, stock structure and migratory rate of prespawning short-finned squid *Illex argentinus* based on statolith ageing investigations. Fish. Res. 16, 313–338.

Arkhipkin, A., 2000. Intrapopulation structure of winter-spawned Argentine shortfin squid, *Illex argentinus* (Cephalopoda, Ommastrephidae), during its feeding period over the Patagonian Shelf. Fish. Bull. 98, 1–13.

Arkhipkin, A.I., 2013. Squid as nutrient vectors linking Southwest Atlantic marine ecosystems. Deep Sea Res. II 95, 7–20.
Arkhipkin, A., Laptikhovsky, V., 1994. Seasonal and interannual variability in growth and maturation of winter-spawning *Illex argentinus* (Cephalopoda, Ommastrephidae) in the Southwest Atlantic. Aquat. Living Resour. 7, 221–232.
Arkhipkin, A.I., Laptikhovsky, V.V., 2012. Impact of ocean acidification on plankton larvae as a cause of mass extinctions in ammonites and belemnites. Neu. Jahr. Geol. Palaont. Abh. 266, 39–50.
Arkhipkin, A.I., Middleton, D.A., 2002a. Sexual segregation in ontogenetic migrations by the squid *Loligo gahi* around the Falkland Islands. Bull. Mar. Sci. 71, 109–127.
Arkhipkin, A.I., Middleton, D.A., 2002b. Inverse patterns in abundance of *Illex argentinus* and *Loligo gahi* in Falkland waters: possible interspecific competition between squid? Fish. Res. 59, 181–196.
Arkhipkin, A.I., Pérez, A.A., 1998. Life-history reconstruction. In: Rodhouse, P.G., Dawe, E.G., O'Dor, R.K. (Eds.), Squid Recruitment Dynamics: The Genus *Illex* as a Model, the Commercial *Illex* Species and Influences on Variability. FAO, Rome, pp. 155–178.
Arkhipkin, A.I., Roa-Ureta, R., 2005. Identification of ontogenetic growth models for squid. Mar. Freshw. Res. 56, 371–386.
Arkhipkin, A., Jereb, P., Ragonese, S., 2000. Growth and maturation in two successive seasonal groups of the short-finned squid, *Illex coindetii* from the Strait of Sicily (central Mediterranean). ICES J. Mar. Sci. 57, 31–41.
Arkhipkin, A.I., Middleton, D.A.J., Sirota, A.M., Grzebielec, R., 2004. The effect of Falkland Current inflows on offshore ontogenetic migrations of the squid *Loligo gahi* on the southern shelf of the Falkland Islands. Estuar. Coast. Shelf Sci. 60, 11–22.
Arkhipkin, A.I., Laptikhovsky, V.V., Sirota, A.M., Grzebielec, R., 2006. The role of the Falkland Current in the dispersal of the squid *Loligo gahi* along the Patagonian Shelf. Estuar. Coast. Shelf Sci. 67, 198–204.
Arkhipkin, A., Barton, J., Wallace, S., Winter, A., 2013. Close cooperation between science, management and industry benefits sustainable exploitation of the Falkland Islands squid fisheries. J. Fish Biol. 83, 905–920.
Arreguín-Sánchez, F., 1996. Catchability: a key parameter for fish stock assessment. Rev. Fish Biol. Fish. 6, 221–242.
Arreguín-Sánchez, F., Solís-Ramírez, M.J., González de la Rosa, M.E., 2000. Population dynamics and stock assessment for *Octopus maya* (Cephalopoda: Octopodidae) fishery in the Campeche Bank, Gulf of Mexico. Rev. Biol. Trop. 48 (2–3), 323–331.
Augustyn, C.J., 1989. Systematics, life cycle and fisheries potential of the chokker squid *Loligo vulgaris reynaudii*. PhD Thesis, University of Port Elizabeth, South Africa.
Augustyn, C.J., Lipiński, M., Sauer, W., 1992. Can the *Loligo* squid fishery be managed effectively? A synthesis of research on *Loligo vulgaris reynaudii*. S. Afr. J. Mar. Sci. 12, 903–918.
Augustyn, C.J., Roel, B.A., Cochrane, K.L., 1993. Stock assessment in the chokka squid *Loligo vulgaris reynaudii* fishery off the coast of South Africa. In: Okutani, T., O'Dor, R.K., Kubodera, T. (Eds.), Recent Advances in Fisheries Biology. Tokai University Press, Tokyo, pp. 3–14.
Augustyn, C.J., Lipinski, M., Sauer, W., Roberts, M., Mitchell-Innes, B., 1994. Chokka squid on the Agulhas Bank: life history and ecology. S. Afr. J. Sci. 90, 143–154.
Baldaque da Silva, A.A., 1891. Estado Actual das Pescas em Portugal. Imprensa Nacional, Lisboa.
Bañón Diaz, R., Campelos Álvarez, J., Garcia Tasende, M., 2006. A pesca de polbo comun con nasas na costa Galega, 1999-2004. Os Recursos Mariños de Galicia, Serie técnica 6. Xunta de Galicia.

Barahona, N., Aguilera, A. Olguín y V. Pezo., A. 2010. Caracterización biológico pesquera de las actividades extractivas del recurso pulpo en la X Región. FIP N°2008-40. Informe Final. Instituto de Fomento Pesquero.
Baron, P.J., 2003. The paralarvae of two South American sympatric squid: *Loligo gahi* and *Loligo sanpaulensis*. J. Plankton Res. 25, 1347–1358.
Basson, M., Beddington, J.R., Crombie, J.A., Holden, S.J., Purchase, L.V., Tingley, G.A., 1996. Assessment and management techniques for migratory annual squid stocks: the *Illex argentinus* fishery in the Southwest Atlantic as an example. Fish. Res. 28, 3–27.
Bayse, S.M., He, P., Pol, M.V., Chosid, D.M., 2014. Quantitative analysis of the behavior of longfin inshore squid (*Doryteuthis pealeii*) in reaction to a species separation grid of an otter trawl. Fish. Res. 152, 55–61.
Beddington, J.R., Rosenberg, A.A., Crombie, J.A., Kirkwood, G.P., 1990. Stock assessment and the provision of management advice for the short fin squid fishery in Falkland Islands waters. Fish. Res. 8, 351–365.
Belcari, P., Tserpes, G., González, M., Lefkaditou, E., Marceta, B., Piccinetti Manfrin, G., Souplet, A., 2002. Distribution and abundance of *Eledone cirrhosa* (Lamarck, 1798) and *E. moschata* (Lamarck, 1798) (Cephalopoda: Octopoda) in the Mediterranean Sea. Sci. Mar. 66 (S2), 143–155.
Belcher, E., Matsuyama, B., Trimble, G., 2001. Object identification with acoustic lenses. In: An Ocean Odyssey—Oceans 2001 MTS/IEEE Conference Proceedings. Marine Technology Society, Washington, DC, pp. 6–11.
Bellido, J., Pierce, G., Wang, J., 2001. Modelling intra-annual variation in abundance of squid *Loligo forbesi* in Scottish waters using generalised additive models. Fish. Res. 52, 23–39.
Benoit-Bird, K.J., Gilly, W.F., Au, W.W.L., Mate, B., 2008. Controlled and *in situ* target strengths of the jumbo squid *Dosidicus gigas* and identification of potential acoustic scattering sources. J. Acoust. Soc. Am. 123, 1318–1328.
Berg, P.E., Davies, S., 2002. Fishery monitoring, control and surveillance. In: Cochrane, K. (Ed.), A Fishery Manager's Guidebook. Management Measures and Their Application. In: FAO Fisheries Technical Paper No. 424, FAO, Rome.
Bethoney, N.D., Schondelmeier, B.P., Stokesbury, K.D.E., Hoffman, W.S., 2013. Developing a fine-scale system to address river herring (*Alosa pseudoharengus*, *A. aestivalis*) and American shad (*A. sapidissima*) bycatch in the U.S. Northwest Atlantic mid-water trawl fishery. Fish. Res. 141, 79–87.
Bettencourt, V., Guerra, Á., 2001. Age studies based on daily growth increments in statoliths and growth lamellae in cuttlebone of cultured *Sepia officinalis*. Mar. Biol. 139, 327–334.
Beverton, R.J.H., Holt, S.J., 1957. On the Dynamics of Exploited Fish Populations. Fisheries Investment Series 2, vol. 16, U.K. Ministry of Agriculture, Fisheries and Food, London.
Binet, T., 2012. Rebuilding fisheries in developing countries. Economics of Rebuilding Fisheries: Case Studies of Fisheries Rebuilding Plans. OECD.http://www.oecd.org/tad/fisheries/rebuildingfisheriesthewayforward.htm.
Bloor, I.S.M., Attrill, M.J., Jackson, M.L., 2013a. A review of the factors influencing spawning, early life stage survival and recruitment variability in the common cuttlefish (*Sepia officinalis*). Adv. Mar. Biol. 65, 1–65.
Bloor, I.S.M., Wearmouth, V.J., Cotterell, V.P., McHugh, M.J., Humphries, N.E., Jackson, E.J., Attrill, M.J., Sims, D.W., 2013b. Movements and behaviour of European common cuttlefish *Sepia officinalis* in English Channel inshore waters: first results from acoustic telemetry. J. Exp. Mar. Biol. Ecol. 448, 19–27.
Boletzky, S. V., Dohle, W. 1967. Observations sur un capitellidé (Capitella hermaphrodita sp. n.) et d'autres polychètes habitant la ponte de *Loligo vulgaris*. Vie Milieu, A 18(1), 79–98.

Boletzky, S.V., 1983. Sepia Officinalis. In: Boyle, P.R. (Ed.), Cephalopod Life Cycles: Species Account, vol. I. Academic Press, London, pp. 31–52.
Boletzky, S.V., 1987. Fecundity variation in relation to intermittent or chronic spawning in the cuttlefish, *Sepia officinalis* L. (Mollusca, Cephalopoda). Bull. Mar. Sci. 40, 382–388.
Boletzky, S.V., 1988. A new record of long continued spawning of *Sepia officinalis* (Mollusca: Cephalopoda). Rapp. Comm. Int. Mer Medit. 31, 257.
Boletzky, S., 2003. Biology of early life stages in cephalopod molluscs. Adv. Mar. Biol. 44, 143–203.
Bolker, B.M., 2008. Ecological Models and Data in R. Princeton University Press, Princeton.
Borges, T.C., 2001. Cephalopod Resources Dynamics and Fisheries Trends in the Algarve and the Gulf of Cadiz (ALCACEPH): Final Report to the Commission of the European Communities Directorate General of Fisheries. University of the Algarve, Faro, Portugal.
Bouchaud, O., 1991. Energy consumption of the cuttlefish *Sepia officinalis* L. (Mollusca: Cephalopoda) during embryonic development, preliminary results. Bull. Mar. Sci. 49, 333–340.
Bower, J.R., 1996. Estimated paralarval drift and inferred hatching sites for *Ommastrephes bartramii* (Cephalopoda: Ommastrephidae) near the Hawaiian Archipelago. Fish. Bull. 94, 398–411.
Bower, J.R., Nakamura, Y., Mori, K., Yamamoto, J., Isoda, Y., Sakurai, Y., 1999a. Distribution of *Todarodes pacificus* (Cephalopoda: Ommastrephidae) paralarvae near the Kuroshio off southern Kyushu, Japan. Mar. Biol. 135, 99–106.
Bower, J.R., Seki, M.P., Young, R.E., Bigelow, K.A., Hirota, J., Flament, P., 1999b. Cephalopod paralarvae assemblages in Hawaiian Islands waters. Mar. Ecol. Prog. Ser. 185, 203–212.
Boyle, P.R., Boletzky, S.V., 1996. Cephalopod populations: definition and dynamics. Philos. Trans. Biol. Sci. 351, 985–1002.
Boyle, P.R., Knobloch, D., 1983. The female reproductive cycle of the octopus, *Eledone cirrhosa*. J. Mar. Biol. Assoc. U.K. 63, 71–83.
Boyle, P.R., Pierce, G.J. (Eds.), 1994. Fishery biology of Northeast Atlantic squid. Fish. Res. 21(1–2) (special issue): 314 pp.
Boyle, P., Rodhouse, P., 2005. Cephalopods: Ecology and Fisheries. Blackwell Science, Oxford.
Brady, B.C., 2008. Long-term changes in biological characteristics and fishery of *Loligo opalescens*. MSc Thesis, San Jose State University, California, USA.
Bravo de Laguna, J., 1989. Managing an international multispecies fishery: the Saharan trawl fishery for cephalopods. In: Caddy, J.F. (Ed.), Marine Invertebrate Fisheries: Their Assessment and Management. Wiley, New York, pp. 591–612.
Brierley, A.S., Thorpe, J.P., Clarke, M.R., Martins, H.R., 1993. A preliminary biochemical genetic investigation of the population structure of *Loligo forbesi* Steenstrup, 1856 from the British Isles and the Azores. In: Okutani, T., O'Dor, R.K., Kubodera, T. (Eds.), Recent Advances in Fisheries Biology. Tokai University Press, Tokyo, pp. 59–67.
Brill, R.W., Lutcavage, M.E., 2001. Understanding environmental influences on movements and depth distributions of tunas and billfishes can significantly improve population assessments. Am. Fish. Soc. Symp. 25, 179–198.
Brodziak, J., Hendrickson, L., 1999. An analysis of environmental effects on survey catches of squids *Loligo pealei* and *Illex illecebrosus* in the northwest Atlantic. Fish. Bull. 97, 9–24.
Brodziak, J.K.T., Rosenberg, A.A., 1993. A method to assess the squid fisheries in the northwest Atlantic. ICES J. Mar. Sci. 50, 187–194.
Brunetti, N.E., Ivanovic, M.L., 1992. Distribution and abundance of early life stages of squid (*Illex argentinus*) in the south-west Atlantic. ICES J. Mar. Sci. 49, 175–183.

Brunetti, N.E., Elena, B., Rossi, G.R., Ivanovic, M.L., Aubone, A., Guerrero, R., Benavides, H., 1998. Summer distribution, abundance, and population structure of *Illex argentinus* on the Argentine shelf in relation to environmental features. S. Afr. J. Mar. Sci. 20, 175–186.

Buckley, L.J., Buckley, L.B., 2010. Toward linking ocean models to fish population dynamics. Prog. Oceanogr. 84, 85–88.

Caballero-Alfonso, A.M., Ganzedo, U., Trujillo-Santana, A., Polanco, J., Santana del Pinoa, A., Ibarra-Berastegic, G., Castro-Hernández, J.J., 2010. The role of climatic variability on the short-term fluctuations of octopus captures at the Canary Islands. Fish. Res. 102, 258–265.

Cabranes, C., Fernández-Rueda, P., Martínez, J., 2008. Genetic structure of *Octopus vulgaris* around the Iberian Peninsula and Canary Islands as indicated by microsatellite DNA variation. ICES J. Mar. Sci. 65, 12–16.

Caddy, J.F., 1983. The cephalopod: factors relevant to their population dynamics and to the assessment and management of stocks. In: Caddy, J.F. (Ed.), Advances in Assessment of World Cephalopod Resources. Fisheries Technical Paper No. 231, FAO, Rome, pp. 416–452.

Caddy, J.F., 1996. Modelling natural mortality with age in short-lived invertebrate populations: definition of a strategy of gnomonic time division. Aquat. Living Resour. 9, 197–207.

Caddy, J.F., 1998. A Short Review of Precautionary Reference Points and Some Proposals for their Use in Data-Poor Situations. FAO Fisheries Technical Paper No. 379, FAO, Rome.

Caddy, J.F., 2004. Current usage of fisheries indicators and reference points, and their potential application to management of fisheries for marine invertebrates. Can. J. Fish. Aquat. Sci. 61, 1307–1324.

Caddy, J.F., Mahon, R., 1995. Reference Points for Fisheries Management. FAO Fisheries Technical Paper No. 347, FAO, Rome.

Caddy, J.F., Rodhouse, P.G., 1998. Cephalopod and groundfish landings: evidence for ecological change in global fisheries? Rev. Fish Biol. Fish. 8 (4), 431–444.

Cadrin, S.X., Hatfield, E.M.C., 1999. Stock Assessment of Longfin Inshore Squid, *Loligo pealeii*, Northeast Fisheries Science Center Reference Document 99-12. Woods Hole, MA.

Cadrin, S.X., Secor, D.H., 2009. Accounting for spatial population structure in stock assessment: past, present, and future. In: Rothschild, B.J., Beamish, R.J. (Eds.), The Future of Fisheries Science in North America. Springer, Dordrecht, Netherlands, pp. 405–426.

Cadrin, S.X., Clark, S.H., Schick, D.F., Armstrong, M.P., McCarron, D., Smith, B., 1999. Application of catch-survey models to the northern shrimp fishery in the Gulf of Maine. N. Am. J. Fish. Manag. 19, 551–568.

Cadrin, S.X., Kerr, L.A., Mariani, S. (Eds.), 2013. Stock Identification Methods: Applications in Fishery Science. Academic Press, London.

California Department of Fish and Game, 2007. Review of some California fisheries for 2006: coastal pelagic finfish, market squid, Dungeness crab, spot prawn, highly migratory species, ocean salmon, California halibut, nearshore live-fishes, cabezon, surfperches, and leopard shark. CalCoFI Rep. 46, 10–31.

Cao, J., Chen, X., Chen, Y., Liu, B., Ma, J., Li, S., 2011. Generalized linear Bayesian models for standardizing CPUE: an application to a squid-jigging fishery in the northwest Pacific Ocean. Sci. Mar. 75, 679–689.

Castilla, J.C., 2010. Fisheries in Chile: small-pelagics, management, rights and sea zoning. Bull. Mar. Sci. 86, 221–234.

Cavaleiro, C.D., 2006. Segmentação da pescaria de potas desembarcadas no porto de Peniche. Scientific report of the B.Sc. thesis. Escola Superior de Tecnologia do Mar, Instituto Politécnico de Leiria.

Ceriola, L., Ungaro, N., Toteda, F., 2007. A "Traffic" Light approach for the assessment of the broadtail shortfin squid *Illex coindetii* Verany, 1839 in the Southern Adriatic Sea (Central Mediterranean). Rev. Fish Biol. Fish. 17, 145–157.

Challier, L., Royer, J., Robin, J.-P., 2002. Variability in age-at-recruitment and early growth in English Channel *Sepia officinalis* described with statolith analysis. Aquat. Living Resour. 15, 303–311.

Challier, L., Dunn, M.R., Robin, J.-P., 2005a. Trends in age-at-recruitment and juvenile growth of cuttlefish, *Sepia officinalis*, from the English Channel. ICES J. Mar. Sci. 62, 1671–1682.

Challier, L., Royer, J., Pierce, G.J., Bailey, N., Roel, B., Robin, J.-P., 2005b. Environmental and stock effects on recruitment variability in the English Channel squid *Loligo forbesi*. Aquat. Living Resour. 18, 353–360.

Challier, L., Orr, P., Robin, J.-P., 2006a. Introducing inter-individual growth variability in the assessment of a cephalopod population: application to the English Channel squid *Loligo forbesi*. Oecologia 150, 17–28.

Challier, L., Pierce, G.J., Robin, J.-P., 2006b. Spatial and temporal variation in age and growth in juvenile *Loligo forbesi* and relationships with recruitment in the English Channel and Scottish waters. J. Sea Res. 55, 217–229.

Charles, A., 2009. Rights-based fisheries management: the role of use rights in managing access and harvesting. In: Cochrane, K.L., Garcia, S.M. (Eds.), A Fishery Manager's Guidebook. FAO/Wiley-Blackwell.

Chen, C.-S., 2010. Abundance trends of two neon flying squid (*Ommastrephes bartramii*) stocks in the North Pacific. ICES J. Mar. Sci. 67, 1336–1345.

Chen, D.S., Dykhuizen, G., Hodge, J., Gilly, W.F., 1996. Ontogeny of copepod predation in juvenile squid (*Loligo opalescens*). Biol. Bull. 190, 69–81.

Chen, C., Pierce, G., Wang, J., Robin, J.-P., Poulard, J.-C., Pereira, J., Zuur, A., Boyle, P., Bailey, N., Beare, D., 2006. The apparent disappearance of *Loligo forbesi* from the south of its range in the 1990s: trends in *Loligo* spp. abundance in the northeast Atlantic and possible environmental influences. Fish. Res. 78, 44–54.

Chen, C.-S., Huang, W.-B., Chiu, T.-S., 2007. Different spatiotemporal distribution of Argentine short-finned squid (*Illex argentinus*) in the Southwest Atlantic during high-abundance year and its relationship to sea water temperature changes. Zool. Stud. 46, 362–374.

Chen, X., Liu, B., Chen, Y., 2008. A review of the development of Chinese distant-water squid jigging fisheries. Fish. Res. 89, 211–221.

Cheng, S.H., Anderson, F.E., Bergman, A., Mahardika, G.N., Muchlisin, Z.A., Dang, B.T., Calumpong, H.P., Mohamed, K.S., Sasikumar, G., Venkatesan, V., Barber, P.H., 2014. Molecular evidence for co-occurring cryptic lineages within the *Sepioteuthis cf. lessoniana* species complex in the Indian and Indo-West Pacific Oceans. Hydrobiologia 725, 165–188.

Chilvers, B.L., 2008. New Zealand sea lions *Phocarctos hookeri* and squid trawl fisheries: bycatch problems and management options. Endanger. Species Res. 5, 193–204.

Christofferson, J.P., Foss, A., Lambert, W.E., Welge, B., 1978. An electrophoretic study of select proteins from the market squid, *Loligo opalescens* Berry. Calif. Dept. Fish & Game Fish Bull. 169, 123–133.

Cinti, A., Baron, P.J., Rivas, A.L., 2004. The effect of environmental factors on the embryonic survival of the Patagonian squid *Loligo gahi*. J. Exp. Mar. Biol. Ecol. 313, 225–240.

Clarke, M.R., 1996. Cephalopods as prey. III Cetaceans. Philos. Trans. R Soc. Lond. B 351, 1053–1065.

Cochrane, K.L. (Ed.), 2002. A Fishery Manager's Guidebook. Management Measures and Their Application. FAO Fisheries Technical Paper No. 424, FAO, Rome.

Cochrane, K.L., Garcia, S.M. (Eds.), 2009. A Fishery Manager's Guidebook. FAO/Wiley-Blackwell, United Kingdom.

Coelho, M., Quintela, J., Bettencourt, V., Olavo, G., Villa, H., 1994. Population structure, maturation patterns and fecundity of the squid *Loligo vulgaris* from southern Portugal. Fish. Res. 21, 87–102.

Coffey, C., 2005. What role for public participation in fisheries governance? In: Gray, T.S. (Ed.), Participation in Fisheries Governance. Springer Netherlands, Dordrecht, The Netherlands, pp. 27–44.

Collie, J.S., Kruse, G.H., 1998. Estimating king crab *Paralithodes camtschaticus* abundance from commercial catch and research survey data. Can. Spec. Publ. Fish. Aquat. Sci. 125, 73–83.

Collie, J.S., Sissenwine, M.P., 1983. Estimating population size from relative abundance data measured with error. Can. J. Fish. Aquat. Sci. 40, 1871–1879.

Collins, M., Burnell, G., Rodhouse, P., 1995. Recruitment, maturation, and spawning of *Loligo forbesi* Steenstrup (Cephalopoda: Loliginidae) in Irish waters. ICES J. Mar. Sci. 52, 127–137.

Collins, M., Pierce, G., Boyle, P., 1997. Populations indices of reproduction and recruitment in *Loligo forbesi* (Cephalopoda: Loliginidae) in Scottish and Irish waters. J. Appl. Ecol. 34, 778–786.

Collins, M., Boyle, P., Pierce, G., Key, L., Hughes, S., Murphy, J., 1999. Resolution of multiple cohorts in the *Loligo forbesi* population from the west of Scotland. ICES J. Mar. Sci. 56, 500–509.

Collins, M.A., Yau, C., Boyle, P.R., Friese, D., Piatkowski, U., 2002. Distribution of cephalopods from plankton surveys around the British Isles. Bull. Mar. Sci. 71, 239–254.

Conser, R.J., 1991. A DeLury method for scallops incorporating length based selectivity of the recruiting year-class to the survey gear and partial recruitment to the commercial fishery. In: Northeast Fisheries Science Center, Report of the Twelfth Northeast Regional Stock Assessment Workshop (12th SAW), Northeast Fisheries Science Center Reference Document 91-03. Woods Hole, MA. Appendix.

Conser, R.J., Idoine, J., 1992. A modified DeLury model for estimating mortality rates and stock sizes of American lobster populations. In: Northeast Fisheries Science Center, Report of the Fourteenth Northeast Regional Stock Assessment Workshop (14th SAW), Northeast Fisheries Science Center Reference Document 92-07. Woods Hole, MA. Appendix.

Crespi-Abril, A.C., Baron, P.J., 2012. Revision of the population structuring of *Illex argentinus* (Castellanos, 1960) and a new interpretation based on modelling the spatio-temporal environmental suitability for spawning and nursery. Fish. Oceanogr. 21, 199–214.

Crespi-Abril, A.C., Morsan, E.M., Williams, G.N., Gagliardini, D.A., 2013. Spatial distribution of *Illex argentinus* in San Matias Gulf (Northern Patagonia, Argentina) in relation to environmental variables: a contribution to the new interpretation of the population structuring. J. Sea Res. 77, 22–31.

Crespi-Abril, A., Villanueva Gomila, G., Venerus, L., Baroń, P., 2014. Spatial distribution of cephalopod paralarvae in San Jose′ Gulf (Northern Patagonia, Argentina): the role of tidal circulation in larval dispersal. Fish. Res. 152, 13–20.

Croxall, J.P., Prince, P.A., 1996. Cephalopods as prey. I. Seabirds. Philos. Trans. R Soc. Lond. 351, 1023–1043.

Cuccu, D., Mereu, M., Porcu, C., Follesa, M.C., Cau, Al, Cau, A., 2013. Development of sexual organs and fecundity in *Octopus vulgaris* Cuvier, 1797 from the Sardinian waters (Mediterranean Sea). Mediterranean. Mar. Sci. 14, 270–277.

Dawe, E.G., 1988. Length-weight relationships of the shortfinned squid in Newfoundland and the effect of the diet on condition and growth. Trans. Am. Fish. Soc. 117, 591–599.

Dawe, E.G., Brodziak, J.K.T., 1998. Trophic relationships, ecosystem variability and recruitment. In: Rodhouse, P.G., Dawe, E.G., O'Dor, R.K. (Eds.), Squid Recruitment Dynamics. The Genus *Illex* as a Model, the Commercial *Illex* Species and Influence on Variability. In: FAO Fisheries Technical Paper No. 376, FAO, Rome, pp. 125–156.

Dawe, E.G., Hendrickson, L., 1998. A review of the biology, population dynamics, and exploitation of short-finned squid in the northwest Atlantic Ocean, in relation to assessment and management of the resource. NAFO SCR Doc. 98/59.

Dawe, E.G., Warren, W.G., 1993. Recruitment of short-finned squid in the northwest Atlantic Ocean and some environmental relationships. J. Cephalopod Biol. 2 (2), 1–21.

Dawe, E.G., Colbourne, E.B., Drinkwater, K.F., 2000. Environmental effects on recruitment of short-finned squid (*Illex illecebrosus*). ICES J. Mar. Sci. 57, 1002–1013.

Dawe, E.G., Hendrickson, L.C., Colbourne, E.B., Drinkwater, K.F., Showell, M.A., 2007. Ocean climate effects on the relative abundance of short-finned (*Illex illecebrosus*) and long-finned (*Loligo pealeii*) squid in the northwest Atlantic Ocean. Fish. Oceanogr. 16, 303–316.

De Angelis, P., 2012. Assessing the impact of international trade on chambered nautilus. Geobios 45, 5–11.

De Lury, D.B., 1947. On the estimation of biological populations. Biometrics 3, 145–167.

Demarcq, H., Faure, V., 2013. Coastal upwelling and associated retention indices derived from satellite SST. Application to *Octopus vulgaris* recruitment. Oceanol. Acta 23, 391–408.

DfID (Department for International Development, UK), 2007. Programme of Support to Tackling IUU Fishing in Southern Africa: Programme Brief and Update no. 1. p. 11.

Diallo, M., Ortiz, M., 2002. Estimation of standardized index of abundance of octopus (*O. vulgaris*) from the Senegalese artisanal fishery (1989-1994). In: Caverivière, A., Thiam, M., Jouffre, D. (Eds.), Le Poulpe *Octopus vulgaris*. IRD, Paris, pp. 223–232.

Diaz-De-León, A.J., Seijo, J.C., 1992. A multi-criteria non-linear optimization model. Mar. Resour. Econ. 7, 23–40.

Diekmann, R., Nellen, W., Piatkowski, U., 2006. A multivariate analysis of larval fish and paralarval cephalopod assemblages at Great Meteor Seamount. Deep Sea Res. I 53, 1635–1657.

Dillane, E., Galvin, P., Coughlan, J., Rodhouse, P., Cross, T., 2000. Polymorphic variable number of tandem repeat (VNTR) loci in the ommastrephid squid, *Illex coindetii* and *Todaropsis eblanae*. Mol. Ecol. 9, 1002–1004.

Dillane, E., Galvin, P., Coughlan, J., Lipinski, M., Cross, T., 2005. Genetic variation in the lesser flying squid *Todaropsis eblanae* (Cephalopoda, Ommastrephidae) in east Atlantic and Mediterranean waters. Mar. Ecol. Prog. Ser. 292, 225–232.

Domain, F., Jouffre, D., Caverivière, A., 2000. Growth of *Octopus vulgaris* from tagging in Senegalese waters. J. Mar. Biol. Assoc. U.K. 80, 699–705.

Dorval, E., Crone, P.R., McDaniel, J.D., 2013. Variability of egg escapement, fishing mortality and spawning population in the market squid fishery in the California Current Ecosystem. Mar. Freshw. Res. 64, 80–90.

Dunn, M.R., 1999. The exploitation of selected non-quota species in the English channel. PhD Thesis, University of Portsmouth, United Kingdom.

Dunstan, A., Alanis, O., Marshall, J., 2010. *Nautilus pompilius* fishing and population decline in the Philippines: a comparison with an unexploited Australian *Nautilus* population. Fish. Res. 106, 239–247.

Dunstan, A.J., Ward, P.D., Marshall, N.J., 2011. Vertical distribution and migration patterns of *Nautilus pompilius*. PLoS One 6 (2), e16311.

Dupavillon, J.L., Gillanders, M.L., 2009. Impacts of seawater desalination on the giant Australian cuttlefish *Sepia apama* in the upper Spencer Gulf, South Australia. Mar. Environ. Res. 67, 207–218.

Ehrhardt, N.M., Jacquemin, P.S., Garcia, F.B., Gonzalez, G.D., Lopez, J.M.B., Ortiz, J.C., Solis, A.N., 1983. On the fishery and biology of the giant squid *Dosidicus gigas* in the Gulf of California, Mexico. In: Caddy, J.F. (Ed.), Advances in Assessment of World Cephalopod Resources. In: Fisheries Technical Paper No. 231, FAO, Rome, pp. 306–340.

Emeis, K.-C., Brüchert, V., Currie, B., Endler, R., Ferdelman, T., Kiessling, A., Leipe, T., Noli-Peard, K., Struck, U., Vogt, T., 2004. Shallow gas in shelf sediments of the Namibian coastal upwelling ecosystem. Cont. Shelf Res. 24, 627–642.

Epps, M., 2007. A socioeconomic baseline assessment: implementing the socioeconomic monitoring guidelines in southwest Madagascar. Blue Ventures Conservation Report.

Erzini, K., Inejih, C.A.O., Stobberup, K.A., 2005. An application of two techniques for the analysis of short, multivariate non-stationary time-series of Mauritanian trawl survey data. ICES J. Mar. Sci. 62, 353–359.

FAO, 1996. Precautionary Approach to Capture Fisheries and Species Introductions, Elaborated by the Technical Consultation on the Precautionary Approach to Capture Fisheries (Including species introductions) (6-13 June 1995, Lysekil, Sweden). FAO Technical Guidelines for Responsible Fisheries, No. 2, FAO, Rome.

FAO, 2003. Technical Guidelines for Responsible Fisheries, No. 4, FAO, Rome, Suppl. 2.

FAO, 2004. The State of World Fisheries and Aquaculture. FAO Fisheries Department, Rome.

FAO, 2011. FishStatJ: Universal Software for Fishery Statistical Time Series. FAO Fisheries and Aquaculture Department. Statistics and Information Service, Rome.

Faure, V., Inejih, C.A., Demarcq, H., Cury, P., 2000. The importance of retention processes in upwelling areas for recruitment of *Octopus vulgaris*: the example of the Arguin Bank (Mauritania). Fish. Oceanogr. 9, 343–355.

Fernández-Rueda, P., García-Flórez, L., 2007. *Octopus vulgaris* (Mollusca: Cephalopoda) fishery management assessment in Asturias (north-west Spain). Fish. Res. 83, 351–354.

Field, J.C., Baltz, K., Phillips, A.J., Walker, W.A., 2007. Range expansion and trophic interactions of the jumbo squid, *Dosidicus gigas*, in the California Current. CalCoFI Rep. 48, 131–146.

Fields, W.G., 1962. The structure, development, food relations, reproduction and life history of the squid *Loligo opalescens* Berry. PhD Thesis, Department of Biological Sciences, Stanford University.

FIGIS, 2011. FAO Fisheries Statistics. Global Capture Production 1950-2009. http://www.fao.org/fishery/statistics/global-capture-production.

Fletcher, D., MacKenzie, D., Villouta, E., 2005. Modelling skewed data with many zeros: a simple approach combining ordinary and logistic regression. Environ. Ecol. Stat. 12, 45–54.

Flewwelling, P., 2001. Fisheries Management and MCS in South Asia. FAO/FISHCODE Project: GCP/INT/648/NOR: Field Report C-6 (En). FAO, Rome.

Flood, M., Stobutzki, I., Andrews, J., Begg, G., Fletcher, W., Gardner, C., Kemp, J., Moore, A., O'Brien, A., Quinn, R., Roach, J., Rowling, K., Sainsbury, K., Saunders, T., Ward, T., Winning, M. (Eds.), 2012. Status of Key Australian Fish Stocks Reports 2012. Fisheries Research and Development Corporation, Canberra.

Fonseca, P., Campos, A., Garcia, A., 2002. Bottom trawl codend selectivity for cephalopods in Portuguese continental waters. Fish. Res. 59, 263–271.

Forsythe, J.W., 1993. A working hypothesis of how seasonal temperature change may impact the field growth of young cephalopods. In: Okutani, T., O'Dor, R.K., Kubodera, T.

(Eds.), Recent Advances in Fisheries Biology. Tokai University Press, Tokyo, pp. 133–143.
Forsythe, J.W., 2004. Accounting for the effect of temperature on squid growth in nature: from hypothesis to practice. Mar. Freshw. Res. 55, 331–339.
Forsythe, J.W., Walsh, L.S., Turk, P.E., Lee, P.G., 2001. Impact of temperature on juvenile growth and age at the first egg-laying of the Pacific reef squid *Sepioteuthis lessoniana* reared in captivity. Mar. Biol. 138, 103–112.
Fox, W.W.J., 1970. An exponential surplus-yield model for optimizing exploited fish populations. Trans. Am. Fish. Soc. 99, 80–88.
Fréon, P., Misund, O.A., 1999. Dynamics of Pelagic Fish Distribution and Behavior: Effects on Fisheries and Stock Assessment. Fishing News Books, Oxford.
Fréon, P., Cury, P., Shannon, L., Roy, C., 2005. Sustainable exploitation of small pelagic fish stocks challenged by environmental and ecosystem changes: a review. Bull. Mar. Sci. 76, 385–462.
Fries, A.S., 2010. A gap analysis of the distributions of cephalopod species worldwide with a focus on commercially important species. Master's Thesis, Nicholas School of the Environment and Earth Sciences, Duke University.
Froerman, Y.M., 1986. Ecology and mechanism of abundance dynamics of the short-finned squid *Illex illecebrosus*. PhD Thesis, Shirshov Institute of Oceanography, Moscow.
Fulton, E.A., Parslow, J.S., Smith, A.D., Johnson, C.R., 2004. Biogeochemical marine ecosystem models II: the effect of physiological detail on model performance. Ecol. Model. 173, 371–406.
Gaichas, S.K., Aydin, K.Y., Francis, R.C., 2010. Using food web model results to inform stock assessment estimates of mortality and production for ecosystem-based fisheries management. Can. J. Fish. Aquat. Sci. 67, 1490–1506.
Garcia, S.M., 2003. A review of the ecosystem approach to fisheries. In: Ba, M., Chavance, P., Gascuel, D., Pauly, D., Vakily, M., Ba, M., Chavance, P., Gascuel, D., Pauly, D., Vakily, M. (Eds.), Pêcheries Maritimes, Écosystèmes et Sociétés en Afrique de l'Ouest: un Demi-siècle de Changement, Actes du Symposium International de Dakar (Sénégal). ACP-UE.
Garcia, S.M., Cochrane, K.L., 2005. Ecosystem approach to fisheries: a review of implementation guidelines. ICES J. Mar. Sci. 62, 311–318.
García-Rodriguez, M., Fernandez, A.M., Esteban, A., 2006. Characterisation, analysis and catch rates of the small scale fisheries of the Alicante Gulf (SE Spain) over a 10 year time series. Fish. Res. 77, 226–238.
Garofalo, G., Ceriola, L., Gristina, M., Fiorentino, F., Pace, R., 2010. Nurseries, spawning grounds and recruitment of *Octopus vulgaris* in the Strait of Sicily, central Mediterranean Sea. ICES J. Mar. Sci. 67, 1363–1371.
Gasalla, M.A., Rodrigues, A.R., Postuma, F.A., 2010. The trophic role of the squid *Loligo plei* as a keystone species in the South Brazil Bight ecosystem. ICES J. Mar. Sci. 67, 1413–1424.
Georgakarakos, S., Koutsoubas, D., Valavanis, V., 2006. Time series analysis and forecasting techniques applied on loliginid and ommastrephid landings in Greek waters. Fish. Res. 78, 55–71.
Gilly, W., Elliger, C., Salinas, C., Camarilla-Coop, S., Bazzino, G., Beman, M., 2006. Spawning by jumbo squid *Dosidicus gigas* in San Pedro Mártir Basin, Gulf of California, Mexico. Mar. Ecol. Prog. Ser. 313, 125–133.
Glass, C.W., Sarno, B., Morris, G.D., Milliken, H.O., Carr, H.A., 1999. Bycatch reduction in Massachusetts inshore squid (*Loligo pealeii*) trawl fisheries. Mar. Technol. Soc. J. 33 (2), 35–42.

Glazer, J.P., Butterworth, D.S., 2002. GLM-based standardization of the catch per unit effort series for South African west coast hake, focusing on adjustments for targeting other species. S. Afr. J. Mar. Sci. 24, 323–339.

Glazer, J.P., Butterworth, D.S., 2006. Some refinements of the assessment of the South African squid resource, *Loligo vulgaris reynaudii*. Fish. Res. 78, 14–25.

Goethel, D.R., Quinn, T.J., Cadrin, S.X., 2011. Incorporating spatial structure in stock assessment: movement modeling in marine fish population dynamics. Rev. Fish. Sci. 19, 119–136.

Gonzalez, A., Guerra, A., 2013. *Illex coindeti*, Broadtail short-fin squid. In: Rosa, R., Pierce, G., O'Dor, R. (Eds.), Advances in Squid Biology, Ecology and Fisheries. Part II Oegopsid Squids. Nova Biomedical, New York, pp. 49–72.

González, A.F., Otero, J., Guerra, A., Prego, R., Rocha, F.J., Dale, A.W., 2005. Distribution of common octopus and common squid paralarvae in a wind-driven upwelling area (Ria of Vigo, northwestern Spain). J. Plankton Res. 27, 271–277.

Goss, C., Middleton, D., Rodhouse, P., 2001. Investigations of squid stocks using acoustic survey methods. Fish. Res. 54, 111–121.

Goto, T., 2002. Paralarval distribution of the Ommastrephid squid *Todarodes pacificus* during fall in the southern Sea of Japan, and its implication for locating spawning grounds. Bull. Mar. Sci. 71, 299–312.

Gowland, F.C., Moltschaniwskyj, N.A., Steers, M.A., 2002. Description and quantification of developmental abnormalities in a natural *Sepioteuthis australis* spawning population (Mollusca: Cephalopoda). Mar. Ecol. Prog. Ser. 243, 133–141.

Graham, M., 1935. Modern theory of exploiting a fishery, and application to North Sea trawling. J. Cons. Int. Explor. Mer. 10, 264–274.

Gras, M., Roel, B. A., Coppin, F., Foucher, E., Robin, J.-P. (2014). A two-stage biomass model to assess the English Channel cuttlefish (*Sepia officinalis L.*) stock. ICES journal of Marine Science in press.

Greatorex, E., Jones, C., Murphy, J., Key, L., Emery, A., Boyle, P., 2000. Microsatellite markers for investigating population structure in *Octopus vulgaris* (Mollusca: Cephalopoda). Mol. Ecol. 9, 641–642.

Grist, E.P., Jackson, G.D., 2004. Energy balance as a determinant of two-phase growth in cephalopods. Mar. Freshw. Res. 55, 395–401.

Guard, M., 2003. Assessment of the artisanal fishery of *Octopus cyanea* Gray, 1929 in Tanzania: catch dynamics, fisheries biology, socio-economics and implications for management. PhD Thesis, University of Aberdeen, Scotland.

Guard, M., 2009. Terms of Reference Report for the Stock Assessment and Management of Octopus Fisheries in South-west Madagascar: a Proposed Framework, Regional Programme for the Sustainable Management of the Coastal Zone of the Countries of the Indian Ocean (Recomap).

Guerra, Á., 2006. Ecology of *Sepia officinalis*. Vie Milieu 56 (2), 97–107.

Guerra, A., Rocha, F., 1994. The life history of *Loligo vulgaris* and *Loligo forbesi* (Cephalopoda: Loliginidae) in Galician waters (NW Spain). Fish. Res. 21, 43–69.

Guidetti, P., Claudet, J., 2010. Comanagement practices enhance fisheries in marine protected areas. Conserv. Biol. 24, 312–318.

Gutiérrez, N.L., Hilborn, R., Defeo, O., 2011. Leadership, social capital and incentives promote successful fisheries. Nature 470, 386–389.

Haimovici, M., Brunetti, N., Rodhouse, P., Csirke, J., Leta, R., 1998. Illex argentinus. In: Rodhouse, P.G., Dawe, E.G., O'Dor, R.K. (Eds.), Squid Recruitment Dynamics. The Genus *Illex* as a Model, the Commercial *Illex* Species and Influence on Variability. In: FAO Fisheries Technical Paper No. 376, FAO, Rome, pp. 27–58.

Haimovici, M., Piatkowski, U., dos Santos, R.A., 2002. Cephalopod paralarvae around tropical seamounts and oceanic islands off the north-eastern coast of Brazil. Bull. Mar. Sci. 71, 313–330.

Haimovici, M., Leite, T.S., Marinho, R.A., Batista, B., Madrid, R.M., Oliveira, J.E.L., Lima, F. D., Candice, L., in press. As pescarias de polvos do Nordeste do Brasil. In: Haimovici, Andriguetto, Sunye (Eds.), A pesca marinha e estuarina no Brasil: abordagem multidisciplinar aplicada a estudos de caso. Editora da Furg.
Hall, K., Hanlon, R., 2002. Principal features of the mating system of a large spawning aggregation of the giant Australian cuttlefish *Sepia apama* (Mollusca: Cephalopoda). Mar. Biol. 140, 533–545.
Hall, K.C., Fowler, A.J., Geddes, M.C., 2007. Evidence for multiple year classes of the giant Australian cuttlefish *Sepia apama* in northern Spencer Gulf, South Australia. Rev. Fish Biol. Fish. 17, 367–384.
Hanlon, R.T., 1998. Mating systems and sexual selection in the squid *Loligo*: how might commercial fishing on spawning grounds affect them? CalCOFI Rep. 39, 92–100.
Hanlon, R.T., Messenger, J.B., 1996. Cephalopod Behaviour. Cambridge University Press, Cambridge.
Hare, S.R., Mantua, N.J., 2000. Empirical evidence for North Pacific regime shifts in 1977 and 1989. Prog. Oceanogr. 47, 103–145.
Hastie, L.C., 1996. Estimation of trawl codend selectivity for squid (*Loligo forbesi*), based on Scottish research vessel survey data. ICES J. Mar. Sci. 53, 741–744.
Hastie, T., Tibshirani, R.A., 1986. Generalized additive models. Stat. Sci. 1, 297–318.
Hastie, L., Joy, J., Pierce, G., Yau, C., 1994. Reproductive biology of *Todaropsis eblanae* (Cephalopoda: Ommastrephidae) in Scottish waters. J. Mar. Biol. Assoc. U.K. 74, 367–382.
Hastie, L., Pierce, G., Pita, C., Viana, M., Smith, J., Wangvoralak, S., 2009. Squid Fishing in UK Waters: A Report to SEAFISH Industry Authority. University of Aberdeen.
Hatanaka, H., 1979. Studies on the fisheries biology of common octopus off the northwest coast of Africa. Bull. Far Seas Fish. Res. Lab. 17, 13–124.
Hatfield, E.M., 1996. Towards resolving multiple recruitment into loliginid fisheries: *Loligo gahi* in the Falkland Islands fishery. ICES J. Mar. Sci. 53, 565–575.
Hatfield, E., Des Clers, S., 1998. Fisheries management and research for *Loligo gahi* in the Falkland Islands. CalCoFI Rep. 39, 81–91.
Hatfield, E.M.C., Rodhouse, P.G., 1994. Distribution and abundance of juvenile *Loligo gahi* in Falkland Island waters. Mar. Biol. 121, 267–272.
Hauck, M., Sweijd, N.A., 1999. A case study of abalone poaching in South Africa and its impact on fisheries management. ICES J. Mar. Sci. 56, 1024–1032.
Hendrickson, L.C., 2004. Population biology of northern shortfin squid (*Illex illecebrosus*) in the Northwest Atlantic Ocean and initial documentation of a spawning area. ICES J. Mar. Sci. 61, 252–266.
Hendrickson, L.C., 2011. Effects of a codend mesh size increase on size selectivity and catch rates in a small-mesh bottom trawl fishery for longfin inshore squid, *Loligo pealeii*. Fish. Res. 108, 42–51.
Hendrickson, L.C., Hart, D.R., 2006. An age-based cohort model for estimating the spawning mortality of semelparous cephalopods with an application to per-recruit calculations for the northern shortfin squid, *Illex illecebrosus*. Fish. Res. 78, 4–13.
Hibberd, T., Pecl, G.T., 2007. Effects of commercial fishing on the population structure of spawning southern calamary (*Sepioteuthis australis*). Rev. Fish Biol. Fish. 17, 207–221.
Hilborn, R., Walters, C., 1992. Quantitative Fisheries Stock Assessment: Choice, Dynamics and Uncertainty. Springer, Chapman and Hall, New York.
Hill, S., Agnew, D.J., 2002. Optimal harvest strategies for single-cohort squid populations. Bull. Mar. Sci. 71, 371–381.
Hiramatsu, K., 2010. Statistical problems of estimating stock and recruitment relationship. Kaiyo 42, 204–208.
Hochberg, F.G., 1983. The parasites of cephalopods: a review. Mem. Natl. Mus. Vict. 44, 109–145.

Holme, N., 1974. The biology of *Loligo forbesi* Steenstrup (Mollusca: Cephalopoda) in the Plymouth area. J. Mar. Biol. Assoc. U.K. 54, 481–503.
Hoving, H.J., Gilly, W.F., Markaida, U., Benoit-Bird, K.J., Brown, Z.W., Daniel, P., Field, J.C., Parissenti, L., Liu, B., Campos, B., 2013. Extreme plasticity in life-history strategy allows a migratory predator (jumbo squid) to cope with a changing climate. Glob. Change Biol. 19, 2089–2103.
Hunsicker, M.E., Essington, T.E., Watson, R., Sumaila, U.R., 2010. The contribution of cephalopods to global marine fisheries: can we have our squid and eat them too? Fish Fish. 11, 421–438.
Ibáñez, C.M., Chong, J., 2008. Feeding ecology of *Enteroctopus megalocyathus* (Gould 1852) (Cephalopoda: Octopodidae). J. Mar. Biol. Assoc. U.K. 88, 793–798.
Ibáñez, C.M., Keyl, F., 2010. Cannibalism in cephalopods. Rev. Fish Biol. Fish. 20, 123–136.
Ichii, T., Mahapatra, K., Sakai, M., Wakabayashi, T., Okamura, H., Igarashi, H., Inagaki, D., Okada, Y., 2011. Changes in abundance of the neon flying squid *Ommastrephes bartramii* in relation to climate change in the central North Pacific Ocean. Mar. Ecol. Prog. Ser. 441, 151–164.
Iida, K., Takahashi, R., Tang, Y., Mukai, T., Sato, M., 2006. Observation of marine animals using underwater acoustic camera. Jpn. J. Appl. Phys. 45, 4875–4881.
Imaizumi, T., Furusawa, M., Akamatsu, T., Nishimori, Y., 2008. Measuring the target strength spectra of fish using dolphin-like short broadband sonar signals. J. Acoust. Soc. Am. 124, 3440–3449.
Instituto de Fomento Pesquero, 2010. Caracterización biológico pesquera de las actividades extractivas del recurso pulpo en la X Región. Informe Final Corregido. FIP No 2008-40. Instituto de Fomento Pesquero, Valparaíso, Chile. 485 pp.
Ish, T., Dick, E.J., Switzer, P.V., Mangel, M., 2004. Environment, krill and squid in the Monterey Bay: from fisheries to life histories and back again. Deep Sea Res. II 51, 849–862.
IUCN, 1996. Resolutions and recommendations, World Conservation Congress, Montreal (Canada), 13–23 October, 1996, 164p, ISBN 2-8317-0391-3. http://iucn.org/wcc/resolutions/resrecen.pdf.
Iwata, Y., Ito, K., Sakurai, Y., 2010. Is commercial harvesting of spawning aggregations sustainable? The reproductive status of the squid *Loligo bleekeri*. Fish. Res. 102, 286–290.
Jackson, G.D., 1998. Research into the life history of *Loligo opalescens*: where to from here? CalCoFI Rep. 39, 101–107.
Jackson, G., Domeier, M., 2003. The effects of an extraordinary El Niño/La Niña event on the size and growth of the squid *Loligo opalescens* off Southern California. Mar. Biol. 142, 925–935.
Jackson, G.D., Pecl, G., 2003. The dynamics of the summer-spawning population of the loliginid squid *Sepioteuthis australis* in Tasmania, Australia—a conveyor belt of recruits. ICES J. Mar. Sci. 60, 290–296.
Jackson, G., Wotherspoon, S., McGrath-Steer, B., 2005. Temporal population dynamics in arrow squid *Nototodarus gouldi* in southern Australian waters. Mar. Biol. 146, 975–983.
Jackson, G., Bustamante, P., Cherel, Y., Fulton, E., Grist, E., Jackson, C., Nichols, P., Pethybridge, H., Phillips, K., Ward, R., 2007. Applying new tools to cephalopod trophic dynamics and ecology: perspectives from the Southern Ocean Cephalopod Workshop, February 2–3, 2006. Rev. Fish Biol. Fish. 17, 79–99.
Jelizarov, A.A. (Ed.), 1996. Commercial Aspects of Biology of Commander Squid *Berryteuthis magister* and of Fishes of Slope Communities in the Western Part of the Bering Sea. VNIRO Publishing House, Moscow.
Jentoft, S., Kristoffersen, T., 1989. Fishermen's co-management: the case of the Lofoten fishery. Hum. Organ. 48, 355–365.

Jentoft, S., McCay, B., 1995. User participation in fisheries management: lessons drawn from international experiences. Mar. Policy 19, 227–246.
Jereb, P., Roper, C.F. (Eds.), 2005. Cephalopods of the World. An Annotated and Illustrated Catalogue of Cephalopod Species Known to Date. Volume 1. Chambered Nautiluses and Sepioids (Nautilidae, Sepiidae, Sepiolidae, Sepiadariidae, Idiosepiidae and Spirulidae). FAO, Rome.
Jereb, P., Roper, C.F.E. (Eds.), 2010. Cephalopods of the World. An Annotated and Illustrated Catalogue of Cephalopod Species Known to Date. Vol 2. Myopsid and Oegopsid Squids. FAO Species Catalogue for Fishery Purposes, Vol. 2. FAO, Rome, No. 4.
Johnson, T.R., 2011. Fishermen, scientists, and boundary spanners: cooperative research in the U.S. *Illex* squid fishery. Soc. Nat. Resour. 24, 242–255.
Johnson, T.R., van Densen, W.L.T., 2007. The benefits and organization of cooperative research for fisheries management. ICES J. Mar. Sci. 64, 834–840.
Jones, R., 1976. Growth of fishes. In: Cushing, D.H., Walsh, J.J. (Eds.), The Ecology of the Seas. Blackwell Scientific Publications, Oxford, pp. 251–279.
Jorgensen, E.M., 2007. Identification, distribution and relative abundance of paralarval gonatid squids (Cephalopoda: Oegopsida: Gonatidae) from the Gulf of Alaska, 2001-2003. J. Molluscan Stud. 73, 155–165.
Jouffre, D., Lanco, S., Gascuel, D., Caverivière, A., 2002. Évaluation par modélisation analytique des effets de périodes de fermeture de la pêche du poulpe au Sénégal. In: Caverivière, A., Thiam, M., Jouffre, D. (Eds.), Le Poulpe *Octopus vulgaris*. IRD, Paris, pp. 297–316.
Kang, M., Furusawa, M., Miyashita, K., 2002. Effective and accurate use of difference in mean volume backscattering strength to identify fish and plankton. ICES J. Mar. Sci. 59, 794–804.
Kang, D., Mukai, T., Iida, K., Hwang, D., Myoung, J., 2005. The influence of tilt angle on the acoustic target strength of the Japanese common squid (*Todarodes pacificus*). ICES J. Mar. Sci. 62, 779–789.
Kaplan, M.B., Mooney, T.A., McCorkle, D.C., Cohen, A.L., 2013. Adverse effects of ocean acidification on early development of squid (*Doryteuthis pealeii*). PLoS One 8 (5), e63714. http://dx.doi.org/10.1371/journal.pone.006371.
Karnik, N.S., Chakraborty, S., Jaiswar, A., Swamy, R., Rajaprasad, R., Boomireddy, S., Rizvi, A., 2003. Growth and mortality of Indian squid, *Loligo duvauceli* (d'Orbigny) (Mollusca: Cephalopoda: Teuthoidea) from Mumbai waters, India. Ind. J. Mar. Sci. 32, 67–70.
Kasahara, S., 1978. Descriptions of offshore squid angling in the Sea of Japan, with special reference to the distribution of common squid (*Todarodes pacificus* Steenstrup); and on the techniques for forecasting fishing conditions. Bull. Jpn Sea Reg. Fish. Res. Lab. 28, 179–199 (in Japanese).
Kashiwada, J., Recksiek, C.W., 1978. Possible morphological indicators of population structure in the market squid, *Loligo opalescens*. Calif. Dept. Fish Game Fish Bull. 169, 99–111.
Kato, S., Hardwick, J.E., 1975. The California squid fishery. FAO Fish. Rep. 170, 107–127.
Katsanevakis, S., Verriopoulos, G., 2006. Seasonal population dynamics of *Octopus vulgaris* in the eastern Mediterranean. ICES J. Mar. Sci. 63, 151–160.
Katugin, O., 2002. Patterns of genetic variability and population structure in the North Pacific squids *Ommastrephes bartramii*, *Todarodes pacificus*, and *Berryteuthis magister*. Bull. Mar. Sci. 71, 383–420.
Katugin, O.N., Shevtsov, G.A., Zuev, M.A., Didenko, V.D., Kulik, V.V., Vanin, N.S., 2013. *Berryteuthis magister*, Schoolmaster gonate squid. In: Rosa, R., O'Dor, R., Pierce, G. (Eds.), Advances in Squid Biology, Ecology and Fisheries. Part II—Oegopsid Squids. Nova Science Publishers, New York, pp. 1–48.
Kawabata, A., 2005. Target strength measurements of suspended live ommastrephid squid, *Todarodes pacificus*, and its application in density estimations. Fish. Sci. 71, 63–72.

Kawabata, A., Yatsu, A., Ueno, Y., Suyama, S., Kurita, Y., 2006. Special distribution of the Japanese common squid, *Todarodes pacificus*, during its northward migration in the western North Pacific Ocean. Fish. Oceanogr. 15, 113–124.

Keyl, F., Wolff, M., 2008. Environmental variability and fisheries: what can models do? Rev. Fish Biol. Fish. 18, 273–299.

Keyl, F., Argüelles, J., Mariátegui, L., Tafur, R., Wolff, M., Yamashiro, C., 2008. A hypothesis on range expansion and spatio-temporal shifts in size-at-maturity of jumbo squid (*Dosidicus gigas*) in the Eastern Pacific Ocean. CalCOFI Rep. 49, 119–128.

Keyl, F., Argüelles, J., Tafur, R., 2011. Interannual variability in size structure, age, and growth of jumbo squid (*Dosidicus gigas*) assessed by modal progression analysis. ICES J. Mar. Sci. 68, 507–518.

Khoufi, W., Jabeur, C., Bakhrouf, A., 2012. Stock assessment of the Common Octopus (*Octopus vulgaris*) in Monastir; the Mid-eastern Coast of Tunisia. Int. J. Mar. Sci. 2 (8), 57–61.

Kidokoro, H., 2009. Impact of climatic changes on the distribution, migration pattern and stock abundance of the Japanese common squid, *Todarodes pacificus* in the sea of Japan. Bull. Fish. Res. Agency 27, 95–189.

Kidokoro, H., Mori, K., 2004. Stock assessment method and management procedure used for the *Todarodes pacificus* fishery in Japan. ICES CM2004/CC:02.

Kidokoro, H., Goto, T., Nagasawa, T., Nishida, H., Akamine, T., Sakurai, Y., 2010. Impacts of a climate regime shift on the migration of Japanese common squid (*Todarodes pacificus*). ICES J. Mar. Sci. 67, 1314–1322.

Kidokoro, H., Goto, T., Matsukura, R., 2013. Stock assessment and evaluation for autumn spawning stock of Japanese common squid. In: Marine Fisheries Stock Assessment and Evaluation for Japanese Waters. Fisheries Agency and Fisheries Research Agency of Japan, Tokyo, pp. 605–635.

Kidokoro, H., Shikata, T., Kitagawa, S., 2014. Forecasting the stock size of the autumn cohort of Japanese common squid (Todarodes pacificus) based on the abundance of trawl-caught juveniles. Hidrobiologica 24, in press.

Kiyofuji, H., Saitoh, S., 2004. Use of night-time visible images to detect Japanese common squid *Todarodes pacificus* fishing areas and potential migration routes in the Sea of Japan. Mar. Ecol. Prog. Ser. 276, 173–186.

Klages, N.T.W., 1996. Cephalopods as prey. II. Seals. Philos. Trans. R Soc. Lond. B 351, 1045–1052.

Koronkiewicz, A., 1995. Size and maturity differences between trawl and jigger caught short-finned squid *Illex argentinus* (Cephalopoda: Ommastrephidae). Acta Ichthyol. Piscat. 25, 97–112.

Koslow, J.A., 2009. The role of acoustics in ecosystem-based fishery management. ICES J. Mar. Sci. 66, 966–973.

Koueta, N., Boucaud-Camou, E., 2003. Combined effects of photoperiod and feeding frequency on survival and growth of juvenile cuttlefish *Sepia officinalis* L. in experimental rearing. J. Exp. Mar. Biol. Ecol. 296, 215–226.

Krstulović Šifner, S., Vrgoč, N., 2004. Population structure, maturation and reproduction of the European squid, *Loligo vulgaris*, in the Central Adriatic Sea. Fish. Res. 69, 239–249.

Kuipers, M.R., 2012. Growth and reproduction of a short-lived Cephalopod: mechanisms that facilitate population success in a highly variable environment. PhD Thesis, University of Tasmania.

Lange, A.M., Waring, G.T., 1992. Fishery interactions between long-finned squid (*Loligo pealei*) and butterfish (*Peprilus triacanthus*) off the Northeast USA. J. Northw. Atl. Fish. Sci. 12, 49–62.

Laptikhovsky, V., 2000. Fecundity of the squid *Loligo vulgaris* Lamarck, 1798 (Myopsida, Loliginidae) of northwest Africa. Sci. Mar. 64, 275–278.

Laptikhovsky, V., 2006. Latitudinal and bathymetric trends in egg size variation: a new look at Thorson's and Rass's rules. Mar. Ecol. 27, 7–14.

Laptikhovsky, V., 2013. Chapter VIII. Todarodes sagittatus. In: Rosa, R., Pierce, G., O'Dor, R. (Eds.), Advances in squid biology, ecology and fisheries. Part II Oegopsid squids. Nova Biomedical, NY, pp. 223–247.

Laptikhovsky, V.V., Nigmatullin, C.M., 1999. Egg size and fecundity in females of the subfamilies Todaopsinae and Todarodinae (Cephalopoda: Ommastrephidae). J. Mar. Biol. Assoc. U.K. 79, 569–570.

Laptikhovsky, V., Salman, A., Onsoy, B., Katagan, T., 2003. Fecundity of the common cuttlefish, *Sepia officinalis* L. (Cephalopoda, Sepiida): a new look on an old problem. Sci. Mar. 67, 279–284.

Laptikhovsky, V.V., Rogov, M.A., Nikolaeva, S.V., Arkhipkin, A.I., 2013. Environmental impact on ecotochleate cephalopod reproductive strategies and the evolutionary significance of cephalopod egg size. Bull. Geosci. 88, 83–93.

Laurans, M., Gascuel, D., Caverivière, A., 2002. Application d'un modèle global avec effet de l'environnement au stock de poulpe du Sénégal. In: Caverivière, A., Thiam, M., Jouffre, D. (Eds.), Le Poulpe *Octopus vulgaris*. IRD, Paris, pp. 255–267.

Le Goff, R., Gauvrit, E., Pinczon du Sel, G., Daguzan, J., 1998. Age group determination by analysis of the cuttlebone of the cuttlefish *Sepia officinalis* L. in reproduction in the Bay of Biscay. J. Molluscan Stud. 64, 183–193.

Le Pape, O., Chauvet, F., Mahévas, S., Lazure, P., Guérault, D., Désaunay, Y., 2003. Quantitative description of habitat suitability for the juvenile common sole (*Solea solea*, L.) in the Bay of Biscay (France) and the contribution of different habitats to the adult population. J. Sea Res. 50, 139–149.

Le Pape, O., Baulier, L., Cloarec, A., Martin, J., Le Loc'h, F., Désaunay, Y., 2007. Habitat suitability for juvenile common sole (*Solea solea* L.) in the Bay of Biscay (France): a quantitative description using indicators based on epibenthic fauna. J. Sea Res. 57, 126–136.

Lee, P.G., Turk, P.E., Yang, W.T., Hanlon, R.T., 1994. Biological characteristics and biomedical applications of the squid *Sepioteuthis lessoniana* cultured through multiple generations. Biol. Bull. 186, 328–341.

Lee, W.-J., Lavery, A.C., Stanton, T.K., 2012. Orientation dependence of broadband acoustic backscattering from live squid. J. Acoust. Soc. Am. 131, 4461–4475.

Lefkaditou, E., Kavadas, S., Papaconstantinou, C., 2002. Cephalopod fisheries statistics in Greek waters (NE Mediterranean), ICES Document CM 1999/G:4.

Lehodey, P., Murtugudde, R., Senina, I., 2010. Bridging the gap from ocean models to population dynamics of large marine predators: a model of mid-trophic functional groups. Prog. Oceanogr. 84 (1), 69–84.

Leite, T.S., Haimovici, M., Lins, J., 2008. Uma proposta de manejo para a pesca de polvo (Mollusca: Cephalopoda) na Área de Preservação Ambiental do Arquipélago de Fernando de Noronha, Brasil. Arq. Ciênci. Mar 41 (1), 81–89.

Leporati, S., 2008. Fisheries biology and population dynamics of the pale octopus (*Octopus pallidus*). PhD Thesis, University of Tasmania.

Leporati, S.C., Ziegler, P.E., Semmens, J.M., 2009. Assessing the stock status of holobenthic octopus fisheries: is catch per unit effort sufficient? ICES J. Mar. Sci. 66, 478–487.

Leslie, P.H., Davis, D.H.S., 1939. An attempt to determine the absolute number of rats on a given area. J. Anim. Ecol. 8, 94–113.

Link, J.S., Nye, J.A., Hare, J.A., 2011. Guidelines for incorporating fish distribution shifts into a fisheries management context. Fish Fish. 12, 461–469.

Lipiński, M.R., 1987. Food and feeding of *Loligo vulgaris reynaudii* from St Francis Bay, South Africa. S. Afr. J. Mar. Sci. 5, 557–564.

Lipiński, M.R., 1992. Cephalopods and the Benguela ecosystem: trophic relationships and impact. S. Afr. J. Mar. Sci. 12, 791–802.

Lipiński, M.R., 1998. Cephalopod life cycles: patterns and exceptions. S. Afr. J. Mar. Sci. 20, 439–447.

Lipinski, M.R., 2001. Growth of cephalopods: conceptual model. Abh. Geol. Bundesanstalt 57, 133–138.

Lipinski, M.R., David, J., 1990. Cephalopods in the diet of the South African fur seal (*Arctocephalus pusillus pusillus*). J. Zool. 221, 359–374.

Lipinski, M.R., Jackson, S., 1989. Surface-feeding on cephalopods by procellariiform seabirds in the southern Benguela region, South Africa. J. Zool. 218, 549–563.

Lipinski, M.R., Soule, M.A., 2007. A new direct method of stock assessment of the loliginid squid. Rev. Fish Biol. Fish. 17, 437–453.

Lipinski, M.R., Roeleveld, M.A., Augustyn, C.J., 1991. Feeding studies on *Sepia australis* with an assessment of its significance in the Benguela ecosystem. In: La seiche—The Cuttlefish. Centre de Publications de l'Universite de Caen, France, pp. 117–129.

Lipiński, M.R., Payne, A., Rose, B., 1992. The importance of cephalopods as prey for hake and other groundfish in South African waters. S. Afr. J. Mar. Sci. 12, 651–662.

Lipinski, M.R., Butterworth, D., Augustyn, C., Brodziak, J., Christy, G., Des Clers, S., Jackson, G., O'Dor, R., Pauly, D., Purchase, L., 1998a. Cephalopod fisheries: a future global upside to past overexploitation of living marine resources? Results of an International Workshop, 31 August–2 September 1997, Cape Town, South Africa. S. Afr. J. Mar. Sci. 20, 463–469.

Lipinski, M.R., Hampton, I., Sauer, W.H.H., Augustyn, C.J., 1998b. Daily net emigration from a spawning concentration of chokka squid (*Loligo vulgaris reynaudii* d'Orbigny, 1845) in Kromme Bay, South Africa. ICES J. Mar. Sci. 55, 258–270.

Litz, M.N., Phillips, A.J., Brodeur, R.D., Emmett, R.L., 2011. Seasonal occurrences of Humboldt squid (*Dosidicus gigas*) in the northern California Current System. CalCoFI Rep. 52, 97–108.

Lordan, C., Burnell, G., Cross, T., 1998. The diet and ecological importance of *Illex coindetii* and *Todaropsis eblanae* (Cephalopoda: Ommastrephidae) in Irish waters. S. Afr. J. Mar. Sci. 20, 153–163.

Lourenço, S., Pereira, J., 2006. Estimating standard landings per unit effort for an octopus mixed components fishery. Fish. Res. 78, 89–95.

Lu, H., Chen, X., Li, G., Cao, J., 2013. Stock assessment and management for *Illex argentinus* in southwest Atlantic Ocean based on Bayesian Schaefer model. Chin. J. Appl. Ecol. 24, 2007–2014.

Macewicz, B.J., Hunter, J.R., Lo, N.C.H., LaCasella, E.L., 2004. Fecundity, egg deposition, and mortality of market squid (*Loligo opalescens*). Fish. Bull. 102, 306–327.

Maltagliati, F., Belcari, P., Casu, D., Casu, M., Sartor, P., Vargiu, G., Castelli, A., 2002. Allozyme genetic variability and gene flow in *Octopus vulgaris* (Cephalopoda, Octopodidae) from the Mediterranean Sea. Bull. Mar. Sci. 71, 473–486.

Manderson, J., Palamara, L., Kohut, J., Oliver, M.J., 2011. Ocean observatory data are useful for regional habitat modeling of species with different vertical habitat preferences. Mar. Ecol. Prog. Ser. 438, 1–17.

Mangold, K., 1983. Octopus vulgaris. In: Boyle, P.B. (Ed.), Cephalopod Life Cycles. In: Species Accounts, vol. I. Academic Press, London, pp. 335–364.

Mangold, K., Boletzky, S.V., 1973. New data on reproductive biology and growth of *Octopus vulgaris*. Mar. Biol. 19, 7–12.

Marques Moreno, A.C.A., 2012. Environmental drivers on the life cycle strategies and distribution of cephalopods in the Portuguese coast. PhD Thesis, Universidade de Lisboa.

Martins, H.R., 1982. Biological studies of the exploited stock of *Loligo forbesi* (Mollusca: Cephalopoda) in the Azores. J. Mar. Biol. Assoc. U.K. 62, 799–808.

Martins, R.S., Roberts, M.J., Lett, C., Chang, N., Moloney, C.L., Camargo, M.G., Vidal, E.A., 2014. Modelling transport of chokka squid (*Loligo reynaudii*) paralarvae off South Africa: reviewing, testing and extending the 'Westward Transport Hypothesis'. Fish. Oceanogr, 23, 116–131.

Masuda, S., Yokawa, K., Yatsu, A., Kawahara, S., 1998. Growth and population structure of *Dosidicus gigas* in the Southeastern Pacific Ocean. In: Okutani, T. (Ed.), Large Pelagic Squids. Japan Marine Fishery Resources Research Center, Tokyo, pp. 107–118.

Maunder, M.N., Watters, G.M., 2003. A general framework for integrating environmental time series into stock assessment models: model description, simulation testing, and example. Fish. Bull. 101, 89–99.

Maxwell, M.R., Jacobson, L.D., Conser, R.J., 2005. Eggs-per-recruit model for management of the California market squid (*Loligo opalescens*) fishery. Can. J. Fish. Aquat. Sci. 62, 1640–1650.

McCullagh, P., Nelder, J.A., 1989. Generalized Linear Models. Chapman and Hall, London.

Meiyappan, M.M., Srinath, M., Prabhakaran Nair, K., Satyanarayana Rao, K., Sarvesan, R., Syda Rao, G., SunilkumarMohamed, K., Vidhyasagar, K., Sundraram, K.S., Lipton, A.P., Natarajan, P., Radhakrishnan, G., Narasimham, K.A., Balan, K., Kripa, V., Sathianandan, T.V., 1993. Stock assessment of the Indian squid Loligo duvauceli Orbigny. Ind. J. Fish. 40 (1–2), 74–84.

Meiyappan, M.M., Mohamed, K.S., Vidyasagar, K., Nair, K.P., Ramachandran, N., Lipton, A.P., Rao, G.S., Kripa, V., Joshi, K.K., Abdussamad, E.M., Sarvesan, R., Achari, G.P., 2000. A review on cephalopod resources, biology and stock assessment in Indian seas. In: Pillai, V.N., Menon, N.G. (Eds.), Marine Fisheries Research and Management. CMFRI, Kochi, India, pp. 546–562.

Melo, Y.C., Sauer, W.H.H., 1999. Confirmation of serial spawning in the chokka squid *Loligo vulgaris reynaudii* off the coast of South Africa. Mar. Biol. 135, 307–313.

Melo, Y.C., Sauer, W.H.H., 2007. Determining the daily spawning cycle of the chokka squid, Loligo reynaudii off the South African coast. Rev. Fish Biol. Fish. 17, 247–257.

Meltzner, F., Bock, C., Pörtner, H.-O., 2007. Allometry of thermal limitation in the cephalopod *Sepia officinalis*. Comp. Biochem. Physiol. A 146, 149–154.

Mercer, R.W., Bucy, M., 1983. Experimental squid jigging off the Washington coast. Mar. Fish. Rev. 45, 7–8.

Mesnil, B., 1977. Growth and life cycle of squid, *Loligo pealei* and *Illex illecebrosus*, from the Northwest Atlantic. Int. Comm. Northw. Atl. Fish. (ICNAF) Sel. Pap. 2, 55–69.

Mesnil, B., 2003. The Catch-Survey Analysis (CSA) method of fish stock assessment: an evaluation using simulated data. Fish. Res. 63, 193–212.

Mid-Atlantic Fishery Management Council, 2011. 2012 Atlantic Mackerel, Squid and Butterfish Specifications and Management Measures. http://www.nero.noaa.gov/nero/regs/frdoc/11/11smb2012-2014specsEA.pdf.

Mikalsen, K.H., Jentoft, S., 2001. From user-groups to stakeholders? The public interest in fisheries management. Mar. Policy 25, 281–292.

Mikalsen, K.H., Jentoft, S., 2008. Participatory practices in fisheries across Europe: making stakeholders more responsible. Mar. Policy 32, 169–177.

Miksis-Olds, J.L., Stokesbury, K.D.E., 2007. Measuring fish abundance in a weir trap using an acoustical-optical platform. J. Acoust. Soc. Am. 122, 2431–2438.

Mills, K.E., Pershing, A.J., Brown, C.J., Chen, Y., Chiang, F.-S., Holland, D.S., Lehuta, S., Nye, J.A., Sun, J.C., Thomas, A.C., Wahle, R.A., 2013. Fisheries management in a changing climate: lessons from the 2012 ocean heat wave in the Northwest Atlantic. Oceanography 26 (2), 191–195.

Mohamad Kasim, H., 1985. Population dynamics of the squid *Loligo duvaucelii* D'orbigny (Cephalopoda) in Saurashtra waters. J. Mar. Biol. Assoc. Ind. 27 (1–2), 103–112.

Mohamad Kasim, H., 1993. Population dynamics of the cuttlefish *Sepia elliptica* Hoyle in Saurashtra waters. J. Mar. Biol. Assoc. Ind. 35 (1–2), 80–86.
Mohn, R.K., 1982. Yield-per-recruit analysis of short-finned squid on the Scotian Shelf by simulation. NAFO Sci. Coun. Stud. 3, 39–45.
Moltschaniwskyj, N., 1994. Muscle tissue growth and muscle fibre dynamics in the tropical loliginid squid Photololigo sp. (Cephalopoda: Loliginidae). Can. J. Fish. Aquat. Sci. 51, 830–835.
Moltschaniwskyj, N.A., 2004. Understanding the process of growth in cephalopods. Mar. Freshw. Res. 55, 379–386.
Morales-Bojórquez, E., Cisneros-Mata, M.A., Nevárez-Martı´nez, M.O., Hernández-Herrera, A.n., 2001a. Review of stock assessment and fishery biology of *Dosidicus gigas* in the Gulf of California, Mexico. Fish. Res. 54, 83–94.
Morales-Bojórquez, E., Hernández-Herrera, A., Nevárez-Martínez, M.O., Cisneros-Mata, M.Á., Guerrero-Escobedo, F.J., 2001b. Population size and exploitation of giant squid (*Dosidicus gigas* D'Orbigny, 1835) in the Gulf of California, Mexico. Sci. Mar. 65, 75–80.
Morales-Bojórquez, E., Hernández-Herrera, A., Ángel Cisneros-Mata, M., Nevárez-Martínez, M.O., 2008. Improving estimates of recruitment and catchability of jumbo squid *Dosidicus gigas* in the Gulf of California, Mexico. J. Shellfish Res. 27, 1233–1237.
Morales-Bojorquez, E., Hernandez-Herrera, A., Nevarez-Martinez, M.O., Diaz-Uribe, J.G., 2012. Population size of the jumbo squid *Dosidicus gigas* in the central Gulf of California, Mexico, based on mark-recapture data. Aquat. Biol. 15, 27–34.
Moreira, A.A., Tomás, A.R.G., Hilsdorf, A.W.S., 2011. Evidence for genetic differentiation of *Octopus vulgaris* (Mollusca, Cephalopoda) fishery populations from the southern coast of Brazil as revealed by microsatellites. J. Exp. Mar. Biol. Ecol. 407, 34–40.
Moreno, A., Morais da Cunha, M., Pereira, J.M., 1994. Population biology of veined squid (*Loligo forbesi*) and European squid (*Loligo vulgaris*) from the Portuguese coast. Fish. Res. 21, 71–86.
Moreno, A., Pereira, J., Arvanitidis, C., Robin, J., Koutsoubas, D., Perales-Raya, C., Cunha, M., Balguerias, E., Denis, V., 2002. Biological variation of *Loligo vulgaris* (Cephalopoda: Loliginidae) in the eastern Atlantic and Mediterranean. Bull. Mar. Sci. 71, 515–534.
Moreno, A., Pereira, J., Cunha, M., 2005. Environmental influences on age and size at maturity of *Loligo vulgaris*. Aquat. Living Resour. 18, 377–384.
Moreno, A., dos Santos, A., Piatkowski, U., Santos, A.M.P., Cabral, H., 2009. Distribution of cephalopod paralarvae in relation to the regional oceanography of the western Iberia. J. Plankton Res. 31, 73–91.
Moustahfid, H., Tyrrell, M.C., Link, J.S., 2009. Accounting explicitly for predation mortality in surplus production models: an application to longfin inshore squid. N. Am. J. Fish Manag. 29, 1555–1566.
MRAG, 2005. Review of Impacts of Illegal, Unreported and Unregulated Fishing on Developing Countries. MRAG, London.http://www.dfid.gov.uk/pubs/files/illegal-fishing-mrag-report.pdf.
MRAG, 2010. Towards Sustainable Fisheries Management: International Examples of Innovation. MRAG, London.
Mueller, L.D., Joshi, A., 2000. Stability in Model Populations. Princeton University Press, Princeton, NJ.
Mukai, T., Iida, K., Sakaguchi, K., Abe, K., 2000. Estimations of squid target strength using a small cage and theoretical scattering models. Proc. JSPS-DGHE Int. Symp. 10, 135–140.
Murata, M., 1989. Population assessment, management and fishery forecasting for the Japanese common squid, *Todarodes pacificus*. In: Caddy, J.F. (Ed.), Marine Invertebrate Fisheries: Their Assessment and Management. Wiley, New York, pp. 613–636.

Murphy, J., Balguerías, E., Key, L., Boyle, P., 2002. Microsatellite DNA markers discriminate between two *Octopus vulgaris* (Cephalopoda: Octopoda) fisheries along the northwest African coast. Bull. Mar. Sci. 71, 545–553.

Murray, F.C., 1999. Developmental abnormalities in the cephalopod embryo. BSc Hon. thesis, University of Aberdeen.

Myers, R.A., 2000. The synthesis of dynamic and historical data on marine populations and communities; putting dynamics into the Ocean Biogeographical Information System (OBIS). Oceanography 13 (3), 56–59.

Nabhitabhata, J., Asawangkune, P., Amornjaruchit, S., Promboon, P., 2001. Tolerance of eggs and hatchlings of neritic cephalopods to salinity changes. Phuket Marine Biological Center Special Publication 25 (1), 91–99.

Nagasawa, K., Mori, J., Okamura, H., 1998. Parasites as biological tags of stocks of neon flying squid (*Ommastrephes bartramii*) in the North Pacific Ocean. In: Okutani, T. (Ed.), Large Pelagic Squids. Japan Marine Fishery Resources Research Center, Tokyo, pp. 49–64.

Nakata, J., 1993. Long-term changes in catch and biological features of Japanese common squid (*Todarodes pacificus*) in waters off the east coast of Hokkaido. In: Okutani, T., O'Dor, R.K., Kubodera, T. (Eds.), In: Recent Advances in Fisheries Biology, Tokai University Press, Tokyo, pp. 343–350.

Narvarte, M., González, R., Filippo, P., 2007. Artisanal mollusk fisheries in San Matías Gulf (Patagonia, Argentina): an appraisal of the factors contributing to unsustainability. Fish. Res. 87, 68–76.

Natsukari, Y., Tashiro, M., 1991. Neritic squid resources and cuttlefish resources in Japan. Mar. Freshw. Behav. Physiol. 18, 149–226.

NEFSC (Northeast Fisheries Science Center), 2011. 51st Northeast Regional Stock Assessment Workshop (51st SAW) Assessment Report, Northeast Fisheries Science Center Reference Document 11-02. Woods Hole, MA.

Nesis, K.N., 1985. Oceanic Cephalopods, Distribution, Life Forms and Evolution. Nauka, Moscow.

Nesis, K., 1993. Population structure of oceanic ommastrephids, with particular reference to *Sthenoteuthis oualaniensis*: a review. In: Okutani, T., O'Dor, R.K., Kubodera, (Eds.), Recent Advances in Fisheries Biology. Tokai University Press, Tokyo, pp. 375–383.

Nevárez-Martínez, M.O., Hernández-Herrera, A., Morales-Bojórquez, E., Balmori-Ramírez, A., Cisneros-Mata, M.A., Morales-Azpeitia, R., 2000. Biomass and distribution of the jumbo squid (Dosidicus gigas; d'Orbigny, 1835) in the Gulf of California, Mexico. Fish. Res. 49, 129–140.

Nevárez-Martínez, M.O., Méndez-Tenorio, F.J., Cervantes-Valle, C., López-Martínez, J., Anguiano-Carrasco, M.L., 2006. Growth, mortality, recruitment, and yield of the jumbo squid (*Dosidicus gigas*) off Guaymas, Mexico. Fish. Res. 79, 38–47.

Nevárez-Martínez, M.O., Morales-Bojórquez, E., Cervantes-Valle, C., Santos-Molina, J.P., López- Martínez, J., 2010. Population dynamics of the jumbo squid (*Dosidicus gigas*) in the 2002–2008 fishing seasons off Guaymas, Mexico. Fish. Res. 106, 132–140.

Nigmatullin, C.M., Nesis, K.N., Arkhipkin, A.I., 2001. Biology of the jumbo squid *Dosidicus gigas* (Cephalopoda: Ommastrephidae). Fish. Res. 54, 9–19.

O'Dor, R.K., 1998. Squid life-history strategies. In: Rodhouse, P.G., Dawe, E.G., O'Dor, R.K. (Eds.), Squid Recruitment Dynamics. The Genus *Illex* as a Model, the Commercial *Illex* Species and Influence on Variability. In: FAO Fisheries Technical Paper No. 376FAO, Rome, pp. 233–254.

O'Dor, R.K., Coelho, M.L., 1993. Big squid, big currents and big fisheries. In: Okutani, T., O'Dor, R.K., Kubodera, T. (Eds.), Recent Advances in Cephalopod Fisheries Biology. Tokai University Press, Tokyo, pp. 385–396.

O'Dor, R.K., Dawe, E.G., 2013. *Illex illecebrosus*, Northern short-finned squid. In: Rosa, R., Pierce, G., O'Dor, R. (Eds.), Advances in Squid Biology, Ecology and Fisheries. Part II. Oegopsid Squids. Nova Biomedical, New York, pp. 73–108.

O'Dor, R., Wells, M., 1987. Energy and nutrient flow. In: Boyle, P.R. (Ed.), In: Cephalopod Life Cycles, vol. II. Academic Press, London, pp. 109–133.

O'Dor, R.K., 1983. Illex illecebrosus. In: Boyle, P.B. (Ed.), Cephalopod Life Cycles. In: Species Accounts, vol. I. Academic Press, London, pp. 175–200.

O'Dor, R., 1998. Can understanding squid life-history strategies and recruitment improve management? S. Afr. J. Mar. Sci. 20, 193–206.

O'Dor, R., Dawe, E., 1998. Illex illecebrosus. In: Rodhouse, P.G., Dawe, E.G., O'Dor, R.K. (Eds.), Squid Recruitment Dynamics. The Genus *Illex* as a Model, the Commercial *Illex* Species and Influence on Variability. In: FAO Fisheries Technical Paper No. 376, FAO, Rome, pp. 77–104.

O'Dor, R., Durward, R., Vessey, E., Amaratunga, T., 1980. Feeding and growth in captive squid, *Illex illecebrosus*, and the influence of food availability on growth in the natural population. ICNAF Selected Papers 6, 15–21.

O'Dor, R., Aitken, J., Jackson, G.D., 2005. Energy balance growth models: applications to cephalopods. Phuket Mar. Biol. Center Res. Bull. 66, 329–336.

Okutani, T., 1977. Stock Assessment of Cephalopod Resources Fished by Japan. FAO Fisheries Technical Paper No. 173, FAO, Rome.

Okutani, T., 1983. Todarodes pacificus. In: Boyle, P.R. (Ed.), Cephalopod Life Cycles, vol. I. Academic Press, London, pp. 201–214.

Okutani, T., Watanabe, T., 1983. Stock assessment by larval surveys of the winter population of *Todarodes pacificus* Steenstrup (Cephalopoda: Ommastrephidae), with a review of early works. Biol. Oceanogr. 2, 401–431.

Olyott, L., Sauer, W., Booth, A.J., 2006. Spatio-temporal patterns in maturation of the chokka squid (*Loligo vulgaris reynaudii*) off the coast of South Africa. ICES J. Mar. Sci. 63, 1649–1664.

Olyott, L., Sauer, W., Booth, A., 2007. Spatial patterns in the biology of the chokka squid, *Loligo reynaudii* on the Agulhas Bank, South Africa. Rev. Fish Biol. Fish. 17, 159–172.

Oosthuizen, A., 2004. Economic feasibility of an experimental octopus fishery in South Africa. S. Afr. J. Sci. 100, 595–602.

Oosthuizen, A., Smale, M.J., 2003. Population biology of *Octopus vulgaris* on the temperate south-eastern coast of South Africa. J. Mar. Biol. Assoc. U.K. 83, 535–541.

Oosthuizen, A., Roberts, M.J., Sauer, W.H.H., 2002. Temperature effects on the embryonic development and hatching success of the squid *Loligo vulgaris reynaudi*. Bull. Mar. Sci. 71, 619–632.

Oosthuizen, A., Jiwaji, M., Shaw, P., 2004. Genetic analysis of the *Octopus vulgaris* population on the coast of South Africa. S. Afr. J. Sci. 100, 603–607.

Ordines, F., Massutí, E., Guijarro, B., Mas, R., 2006. Diamond vs. square mesh codend in a multi-species trawl fishery of the western Mediterranean: effects on catch composition, yield, size selectivity and discards. Aquat. Living Resour. 19, 329–338.

Ortiz, N., Ré, M.E., Márquez, F., Glembocki, N.G., 2011. The reproductive cycle of the red octopus *Enteroctopus megalocyathus* in fishing areas of the North Atlantic coast. Fish. Res. 110, 217–223.

Otero, J., Álvarez-Salgado, J.A., González, A.F., Miranda, A., Groom, S.B., Cabanas, J.M., Casas, G., Wheatley, B., Guerra, A., 2008. Bottom-up control of common octopus *Octopus vulgaris* in the Galician upwelling system, northeast Atlantic Ocean. Mar. Ecol. Prog. Ser. 362, 181–192.

Otieno, M.J., 2011. Fishery Value Chain Analysis: Background Report—Kenya.

Overholtz, W., Link, J., Suslowicz, L., 2000. Consumption of important pelagic fish and squid by predatory fish in the northeastern USA shelf ecosystem with some fishery comparisons. ICES J. Mar. Sci. 57, 1147–1159.

Overholtz, W., Jacobson, L., Link, J., 2008. An ecosystem approach for assessment advice and biological reference points for the Gulf of Maine–Georges Bank Atlantic herring complex. N. Am. J. Fish Manag. 28, 247–257.
Parmesan, C., Yohe, G., 2003. A globally coherent fingerprint of climate change impacts across natural systems. Nature 421, 37–42.
Pascual, S., Guerra, A., 2001. Vexing question on fisheries research: the study of cephalopods and their parasites. Iberus 19, 87–95.
Pascual, S., González, A., Guerra, A., 1998. Effect of parasitism on the productivity of the ommastrephid stocks in Galician waters (NW Spain): economic loss. Iberus 16, 95–98.
Pascual, S., González, A., Guerra, A., 2005. The recruitment of gill-infesting copepods as a categorical predictor of size-at-age data in squid populations. ICES J. Mar. Sci. 62, 629–633.
Pascual, S., González, A., Guerra, A., 2007. Parasites and cephalopod fisheries uncertainty: towards a waterfall understanding. Rev. Fish Biol. Fish. 17, 139–144.
Patterson, K., 1988. Life history of Patagonian squid *Loligo gahi* and growth parameter estimates using least-squares fits to linear and von Bertalanffy models. Mar. Ecol. Prog. Ser. 47, 65–74.
Paulij, W.P., Bogaards, R.H., Denucé, J.M., 1990. Influence of salinity on embryonic development and the distribution of *Sepia officinalis* in the Delta Area (South Western part of The Netherlands). Mar. Biol. 107, 17–23.
Pauly, D., 1985. Population dynamics of short-lived species, with emphasis on squids. NAFO Sci. Coun. Stud. 9, 143–154.
Pauly, D., 1998. Why squids, though not fish, may be better understood by pretending they are. S. Afr. J. Mar. Sci. 20, 47–58.
Pauly, D., Christensen, V., Dalsgaard, J., Froese, R., Torres, F., 1998. Fishing down marine food webs. Science 279, 860–863.
Payne, A.G., Agnew, D.J., Pierce, G.J., 2006. Trends and assessment of cephalopod fisheries. Fish. Res. 78, 1–3.
Pecl, G.T., 2000. Comparative life history of tropical and temperate *Sepioteuthis* squids in Australian waters. PhD Thesis James Cook University.
Pecl, G.T., Jackson, G.D., 2008. The potential impacts of climate change on inshore squid: biology, ecology and fisheries. Rev. Fish Biol. Fish. 18, 373–385.
Pecl, G.T., Tracey, S.R., Semmens, J.M., Jackson, G.D., 2006. Use of acoustic telemetry for spatial management of southern calamary *Sepioteuthis australis*, a highly mobile inshore squid species. Mar. Ecol. Prog. Ser. 328, 1–15.
Pereira, J.M.F., 1999. Control of the Portuguese artisanal octopus fishery. In: Proceedings of the International Conference on Integrated Fisheries Monitoring. Sydney, Australia, 1–5 February 1999. FAO, Rome, pp. 369–378.
Pérez-Losada, M., Guerra, Á., Sanjuan, A., 1999. Allozyme differentiation in the cuttlefish *Sepia officinalis* (Mollusca: Cephalopoda) from the NE Atlantic and Mediterranean. Heredity 83, 280–289.
Pérez-Losada, M., Guerra, A., Carvalho, G., Sanjuan, A., Shaw, P., 2002. Extensive population subdivision of the cuttlefish *Sepia officinalis* (Mollusca: Cephalopoda) around the Iberian Peninsula indicated by microsatellite DNA variation. Heredity 89, 417–424.
Pérez-Losada, M., Nolte, M.J., Crandall, K.A., Shaw, P.W., 2007. Testing hypotheses of population structuring in the Northeast Atlantic Ocean and Mediterranean Sea using the common cuttlefish *Sepia officinalis*. Mol. Ecol. 16, 2667–2679.
Piatkowski, U., Pierce, G.J., Morais da Cuhna, M., 2001. Impact of cephalopods on food chain and their interaction with environment and fisheries: an overview. Fish. Res. 52, 5–10.
Pierce, G.J., Boyle, P.R., 2003. Empirical modelling of interannual trends in abundance of squid (*Loligo forbesi*) in Scottish waters. Fish. Res. 59, 305–326.

Pierce, G.J., Guerra, Á., 1994. Stock assessment methods used for cephalopod fisheries. Fish. Res. 21, 255–285.

Pierce, G.J. (Ed.), 1999. Data collection for assessment of cephalopod fished stocks: Final Report to the Commission of the European Communities Directorate General for Fisheries on Study No. 96/081. University of Aberdeen, UK.

Pierce, G.J., Santos, M.B., 1996. Trophic interactions of squid *Loligo forbesi* in Scottish waters. In: Tasker, M.L., Greenstreet, S.P.R. (Eds.), Aquatic Predators and Their Prey. Fishing News Books, Oxford, UK, pp. 58–64.

Pierce, G.J., Boyle, P.R., Hastie, L.C., Key, L., 1994. The life history of *Loligo forbesi* (Cephalopoda: Loliginidae) in Scottish waters. Fish. Res. 21, 17–41.

Pierce, G.J., Bailey, N., Stratoudakis, Y., Newton, A., 1998. Distribution and abundance of the fished population of *Loligo forbesi* in Scottish waters: analysis of research cruise data. ICES J. Mar. Sci. 55, 14–33.

Pierce, G.J., Zuur, A.F., Smith, J.M., Santos, M.B., Bailey, N., Chen, C., Boyle, P.R., 2005. Interannual variation in life-cycle characteristics of the veined squid (*Loligo forbesi*) in Scottish (UK) waters. Aquat. Living Resour. 18, 327–340.

Pierce, G.J., Valavanis, V.D., Guerra, Á., Jereb, P., Orsi-Relini, L., Bellido, J.M., Katara, I., Piatkowski, U., Pereira, J., Balguerias, E., Sobrino, I., Lefkaditou, E., Wang, J., Santurtun, M., Boyle, P.R., Hastie, L.C., MacLeod, C.D., Smith, J.M., Viana, M., González, Á.F., Zuur, A.F., 2008. A review of cephalopod-environment interactions in European Seas. Hydrobiologia 612, 49–70.

Pierce, G.J., Allcock, L., Bruno, I., Bustamante, P., González, Á., Guerra, Á., Jereb, P., 2010. Cephalopod Biology and Fisheries in Europe: ICES Coop. Res. Rep. No. 303.

Pilling, G.M., Apostolaki, P., Failler, P., Floros, C., Large, P.A., Morales-Nin, B., Reglero, P., Stergiou, K.I., Tsikliras, A.C., 2008. Assessment and management of data-poor fisheries. In: Payne, A., Cotter, J., Potter, T. (Eds.), Advances in Fisheries Science: 50 Years on From Beverton and Holt. Blackwell Publishing, CEFAS, pp. 280–305.

Pinsky, M.L., Fogarty, M.J., 2012. Lagged social-ecological responses to climate and range shifts in fisheries. Clim. Change 115, 883–891.

Pinsky, M.L., Worm, B., Fogarty, M.J., Sarmiento, J.L., Levin, S.A., 2013. Marine taxa track local climate velocities. Science 341, 1239–1242.

Pita, C., Pierce, G.J., Theodossiou, I., 2010. Stakeholders' participation in the fisheries management decision-making process: fishers' perceptions of participation. Mar. Policy 34, 1093–1102.

Pomeroy, C., Fitzsimmons, M., 1998. Information needs for effective management of the Californian market squid fishery: the role of social science research. CalCoFI Rep. 39, 108–114.

Pope, J., 2009. Input and output controls: the practice of fishing effort and catch management in responsible fisheries. In: Cochrane, K.L., Garcia, S.M. (Eds.), A Fishery Manager's Guidebook. FAO/Wiley-Blackwell.

Punt, A.E., 2006. The FAO precautionary approach after almost 10 years: have we progressed towards implementing simulation-tested feedback-control management systems for fisheries management? Nat. Resour. Model. 19, 441–464.

Quetglas, A., Ordines, F., Hidalgo, M., Monserrat, S., Ruiz, S., Amores, Á., Moranta, J., Massutí, E., 2013. Synchronous combined effects of fishing and climate within a demersal community. ICES J. Mar. Sci. 70, 319–328.

Quinn, T., Deriso, R.B., 1999. Quantitative Fish Dynamics. Oxford University Press, New York.

Raberinary, D., Peabody, S., 2011. Octopus Fishery Management. Indian Ocean Community Conservation Handbooks, Blue Ventures.

Ranta, E., Kaitala, V., Lundberg, P., 1997. The spatial dimension in population fluctuations. Science 278, 1621–1623.

Rasero, M., Gonzalez, A., Castro, B.G., Guerra, A., 1996. Predatory relationships of two sympatric squid, *Todaropsis eblanae* and *Illex coindetii* (Cephalopoda: Ommastrephidae) in Galician waters. J. Mar. Biol. Assoc. U.K. 76, 73–87.
Recksiek, C.W., Frey, H.W. (Eds.), 1978. Biological, oceanographic, and acoustic aspects of the market squid, *Loligo opalescens* Berry, California Department of Fish and Game Fish Bulletin 169.
Regueira, M., González, A.F., Guerra, A., Soares, A., 2013. Reproductive traits of horned octopus *Eledone cirrhosa* in Atlantic Iberian waters. J. Mar. Biol. Assoc. U.K. 93, 1641–1652.
Reichow, D., Smith, M.J., 1999. Highly variable microsatellites in the California market squid *Loligo opalescens*. Mar. Biotechnol. 1 (4), 403–406.
Reichow, D., Smith, M., 2001. Microsatellites reveal high levels of gene flow among populations of the California squid *Loligo opalescens*. Mol. Ecol. 10, 1101–1109.
Reiss, C.S., Maxwell, M.R., Hunter, J.R., Henry, A., 2004. Investigating environmental effects on population dynamics of *Loligo opalescens* in the Southern California Bight. CalCoFI Rep. 45, 87–97.
Richard, A., 1975. L'élevage de la seiche (*Sepia officinalis* L., Mollusque). In: 10th European Symposium on Marine Biology, Ostend, Belgium, pp. 359–380.
Ricker, W.E., 1975. Computation and interpretation of biological statistics of fish populations. Bull. Fish. Res. Board Can. 119, 1–382.
Rigby, P.R., Sakurai, Y., 2004. Temperature and feeding related growth efficiency of immature octopuses *Enteroctopus dofleini*. Suisan Zoshoku 52, 29–36.
Roa-Ureta, R.H., 2012. Modelling in-season pulses of recruitment and hyperstability-hyperdepletion in the *Loligo gahi* fishery around the Falkland Islands with generalized depletion models. ICES J. Mar. Sci. 69, 1403–1415.
Roa-Ureta, R., Arkhipkin, A.I., 2007. Short-term stock assessment of *Loligo gahi* at the Falkland Islands: sequential use of stochastic biomass projection and stock depletion models. ICES J. Mar. Sci. 64, 3–17.
Robert, M., Faraj, A., McAllister, M.K., Rivot, E., 2010. Bayesian state-space modeling of the De Lury depletion model: strengths and limitations of the method, and applications to the Moroccan octopus fishery. ICES J. Mar. Sci. 67, 1272–1290.
Roberts, M.J., 2005. Chokka squid (*Loligo vulgaris reynaudii*) abundance linked to changes in South Africa's Agulhas Bank ecosystem during spawning and the early life cycle. ICES J. Mar. Sci. 62, 33–55.
Roberts, M.J., Sauer, W., 1994. Environment: the key to understanding the South African chokka squid (*Loligo vulgaris reynaudii*) life cycle and fishery? Antarctic Sci. 6, 249–258.
Roberts, M.J., Van Den Berg, M., 2002. Recruitment variability of chokka squid (*Loligo vulgaris reynaudii*) role of currents on the Agulhas Bank (South Africa) in paralarvae distribution and food abundance. Bull. Mar. Sci. 71, 691–710.
Robinson, C.J., Gómez-Gutiérrez, J., de León, D.A.S., 2013. Jumbo squid (*Dosidicus gigas*) landings in the Gulf of California related to remotely sensed SST and concentrations of chlorophyll a (1998–2012). Fish. Res. 137, 97–103.
Rocha, F.J., Vega, M.A., 2003. Overview of cephalopod fisheries in Chilean waters. Fish. Res. 60, 151–159.
Rochet, M.-J., 2000. Does the concept of spawning per recruit make sense? ICES J. Mar. Sci. 57, 1160–1174.
Rochette, S., Rivot, E., Morin, J., Mackinson, S., Riou, P., Le Pape, O., 2010. Effect of nursery habitat degradation on flatfish population: application to *Solea solea* in the Eastern Channel (Western Europe). J. Sea Res. 64, 34–44.
Rodhouse, P.G., 2001. Managing and forecasting squid fisheries in variable environments. Fish. Res. 54, 3–8.

Rodhouse, P.G., Hatfield, E.M.C., 1990. Dynamics of growth and maturation in the cephalopod *Illex argentinus* de Castellanos, 1960 (Teuthoidea: Ommastrephidae). Philos. Trans. R Soc. B 329, 229–241.

Rodhouse, P.G., Nigmatullin, C.M., 1996. Role as consumers. Philos. Trans. R Soc. B 351, 1003–1022.

Rodhouse, P.G., Elvidge, C.D., Trathan, P.N., 2001. Remote sensing of the global light-fishing fleet: an analysis of interactions with oceanography, other fisheries and predators. Adv. Mar. Biol. 39, 261–303.

Rodhouse, P.G.K., Arkhipkin, A.I., Laptikhovsky, V., Nigmatullin, C., Waluda, C.M., 2013. *Illex argentinus*, Argentine shortfin squid. In: Rosa, R., Pierce, G., O'Dor, R. (Eds.), Advances in Squid Biology, Ecology and Fisheries. Part II—Oegopsid squids. Nova Science Publishers, New York, pp. 109–148.

Roel, B.A., 1998. Stock assessment of the chokka squid. PhD Thesis, University of Cape Town, South Africa.

Roel, B.A., Butterworth, D.S., 2000. Assessment of the South African chokka squid *Loligo vulgaris reynaudii*: is disturbance of aggregations by the recent jig fishery having a negative effect on recruitment? Fish. Res. 48, 213–228.

Roel, B.A., Cochrane, K.L., Field, J.G., 2000. Investigation into the declining trend in chokka squid *Loligo vulgaris reynaudii* catches made by South African trawlers. S. Afr. J. Mar. Sci. 22, 121–135.

Roper, C.F.E., Young, R.E., 1975. Vertical distribution of pelagic cephalopods. Smithson. Contrib. Zool. 209, 1–51.

Roper, C.L., Sweeney, M.J., Nauen, C.E., 1984. In: FAO Species Catalogue, Vol. 3. Cephalopods of the World: An Annotated and Illustrated Catalogue of Species of Interest to Fisheries. FAO, Rome, FAO Fisheries Synopsis No. 125: 277 pp.

Rosa, R., Seibel, B.A., 2008. Synergistic effects of climate-related variables suggest future physiological impairment in a top oceanic predator. Proc. Natl. Acad. Sci. U.S.A. 105, 20776–20780.

Rosa, R., Seibel, B.A., 2010. Metabolic physiology of the Humboldt squid, *Dosidicus gigas*: implications for vertical migration in a pronounced oxygen minimum zone. Prog. Oceanogr. 86, 72–80.

Rosa, A.L., Yamamoto, J., Sakurai, Y., 2011. Effects of environmental variability on the spawning areas, catch, and recruitment of the Japanese common squid, *Todarodes pacificus* (Cephalopoda: Ommastrephidae), from the 1970s to the 2000s. ICES J. Mar. Sci. 68, 1114–1121.

Rosa, R., Yamashiro, C., Markaida, U., Rodhouse, P.G.K., Waluda, C.M., Salinas-Zavala, C.A., Keyl, F., O'Dor, R.K., Stewart, J.S., Gilly, W.F., 2013. *Dosidicus gigas*, Humbolt squid. In: Rosa, R., Pierce, G., O'Dor, R. (Eds.), Advances in Squid Biology, Ecology and Fisheries. Part II Oegopsid Squids. Nova Biomedical, New York, pp. 169–206.

Rosas-Luis, R., Tafur-Jimenez, R., Alegre-Norza, A.R., Castillo-Valderrama, P.R., Cornejo-Urbina, R.M., Salinas-Zavala, C.A., Sánchez, P., 2011. Trophic relationships between the jumbo squid (*Dosidicus gigas*) and the lightfish (*Vinciguerria lucetia*) in the Humboldt Current System off Peru. Sci. Mar. 75, 549–557.

Rosenberg, A.A., Kirkwood, G.P., Crombie, J.A., Beddington, J.R., 1990. The assessment of annual stocks of squid species. Fish. Res. 8, 335–350.

Rowell, T.W., Trites, R.W., Dawe, E.G., 1985. Distribution of shortfinned squid (*Illex illecebrosus*) larvae and juveniles in relation to the Gulf Stream Frontal Zone between Florida and Cape Hatteras. NAFO Sci. Coun. Stud. 9, 77–92.

Royer, J., Périès, P., Robin, J.P., 2002. Stock assessments of English Channel loliginid squids: updated depletion methods and new analytical methods. ICES J. Mar. Sci. 59, 445–457.

Royer, J., Pierce, G.J., Foucher, E., Robin, J.-P., 2006. The English Channel stock of *Sepia officinalis*: modeling variability in abundance and impact of the fishery. Fish. Res. 78, 96–106.

Sacau, M., Pierce, G.J., Wang, J., Arkhipkin, A.I., Portela, J., Brickle, P., Santos, M.B., Zuur, A.F., Cardoso, X., 2005. The spatio-temporal pattern of Argentine shortfin squid *Illex argentinus* abundance in the southwest Atlantic. Aquat. Living Resour. 18, 361–372.
Sakurai, Y., Bower, J.R., Nakamura, Y., Yamamoto, S., Watanabe, K., 1996. Effect of temperature on development and survival of *Todarodes pacificus* embryos and paralarvae. Am. Malacol. Bull. 13, 89–95.
Sakurai, Y., Kiyofuji, H., Saitoh, S., Goto, T., Hiyama, Y., 2000. Changes in inferred spawning areas of *Todarodes pacificus* (Cephalopoda: Ommastrephidae) due to changing environmental conditions. ICES J. Mar. Sci. 57, 24–30.
Sakurai, Y., Kiyofuji, H., Saitoh, S., Yamamoto, J., Goto, T., Mori, K., Kinoshita, T., 2002. Stock fluctuations of the Japanese common squid, *Todarodes pacificus*, related to recent climate changes. Fish. Sci. 68 (Suppl. I), 226–229.
Sakurai, Y., Kidokoro, H., Yamashita, N., Yamamoto, J., Uchikawa, K., Takahara, H., 2013. *Todarodes pacificus*, Japanese common squid. In: Rosa, R., Pierce, G., O'Dor, R. (Eds.), Advances in Squid Biology, Ecology and Fisheries. Part II Oegopsid Squids. Nova Biomedical, New York, pp. 249–272.
Sanchez, P., Obarti, R., 1993. The biology and fishery of *Octopus vulgaris* caught with clay pots on the Spanish Mediterranean coast. In: Okutani, T., O'Dor, R.K., Kubodera, T. (Eds.), Recent Advances in Cephalopod Fishery Biology. Tokai University Press, Shimizu, Japan, pp. 477–487.
Santos, M.B., Clarke, M.R., Pierce, G.J., 2001. Assessing the importance of cephalopods in the diets of marine mammals and other top predators: problems and solutions. Fish. Res. 52, 121–139.
Sasaki, M., 1921. On the life history of an economic cuttlefish of Japan, *Ommastrephes sloani pacificus*. 9, Transactions of Wagner Free Institute of Science in Philadelphia, Philadelphia, USA, pp. 1–25.
Sasaki, M., 1929. Supplementary notes on the life history of an economic cuttlefish, *Ommastrephes sloani pacificus*, Steenstrup. Jap. J. Zool. 2, 199–211.
Sato, T., Hatanaka, H., 1983. A review of the assessments of Japanese distant-water fisheries for cephalopods. In: Caddy, J.F. (Ed.), In: Advances in Assessment of World Cephalopod Resources, Fisheries Technical Paper No. 231, FAO, Rome, pp. 145–203.
Sauer, W.H., 1991. Aspects of the ecology of the chokka squid *Loligo vulgaris reynaudii* (D'Orbigny) in Algoa Bay and St. MSc Thesis, University of Port Elizabeth (South Africa), Francis Bay.
Sauer, W.H.H., 1993. The ecology of spawning squid, *Loligo vulgaris reynaudii*, in the inshore areas of the Eastern Cape, South Africa. PhD Thesis, University of Port Elizabeth, South Africa.
Sauer, W.H.H., 1995. South Africa's Tsitsikamma National Park as a protected breeding area for the commercially exploited chokka squid *Loligo vulgaris reynaudii*. S. Afr. J. Mar. Sci. 6, 365–371.
Sauer, W.H.H., Smale, M.J., 1993. Spawning behaviour of *Loligo vulgaris reynaudii* in shallow waters of the south eastern Cape, South Africa. In: Okutani, T., O'Dor, R.K., Kubodera, T. (Eds.), Recent Advances in Cephalopod Fishery Biology. Tokai University Press, Shimizu, Japan, pp. 489–498.
Sauer, W.H.H., Smale, M.J., Lipinski, M.R., 1992. The location of spawning grounds, spawning and schooling behaviour of the squid *Loligo vulgaris reynaudii* (Cephalopoda: Myopsida) off the Eastern Cape Coast, South Africa. Mar. Biol. 114, 97–107.
Sauer, W.H., Roberts, M.J., Lipinski, M.R., Smale, M.J., Hanlon, R.T., Webber, D.M., O'Dor, R.K., 1997. Choreography of the squid's nuptial dance. Biol. Bull. 192, 203–207.
Sauer, W., Lipinski, M., Augustyn, C., 2000. Tag recapture studies of the chokka squid *Loligo vulgaris reynaudii* D' Orbigny, 1845 on inshore spawning grounds on the south-east coast of South Africa. Fish. Res. 45, 283–289.

Sauer, W.H.H., Potts, W., Raberinary, D., Anderson, J., Sylvio Perrine, M.J., 2011. Assessment of current data for the octopus resource in Rodrigues, western Indian Ocean. Afr. J. Mar. Sci. 33, 181–187.

Sauer, W.H.H., Downey, N.J., Lipinski, M., Roberts, M.J., Smale, M.J., Shaw, P., Glazer, J., Melo, Y., 2013. Loligo reynaudi. Chokka squid. In: Rosa, R., Pierce, G., O'Dor, R. (Eds.), Advances in Squid Biology, Ecology and Fisheries. Part I. Myopsid Squids. Nova Biomedical, New York, pp. 33–72.

Saunders, W.B., Spinosa, C., 1979. *Nautilus* movement and distribution in Palau, western Caroline Islands. Science 204, 1199–1201.

Saville, A., 1987. Comparisons between cephalopods and fish of those aspects of the biology related to stock management. In: Boyle, P.R. (Ed.), Cephalopod Life Cycles, vol. 2. Academic Press, London, pp. 277–290.

Sawada, K., Takahashi, H., Takao, Y., Horne, J.K., McClatchie, S., Abe, K., 2004. Development of an acoustic-optical system to estimate target strengths and tilt angles from fish aggregations. In: Oceans 04/Tech-Ocean 04 Conference Proceedings. Marine Technology Society, Washington, DC, pp. 395–400.

Schaefer, M.B., 1954. Some aspects of the dynamics of populations important to the management of the commercial marine fisheries. Bull. Int. Am. Trop. Tuna Comm. 1, 27–56.

Scheel, D., Bisson, L., 2012. Movement patterns of giant Pacific octopuses, *Enteroctopus dofleini* (Wülker, 1910). J. Exp. Mar. Biol. Ecol. 416, 21–31.

Schmidt, C.C., 2004. Addressing illegal, unreported and unregulated (IUU) fishing. In: International Fisheries Compliance 2004 Conference, pp. 29–30.

Schön, P.J., Sauer, W.H.H., Roberts, M.J., 2002. Environmental influences on spawning aggregations and jig catches of chokka squid *Loligo vulgaris reynaudii*: a 'black box' approach. Bull. Mar. Sci. 71, 783–800.

Schwarz, R., Perez, J.A.A., 2010. Growth model identification of short-finned squid *Illex argentinus* (Cephalopoda: Ommastrephidae) off southern Brazil using statoliths. Fish. Res. 106, 177–184.

Schwarz, R., Perez, J.A.A., 2013. Age structure and life cycles of the Argentine shortfin squid *Illex argentinus* (Cephalopoda: Ommastrephidae) in southern Brazil. J. Mar. Biol. Assoc. U.K. 93, 557–565.

Segawa, S., 1995. Effect of temperature on oxygen consumption of juvenile oval squid *Sepioteuthis lessoniana*. Fish. Sci. 61, 743–746.

Seibel, B.A., Hochberg, F.G., Carlini, D.B., 2000. Life history of Gonatus onyx (Cephalopoda: Teuthoidea): deep-sea spawning and post-spawning egg care. Mar. Biol. 137, 519–526.

Seibel, B.A., 2007. On the depth and scale of metabolic rate variation: scaling of oxygen consumption rates and enzymatic activity in the Class Cephalopoda (Mollusca). J. Exp. Biol. 210, 1–11.

Seibel, B.A., Drazen, J.C., 2007. The rate of metabolism in marine animals: environmental constraints, ecological demands and energetic opportunities. Phil. Trans. R. Soc. B. 362, 2061–2078.

Semmens, J.M., Pecl, G.T., Gillanders, B.M., Waluda, C.M., Shea, E.K., Jouffre, D., Ichii, T., Zumholz, K., Katugin, O.N., Leporati, S.C., Shaw, P.W., 2007. Approaches to resolving cephalopod movement and migration patterns. Rev. Fish Biol. Fish. 17, 401–423.

Semmens, J., Doubleday, Z., Hoyle, K., Pecl, G., 2011. A multilevel approach to examining cephalopod growth using *Octopus pallidus* as a model. J. Exp. Biol. 214, 2799–2807.

Shaw, R., 1994. An economic description and assessment of the squid catching industry in the UK, Portugal and Spain. Fish. Res. 21, 287–303.

Shaw, P.W., 2002. Past, present and future applications of DNA-based markers in cephalopod biology: workshop report. Bull. Mar. Sci. 71, 67–78.

Shaw, P., Pierce, G., Boyle, P., 1999. Subtle population structuring within a highly vagile marine invertebrate, the veined squid *Loligo forbesi*, demonstrated with microsatellite DNA markers. Mol. Ecol. 8, 407–417.

Shaw, P., Arkhipkin, A., Adcock, G., Burnett, W., Carvalho, G., Scherbich, J., Villegas, P., 2004. DNA markers indicate that distinct spawning cohorts and aggregations of Patagonian squid, *Loligo gahi*, do not represent genetically discrete subpopulations. Mar. Biol. 144, 961–970.

Shaw, P.W., Hendrickson, L., McKeown, N.J., Stonier, T., Naud, M.-J., Sauer, W.H.H., 2010. Discrete spawning aggregations of loliginid squid do not represent genetically distinct populations. Mar. Ecol. Prog. Ser. 408, 117–127.

Shibata, K., Flores, E.E.C., 1972. Echo-traces typical of squids in waters surrounding Japan. FAO Fish. Circ. 142, 7–13.

Simmonds, J., MacLennan, D., 2005. Fisheries Acoustics Theory and Practice, second ed. Blackwell Science, Oxford, UK.

Smale, M.J., 1996. Cephalopods as prey. IV. Fishes. Philos. Trans. R Soc. Lond. B 351, 1067–1081.

Smale, M., Buchan, P., 1981. Biology of *Octopus vulgaris* off the east coast of South Africa. Mar. Biol. 65, 1–12.

Sobrino, I., Silva, L., Bellido, J.M., Ramos, F., 2002. Rainfall, river discharges and sea temperature as factors affecting abundance of two coastal benthic cephalopod species in the Gulf of Cadiz (SW Spain). Bull. Mar. Sci. 71, 851–865.

Solow, A.R., 2002. Fisheries recruitment and the North Atlantic Oscillation. Fish. Res. 54, 295–297.

Song, H., Yamashita, N., Kidokoro, H., Sakurai, Y., 2012. Comparison of growth histories of immature Japanese common squid *Todarodes pacificus* between the autumn and winter spawning cohorts based on statolith and gladius analyses. Fish. Sci. 78, 785–790.

Soule, M.A., Hampton, I., Lipinski, M.R., 2010. Estimating the target strength of live, free-swimming chokka squid *Loligo reynaudii* at 38 and 120 kHz. ICES J. Mar. Sci. 67, 1381–1391.

Stanton, T.K., 2012. 30 years of advances in active bioacoustics: a personal perspective. Methods Oceanogr. 1–2, 49–77.

Stanton, T.K., Chu, D., Jech, J.M., Irish, J.D., 2010. New broadband methods for resonance classification and high-resolution imagery of fish with swimbladders using a modified commercial broadband echosounder. ICES J. Mar. Sci. 67, 365–378.

Starr, R.M., Thorne, R.E., 1998. Acoustic assessment of squid stocks. In: Rodhouse, P.G., Dawe, E.G., O'Dor, R.K. (Eds.), Squid Recruitment Dynamics: The Genus *Illex* as a Model, the Commercial *Illex* Species and Influences on Variability. In: FAO Fisheries Technical Paper No. 376FAO, Rome, pp. 181–198.

Starr, R.M., Culver, C.S., Pomeroy, C., 2010. Managing Data-poor Fisheries Workshop: Case Studies, Models, and Solutions. California Sea Grant College Program.

Steer, M.A., Moltschaniwskyj, N.A., 2007. The effects of egg position, egg mass size, substrate and biofouling on embryo mortality in the squid *Sepioteuthis australis*. Rev. Fish Biol. Fish. 17, 173–182.

Steer, M.A., Moltschaniwskyj, N.A., Nichols, D.S., Miller, M., 2004. The role of temperature and maternal ration in embryo survival: using the dumpling squid *Euprymna tasmanica* as a model. J. Exp. Mar. Biol. Ecol. 307, 73–89.

Stefansson, G., 1996. Analysis of groundfish survey abundance data: combining the GLM and delta approaches. ICES J. Mar. Sci. 53, 577–588.

Stonier, T., 2012. Microsatellite and morphometric analysis of chokka squid (*Loligo reynaudii*) from different spawning aggregations around the southern African coast. MSc Thesis, Rhodes University, Grahamstown (South Africa).

Sundaram, S., Khan, M.Z., 2009. Stock assessment of *Sepiella inermis* (Orbigny, 1848) from Mumbai waters. J. Mar. Biol. Assoc. Ind. 51 (2), 178–183.

Sunday, J.M., Bates, A.E., Dulvy, N., 2012. Thermal tolerance and the global redistribution of animals. Nat. Clim. Change 2, 686–690.

Suzuki, T., Tashiro, M., Yamagishi, Y., 1974. Studies on the swimming layer of squid *Todarodes pacificus* Steenstrup as observed by a fish finder in the offshore region of the northern part of the Japan Sea. Bull. Facult. Fish. Hokkaido Univ. 25, pp. 238–246.

Swartzman, G., 1997. Analysis of the summer distribution of fish schools in the Pacific Eastern Boundary Current. ICES J. Mar. Sci. 54, 105–116.

Syrjala, S.E., 2000. Critique on the use of the delta distribution for the analysis of trawl survey data. ICES J. Mar. Sci. 57, 831–842.

Takayanagi, S., 1993. Changes in growth and maturity of Japanese common squid (*Todarodes pacificus*) related to differences in stock size in the Tsugaru Strait, Northern Japan. In: Okutani, T., O'Dor, R.K., Kubodera, T. (Eds.), Recent Advances in Fisheries Biology. Tokai University Press, Tokyo, pp. 545–553.

Teske, P., Oosthuizen, A., Papadopoulos, I., Barker, N., 2007. Phylogeographic structure of *Octopus vulgaris* in South Africa revisited: identification of a second lineage near Durban harbour. Mar. Biol. 151, 2119–2122.

Thiaw, M., Gascuel, D., Thiao, D., Thiaw, O.T., Jouffre, D., 2011. Analysing environmental and fishing effects on a short-lived species stock: the dynamics of the octopus *Octopus vulgaris* population in Senegalese waters. Afr. J. Mar. Sci. 33, 209–222.

Thomas, S., Kizhakudan, S.J., 2011. Cephalopod fishery and population dynamics of *Loligo duvauceli* (Orbigny) off Saurashtra region, Gujarat. Ind. J. Fish. 53 (4), 425–430.

Tian, S., Chen, X., Chen, Y., Xu, L., Dai, X., 2009a. Impacts of spatial scales of fisheries and environmental data on catch per unit effort standardisation. Mar. Freshw. Res. 60, 1273–1284.

Tian, S., Chen, X., Chen, Y., Xu, L., Dai, X., 2009b. Standardizing CPUE of *Ommastrephes bartramii* fishery in Northwest Pacific Ocean. Chin. J. Oceanol. Limnol. 27, 729–739.

Tomas, A.R.G., Petrere, M., 2012. Stock assessment of Octopus vulgaris Cuvier, 1797 (Cephalopoda: Octopodidae) exploited by trawling fisheries in southeast Brazil, Abstracts, CIAC Symposium, 27 October – 02 November 2012, Florianopolis, Brazil.

Tosunoğlu, Z., Aydın, C., Salman, A., Fonseca, P., 2009. Selectivity of diamond, hexagonal and square mesh codends for three commercial cephalopods in the Mediterranean. Fish. Res. 97, 95–102.

Triantafillos, L., Adams, M., 2001. Allozyme analysis reveals a complex population structure in the southern calamary *Sepioteuthis australis* from Australia and New Zealand. Mar. Ecol. Prog. Ser. 212, 193–209.

Triantafillos, L., Adams, M., 2005. Genetic evidence that the northern calamary, *Sepioteuthis lessoniana*, is a species complex in Australian waters. ICES J. Mar. Sci. 62, 1665–1670.

Tserpes, G., Peristeraki, P., 2002. Trends in the abundance of demersal species in the southern Aegean Sea. Sci. Mar. 66, 243–252.

Turan, C., Yaglioglu, D., 2010. Population identification of common cuttlefish (*Sepia officinalis*) inferred from genetic, morphometric and cuttlebone chemistry data in the NE Mediterranean Sea. Sci. Mar. 74, 77–86.

Turchin, P., 2003. Complex Population Dynamics: A Theoretical/Empirical Synthesis. Princeton University Press, Princeton, New Jersey.

Tyrrell, M.C., Link, J.S., Moustahfid, H., Overholtz, W.J., 2008. Evaluating the effect of predation mortality on forage species population dynamics in the Northeast US continental shelf ecosystem using multispecies virtual population analysis. ICES J. Mar. Sci. 65, 1689–1700.

Tyrrell, M., Link, J., Moustahfid, H., 2011. The importance of including predation in fish population models: implications for biological reference points. Fish. Res. 108, 1–8.
Ulrich, C., Le Gallic, B., Dunn, M.R., Gascuel, D., 2002. A multi-species multi-fleet bioeconomic simulation model for the English Channel artisanal fisheries. Fish. Res. 58, 379–401.
Uozumi, Y., 1998. Fishery biology of arrow squids, *Nototodarus gouldi* and *N. sloanii* in New Zealand waters. Bull. Natl. Inst. Far Seas Fish. 35, 1–111.
Uozumi, Y., Shiba, C., 1993. Growth and age composition of *Illex argentinus* (Cephalopoda: Oegopsida) based on daily increment counts in statoliths. In: Okutani, T., O'Dor, R.K., Kubodera, T. (Eds.), Recent Advances in Fisheries Biology. Tokai University Press, Tokyo, pp. 591–605.
van der Vyver, F., 2013. The geographic stock structure of chokka (*Loligo reynaudii*) and its implications for management of the fishery. MSc Thesis, Rhodes University, Grahamstown (South Africa).
Vargas-Yáñez, M., Moya, F., García-Martínez, M., Rey, J., González, M., Zunino, P., 2009. Relationships between *Octopus vulgaris* landings and environmental factors in the northern Alboran Sea (Southwestern Mediterranean). Fish. Res. 99, 159–167.
Vasconcellos, M., Haimovici M., Ramos, K. L., In press. Pesca de emalhe demersal no sul do Brasil: evolução, conflitos e (des)ordenamento. In: M. Haimovici, J. M. Andriguetto & P. Sunye (Orgs.). A Pesca Marinha e Estuarina no Brasil: Abordagem Multidisciplinar Aplicada a Estudos de Caso. Editora da Universidade Federal do Rio Grande, Rio Grande.
Vecchione, M., 1987. Juvenile ecology. In: Boyle, P.R. (Ed.), Cephalopod Life Cycles, vol. 2. Academic Press Inc., London, pp. 61–84.
Venables, W.N., Dichmont, C.M., 2004. GLMs, GAMs and GLMMs: an overview of theory for applications in fisheries research. Fish. Res. 70, 319–337.
Vidal, E.A.G., DiMarco, F.P., Wormuth, J.H., Lee, P.G., 2002. Influence of temperature and food availability on survival, growth, and yolk utilization in hatchling squid. Bull. Mar. Sci. 71, 915–931.
Vidal, E.A.G., Roberts, M.J., Martins, R.S., 2005. Yolk utilization, metabolism and growth in reared Loligo vulgaris paralarvae Aquat. Living Resour. 18, 385–393.
Vidal, E.A.G., DiMarco, P., Lee, P., 2006. Effects of starvation and recovery on the survival, growth and RNA/DNA ratio in loliginid squid paralarvae. Aquaculture 260, 94–105.
Vidal, E.A.G., Fuentes, L., da Silva, L.B., 2010a. Defining *Octopus vulgaris* populations: a comparative study of the morphology and chromatophore pattern of paralarvae from Northeastern and Southwestern Atlantic. Fish. Res. 106, 199–208.
Vidal, E.A.G., Haimovici, M., Hackbart, V.C.S., 2010b. Distribution of paralarvae and small juvenile cephalopods in relation to primary production in an upwelling area off southern Brazil. ICES J. Mar. Sci. 67, 1346–1352.
Vidal, E.A.G., Boletzky, S.V., 2014. Loligo vulgaris and Doryteuthis opalescens. In: Iglesias, J., Fuentes, L., Villanueva, R. (Eds.), Cephalopod Culture. Springer-Verlag, London.
Vidyasagar, K., Deshmukh, V., 1992. Stock assessment of *Loligo duvaucelii* (D'Orbigny) in Bombay waters. J. Mar. Biol. Assoc. Ind. 34 (1–2), 14–17.
Vila, Y., Silva, L., Torres, M., Sobrino, I., 2010. Fishery, distribution pattern and biological aspects of the common European squid *Loligo vulgaris* in the Gulf of Cadiz. Fish. Res. 106, 222–228.
Villanueva, R., Quintana, D., Petroni, G., Bozzano, A., 2011. Factors influencing the embryonic development and hatchling size of the oceanic squid *Illex coindetii* following in vitro fertilization. J. Exp. Mar. Biol. Ecol. 407, 54–62.
Villegas, P., 2001. Growth, life cycle and fishery biology of *Loligo gahi* (d'Orbigny, 1835) off the Peruvian coast. Fish. Res. 54, 123–131.

Vojkovich, M., 1998. The California fishery for market squid (*Loligo opalescens*). CalCOFI Rep. 39, 55–60.

Wakabayashi, T., Suzuki, N., Sakai, M., Ichii, T., Chow, S., 2006. Identification of ommastrephid squid paralarvae collected in northern Hawaiian waters and phylogenetic implications for the family Ommastrephidae using mtDNA analysis. Fish. Sci. 72, 494–502.

Walsh, L., Turk, P., Forsythe, J., Lee, P., 2002. Mariculture of the loliginid squid *Sepioteuthis lessoniana* through seven successive generations. Aquaculture 212, 245–262.

Walters, C.J., Martell, S.J.D., 2004. Fisheries Ecology and Management. Princeton University Press, Princeton, NJ.

Waluda, C.M., Rodhouse, P.G., 2006. Remotely sensed mesoscale oceanography of the central Eastern Pacific and recruitment variability in *Dosidicus gigas*. Mar. Ecol. Prog. Ser. 310, 25–32.

Waluda, C.M., Trathan, P.N., Rodhouse, P.G., 1999. Influence of oceanographic variability on recruitment in the *Illex argentinus* (Cephalopoda: Ommastrephidae) fishery in the South Atlantic. Mar. Ecol. Prog. Ser. 183, 159–167.

Waluda, C.M., Rodhouse, P.G., Podestá, G.P., Trathan, P.N., 2001a. Surface oceanography of the inferred hatching grounds of *Illex argentinus* (Cephalopoda: Ommastreshidae) and influences on recruitment variability. Mar. Biol. 139, 671–679.

Waluda, C.M., Rodhouse, P.G., Trathan, P.N., Pierce, G.J., 2001b. Remotely sensed mesoscale oceanography and the distribution of *Illex argentinus* in the South Atlantic. Fish. Oceanogr. 10, 207–216.

Waluda, C.M., Trathan, P.N., Rodhouse, P.G., 2004. Synchronicity in southern hemisphere squid stocks and the influence of the Southern Oscillation and Trans Polar Index. Fish. Oceanogr. 13, 255–266.

Wang, J.G., Pierce, G.J., Boyle, P.R., Denis, V., Robin, J.-P., Bellido, J.M., 2003. Spatial and temporal patterns of cuttlefish (*Sepia officinalis*) abundance and environmental influences—a case study using trawl fishery data in French Atlantic coastal, English Channel, and adjacent waters. ICES J. Mar. Sci. 60, 1149–1158.

Wangvoralak, S., 2011. Life history and ecological importance of the veined squid *Loligo forbesii* in Scottish waters. PhD Thesis, University of Aberdeen, Aberdeen.

Warner, R.R., Hamilton, S.L., Sheehy, M.S., Zeidberg, L.D., Brady, B.C., Caselle, J.E., 2009. Geographic variation in natal and early larval trace-elemental signatures in the statoliths of the market squid *Doryteuthis* (formerly *Loligo*) *opalescens*. Mar. Ecol. Prog. Ser. 379, 109–121.

Wells, M.J., Clarke, A., 1996. Energetics: the costs of living and reproducing for an individual cephalopod. Philos. Trans. R Soc. B 35, 1083–1104.

Wilberg, M.J., Thorson, J.T., Linton, B.C., Berkson, J., 2010. Incorporating time-varying catchability into population dynamic stock assessment models. Rev. Fish. Sci. 18, 7–24.

Willcox, S., Lyle, J., Steer, M., 2001. Tasmanian Arrow Squid Fishery—Status Report 2001. Tasmanian Aquaculture and Fisheries Institute, University of Tasmania.

Wolfram, K., Mark, F.C., John, U., Lucassen, M., Pörtner, H.O., 2006. Microsatellite DNA variation indicates low levels of genetic differentiation among cuttlefish (*Sepia officinalis*) populationsin the English Channel and the Bay of Biscay. Comp. Biochem. Physiol. D Genom. Proteom. 1, 375–383.

Worm, B., Hilborn, R., Baum, J.K., Branch, T.A., Collie, J.S., Costello, C., Fogarty, M.J., Fulton, E.A., Hutchings, J.A., Jennings, S., Jensen, O.P., Lotze, H.K., Mace, P.M., McClanahan, T.R., Minto, C., Palumbi, S.R., Parma, A.M., Ricard, D., Rosenberg, A.A., Watson, R., Zeller, D., 2009. Rebuilding global fisheries. Science 325, 578–585.

Wormuth, J.H., Roper, C.F.E., 1983. Quantitative sampling of oceanic cephalopods by nets, problems and recommendations. Biol. Oceanogr. 2, 357–377.

Yamamoto, J., Shimura, T., Uji, R., Watanabe, S., Sakurai, Y., 2007. Vertical distribution of *Todarodes pacificus* (Cephalopoda: Ommastrephidae) paralarvae near the Oki Islands, southwestern Sea of Japan. Mar. Biol. 153, 7–13.

Yamashita, N., Kaga, T., 2013. Stock assessment and evaluation for winter spawning stock of Japanese common squid. Marine Fisheries Stock Assessment and Evaluation for Japanese Waters. Fisheries Agency and Fisheries Research Agency of Japan, Tokyo, pp. 571–604.

Yatsu, A., Tanaka, H., Mori, J., 1998. Population structure of the neon flying squid, *Ommastrephes bartramii*, in the North Pacific Ocean. In: Okutani, T. (Ed.), Large Pelagic Squids. Japan Marine Fishery Resources Research Center, Tokyo, pp. 31–48.

Ye, Y., 2000. Is recruitment related to spawning stock in penaeid shrimp fisheries? ICES J. Mar. Sci. 57, 1103–1109.

Yeatman, J.M., Benzie, J.A.H., 1993. Cryptic speciation in *Loligo* from Northern Australia. In: Okutani, T., O'Dor, R.K., Kubodera, T. (Eds.), Recent Advances in Fisheries Biology. Tokai University Press, Tokyo, pp. 641–652.

Young, I.A.G., Pierce, G.J., Daly, H.I., Santos, M.B., Key, L.N., Bailey, N., Robin, J.-P., Bishop, A.J., Stowasser, G., Nyegaard, M., Cho, S.K., Rasero, M., Pereira, J.M.F., 2004. Application of depletion methods to estimate stock size in the squid *Loligo forbesi* in Scottish waters (UK). Fish. Res. 69, 211–227.

Young, I.A.G., Pierce, G.J., Murphy, J., Daly, H.I., Bailey, N., 2006. Application of the Gómez-Muñoz model to estimate catch and effort in squid fisheries in Scotland. Fish. Res. 78, 26–38.

Young, M.A., Kvitek, R.G., Iampetro, P.J., Garza, C.D., Maillet, R., Hanlon, R.T., 2011. Seafloor mapping and landscape ecology analyses used to monitor variations in spawning site preference and benthic egg mop abundance for the California market squid (*Doryteuthis opalescens*). J. Exp. Mar. Biol. Ecol. 407, 226–233.

Zavala, C.A.S., Ferreri, G.A.B., Keyl, F., 2012. Eulerian approach to the estimation of growth rates and population structure of jumbo squid (*Dosidicus gigas*) in the Central Gulf of California. J. Shellfish Res. 31, 801–807.

Zeidberg, L.D., Robison, B.H., 2007. Invasive range expansion by the Humboldt squid, *Dosidicus gigas*, in the eastern North Pacific. Proc. Natl. Acad. Sci. U.S.A. 104, 12948–12950.

Zeidberg, L.D., Hamner, W.M., Nezlin, N.P., Henry, A., 2006. The fishery for California market squid (*Loligo opalescens*) (Cephalopoda: Myopsida), from 1981 through 2003. Fish. Bull. 104, 46–59.

Zeidberg, L.D., Isaac, G., Widmer, C.L., Neumeister, H., Gilly, W.H., 2011. Egg capsule hatch rate and incubation duration of the California market squid, *Doryteuthis* (= *Loligo*) *opalescens*: insights from laboratory manipulations. Mar. Ecol. 32, 468–479.

Zeidberg, L.D., Butler, J.L., Ramon, D., Cossio, A., Stierhoff, K.L., Henry, A., 2012. Estimation of spawning habitats of market squid (*Doryteuthis opalescens*) from field surveys of eggs off Central and Southern California. Mar. Ecol. 33, 326–336.

Zheng, J., Murphy, M.C., Kruse, G.H., 1997. Application of a catch-survey analysis to blue king crab stocks near Pribilof and St. Matthew Islands. Alaska Fishery Res. Bull. 4, 62–74.

Zuev, G.V., Nigmatullin, C.M., Nikolski, G.V., 1985. Nektonic Oceanic Squids. Agropromizdat, Moscow.

Zumholz, K., Piatkowski, U., 2005. Research cruise data on the biology of the lesser flying squid, *Todaropsis eblanae*, in the North Sea. Aquat. Living Resour. 18, 373–376.

CHAPTER THREE

The Study of Deep-Sea Cephalopods

Henk-Jan T. Hoving[*,1], Jose Angel A. Perez[†], Kathrin S.R. Bolstad[‡], Heather E. Braid[‡], Aaron B. Evans[‡], Dirk Fuchs[§], Heather Judkins[¶], Jesse T. Kelly[‡], José E.A.R. Marian[||], Ryuta Nakajima[#], Uwe Piatkowski[*], Amanda Reid[**], Michael Vecchione[††], José C.C. Xavier[‡‡,§§]

[*]GEOMAR, Helmholtz Centre for Ocean Research Kiel, Kiel, Germany
[†]Centro de Ciências Tecnológicas da Terra e do Mar Universidade do Vale do Itajaí, Itajaí, Santa Catarina, Brazil
[‡]Institute for Applied Ecology New Zealand, Auckland University of Technology, Auckland, New Zealand
[§]Freie Universität Berlin, Institute of Geological Sciences, Branch Paleontology, Berlin, Germany
[¶]Department of Biological Sciences, University of South Florida St. Petersburg, St. Petersburg, Florida, USA
[||]Departamento de Zoologia, Instituto de Biociências, Universidade de São Paulo, Sao Paulo, Brazil
[#]Department of Art and Design, University of Minnesota Duluth, Duluth, Minnesota, USA
[**]Australian Museum Research Institute, Sydney, New South Wales, Australia
[††]NMFS National Systematics Laboratory, National Museum of Natural History, Washington, DC, USA
[‡‡]Institute of Marine Research, Department of Life Sciences, University of Coimbra, Coimbra, Portugal
[§§]British Antarctic Survey, NERC, Cambridge, United Kingdom
[1]Corresponding author: e-mail address: hhoving@geomar.de

Contents

Advances in Marine Biology, Volume 67
ISSN 0065-2881
http://dx.doi.org/10.1016/B978-0-12-800287-2.00003-2

Abstract

"Deep-sea" cephalopods are here defined as cephalopods that spend a significant part of their life cycles outside the euphotic zone. In this chapter, the state of knowledge in several aspects of deep-sea cephalopod research are summarized, including information sources for these animals, diversity and general biogeography and life cycles, including reproduction. Recommendations are made for addressing some of the remaining knowledge deficiencies using a variety of traditional and more recently developed methods. The types of oceanic gear that are suitable for collecting cephalopod specimens and images are reviewed. Many groups of deep-sea cephalopods require taxonomic reviews, ideally based on both morphological and molecular characters. Museum collections play a vital role in these revisions, and novel (molecular) techniques may facilitate new use of old museum specimens. Fundamental life-cycle parameters remain unknown for many species; techniques developed for neritic species that could potentially be applied to deep-sea cephalopods are discussed. Reproductive tactics and strategies in deep-sea cephalopods are very diverse and call for comparative evolutionary and experimental studies, but even in the twenty-first century, mature individuals are still unknown for many species. New insights into diet and trophic position have begun to reveal a more diverse range of feeding strategies than the typically voracious predatory lifestyle known for many cephalopods. Regular standardized deep-sea cephalopod surveys are necessary to provide insight into temporal changes in oceanic cephalopod populations and to forecast, verify and monitor the impacts of global marine changes and human impacts on these populations.

Keywords: Cephalopoda, Deep sea, Diversity, Collection methods, Reproduction, Longevity, Trophic ecology, Evolution, Biogeography

1. INTRODUCTION

The deep sea, comprising both the sea floor and the overlying water column below the photic zone, is the largest and least-known environment of the Earth. According to recent calculations, less than 0.0001% of the deep-sea floor has been investigated, and even less for the deep pelagic (Ramirez-Llodra et al., 2010), yet 32 of the 33 known phyla of plants and animals are found in the oceans (most with representatives in the deep sea), and 15 of these are exclusively marine (Beaumont et al., 2006), and most of these have representatives in the deep sea. This chapter focuses on one particular class of Mollusca: the cephalopods, which form an important and fascinating component of this vast but unexplored environment.

Although the cephalopod families inhabiting the oceanic and deep-sea environments are numerous (42 of the 50 known families), they tend to be less speciose than coastal families (Young et al., 1998). Cephalopods occur not only in a wide range of deep-sea habitats, for example, seamounts, open water and continental slopes, but also in extreme environments such as hydrothermal vents and oxygen minimum zones (e.g. Clarke, 2008; Gilly et al., 2013; Ramirez-Llodra et al., 2010; Roeleveld et al., 1992; Roper and Young, 1975; Voight, 2000). Cephalopod body forms throughout these waters vary widely (Figures 3.1 and 3.2); squids range in size from small pyroteuthids of several grams to giants and colossals (e.g. *Mesonychoteuthis hamiltoni* Robson, 1925) weighing up to 500 kg (Jereb and Roper, 2010) and from gelatinous drifters (e.g. Bolitaenidae) to swift, agile hunters (e.g. Gonatidae). Deep-sea octopod sizes at maturity also range from just a few hundred grams (e.g. *Bathypolypus arcticus* Prosch, 1849, O'Dor and Malacaster, 1983) to an impressive 75 kg for *Haliphron atlanticus* Steenstrup, 1861 (O'Shea, 2004b). Habitat usage is similarly diverse; while most deep-sea squid species are pelagic and most octopods benthic, examples of each group can be found from the midwater down to or near the sea floor.

Many oceanic cephalopod species, especially squids, have wide geographical ranges. Given their broad distribution patterns and the large volume of their habitat, these oceanic squids are likely to form the bulk of the worldwide deep-sea cephalopod biomass, similarly to what has been shown for midwater fishes (Irigoien et al., 2014). It is therefore not surprising that pelagic cephalopods are among the dominant prey items for many valuable commercial fish species (e.g. tunas and billfishes) and are very important in the diets of many endangered marine animals like whales, seals and sharks (e.g. Clarke, 1996; Klages, 1996; Smale, 1996). The few studies that estimate the abundance of deep-sea squids suggest enormous population sizes and biomass—for example, sperm whales alone have been estimated to consume more than 110 million tonnes of oceanic squids annually (Clarke, 1977a), which considerably exceeds the annual total capture of marine and freshwater organisms harvested by humans (90 million tonnes) (FAO, 2012).

Given that some cephalopod species appear to have cosmopolitan distributions, while others are restricted to narrow geographical ranges, it is not surprising that understanding the systematics of deep-sea taxa in particular remains a work in progress. Specimens of many deep-sea species remain rare

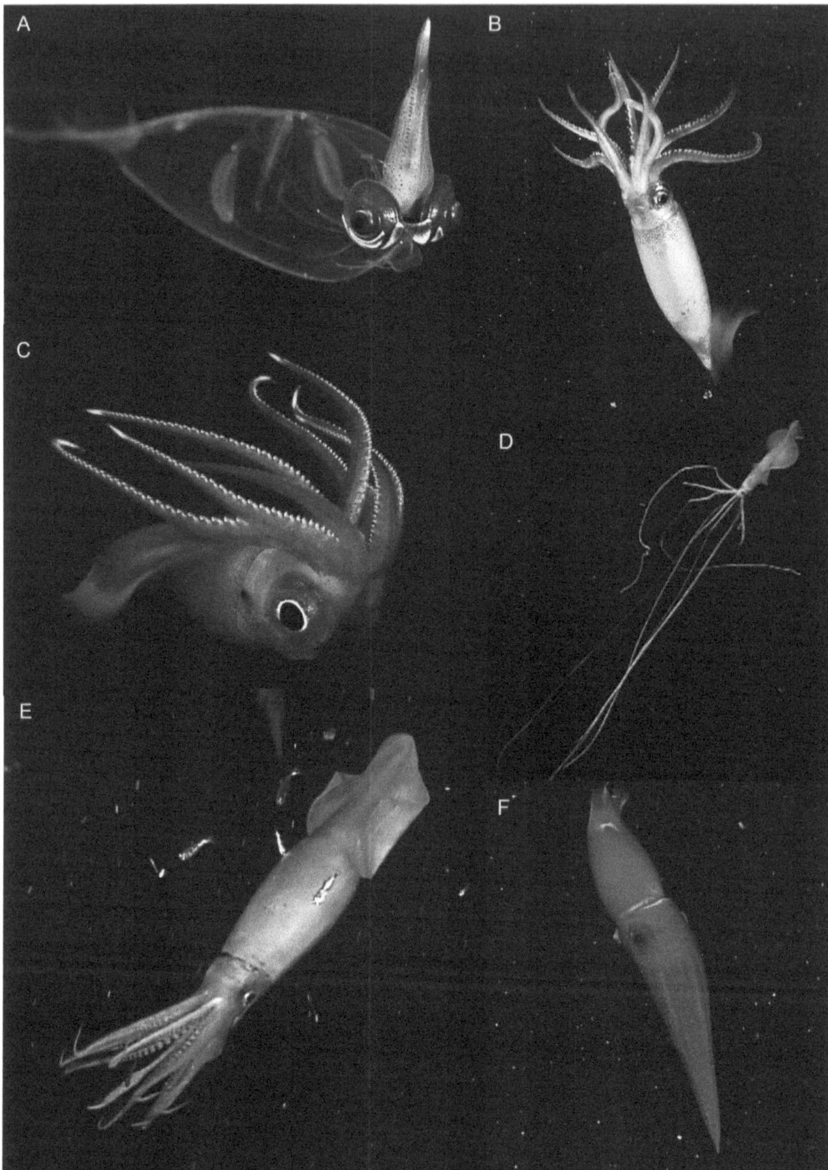

Figure 3.1 Examples of deep-sea squids as observed by the ROV programme of the Monterey Bay Aquarium Research Institute. (A) *Galiteuthis phyllura*. (B) *Octopoteuthis deletron*. (C) *Dosidicus gigas*. (D) *Gonatus* sp. (E) *Magnapinna* sp. (F) *Histioteuthis heteropsis*. *(A) © 2009 MBARI. (B) © 2008 MBARI. (C) © 2006 MBARI. (D) © 2001 MBARI. (E) © 2009 MBARI. (F) © 2011 MBARI.*

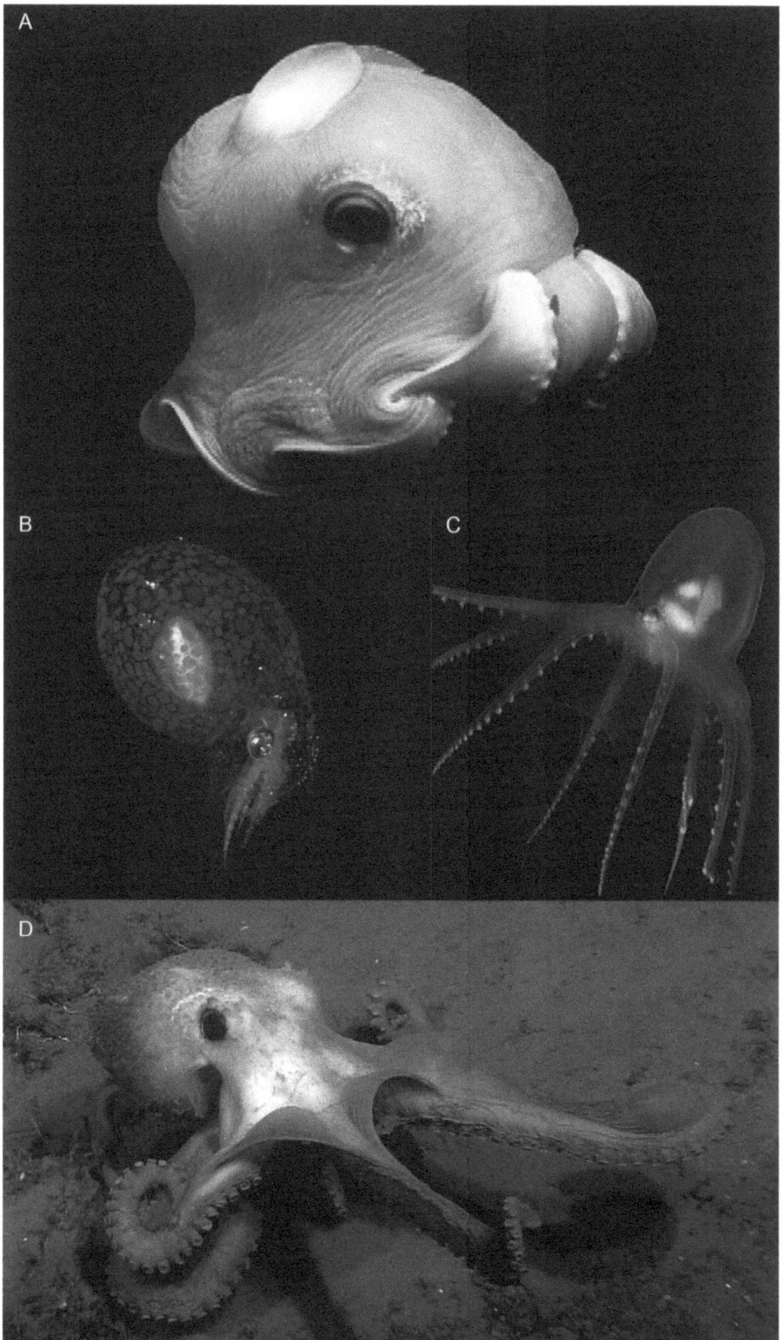

Figure 3.2 Examples of deep-sea octopods as observed by the ROV program of the Monterey Bay Aquarium Research Institute. (A) *Opisthoteuthidae*. (B) *Bolitaenidae*. (C) *Vitreledonella richardi*. (D) *Graneledone*. *(A) © 1999 MBARI. (B) © 2009 MBARI. (C) © 2001 MBARI. (D) © 2007 MBARI.*

in collections, with specimens in suitable condition to permit detailed morphological descriptions in comparison even rarer. Fresh tissue for molecular study is likewise difficult to obtain, so meaningful comparison among taxa and even entire faunas from different geographical region can be quite challenging.

Taxonomy aside, in many squid families such as Histioteuthidae ("jeweled" or "violet" squids) and Cranchiidae ("glass" squids), which are abundant in the diets of cetaceans (e.g. Clarke, 1996), even basic biological knowledge, such as information on age and size at sexual maturity remains unknown for many species. By contrast, the muscular ommastrephid squids (another cephalopod family consumed by cetaceans and comprising 10% of all known oceanic squid species) are relatively well studied (e.g. Rosa et al., 2013a), partly because they form the bulk of all globally fished cephalopods (Caddy and Rodhouse, 1998). While some ommastrephid species primarily inhabit euphotic waters, others migrate regularly into the deep sea. However, their life-history traits (being strong, agile hunters and voracious carnivores, with high growth rates and life cycles of about 1 year) are not necessarily representative of most deep-sea squids and cannot always be used to infer information about lesser known oceanic cephalopods.

Although our knowledge of the life cycle of most oceanic and deep-sea cephalopods is still very limited and built upon generalizations (often from relatively few species), the current understanding of the coleoid cephalopod life cycle is that these animals grow quickly and have relatively short life cycles and die after a single period of reproduction (semelparity) (e.g. Boyle and Rodhouse, 2005). However, accumulating evidence indicates a diversity of semelparous reproductive strategies (Hoving, 2008; Laptikhovsky et al., 2007; Rocha et al., 2001); for example, some species spawn a single egg mass, while others produce several over a period of time. The limited timespan of the cephalopod life and reproductive cycles must heavily influence the development and optimization of the single reproductive cycle. An increasing body of evidence does suggest that the life cycles of deeper-living cephalopods are longer than those of shallow-dwelling species (e.g. Arkhipkin, 1997; Hoving and Lipinski, 2009; Laptikhovsky, 2013; Seibel et al., 2005), but evidence is still lacking for most taxa. Existing knowledge on longevity in deep-sea cephalopods will therefore be reviewed, and available techniques to determine the ages of individual cephalopods will be discussed.

It has been argued that the semelparous life-history strategy requires a highly opportunistic lifestyle, where prey from a wide variety of organisms is taken, often in large amounts, to fuel the energetic costs associated with semelparity and nektonic lifestyle (Rodhouse and Nigmatullin, 1996). It is not surprising that such cephalopods (e.g. families Gonatidae and Ommastrephidae) are insatiable carnivores (Rodhouse and Nigmatullin, 1996). Unfortunately, despite the importance of oceanic and deep-sea cephalopods in marine food webs, data on their diets, feeding strategies and trophic positions are scarce. Recent developments show, however, that the typical "voracious, versatile predatory" feeding strategy (sensu Rodhouse and Nigmatullin, 1996) does not apply to all cephalopods. For example, vampire squids (*Vampyroteuthis infernalis* Chun, 1903) squids have recently been reported to ingest a wide variety of food including detritus (Hoving and Robison, 2012). Since the publication of Rodhouse and Nigmatullin's (1996) comprehensive review, new information has been collected for a wider variety of cephalopods, including oceanic and deep-sea species. New techniques have also provided more detailed insight into the trophic position and temporal feeding variability of a number of species (e.g. Cherel and Hobson, 2005; Jackson et al., 2007; Phillips et al., 2003a,b). For this chapter, literature published since 1996 has been prioritized in order to summarize and describe new information on dietary analysis and the feeding strategies and trophic positions of deep-sea cephalopods.

The primary reason for the absence of general biological knowledge on many deep-sea cephalopods is the inaccessibility of their environment. The first worldwide study of oceanography was carried out during 1872–1876 by the British naval vessel HMS *Challenger.* The best equipment of the day—nets, trawls, dredges, submarine thermometers, instruments to collect sediment samples, and copious quantities of rope—were used to collect vast amounts of information and thousands of specimens that are still being studied today. Some specimens came from depths as great as 5700 m in the Pacific, providing the first evidence that life was indeed to be found in the deep sea. Many new cephalopod species were collected on this expedition, including several deep-sea taxa (Hoyle, 1886) (Figure 3.3).

The collection of deep-sea organisms remains expensive because it requires dedicated expeditions and specialized gear. In addition, oceanic and deep-sea cephalopods are difficult to catch. Being highly mobile and visually attuned animals, some squids are able to avoid many types of oceanographic gear. As a result, there can be considerable differences between the composition of trawl catches and what is found among stomach contents of predators (which are generally capable of capturing large and sexually mature individuals, while human collection methods tend to miss these life stages) (Clarke, 1977b; Rodhouse, 1990; Xavier et al., 2003a). However, when dedicated deep-sea expeditions are able to deploy a wide variety of gear, they may provide a wealth of specimens and information on deep-sea cephalopods. A good example of a successful twenty-first-century expedition to sample cephalopods with traditional gears and modern techniques was been the MAR-ECO cruise of the Norwegian R/V *G.O. Sars* to the Mid-Atlantic Ridge in 2004 (Vecchione et al., 2010b). During this expedition, which formed part of the Census of Marine Life project (Vecchione et al., 2010c), a total number of 1295 cephalopods were caught, representing 56 species of which the majority were oceanic and deep-sea species.

Continuing application development of modern ocean-sampling technology (such as ROVs, manned submersibles and autonomous camera systems) for deep-sea exploration has provided many new insights into the biology and ecology of poorly known deep-sea cephalopods (e.g. Bush et al., 2009; Hoving and Robison, 2012; Kubodera and Mori, 2005; Kubodera et al., 2007; Seibel et al., 2005; Vecchione et al., 2001; Widder, 2013). Spectacular high-definition imagery (even 3D) is now available for species that were previously only known from mangled net-captured specimens. Footage of these animals often (understandably) captures the attention of the general public. Deep-sea cephalopods are charismatic and publications on new behaviour or other discoveries are increasingly covered as news items. Therefore, they may function as emblematic organisms for deep-sea conservation efforts (e.g. *Architeuthis dux* Steenstrup, 1860, and Guerra et al., 2011).

In the recent years, evidence has increasingly shown that populations of (oceanic) cephalopods may respond to or be impacted by changes in the marine environment and human activities including oceanic warming, expanding hypoxia and fishing (Gilly et al., 2013; Golikov et al., 2013; Rodhouse, 2013; Zeidberg and Robison, 2007). Indeed, cephalopods in general have been described as good climate-change indicators (Pecl and Jackson, 2008), and the degree of plasticity some coastal and oceanic species show in their life-history strategies under different oceanographic regimes suggests that they may be preadapted for climate change (Hoving et al., 2013a; Pecl and Jackson, 2008; Rodhouse, 2013). In areas of overexploitation of commercial fish stocks, increases in cephalopod populations have been reported (Caddy and Rodhouse, 1998; Vecchione et al., 2009a; Zeidberg and Robison, 2007). On the other hand, deeper-living, less mobile cephalopods with low metabolic rates (e.g. *Histioteuthis* spp.;

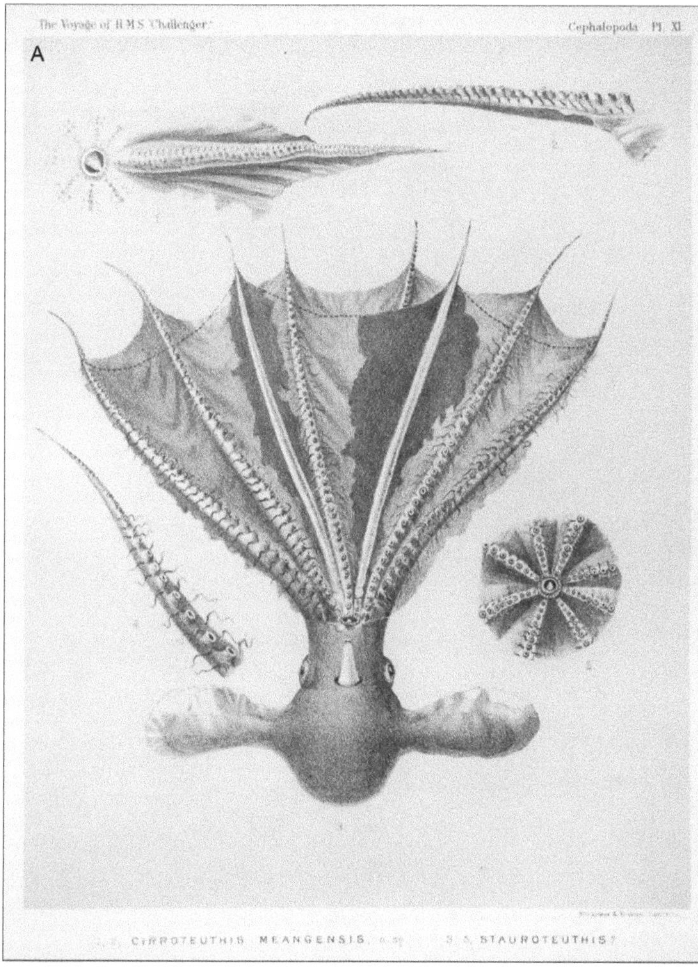

Figure 3.3 (A) Many new deep-sea species were collected during the voyage of the HMS *Challenger* during the years 1873–1876. These are copies of some of the plates from the expedition report (Hoyle, 1886). (A) Plate XI, figs. 1–2, arms of *Cirroteuthis magna* n. sp. (now *Cirrothauma magna*, (Hoyle, 1885)); figs. 3–5, *Stauroteuthis ?*, ventral view of animal (3), distal arm tip (4), and oral view of arm crown (5).

Octopoteuthis deletron Young, 1972; and *Japetella diaphana* Hoyle, 1885; Seibel et al., 1997) may be less able to cope with environmental changes than some of the more opportunistic and adaptable teuthoids. Overall, very little baseline data on life cycles, distribution, diversity, and abundance are available to assist in understanding or predicting the potential responses of deep-sea cephalopods to global marine stressors and change.

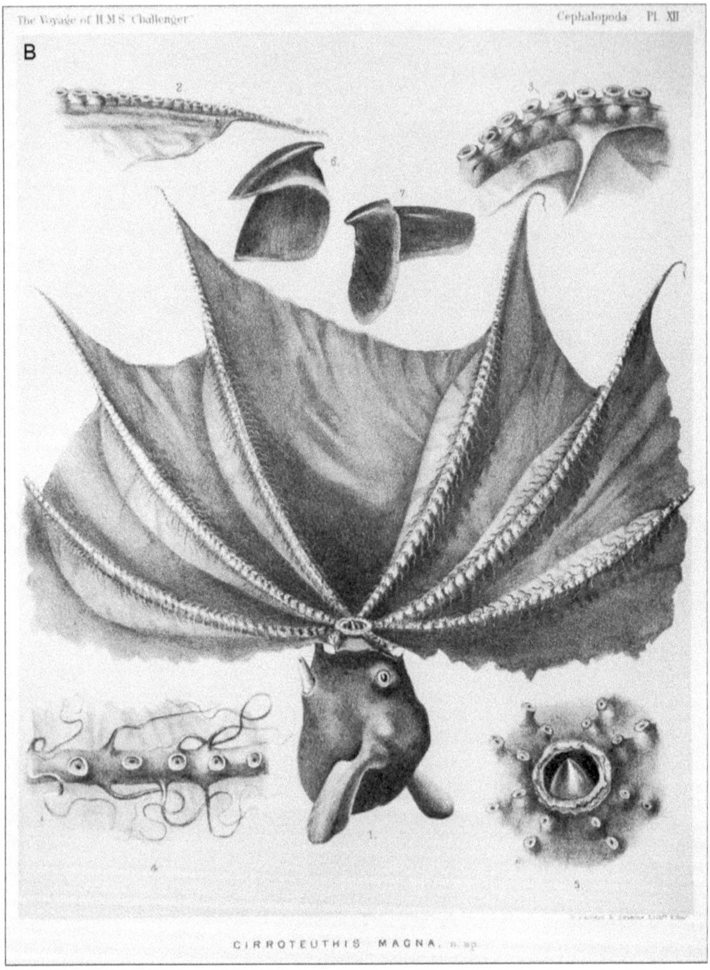

Figure 3.3—Cont'd (B) Plate XII, *Cirroteuthis magna* n. sp. (now *Cirrothauma magna* (Hoyle, 1885)); lateral view of animal (1), distal tip of arm (2), lateral view of right dorsal arm (3), middle of arm (4), centre of arm crown (5), upper beak (6) and lower beak (7).

Because of the discrepancy between the increasing awareness of deep-sea cephalopods' importance and the poor status of our general knowledge on their diversity, biology, and ecology, and our limited ability to collect them, a workshop was held during the Cephalopod International Advisory Council meeting in Florianópolis, Brazil, in October–November 2012. This chapter aims to summarize the discussions that took place, to synthesize what is currently known and to identify knowledge gaps about deep-sea cephalopods. More specifically, we aim to summarize and discuss

- a working definition of a "deep-sea cephalopod";
- the varied sources of information on deep-sea cephalopods that are available;
- the current state of (and gaps in) knowledge on cephalopod diversity, biogeography, evolution, life cycles (i.e. reproduction and longevity) and trophic ecology;
- deep-sea cephalopod populations in the context of ocean change.

We hope that this chapter will enable future researchers studying these animals to better integrate the disparate data collection methods available and to continue to improve worldwide understanding of these important, charismatic residents of the world's largest habitat.

2. WHAT IS A "DEEP-SEA" CEPHALOPOD?

The deep sea is generally defined as the volume of water beneath the euphotic zone. In most environments, this is below 200 m, while in some tropical oligotrophic waters, light may penetrate into ocean layers as deep as 300 m. The deepest unbiased benthic cephalopod record in the literature is a cirrate octopus that was photographed at 5145 m off Barbados (Jahn, 1971). The deepest unbiased record of a squid to date is a "bigfin" squid (Magnapinnidae) observed at 4735 m by DSV *Nautile* in the western Atlantic off Brazil (Vecchione et al., 2001).

One of the difficulties in classifying cephalopods based on their depth distribution is that cephalopods may have quite extreme vertical distributional ranges. For example, *Dosidicus gigas* (d'Orbigny, 1835) commonly ranges from near-surface waters (<25 m, e.g. during nocturnal hunting) to mesopelagic oxygen minimum zones (e.g. Gilly et al., 2006) and has been recorded down to 1447 m (Stewart et al., 2013).

Many other cephalopods migrate from deep waters (below 200 m) up to epipelagic waters at night (e.g. Piatkowski et al., 1994; Roper and Young, 1975; Watanabe et al., 2006). Some cephalopods may occupy different depths during particular phases of their life cycles; the young of many deep-sea cephalopod species spend part of their early lives in epipelagic waters and undergo ontogenetic descent as they grow and mature (e.g. Quetglas et al., 2010; Rodhouse and Piatkowski, 1995; Roper and Young, 1975; Young, 1978).

Therefore dwelling or utilizing the volume outside the euphotic zone for "a significant portion of the life cycle" may constitute a single contiguous period early or late in the life cycle (ontogenetic descent/ascent), or a contiguous proportion of each day (diel vertical migration), or regular but brief forays into deep waters (e.g. diving to hunt). When considering which cephalopods should be treated as "deep-sea" groups, it is important to note that, while this designation can be made at high taxonomic levels in some cases (e.g. the order Vampyromorpha), in others, it may vary among genera within a family or even congeneric species. While cuttlefish are generally considered a coastal, shallow-dwelling group, some species of the genus *Sepia* do occur at great depth (Reid, 2001). Nautiluses are found in tropical locations, where the euphotic zone extends to greater depths, but they have also been reported from true deep-sea strata (Young, 2010). Octopods are more challenging to categorize. Although some species have been well studied and their depth ranges appear reasonably clear (e.g. *Vitreledonella richardi* Joubin, 1918) (Clarke and Lu, 1975; Lu and Clarke, 1975), the combination of taxonomic instability and sparse distribution records leaves others (e.g. *Grimpoteuthis* spp.) with the vague habitat estimate of "benthic," sometimes with a hypothetical depth range. While similar uncertainty also surrounds the exact depth ranges of some squids, in many squid genera, the majority (if not all species) reside below the euphotic zone at some life stage. It appears rare for a squid family to have a mix of deep-dwelling and exclusively shallow-dwelling genera; typically, one deep-sea genus within a family indicates that all other confamilial groups inhabit the deep sea as well. Some notably abundant families with "true" deep-sea members (species that live the majority of their lives below the euphotic zone) are Cranchiidae, Histioteuthidae, Octopoteuthidae, and the "chiroteuthid" families (which contain Chiroteuthidae, Mastigoteuthidae, Joubiniteuthidae, Magnapinnidae, Promachoteuthidae and Batoteuthidae); however, many other families are widely represented in this environment. The onychoteuthids should also be considered important in this category; while little is known about their vertical distribution, and they are often classified as being a "midwater" family of squid (Lefkaditou et al., 2000), specimens have been recorded from depths of over 1200 m (Bolstad, 2010).

Several recent deep-sea observations have suggested that an improved understanding of reproductive strategies could also help characterize species that should be considered "deep-sea" cephalopods. To date, the majority of pelagic squid eggs and egg masses have been found in epipelagic waters, suggesting that squids such as those of the relatively well-reported families

Enoploteuthidae, Ommastrephidae and Thysanoteuthidae, utilize the epipelagic zone for reproduction. However, too little is known about pelagic egg masses from other squid families to assume that all squids that produce egg masses reproduce in the epipelagic zone. Deep-sea squids like Architeuthidae and Octopoteuthidae are also likely to produce egg masses as suggested from their asynchronous ovulation (Hoving et al., 2004, 2008a). Brooding squids have been observed in the deep sea (Bush et al., 2012; Seibel et al., 2005) but also in the epipelagic zone (*Gonatus madokai* Kubodera and Okutani, 1977) (Bower et al., 2012). While additional observations on cephalopod reproductive biology in the deep sea may eventually enable further habitat-use inferences to be drawn across taxa, the body of information presently available on egg masses and spawning in oceanic cephalopods seems insufficient to divide cephalopods into deep-sea species and non-deep-sea species based solely on reproductive strategy.

> *For this chapter, a deep-sea cephalopod is one that depends on the deep-ocean environment (defined as below the euphotic zone) for a significant part of its life cycle*

This working definition is not perfect as it may include some oceanic cephalopods that spawn or spend a large part of the day in epipelagic waters (e.g. Ommastrephidae, Enoploteuthidae and Thysanoteuthidae). However, these species do also depend on the deep sea for a significant part of the day and are therefore included in this chapter. This inclusion allows for comparison among species and families. In several other animal groups, recent information indicates that many animals traditionally categorized as "shallow-water" or "epipelagic" also utilize the deep sea; examples include tiger sharks (Werry et al., 2014), several tuna species (e.g. Block et al., 2001), leatherback sea turtles (Houghton et al., 2008) and emperor penguins (Zimmer et al., 2010). Thus, it appears that, among highly mobile macrofauna, it is not uncommon to range through multiple depth zones even when usually associated with euphotic waters.

In terms of "usual" residents, the neritic zone and deep sea, although comprising quite different physical environments, are thought to contain similar numbers of cephalopod species. However, the genera represented differ greatly between the two environments. According to Young et al. (1998), around 70% of the cephalopod families found in the oceanic pelagic zone are oegopsids, in contrast to neritic environments, which are dominated by myopsid squids, Sepioidea, and incirrate octopods. In fact, cirrate octopods, vampyromorphs and most oegopsid squids (except ommastrephids and enoploteuthids) are rarely found in euphotic waters during their lives, although some oegopsids (e.g. members of the family Cranchiidae) are known to live in euphotic waters during early life phases (Arkhipkin, 1996b,c; Bolstad et al., 2014; Voss, 1985). The results of Young et al. (1998) show two contrasting groups of animals, with little overlap, living in the neritic and oceanic zones of the ocean. Cephalopod taxa represented in the deep sea are outlined in Table 3.1, with references for supporting evidence for their inclusion. Given that 42 cephalopod families (12 families of Octopoda, 29 families of Decapodiformes and the family Nautilidae) are represented in this habitat, it is important to expand our understanding of those cephalopods living at depth.

3. INFORMATION SOURCES FOR DEEP-SEA CEPHALOPODS

Cephalopods are conspicuous components of the deep-sea biota, with some dwelling in the benthic boundary layer (e.g. mastigoteuthid squids and cirrate octopods) and deep pelagic ocean throughout their entire life cycles. Some cephalopods (including *Architeuthis*, some cirrates and some cranchiids) are among the largest species of the deep sea. Early reports of deep-sea cephalopods were often derived from findings of floating or stranded carcasses and the few specimens collected during early expeditions. These reports progressively increased as a growing number of specimens were found in the stomach contents of large deep-sea-feeding predators (e.g. oceanic toothed whales) or captured by commercial and scientific nets (e.g. Clarke, 1980, 1996, 2006; Santos et al., 1999). Morphological and taxonomic analyses of such specimens not only enabled a broader comprehension of the diversity of deep-sea cephalopod forms but also raised many important questions regarding their life histories, behaviour, physiology, abundance, distribution and functional roles in deep-sea communities. These questions were difficult to address due to the shortcomings and relative rarity of available samplers, only sampling a minute fraction of the vast spaces of the deep sea (Herring, 2002). Moreover, because specimens were brought lifeless to the surface (or were partially digested remains), little could be inferred about the lives of these cephalopods.

Modern underwater samplers and optical and acoustic technologies developed to explore the deep sea have not yet fully solved the problem of quantitative sampling of cephalopods. However, optical tools have finally

Table 3.1 Generic and specific diversity of each cephalopod family (alphabetical by order and family)

Order	Family	Known genera	Known species	Revisionary taxonomic works	Deep-sea genera	Depth range sources
Cirrate Octopoda	Cirroctopodidae*	1	4		*Cirroctopus*	Vecchione et al. (2008a,b)
	Cirroteuthidae*	3	6	Guerra et al. (1998); Collins and Henriques (2000)	*Cirroteuthis* *Cirrothauma* *Stauroteuthis*	Vecchione et al. (2008a,b) Vecchione et al. (2008a,b) Collins et al. (2008)
	Grimpoteuthidae*	3	19		*Cryptoteuthis* *Grimpoteuthis* *Luteuthis*	Vecchione et al. (2008a,b)
	Opisthoteuthidae*	1	20		*Opisthoteuthis*	Vecchione et al. (2008a,b)
Incirrate Octopoda	Alloposidae*	1	2		*Haliphron*	Young (2013)
	Amphitretidae*	5	11		*Amphitretus* *Bolitaena* *Japetella* *Vitreledonella*	Young et al. (2013) Young (2008) Young (2008) Young et al. (2010)
	Argonautidae	1	4			
	Bathypolypodidae*	1	7		*Bathypolypus*	Quetglas et al. (2001)
	Eledonidae	2	7	Rochebrune (1884c), Kubodera and Okutani (1994)		
	Enteroctopodidae*	3	11		*Muusoctopus* *Vulcanoctopus*	Kemp et al. (2006) González et al. (1998)

Continued

Table 3.1 Generic and specific diversity of each cephalopod family (alphabetical by order and family)—cont'd

Order	Family	Known genera	Known species	Revisionary taxonomic works	Deep-sea genera	Depth range sources
	Megaleledonidae*	12	40	Lu and Stranks (1994), Allcock et al. (2004), Allcock (2005)	*Graneledone*	Voight (2000)
	Octopodidae*	25	172	Norman et al. (2005), Gleadall (2013)	*Octopus*	Hochberg et al. (2009)
	Ocythoidae	1	1			
	Tremoctopodidae*	1	4	Thomas (1977)	*Tremoctopus*	Mangold et al. (2010)
Myopsida	Australiteuthidae	1	1			
	Loliginidae	11	47	Blainville (1823), Naef (1912a,b), Lu et al. (1985), Borri et al. (1986), Alexeyev (1992), Brakoniecki (1996), Vecchione et al. (1998a)		
Nautilida	Nautilidae*	2	6		*Nautilus*	Young (2010)
Oegopsida	Ancistrocheiridae*	1	1		*Ancistrocheirus*	
	Architeuthidae*	1	1	Winkelmann et al. (2013)	*Architeuthis*	
	Batoteuthidae*	1	1		*Batoteuthis*	Guerra et al. (2012)

Brachioteuthidae*	2	7		*Brachioteuthis*	Lipinski and Young (2008)
				Slosarczykovia	Lipinski (2001)
Chiroteuthidae*	4	19		*Asperoteuthis*	Young and Roper (2011a)
				Chiroteuthis	Roper and Young (2013)
				Grimalditeuthis	Young and Roper (2011b)
				Planctoteuthis	Vecchione et al. (2008a,b)
Cranchiidae*	13	60	Nesis (1972, 1974), Voss (1980)	*Bathothauma*	Young and Mangold (2009) and
				Cranchia	Voss (1980)
				Egea	Voss (1980)
				Galiteuthis	Voss (1980)
				Helicocranchia	Voss (1980)
				Leachia	Young and Mangold (2008)
				Liguriella	Voss (1980)
				Liocranchia	Voss (1980)
				Megalocranchia	Voss (1980)
				Mesonychoteuthis	Voss (1980)
				Sandalops	Voss (1980)
				Taonius	Young and Mangold (1996) and
				Teuthowenia	Voss (1980)
					Young and Mangold (2011) and
					Voss (1980)
					Young and Mangold (2010)
Cycloteuthidae*	2	4	Young and Roper (1969)	*Cycloteuthis*	Young (2012a)
				Discoteuthis	Young (2012a)

Continued

Table 3.1 Generic and specific diversity of each cephalopod family (alphabetical by order and family)—cont'd

Order	Family	Known genera	Known species	Revisionary taxonomic works	Deep-sea genera	Depth range sources
	Enoploteuthidae*	4	43	Nesis (1978, 1987b) and Tsuchiya and Okutani (1998)	*Abralia*	Tsuchiya and Young (2008)
					Abraliopsis	Tsuchiya and Young (2008)
					Enoploteuthis	Tsuchiya and Young (2008)
					Watasenia	Tsuchiya and Young (2008)
	Gonatidae*	4	19		*Berryteuthis*	Kubodera et al. (2013)
					Eogonatus	Kubodera et al. (2013)
					Gonatopsis	Kubodera et al. (2013)
					Gonatus	Kubodera et al. (2013)
	Histioteuthidae*	2	19	Voss (1969) and Voss et al. (1998)	*Histioteuthis*	Horstkotte (2008)
					Stigmatoteuthis	Young and Vecchione (2007)
	Joubiniteuthidae*	1	1		*Joubiniteuthis*	Young (2009)
	Lepidoteuthidae*	1	1	Nesis and Nikitina (1990)	*Lepidoteuthis*	Young and Vecchione (2012)
	Lycoteuthidae*	4	6	Arocha (2003)	*Lampadioteuthis*	Vecchione and Young (1999)
					Lycoteuthis	Vecchione and Young (1999)
					Nematolampas	Vecchione and Young (1999)
					Selenoteuthis	Vecchione and Young (1999)
	Magnapinnidae*	1	3		*Magnapinna*	Vecchione and Young (2013)

	Mastigoteuthidae*	5	~17	Braid et al. (2013)	*Echinoteuthis* *Idioteuthis* *Magnoteuthis* *Mastigopsis* *Mastigoteuthis*	Braid (2013) Vecchione et al. (2007)
	Neoteuthidae*	4	4		*Alluroteuthis* *Narrowteuthis* *Neoteuthis* *Nototeuthis*	Young et al. (1999) Young and Vecchione (2005) Vecchione and Young (2003a) Vecchione and Young (2003b)
	Octopoteuthidae*	2	9	Clarke (1967), Stephen (1985)	*Octopoteuthis* *Taningia*	Young and Vecchione (2013a) Vecchione et al. (2010a)
	Ommastrephidae*	11	20	Nigmatullin (1992), Dunning (1998), Dunning and Förch (1998), Dunning and Wormuth (1998), Roper et al. (1998)	*Dosidicus* *Eucleoteuthis* *Hyaloteuthis* *Illex* *Nototodarus* *Ommastrephes* *Ornithoteuthis* *Sthenoteuthis* *Todarodes* *Todaropsis*	Young and Vecchione (2013b) Watanabe et al. (2006) Young and Vecchione (2010b) Vecchione and Young (2011) Stark (2008) Young and Vecchione (2009) Vecchione and Roper (1991) Moiseev (1991) Quetglas et al. (1998) Clarke (1966)

Continued

Table 3.1 Generic and specific diversity of each cephalopod family (alphabetical by order and family)—cont'd

Order	Family	Known genera	Known species	Revisionary taxonomic works	Deep-sea genera	Depth range sources
	Onychoteuthidae*	7	25	Kubodera et al. (1998), Nesis (2000), Bolstad (2010)	*Ancistroteuthis* *Callimachus* *Filippovia* *Kondakovia* *Notonykia* *Onychoteuthis* *Onykia*	Bolstad (2010) Bolstad (2010) Bolstad et al. (2010) Bolstad (2010) Bolstad (2010) Bolstad (2010) Bolstad (2010)
	Pholidoteuthidae*	1	2		*Pholidoteuthis*	Vecchione and Young (2012)
	Promachoteuthidae*	1	3	Roper and Young (1968)	*Promachoteuthis*	Young and Vecchione (2003)
	Psychroteuthidae*	1	1		*Psychroteuthis*	Piatkowski (2011)
	Pyroteuthidae*	2	6		*Pterygioteuthis* *Pyroteuthis*	Lindgren et al. (2013) Lindgren et al. (2013)
	Thysanoteuthidae*	1	1		*Thysanoteuthis*	Young and Vecchione (2010c)
	Walvisteuthidae	1	1	Nesis et al. (1998)	*Walvisteuthis*	
Sepiida	Sepiadariidae	2	8	Steenstrup (1881)		
	Sepiidae*	3	114	Rochebrune (1884b), Adam (1944), Adam and Rees (1966), Khromov (1987a,b), Khromov et al. (1998), and Lu (1998) [Australia]	*Sepia*	Reid et al. (2005)

	Sepiolidae*	16	70	Joubin (1902), Naef (1912b,c,e), and Berry (1921, 1932)	*Amphorateuthis*	Young et al. (2007)
					Choneteuthis	Lu and Boucher-Rodoni (2006)
					Heteroteuthis	Young et al. (2008a,b)
					Iridoteuthis	Young et al. (2007)
					Nectoteuthis	Young et al. (2007)
					Neorossia	Reid et al. (2005)
					Rossia	Reid et al. (2005)
					Sepiolina	Young et al. (2007)
					Sepioloidea	Reid (2009)
					Stoloteuthis	Young et al. (2007)
Spirulida	Spirulidae*	1	1		*Spirula*	Young (2012b)
Vampyromorpha	Vampyroteuthidae*	1	1		*Vampyroteuthis*	Young (2012c)
(Unassigned) Decapodiformes	Bathyteuthidae*	1	3	Rochebrune (1884a), Roper (1969)	*Bathyteuthis*	Roper (2012)
	Chtenopterygidae*	1	4		*Chtenopteryx*	Young and Vecchione (2010a)
	Idiosepiidae	1	8			
Total	50	174	845			

Families containing deep-sea species are indicated by an asterisk, and those most in need of focused taxonomic revision are highlighted (grey). Data were obtained from WoRMs (WoRMS Editorial Board, 2014), valid taxa were selected (only full genera and species, not subgenera or subspecies) and those that had not been validated by a taxonomic editor were checked individually. For the Octopoda, the recent revised classification of Strugnell et al. (2013) has been followed. Revisionary taxonomic works refer to taxa (either families or genera) that have been revised on the basis of their worldwide representatives and not restricted to a particular geographical area.

allowed live observations in natural habitats (e.g. Kubodera et al., 2007), tracking of active movements with high spatial and temporal resolution (Benoit-Bird and Gilly, 2012), and even live or essentially intact collection (e.g. Robison et al., 2003). These tools, combined with the development of a range of molecular and biochemical techniques, have expanded the scale of scientific approaches to address unanswered questions and have raised new intriguing questions (some of which are described in this chapter). Thus, the information available on deep-sea cephalopods today originates from a variety of sources and serves different purposes. The basic types of records that can be obtained from cephalopods in the deep sea and the information each can potentially produce are summarized in Table 3.2. Additional information can be gained by combining data from multiple sources. For example, the capture of a number of specimens in a defined area of known habitats, geographical positions and depths may provide specimens for a variety of studies in addition to distribution and abundance records. The value of these diverse records for cephalopod and deep-sea scientific research and other purposes, including assessment of human impacts, conservation and society demands, is addressed in Table 3.3.

3.1. Types of gear: Advantages and disadvantages

Diverse methods are useful for obtaining specimens and/or images of cephalopods in the deep sea. Every deep-sea specimen, photo, or video is potentially a valuable source of information about the species, whether revealing aspects of its life history, genetics, biochemical adaptations, role in the food web or other characteristics. However, each sampling method has its own set of problems, especially for quantitative inferences such as distribution and abundance. Logistical considerations, such as size and capability of ships necessary to deploy the gear and distance to the study area, are also important controlling factors in these deep-sea studies. It is important to remember that each of the gear types discussed in the succeeding text is most effective for only a component of the cephalopod assemblage. The greater the variety of gear that can be included in a study, the more completely and confidently inferences can be made about the total cephalopod fauna (Vecchione et al., 2010b). For example, bottom trawling may collect animals on relatively flat substrates, while baited cameras can attract scavengers in rough terrain or midwater, and submersibles can be used to search for animals that are not attracted to bait, in nontrawlable areas.

Table 3.2 Types of deep-sea cephalopod records and the potential information available from each

Record		**Description**	**Potential information**
Organism	Specimen	Individuals brought to surface dead in variable body condition, including body parts	Macro- and microscopic body morphology and morphometrics. If fresh, tissue samples for genetic and biochemical analyses. Potential source of stomach contents, statoliths, gladii, beaks
	Live	Specimens of any deep-sea cephalopod brought to surface alive and potentially fit to live in captivity for some period of time	As for specimens plus observations of live morphology, anatomical structure function and behaviour in captivity. Measurements of physiological processes and experimentation
Observational information	Live animals	Video or still images of live individuals in nature	Observations of natural morphology (e.g. posture, colour) behaviour and other individual characters in nature
		Video or still images of live individuals in aquaria	
	Contextual	Video or still images of deep-sea habitats that contain cephalopod specimens and potentially other organisms	As for specimen images but also observation of natural habitats and species interactions
Abundance records	Species	Numbers or biomass of cephalopod species in a specific area or water volume	Relative abundance indices. Temporal evidence of distribution patterns, e.g. diel or longer-term migrations
	Community	Diversity or biomass of cephalopod species and other organisms in a specific area or water volume	Relative indices of multiple species of a deep-sea community

Continued

Table 3.2 Types of deep-sea cephalopod records and the potential information available from each—cont'd

Record		Description	Potential information
Spatial records	Depth	Cephalopod specimens' position in relation to the sea surface and/or the sea floor (altitude)	Evidence of vertical distribution, movements (e.g. diel, ontogenetic) and habitats (pelagic, benthopelagic)
	Horizontal	Cephalopod specimens' geographical position (latitude, longitude)	Evidence of geographical (horizontal) distribution and movements

3.1.1 Nets and longlines

Many net designs are available, each of which samples only a portion of the total fauna (e.g. Wenneck et al., 2008). Effective sampling of the three-dimensional pelagic environment, with its temporal variability, requires different gear from studies focusing on epibenthic animals. Both pelagic and benthic gear include some nets for which the mouth opening is set by a rigid frame and others for which the mouth shape is determined by hydrodynamic forces.

3.1.1.1 Rigid-frame nets

Rigid-frame epibenthic trawls, such as beam trawls and the Agassiz trawl, have a long history of successful deep-sea sampling (Thistle, 2003). Because they can be fished on a single relatively light wire and because they orient right side up when they contact the bottom, deploying them is relatively simple and winch requirements are modest—a single level-wind winch that will hold enough wire for a scope of at least 3 × the target depth. Such nets have collected important octopod specimens (Vecchione et al., 2005). Rigid-frame epibenthic trawls are available in various sizes but none are wide enough to sweep a broad area like, for example "otter trawl", a double-warp otter trawl can. Generally, the height of the mouth opening is much less than that of an otter trawl. The results of these height/width constraints are both positive and negative. Although the area (or volume) sampled by the rigid-frame net can be calculated with relative confidence, the reduced size of the mouth opening greatly reduces the efficiency of the gear for sampling highly mobile animals like cephalopods.

Many different rigid-frame pelagic nets have been used to collect cephalopods. Examples include bongo nets, Isaacs–Kidd midwater trawls (IKMT), Tucker trawls, rectangular midwater trawls (RMT) (Clarke,

Table 3.3 Value of different types of deep-sea cephalopod records for various scientific applications

	Organism		Images		Abundance		Spatial records	
	Specimen	Live	Live Animals	Contextual	Species	Community	Depth	Horizontal
Cephalopod research								
Taxonomy—systematics								
Phylogeny—evolution								
Life histories								
Population dynamics								
Physiology								
Behaviour								
Abundance								
Distribution								
Feeding								
Predators								
Deep-sea research								
Habitat description								
Biodiversity (richness)								
Community structure								
Energy flow (food chains)								
Biogeography								
Deep-sea adaptations								
Human impacts and conservation								
Fishery assessment								
Fishery impacts								
Oil/mining impacts								
Global changes								
Deep-sea conservation								

Continued

Table 3.3 Value of different types of deep-sea cephalopod records for various scientific applications—cont'd

	Organism		Images		Abundance		Spatial records	
	Specimen	Live	Live Animals	Contextual	Species	Community	Depth	Horizontal
Societal needs								
Deep-sea public awareness	Useful	Useful	Very useful	Very useful	Indifferent	Indifferent	Indifferent	Useful
Education	Useful	Useful	Very useful	Very useful	Indifferent	Indifferent	Useful	Useful
Arts—entertainment	Useful	Useful	Very useful	Very useful	Indifferent	Indifferent	Indifferent	Useful

Very useful
Useful
Indifferent

1977b) and multiple opening/closing nets with environmental sampling system (MOCNESS) (Wiebe et al., 1985). Frame shape and size determine fishing characteristics and mesh size determines what is retained in the cod end. Small nets designed for zooplankton sampling can be effective for collecting paralarvae. The larger rigid-frame gear is generally appropriate for micronekton (e.g. adults of small species such as pyroteuthids, enoploteuthids and juveniles of larger species, such as cranchiids, chiroteuthids and pelagic octopods) (Wiebe et al., 1985). Rigid-frame pelagic nets may be configured to be fished open throughout the tow or rigged so that the net opens and closes at specific depths or times. Although the open configuration is easier to fish, opening/closing nets must be used in order to infer depth distributions with any confidence. A simple mechanical opening/closing device can be effective for targeting a layer or avoiding surface contamination, but multiple samples require multiple deployments, each of which uses precious ship time. Multiple net samplers, such as the RMT and MOCNESS, were therefore developed for determining vertical distributions and migratory patterns. Acoustic triggers and timers can be used for multiple samples during a single deployment. However, the methods that are most effective use conducting cable to send opening and closing signals and can at the same time transmit environmental data such as temperature, pressure (to calculate depth) and conductivity (for salinity).

3.1.1.2 Larger trawls

Nets for which shape is determined by hydrodynamics (e.g. otter trawls) can be much larger than rigid-frame nets and can therefore sample much larger specimens. These types of gear are the best source of adult cephalopods but catch a surprising number of smaller specimens as well. These nets are often similar to gear used in shallow water for fisheries studies but, when used for deep-water nekton, must be rigged with deep-water floats capable of withstanding great pressure. They are effective for sweeping a larger area of the bottom or filtering a larger volume in midwater studies than the rigid-frame nets. The epibenthic versions are usually made of more robust components (heavier trawl doors and bridles, stronger twine, headropes and footropes, etc.) than the pelagic nets. Both epibenthic and pelagic nets can be fished on either a single or a double warp, the latter requiring a much more specialized ship with two large-capacity trawling winches capable of holding great lengths of heavy wire with sophisticated winch-control systems. Few ships exist with such capabilities: larger fisheries research ships such as *G.O. Sars* (Norway) and long-range commercial trawlers are examples. Although the single-warp gear can be fished from a less-specialized ship and is easier to deploy and recover, the double-warp gear can be much larger, again increasing sampled area/volume and capturing larger specimens. It is noteworthy that, because epibenthic trawls are open during deployment and recovery, they sometimes collect important midwater specimens on the way down or back up (e.g. *Magnapinna* sp. A in Vecchione et al., 2010b).

Some otter trawls have uniform mesh size in all parts of the net (e.g. the Norwegian "krill trawl"), but typically, the large nets use graded mesh sizes from as much as 2 m in the wings of the net to just a few millimetres in the cod end. When sampling cephalopods, it is extremely important to search the large mesh carefully as the net, either pelagic or epibenthic, is being retrieved onto the ship. Often, most of the cephalopods are entangled in the large mesh and must be picked by hand before the net is stored on the net reel. This requires careful coordination between the deck crew and the science party.

Pelagic trawls can be rigged with multiple opening/closing cod ends for quasi-discrete depth sampling. However, because just the cod end of the net is opened and closed rather than the entire net, the likelihood of contamination of the sample with specimens that entered the net prior to opening of the cod end is greater than with a rigid-frame net. Another alternative configuration used especially for pelagic sampling is a hard (or "aquarium") cod end that protects the sample from abrasion or crushing. Although this is very useful for obtaining specimens in good condition for some taxa (e.g. fishes and gelatinous megaplankton), it is less useful for cephalopods because of their tendency to entangle in the mesh of the net. Thermally protected cod ends have also been developed; these facilitate recovery of living deep-sea organisms (Childress et al., 1978). Various net sensors can be rigged on the trawl, either using acoustic sensors on the gear itself or using a third wire system. However, the farther the gear is from the ship, the more problematic these sensor configurations become; this is especially troublesome for deep-sea sampling with multi-kilometre depths and wire scope of 2–3 × the depth.

3.1.1.3 Passive collecting gear

Another type of net that has been successfully used in the epipelagic to catch large squids that may migrate vertically into the deep sea during the day, such as ommastrephids and onychoteuthids, is entangling ("gill") nets. The effectiveness of such nets when deployed in the deep sea remains to be seen. Jigging can also be an effective method for catching large vertical migrators (e.g. ommastrephids).

Long lines with many baited hooks are sometimes used in the deep sea both commercially and for research. Although the typical catch of such long lines is almost entirely fishes, sometimes large cephalopods, such as *Ommastrephes bartramii* (Lesueur, 1821), *Mesonychoteuthis hamiltoni* and *Architeuthis dux* are also hooked (M. Vecchione, personal observation). Even recovery of a broken-off tentacle can yield an important tissue sample for DNA and other biochemical analyses (e.g. Kubodera and Mori, 2005; Robison, 1989).

Other baited devices sometimes attract cephalopods as well. The "bait" may be traditional organic bait or some other type of attractant, such as light. The first *in situ* observations of the giant squid (*Architeuthis*) and also of *Taningia danae* Joubin, 1931, were obtained with organic bait (Kubodera et al., 2005) and organic bait and light, respectively (Kubodera et al., 2007). Light traps use a general light source, but recently, lures have been developed that attempt to mimic bioluminescent patterns generated by deep-sea animals. Such a lure, based on the "burglar alarm" bioluminescence signal of a particular medusa, was recently used to obtain the first *in situ* video record of a giant squid, *Architeuthis* (Widder, 2013). A similar lure attracted a large deep-sea squid that could not be identified confidently to any known family (Widder, 2007). The objective of the device using bait or lure may be to catch specimens (a trap) or record photographic or video records (a camera lander). Typically, these devices are deployed by the ship or submersible and left for some period while the ship pursues other objectives. For example, incirrate octopods have been attracted to baited benthic landers that take photos periodically over a period of many hours (Kemp et al., 2006). Similar baited camera systems have been used in midwater layers by lowering the system on a cable from a ship (e.g. Kubodera et al., 2005, 2007).

3.1.2 *Submersibles and related* in situ *mechanisms*

While nets provide samples that integrate distribution over a relatively large area, they are less effective for inferring small-scale distribution or behaviour. Methods based on visual observations are much better for addressing these types of questions. The simplest of these methods are "drop cameras" lowered from a ship to an area of interest. More hydrodynamic camera sleds can also be towed behind a ship. Both drop cameras and camera sleds can be used for video but typically record still images at predetermined intervals, providing semirandom snapshot observations of the target area. Neither observation method collects specimens, which is often an important corroboration tool for confident identification.

Better control and selective targeting of specific fauna, both epibenthic and pelagic, can be achieved by use of submersibles, either manned or unmanned (Robison, 2000, 2004). Although the relative value of manned submersibles versus remotely operated vehicles has been the subject of considerable debate, both have been quite useful for increasing our knowledge of deep-sea cephalopods (e.g. Bush et al., 2009; Hoving and Vecchione, 2012; Robison et al., 2003; Seibel et al., 2005; Vecchione and Roper, 1991). Both manned and unmanned submersibles can record high-quality videos and still images and can search areas not accessible to methods like net sampling. Both can also be configured to collect specimens using suction devices, detritus (or "D") samplers (first developed by Harbor Branch Oceanographic Institution) (essentially large tube segments that can be manoeuvred around an animal in midwater and then closed at both ends) (Figure 3.4b–d) and manipulator arms. While nets can collect many more specimens, these are generally not in very good condition. Specimens collected by submersible are often in excellent condition, with delicate features intact, and many are alive when brought into the laboratory aboard the ship. It has even been possible using such specimens to perform experiments and physiological measurements on deep-sea cephalopods and collect behavioural observations (e.g. Hoving and Robison, 2012; Hunt, 1996; Jacoby et al., 2009; Robison

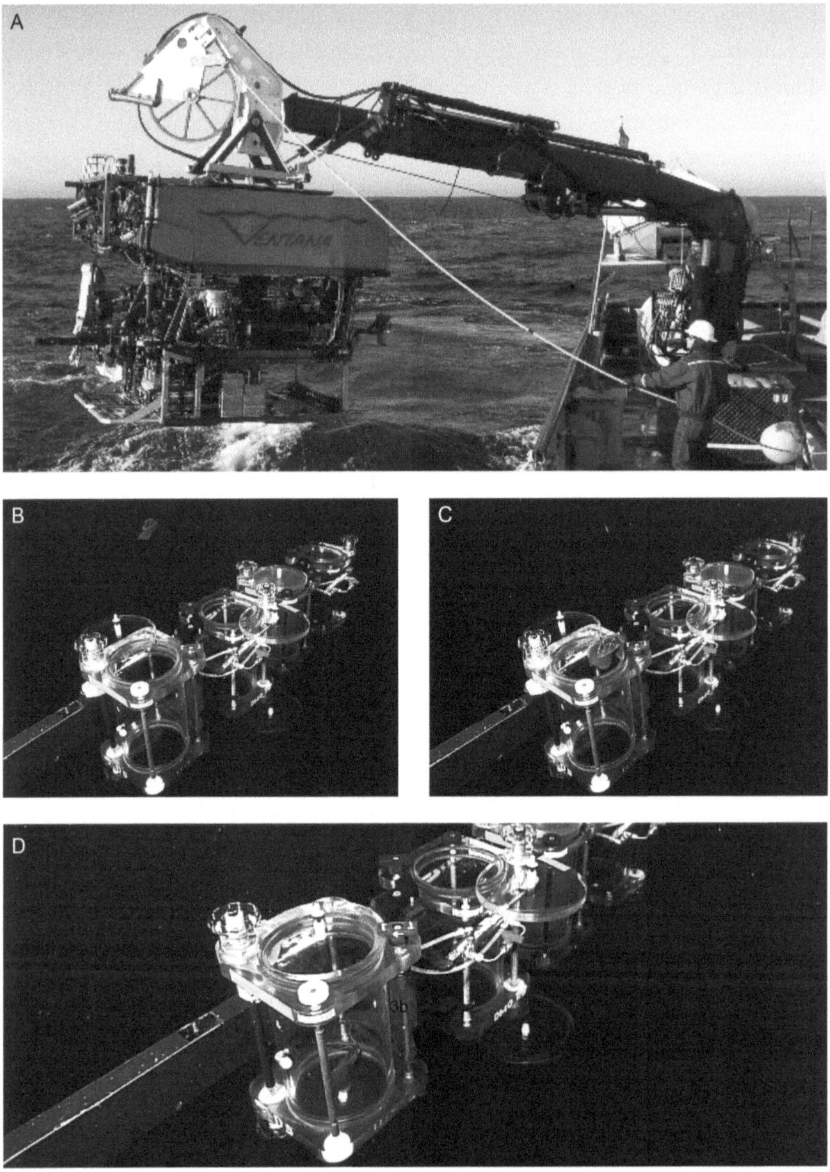

Figure 3.4 (A) The ROV "Ventana" of the Monterey Bay Aquarium Research Institute being deployed in the Monterey Submarine Canyon by R/V *Point Lobos*. (B) *Octopoteuthis deletron* being approached by an opened Detritus sampler (D-sampler) which is mounted on the ROV. (C) The ROV is manoeuvred in such a way that the specimen is positioned over the opened D-sampler. (D) The ROV moves upwards so the specimen is captured inside the D-sampler and the sliding lid of the sampler is closed. A living deep-sea squid is captured. *(A) Photograph Kim Fulton-Bennett © 2004 MBARI. (B–D) © 2013 MBARI.*

et al., 2003). Manned submersibles include those with transparent spheres and those with high-pressure-resistant metal spheres and small portholes. The former allow excellent visibility for the observers—particularly useful for midwater exploration—but have limited depth capability, usually less than 1000 m. The latter, conversely, have much greater depth capability but limited visibility and greater discomfort for the observers.

Unmanned options include remotely operated vehicles (ROVs) (Figure 3.4), which remain connected to the ship by a conducting or fibre-optic cable, allowing control of the vehicle and real-time observations and data, or autonomous underwater vehicles (AUVs), which are pre-programmed for a dive profile prior to release from the ship. Although AUVs have many uses in deep-sea studies, to date they have not been very useful for studying organisms like cephalopods. On the other hand, the ROV has become a standard tool for deep-water biological exploration, resulting in many important cephalopod observations and the collection of excellent specimens (e.g. Robison, 2004). An ROV can remain at depth for prolonged periods, whereas dive duration for manned submersibles is limited by crew endurance and battery capacity. A recent, and expanding, development in ROV operations has been the "telepresence" strategy, in which video, data and two-way communications are linked via satellite between the ship and a shore-based station from which they can be disseminated to scientists and the public at remote command centres (high definition via Internet 2) and any computer connected to the Internet (standard definition via Internet 1). An obvious advantage of this is increased expert participation in the expedition but disadvantages include that specimens generally are not collected and the scientists ashore often cannot dedicate their time to the expedition as intensively as they would if they were aboard. In fact, as this text is being drafted, an ROV is beaming live video ashore from exploration of the Cayman Rise in the Caribbean Sea, including observations of numerous cirrate octopods (http://www.nautiluslive.org).

3.2. Cephalopod collections

Existing institutional collections of cephalopods represent a significant and invaluable resource to the research community. Ranging from small, recent, localized holdings to centuries-old, international repositories, collections contribute to studies in diverse and disparate fields, such as taxonomy, biogeography and reproductive behaviour. In a recent online survey, scientists working in cephalopod research reported accessing specimens from 55 collections around the world (Appendix A). The most commonly accessed were large, well-established collections, such as the US National Museum of Natural History (Smithsonian Institution; Washington, DC, the United States), the Natural History Museum (London, the United Kingdom) and Muséum National d'Histoire Naturelle (Paris, France). However, supplementary material was frequently sourced from smaller institutions as well. In addition, several recent or ongoing collecting programmes were identified around the world, which are equally important as they may help fill gaps in sampling coverage and provide fresh material for studies requiring non-fixed specimens (e.g. genetics). Those programmes open to sampling requests from scientists are given in Appendix B.

3.3. DNA extraction from formalin-fixed tissue

Museum collections hold specimens that are essential to taxonomy. For deep-sea cephalopod species of which very few specimens have ever been caught, museum specimens may be the only possible source of DNA. Unfortunately, most museum specimens are formalin fixed and preserved in ethanol, which makes DNA extraction difficult or impossible (Tang, 2006). Different protocols have been devised for amplifying DNA from museum specimens, such as shown in Fang et al. (2002), who combined gradual dehydration and critical-point drying to successfully amplify high-molecular-weight DNA from formalin-fixed vertebrate tissues. In addition, mitochondrial DNA has been amplified from museum specimens of octopus (Söller et al., 2000) and squid (Carlini et al., 2006). However, there are still many challenges to overcome with formalin-fixed tissue, and failed attempts are generally not reported in the literature (Tang, 2006).

The Barcode of Life Data System (BOLD) is a public compilation of DNA barcodes, which are 652 base pairs from the 5′ end of the mitochondrial cytochrome *c* oxidase subunit I (COI) sequence (Ratnasingham and Hebert, 2007). Strugnell and Lindgren (2007) outlined some considerations and concerns specific to cephalopods in BOLD, such as multiple or nuclear copies of COI, potentially fast rates of evolution, and some uncertainty over whether a single gene is enough to identify species reliably.

Recently, the Barcode Index Number (BIN) system has been introduced, which groups together genetically similar individuals (Ratnasingham and Hebert, 2013). For taxonomy, DNA barcodes can be used to help separate species (Hebert and Gregory, 2005) and have been successful in delineating cephalopod species (Allcock et al., 2011; Dai et al., 2012; Zheng et al., 2012). In addition, DNA barcodes can be used to identify otherwise problematic specimens, such as juveniles (Victor et al., 2009) and badly damaged individuals (St-Onge et al., 2008). Given the great potential for museum material to extend BOLD and the limited success achieved in extracting DNA from formalin-fixed specimens to date (Zhang, 2010), further efforts to improve these techniques should be undertaken, and results reported whether positive or negative.

For example, during a recent review of the Mastigoteuthidae found around New Zealand (Braid, 2013), it became apparent that integrative taxonomy was necessary for working on a group represented mostly by badly damaged specimens, especially as it appears that this family has morphologically similar species with distinct distributions. Therefore, some preliminary tests were conducted on the DNA extraction of formalin-fixed squid tissue using three mitochondrial genes (COI, 16S rRNA and 12S rRNA), trialling combinations of critical-point drying, two DNA extraction protocols and DNA purification, with partial success. Among the most useful protocols were silica-gel column-based extraction, using critical-point-dried tissue combined with a DNA-purification protocol (two sequences recovered out of eight attempted) and alkaline lysis extraction from tissue that had not been critical-point dried (although this only recovered one sequence from 12 samples). Critical-point drying alone was not sufficient to obtain DNA sequences; however, with a DNA-purification step to remove impurities and increase the DNA concentration, some DNA recovery was possible. Overall, it seems that these two steps (critical-point drying of tissue prior to DNA extraction and DNA purification) both assist in sequence recovery from formalin-fixed tissue.

Tissue type may also affect the success of DNA extraction (Tang, 2006), but it appears that DNA extracted from the buccal mass muscle can be successfully amplified (H. Braid, personal observation). The buccal mass is an ideal source of muscle tissue because it is removed during the extraction of beaks for morphological examination. Although DNA extraction is a destructive process, buccal mass tissue can be sampled without damaging important morphological characters. One useful process for providing several different pieces of information and samples is to remove the buccal bulb from specimens prior to fixation, gently extract the beaks (this is usually a simple matter if the specimen has been frozen and then thawed) and store the beaks together with the buccal muscle in 100% ethanol. This ensures that a relatively large piece of muscle tissue is available for molecular work while allowing later confirmation of the parent specimen's identity (should the samples become separated from the whole individual; of course, care should be taken to store subsamples together with parent specimens where possible) by retaining the characteristic beaks together with the tissue.

3.4. Websites, databases and networks

The "Tree of Life" Web Project (ToLweb; http://tolweb.org/), established in 1996, has become one of the leading online repositories of cephalopod information, focussing primarily on systematics, morphology and distribution but also presenting a lot of other biological information. Through the efforts of Richard Young, Michael Vecchione and the late Katharina Mangold (1922–2003), in collaboration with a team of over 20 international researchers, more than 700 pages have been completed for the class Cephalopoda. These remain the most actively updated and informatively complete branches of ToLweb (Maddison et al., 2007). They are particularly useful as taxonomic references, since most taxon pages include images (photos and/or diagrams) and many provide descriptions of key morphological features, as well as general notes on vertical distribution, nomenclature and/or development. In addition to taxonomic information, the ToLweb has many unpublished observations and photographs, in particular for deep-sea cephalopods that are very valuable for biological and ecological studies on these animals. Because of this comprehensive collection of information, which is frequently updated, ToLweb and WoRMS (World Register of Marine Species) (see the succeeding text) were used as the primary references for the current taxonomy of the deep-sea cephalopods discussed herein (Table 3.1).

In addition to web-based resources, advances in ocean science technology have greatly expanded scientific knowledge about the behaviour, ecology and functional morphology of many deep-sea cephalopods; some of this information is also becoming more widely available to the public. The increased feasibility of collecting and analyzing video footage has also necessitated the production and dissemination of field guides to aid in identifying live animals. The Monterey Bay Aquarium Research Institute (MBARI) is a pioneer in this area and maintains a well-established database of deep-sea observations recorded by ROVs during 25 years of operations, a current total of ~19,700 h (S. von Thun, personal communication). From this ROV footage, MBARI's video lab extracts all possible faunal information by annotating the video using their Video Annotation

Reference System (VARS; http://www.mbari.org/vars/). VARS allows biological, geological and experimental observations to be searched for and correlated, together with descriptive and environmental data in the database. This database has revealed unique behaviours of deep-sea cephalopods (e.g. Bush et al., 2009; Hoving and Robison, 2012; Robison et al., 2003), enabled detection of the range expansion of Humboldt squid (*D. gigas*) (Zeidberg and Robison, 2007) and provided unbiased information on distribution and general ecology of deep-sea cephalopods in the Monterey Submarine Canyon (Hunt, 1996). There is a public version of VARS query, and the VARS software is available as an open-source package for use by other institutes. In the northeastern Pacific, *The Marine Life Field Guide iBook* (freely available through the iBook store) has been produced by NEPTUNE Canada (Gervais, 2012) and shows high-quality images of *in situ* observed cephalopod species such as *Graneledone pacifica* Voss and Pearcy, 1990; *Gonatopsis borealis* Sasaki, 1923; and *Dosidicus gigas*. The NEPTUNE Canada cable network gathers live data from instruments deployed and connected on the sea floor off Vancouver Island installed in 800 and 2700 m depth. This tool, installed in area previously lacking an authoritative reference guide, can now be used to help improve the quality of real-time video and photo annotations during cruises and in subsequent studies of archived imagery.

The current ease of communication has facilitated global initiatives and collaborative projects that have increased, or are steadily increasing, our knowledge of deep-sea fauna including cephalopods. For example, following in the footsteps of the Census of Marine Life programme (2000–2010), the INDEEP (International Network for Scientific Investigations of Deep Sea Ecosystems) programme is focusing on determining the global biodiversity and functioning of deep-sea ecosystems. The World Register of Deep-Sea Species (WoRDSS) has recently been launched in collaboration with the WoRMS team (WoRMS Editorial Board, 2014), the Natural History Museum, (London) and the National Oceanographic Centre (Southampton, the United Kingdom) to improve access to deep-sea species information (Higgs et al., 2012). The WoRMS web portal aims to collate the most up-to-date taxonomy with a suite of identification tools (online keys, guides, monographs and papers) recommended by taxonomic experts. Ultimately, it will link directly to a deep-sea field "app" ("Deep Sea ID: A deep-sea field guide"), whose readily available synthesis of information should prove a useful tool for taxonomists, providing a venue for quickly sharing information. It will also be an excellent starting point for people embarking on a studies of new groups or taxa. In linking to an authoritative and updated named database (WoRMS), a consistent taxonomy can also be maintained; WoRDSS will thus be the most up-to-date database of deep-sea taxonomy. Online WoRDSS (http://www.marinespecies.org/deepsea, Glover et al., 2013) lists 419 species of cephalopods as occurring in the deep sea. (The current criterion for inclusion in the WoRDSS database is a sample depth of greater than 500 m, including both pelagic and benthic species; any species recorded below 500 m is included even if it ranges shallower than this depth.)

Although photo imagery is a nondestructive sampling tool for cataloguing the deep sea, identification of species from imagery is often difficult, especially when specimens are not available for confirmation. Organisms can, however, be identified as distinct operational taxonomic units (OTUs) or morphotypes as a basis for future work and detailed species ground truthing through the collection of actual specimens. An ongoing partnership between Ifremer, NOAA, and the University of Plymouth (the United Kingdom) is developing a web-based portal of this kind; initially documenting the Rockall Trough region of the northeast Atlantic, it will soon also include newly collected data from the Bay of Biscay and the western Atlantic (Davies et al., 2012).

3.5. Predators

Large apex predators such as whales, seabirds, seals, sharks and fishes are highly effective samplers, consuming both a great abundance and high diversity of deep-sea cephalopods (Clarke, 1980; Santos et al., 2001; Xavier and Cherel, 2009; Xavier et al., 2003a, 2013). In fact, cephalopods may comprise up to 100% of the total diet in some top predators (Table 3.4), and others are known to feed on more than 35 separate deep-sea cephalopod taxa (e.g. sperm whales; see Gómez-Villota, 2007). Predators are also known to catch larger specimens, and a greater diversity, of cephalopods than are sampled by scientific nets (Clarke, 1977a, 2006; Rodhouse, 1990; Roper, 1974; Staudinger et al., 2014; Wormuth and Roper, 1983; Xavier and Cherel, 2009; Xavier et al., 2006). Furthermore, tracking of cephalopod predators, such as tuna, sharks, seals, sperm whales and seabirds (Block et al., 2011; Croxall et al., 2005; Hindell et al., 2003; Paiva et al., 2010; Phillips et al., 2008; Sims et al., 2008; Xavier et al., 2003b), is currently more feasible than tracking cephalopods directly and is a technique now widely used to support inferences about the vertical and geographical

Table 3.4 Importance of deep-sea cephalopods in the diets of top predators

Top predators	Location	Number of deep-sea cephalopod taxa present in predator's diet	Number of deep-sea cephalopods consumed (%)	Estimated mass of deep-sea cephalopods consumed (%)	Reference
Seabirds					
Wandering albatross *Diomedea exulans*	Subantarctic	>25	93	99	Xavier et al. (2003a)
Grey-headed albatross *Thalassarche chrysostoma*	Subantarctic	>20	100	100	Xavier et al. (2003b)
Black-browed albatross *Thalassarche melanophrys*	Subantarctic	>20	99	99	Xavier et al. (2003b)
Emperor penguin *Aptenodytes forsteri*	Antarctic	>3	100	100	Piatkowski and Pütz (1994)
King penguin *Aptenodytes patagonicus*	Falkland Islands	>9	88	98	Piatkowski et al. (2001)
White-chinned petrel *Procellaria aequinoctialis*	Subantarctic	>5	100	100	Berrow and Croxall (1999)
Whales					
Sperm whale *Physeter macrocephalus*	North Atlantic	>35	100	100	Clarke et al. (1993)
Pygmy sperm whale *Kogia breviceps*	South Pacific	>22	100	100	Beatson (2007)

Continued

Table 3.4 Importance of deep-sea cephalopods in the diets of top predators—cont'd

Top predators	Location	Number of deep-sea cephalopod taxa present in predator's diet	Number of deep-sea cephalopods consumed (%)	Estimated mass of deep-sea cephalopods consumed (%)	Reference
Pilot whale *Globicephala melaena*	South Pacific	>5	70	74	Clarke and Goodall (1994)
Bottlenose whale *Hyperoodon planifrons*	South Pacific	>20	100	100	Clarke and Goodall (1994)
Seals					
Southern elephant seal *Mirounga leonina*	Antarctic	>5	98	98	Daneri et al. (2000)
Southern elephant seal *Mirounga leonina*	Subantarctic	>10	99	99	Rodhouse et al. (1992)
Weddell seal *Leptonychotes weddellii*	Antarctic	>5	65	74	Clarke and MacLeod (1982)
Fishes					
Patagonian toothfish *Dissostichus eleginoides*	Subantarctic	>10	75	98	Xavier et al. (2002)
Conger eel *Conger conger*	North Atlantic Ocean	>1	7	8	Xavier et al. (2010)
Porbeagle shark *Lamna nasus*	South Indian Ocean	>15	100	100	Cherel and Duhamel (2004)

Sleeper shark *Somniosus* cf. *microcephalus*	South Indian Ocean	>15	100	100	Cherel and Duhamel (2004)
Lantern shark *Etmopterus* cf. *granulosus*	South Indian Ocean	>3	100	100	Cherel and Duhamel (2004)
Longnose lancetfish *Alepisaurus ferox*	Central Indian Ocean	>10	85	95	Potier et al. (2007)
	Northwest Pacific Ocean	>15	81	90	Okutani and Kubota (1976)

The number of deep-sea cephalopods consumed (%) and estimated mass (%) corresponds to the deep-sea cephalopods consumed within the cephalopod component in the diets of predators.

distribution of cephalopods and about predator–cephalopod interactions that cannot be obtained by other means (Davis et al., 2007; Ropert-Coudert et al., 2006; Staudinger and Juanes, 2010; Wilson et al., 2002). Thus, dietary data from teuthophagous predators can be used to make inferences on the biology and ecology of deep-sea cephalopods.

Various methods can provide data on the cephalopod component of predator diets (Barrett et al., 2007; Karnovsky et al., 2012; Xavier et al., 2005), including direct observations (e.g. observing an albatross feeding at the sea surface) and the examination of the stomach contents of dead animals and from scats/faeces. Traditionally, the examination of stomach contents has been one of the most widely used methods for studying the diets of predators, having the advantage of potential quantification of prey importance (e.g. the contribution of cephalopod species by frequency of occurrence, by number and by mass) at the time of sampling. Some cephalopod predators are more suitable samplers than others. The longnose lancetfish, *Alepisaurus ferox* Lowe, 1883, for instance, forages on mesopelagic cephalopods. Its stomach contents are extremely useful for biological studies as food is stored in the stomach and digestion takes place in the intestine; relatively intact prey items can therefore be recovered from the stomach (Okutani and Kubota, 1976; Potier et al., 2007). In other instances where less-intact stomach contents are recovered, the original body sizes of digested cephalopods can often be estimated from measurements of their beaks, which are more resistant to digestion than soft tissues and often accumulate in the stomachs of predators (Clarke, 1962, 1986; Lu and Ickeringill, 2002; Xavier and Cherel, 2009). However, beaks can be retained in predator stomachs for 9 months or longer (Xavier et al., 2005); therefore, caution must be used to differentiate among fresh (i.e. with transparent parts or with flesh attached) and highly eroded beaks, to ensure accurate interpretation of the results (Xavier and Cherel, 2009; Xavier et al., 2011). Prey digestion rates also vary widely depending on the size of the predator and of the prey; these variable rates can considerably affect the results obtained from stomach contents for soft-bodied or small prey (e.g. cephalopods, fishes or crustaceans) that can be digested completely. For example, some fishes can be completely digested within 12 h in petrels, seals and penguins, but digestion rates differ according to the fish species consumed (Helm, 1984; Hilton et al., 1998; Jackson and Ryan, 1986; Wilson et al., 1985). The same may be true for different cephalopod species.

Several recent biochemical methods have provided additional novel ways to gain further information about deep-sea cephalopods collected from predator stomachs. Trace elements can be used to determine geographical distribution and to differentiate stocks (e.g. Arbuckle and Wormuth, 2014); stable isotopes of carbon and nitrogen (from fresh cephalopod tissue or tissue recovered from predators) act as ecological tracers of habitat use and trophic level (Cherel and Hobson, 2005); DNA barcoding and fatty-acid analysis can be used to identify cephalopods to species and indicate consumed prey taxa (Braid et al., 2012; Dai et al., 2012; Hoving and Robison, 2012); and chemical pollutants shed light on the accumulation and transfer of persistent anthropogenic pollution through remote deep-ocean communities up to charismatic and endangered megafauna (Unger et al., 2008). In addition, these techniques can help reconstruct seasonal, annual and decadal shifts in oceanographic conditions over varying spatial scales through the analysis of biogenic structures such as beaks or gladii that reflect varying time periods of nutrition and habitat use (e.g. Barrett et al., 2007; Cherel and Hobson, 2005; Karnovsky et al., 2012; Ramos and Gonzalez-Solis, 2012).

4. SYSTEMATICS, DIVERSITY, BIOGEOGRAPHY AND EVOLUTION OF DEEP-SEA CEPHALOPODS

4.1. Evolutionary pathways to modern deep-sea life: Implications from the cephalopod fossil record

In contrast to numerous shallow-water deposits of epicontinental seas, examples of deep-sea deposits and hence records of deep-sea fossil communities are extremely rare; the few fossil deep-sea faunal assemblages known to date have yielded no cephalopods. However, inferences of evolutionary pathways leading to modern deep-sea cephalopods can be made from indirect evidence, including the palaeoenvironment in which fossil cephalopods have been found and/or their functional morphology. By far, the greatest source of fossil cephalopod information is the preserved buoyant shell (phragmocone), which would have imposed a physical habitat depth limit as it does in modern cephalopods such as *Nautilus*, *Spirula* and *Sepia* (Figure 3.5). However, thanks to sedimentary deposits known to exhibit extraordinary soft tissue preservation (called *Konservat-Lagerstätten*, literally "conservation preservation sites"), a relatively good understanding of external and internal morphology can also be gleaned, particularly for coleoids.

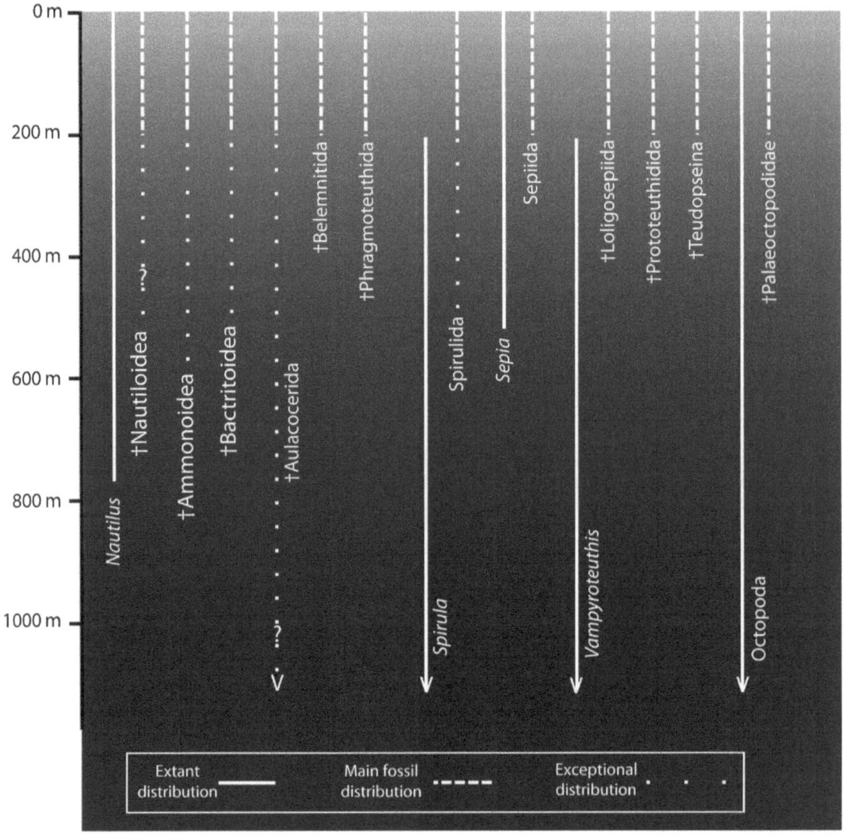

Figure 3.5 Depth distributions of fossil cephalopods (mainly inferred from functional morphologies).

4.1.1 Nautiloidea and Ammonoidea

The large majority of fossil ectocochleate (external shelled) cephalopods were restricted to epicontinental shelf and slope habitats. They first appeared during the Cambrian period (about 500 million years ago) as bottom-associated monoplacophoran-like molluscs (Kröger et al., 2011), and their distributions at any time were determined by transgressions (sea-level rise) and regressions (sea-level decrease). According to depth-range estimates based on shell-strength calculations, only a few of these taxa could dive as deeply as modern *Nautilus* (Daniel et al., 1997; Hewitt and Westermann, 1988; Hewitt et al., 1989; Westermann, 1973); there is no positive evidence from the fossil record that early ectocochleates were adapted to permanent life in the deep sea.

4.1.2 Coleoidea

As with ectocochleates, the fossil record of endocochleate coleoids is confined to shallow inland and shelf areas. Coleoids most likely originated during the Late Devonian or in early Carboniferous times from bactritoid ancestors. Bactritoidea was an extinct group of ectocochleate cephalopods with a straight (orthoconic) phragmocone; this lineage also gave rise to the Ammonoidea. Bactritoids are considered to have been inhabitants of epi- to upper mesopelagic water depths (Westermann, 1973).

4.1.2.1 Belemnoidea

Based on their palaeobiogeographic distribution, belemnoids (Aulacocerida, Phragmoteuthida, Belemnitida and Diplobelida) are generally considered to have been neritic, similar to the recent loliginids (Doyle, 1990; Jeletzky, 1966; Westermann, 1973). Several exceptions may have occurred within the Aulacocerida, where some taxa are known from oceanic deposits (e.g. Triassic of Timor, Indonesia). Westermann (1973) believed that these forms were capable of withstanding higher hydrostatic pressures (maximum estimated implosion depth 1000 m).

4.1.2.2 Decapodiformes

Early predecessors of the recent Decapodiformes probably evolved during the Late Jurassic or Early Cretaceous period from belemnoid diplobelids (Fuchs et al., 2012, 2013; Kröger et al., 2011). These early decapods (either stem-lineage representatives or already within the spirulid lineage), commonly called groenlandibelids, have been found in neritic sediments but most probably drifted there postmortem from oceanic provinces. Therefore, it has been suggested that groenlandibelids—similar to present-day *Spirula*—inhabited mesopelagic waters above the continental slopes (Fuchs et al., 2013; Hewitt et al., 1991).

However, the direct ancestors of mesopelagic *Spirula*, commonly called "Tertiary spirulids," have been found in epicontinental deposits, and the shell morphologies of Eocene–Miocene *Beloptera* and *Spirulirostra* suggest a demersal lifestyle (Young et al., 1998). This indicates that *Spirula* migrated downwards into meso- to bathypelagic waters only 5–10 million years ago. Fuchs (2012) recently concluded that, over the course of evolution, the spirulid shell has progressively retained more early ontogenetic shell characters. These neotenous traits suggest a highly derived (rather than a primitive) morphology and lifestyle for *Spirula*.

Early representatives of the Sepiida (*Ceratisepia*) first appeared in the Late Cretaceous and generally co-occurred with spirulids in Cenozoic neritic deposits (e.g. Eocene *Belosepia*). Since only a handful of modern sepiid species can live in waters deeper than 400 m (Sherrard, 2000), it appears that the bulk of these taxa have retained their approximate original depth distribution.

While spirulids and sepiids are primarily shallow-water inhabitants, Arkhipkin et al. (2012) recently hypothesized that teuthoid squids may have originated in the deep sea. According to this scenario, the squid gladius derived from a belemnitid shell through "...reduction of the dysfunctional rostrum-phragmocone system with subsequent decalcification..." as a physiological adaptation of a life in oxygen-poor deep-water habitats. If this is true, then neritic provinces (e.g. those inhabited by modern loliginid squids) might have been secondarily invaded, unless loliginids are cuttlefishes that have lost the phragmocone rather than oegopsids that moved into shallow waters (Young et al., 1998).

4.1.2.3 Octopodiformes

Octopodiformes (*Vampyroteuthis*, Cirrata, Octopoda) most probably derived from a belemnoid subgroup, the Phragmoteuthida (Jeletzky, 1966). This Middle Triassic divergence must have happened—as in squids—through decalcification of the inner shell and the development of a gladius, which is well known throughout the entire Mesozoic (e.g. Fuchs, 2006; Fuchs and Larson, 2011a,b; Fuchs and Weis, 2008, 2010; Fuchs et al., 2007a,b). In contrast to meso- to bathypelagic *Vampyroteuthis*, these gladius-bearing octopodiforms occupied nektonic or nektobenthic niches in shallow waters; the *Vampyroteuthis* lineage must thus have retreated from shallow waters at some point. However, *Vampyroteuthis* has conserved many ancestral morphological characters.

The Octopoda, comprising Cirrata (Cirroctopoda) and Incirrata, stem from a subgroup of gladius-bearing octopodiforms called Teudopseina (Bizikov, 2004; Fuchs, 2009; Fuchs and Larson, 2011a; Fuchs and Weis, 2010; Fuchs et al., 2009; Haas, 2002). Although still unknown, the hypothetical last common ancestor of the Octopoda (cirrates and incirrates) must have appeared very similar to living cirroctopods. Whereas ancestral cirrates have not been identified with certainty in the fossil record (perhaps the earliest true cirrate was already a deep-sea dweller?), our knowledge about the earliest members of the Incirrata is comparatively good. The first unambiguous octopods with a clearly bipartite gladius vestige appear in Late

Cretaceous epicontinental deposits, indicating moderate water depths (Fuchs et al., 2009); these palaeoctopods already lacked cirri but still possessed fins. In the light of recently proposed phylogenies (e.g. Lindgren et al., 2012; Strugnell et al., 2004, 2009), it appears plausible to assume multiple deep-sea colonizations for various octopod subgroups (Voss, 1988).

In conclusion, expansion into the deep sea from shallow origins is a common and independent occurrence observable in almost all cephalopod subgroups; teuthoid squids, however, may have first arisen and radiated in the deep sea and only later colonized shallow waters.

4.1.3 Misconceptions regarding "living fossils"

The term "living fossil" can only be used with accuracy for *Vampyroteuthis*, since it differs little from its Mesozoic shallow-water-dwelling ancestors. The use of this term for *Nautilus* is still premature, since information about the soft-part morphology of its ancestors remains scarce (e.g. number of arms? number of gills?). Certainly, the *Nautilus* shell exhibits a general conservatism, but in detail (e.g. shape of the conch, course of sutures and position of the siphuncle), it differs significantly from its latest known ancestors, such as *Eutrephoceras* and *Cenoceras* (see Ward, 1984). *Spirula* is definitely not a "living fossil" and must rather be considered as a highly derived cephalopod.

4.2. Diversity and Systematics

Few cephalopod species that spend the majority of their adulthood in the deep sea are well understood, even those considered abundant. New genera are regularly described, and many families are known to contain unnamed species and species complexes. It is perhaps surprising that, in a class with just ~850 extant nominal species (Table 3.1), such systematic uncertainty should remain.

4.2.1 Information sources for deep-sea cephalopod biodiversity

While physical collections form an indispensable resource (as discussed in Section 3.2), their material can only represent a small cross-section of the deep-sea cephalopod fauna. Many of the taxa represented in institution holdings are at the smaller end of the size range for this class and often have soft, gelatinous bodies. Most trawled specimens, after being tumbled in nets with more robust-bodied animals, reach the surface in poor condition, severely limiting their potential for morphological examination. This has resulted in some difficulties in species delineation, with many early species descriptions based only on single and/or badly damaged specimen(s).

A good example is the 21 nominal species of *Architeuthis* that have been described since the first description by Steenstrup in 1857. Genetic data have recently shown that *Architeuthis* comprises a single cosmopolitan species, *A. dux* (Winkelmann et al., 2013), with ~20 synonyms. (Winkelmann et al., 2013). Many deep-sea taxa remain poorly represented in collections because of the relative infrequency of deep sea collecting expeditions, and for some groups (e.g. large species of *Onykia*, *Mesonychoteuthis* and *Architeuthis*), accidental collection as by-catch in commercial fisheries targeting other species remains the main source of available material.

The condition of this material varies greatly, and the geographical distributions of these taxa may appear artificially narrow based on localized commercial fishing effort. Retention (and later, fixation and storage) of this material can pose additional challenges due to its lack of economic value (to fisheries), as well as spatial constraints and the volume of hazardous chemicals required, especially for large specimens.

For some species, extensive museum collections can provide insight into intraspecific variations in morphology, but for many taxa, holdings are restricted to a few individuals. In addition, different life stages and states of maturity for a species might not be represented comprehensively by specimens in collections. However, examination of whole (or partial) physical specimens is no longer the sole means of gaining insight into deep-sea cephalopod diversity. With the recent advances in underwater imagery, and the greater human presence in the deep sea, species are increasingly identified within their natural habitat in their natural (i.e. unpreserved) form. In fact, some species, such as a large magnapinnid known from the Pacific (Vecchione et al., 2001), remain known exclusively from video footage; no physical specimen has ever been positively identified in collections.

4.3. Biogeography

Fries (2010) created a distribution map using primary literature and databases (OBIS and AquaMaps) (Figure 3.6) to illustrate the worldwide distribution of scientific records of the Cephalopoda, including both coastal and deep-sea habitats. It highlighted obvious gaps in cephalopod studies, namely, the Arctic, southwest Atlantic, Indian and Pacific Ocean. A brief summary of studies that have focused specifically on the deep sea (below 200 m) in the major ocean basins follows. They are far from the only studies conducted but are a broad indication of the information available for each region.

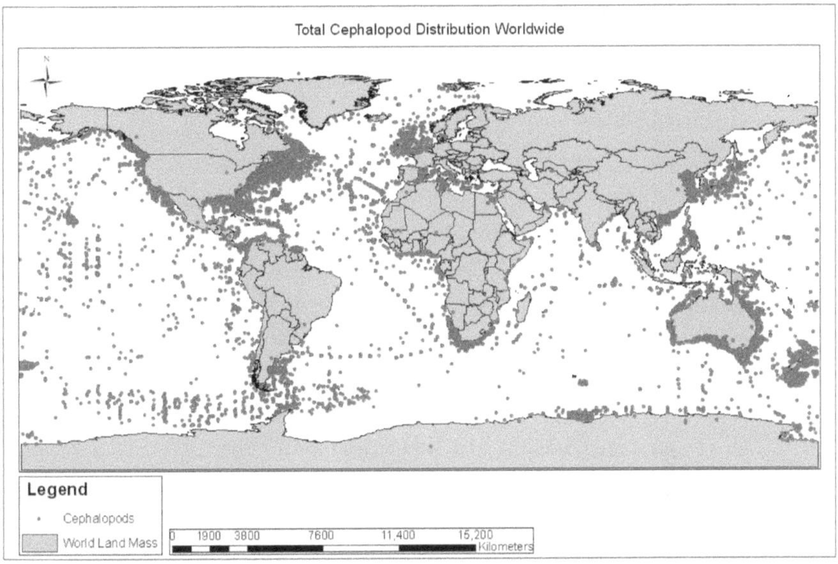

Figure 3.6 Global cephalopod distribution based on meta-analysis study demonstrating worldwide cephalopod distribution and regional gaps requiring future study. *With permission from Fries (2010).*

4.3.1 Atlantic Ocean

Several overviews of the deep-sea fauna have been published, as well as a number of regional reviews. For example, large-scale depth-related diversity patterns of cephalopods in the Atlantic were examined in the open-ocean and deep-sea regions by Rosa et al. (2008) through extensive literature review. Cephalopod diversity was found to peak along continental shelves and within the epipelagic zone and to decline with depth below 200 m. This contrasts with patterns observed for other macrofauna as well as for other molluscan groups such as gastropods and bivalves (Rex et al., 2005). In the southwestern Atlantic, Haimovici and Perez (1991) and Haimovici et al. (2009) have inventoried the deep-sea cephalopod fauna of the Brazilian slope down to 600 m; Guerra et al. (2001) also reported 21 species off the Patagonian Shelf down to 1500 m and showed that benthic octopods penetrate this area within the plume of cold sub-Antarctic waters and are pushed northwards into the South Atlantic by the Falkland/Malvinas current. Future studies were suggested on either side of this plume to investigate the presence of cold water-adapted species. In the northern Atlantic, midwater cephalopods were examined off Nova Scotia during the years 1986–1989 by Vecchione and Pohle (2002), who found 63 species from

28 families, in depths between 169 and 4800 m. Subsequent investigations under the MAR-ECO initiative reported 56 species from along the northern Mid-Atlantic Ridge (Vecchione et al., 2010b) and 50 cephalopod species from the southern portion of the ridge (Perez et al., 2012). In the northeastern Atlantic, cephalopods have been documented from the surface to the sea floor at great depths, including an impressive record of a deep-sea *Muusoctopus* species observed feeding on baitfall (porpoise) at a depth of 2555 m (Kemp et al., 2006). Collins et al. (2001) reviewed the distributions of deep-water (150–4850 m) benthic and benthopelagic cephalopods in this region and identified 36 species of cephalopods from 14 families. Judkins (2009) also reported 129 cephalopod species to inhabit the wider Caribbean, which includes the Gulf of Mexico, Caribbean Sea and a portion of the central western Atlantic. In her study, abundance, distribution and diversity differences were analysed and distribution maps were created for each species, with deep-sea taxa accounting for 45% (58 species) of species documented.

4.3.2 Pacific Ocean

In the southeastern Pacific, a distribution and diversity study conducted off the coast of Chile revealed 71 oceanic species, with the majority found below 200 m (Ibañez et al., 2009). In the northeastern Pacific Ocean, Sinclair and Stabeno (2002) recorded 11 deep-sea cephalopod species from the southeastern Bering Sea, and a field guide to squids and octopodiforms of the eastern North Pacific and Bering Sea was completed in 2009 (Jorgensen, 2009). A deep-sea octopod, *Graneledone* cf. *boreopacifica*, has also been collected near an active hydrothermal vent at 1459 m (Voight, 2000). Given the difficulty of sampling adult cephalopods, cephalopod larvae can also be used to investigate the systematics, distribution and abundances of epiplanktonic squid larvae, such as a study conducted in the California Current (Okutani and McGowan, 1969). Other paralarval studies in the Pacific have focused on the diverse Gonatidae (Hunt and Seibel, 2000; Kubodera and Jefferts, 1984a,b), as well as the ommastrephids (Harman and Young, 1985). A comprehensive study on epipelagic cephalopods was undertaken by Okutani (1974). Young (1972b) also documented 30 species of pelagic deep-sea cephalopods from southern Californian waters, more than a dozen of which have also been observed by MBARI's ROV programme. These cephalopods include *Vampyroteuthis infernalis*, histioteuthids, enoploteuthids, gonatids, cranchiids, octopoteuthids, onychoteuthids and members of the chiroteuthid families (e.g. Bush et al., 2007, 2009; Hoving et al., 2013b;

Hunt, 1996; Robison et al., 2003). In the mid-Pacific, while surveying the Cross Seamount in the mid-Pacific Ocean for mesopelagic micronekton, 22 species of cephalopod were recorded (De Forest and Drazen, 2009). Notable recent findings in the western Pacific along the coast of Japan, both from cruise material and live *in situ* observations category, include those of Kubodera and Mori (2005, for *Architeuthis*), Kubodera et al. (2007, for *Taningia*) and a collaborative expedition yielding the first-ever video footage of a live *Architeuthis dux* at depth (600–900 m, Widder, 2013).

4.3.3 Indian Ocean

Cephalopod captures in the deep Indian Ocean have been scarce, or poorly reported, until recently; specimens were previously only known to be collected as part of other research projects like the International Indian Ocean Expedition in the 1960s. However, at least some regions and environments appear to have quite high species diversity: Nesis (1986) reported 50 species of cephalopods from four tropical Indo-West Pacific-type seamounts, and work conducted by Piatkowski (1991) in the Arabian Sea examined the distribution of pelagic cephalopods to 1000 m. Laptikhovsky et al. (submitted) also recorded at least 68 cephalopod species, from 26 families, from along the Southwest Indian Ridge.

4.3.4 Polar Oceans

Kubodera and Okutani (1994) identified 44 eledonine octopods in the Southern Ocean, describing the distribution and systematics of the group. Allcock et al. (2001) studied octopods in the eastern Weddell Sea to a depth of 1000 m and found eight species of incirrate octopods, three of which were undescribed, and many other subsequent species descriptions have originated from the region (Allcock et al., 2004; Vecchione et al., 2005, 2009b). Xavier et al. (1999) compiled data obtained between 1886 and 1997 in the Southern Ocean and found that 21 species of oegopsid squids had been reported from depths below 200 m. The geographical distributions of each were documented with respect to bathymetry, ocean fronts and sea-ice extent. However, large areas remain from which no data have been reported. Additional studies on cephalopods, using predators as samplers (e.g. Xavier et al., 2006, on albatrosses), have also been conducted in this region, giving further insight into cephalopod distribution and abundance. In the Arctic Ocean, Nesis (2001) discussed the distributional ranges of seven cephalopod species inhabiting the region, concluding that they represented a mixture of circumpolar and non-circumpolar distributions. Another study focusing on distribution and predation was conducted for 36 cephalopod species by Gardiner and Dick (2010) based on past records including museum collections, government reports, international and national databases, published articles and personal communications. Distribution maps were created for each; however, no records could be found for some areas, so coverage of the entire region remained incomplete. Most recently, Golikov et al. (2013) provided new data on the distribution of three Arctic cephalopod species (*Todaropsis eblanae* Ball, 1841; *Todarodes sagittatus* Lamarck, 1798; *Teuthowenia megalops* Prosch, 1849; and *Gonatus fabricii* Lichtenstein, 1818) and showed significant range extensions for these deep-sea cephalopods.

4.4. Combined approaches to cephalopod systematics

Cephalopod taxonomy has advanced markedly in the last few decades. This is partly due to improved exploratory technology and increased sampling but has also resulted from the application of molecular techniques, which provide additional suites of traits that can be used to define species and examine their phylogenetic relationships. While molecular techniques have been utilized primarily in shallower-water taxa to date, due to the relative availability of fresh tissue, they are also increasingly being used as a tool to study deep-sea forms. Although species are traditionally identified by morphology, this approach can underestimate the "true" number of species if cryptic species are present. Similarly, DNA-sequencing techniques can over- or underestimate the "true" number of species when inappropriate markers are used. Thus, a cautionary approach should be taken in determining species boundaries by including as many suites of character traits as possible, ideally both morphological and molecular, and incorporating combined analyses of multiple genetic sequences from each individual (covering both nuclear and mitochondrial DNA sequences).

To distinguish genetically among different species in sympatry, it is essential to independently confirm any phylogenetic hypothesis based on molecular data by either morphological or unlinked molecular characters (Feulner et al., 2007; Milinkovitch et al., 2002). Frankham et al. (2012) have identified approximately 26 definitions of species and, depending upon which concept is followed, different numbers of species and disparate groupings result. These researchers demonstrated that the definition and determination of species have financial, legal, biological and conservation implications and gave an insightful review of this subject.

Recently, considerable advances in our understanding of the higher-level systematics of coleoid cephalopods have occurred. Morphological studies have proven useful in classifying species within subfamilies and genera but less so in determining higher-level relationships due to problems involving the determination of character independence, apomorphic loss of characters or assessment of homology/homoplasy (Young and Vecchione, 1996). This is particularly important for deep-sea cephalopods, where most of the family-level diversity within the class is found. High rates of evolution in cephalopods, coupled with their radiation into numerous habitats, have confounded our ability to recover consistent phylogenetic relationships using both morphological and molecular data (Lindgren et al., 2012).

Historically, molecular studies have produced conflicting results, perhaps due to limited taxon or gene sampling. Strugnell and Nishiguchi (2007) attributed the lack of clear resolution to the different alignment and analysis combinations that have been used in studies of cephalopod phylogenetics. Consistent among these results, however, is the monophyly of the Octopoda, Argonautidae, Oegopsida and Ommastrephidae based on two sequence homology methods and three analysis methods, with the Bathyteuthidae the sister taxon of the Oegopsida in each case. The paper by Strugnell and Nishiguchi (2007) also gave a useful summary of the outcomes of various molecular analyses to 2007 and advocated the use of a variety of different alignment and analysis strategies in phylogenetic analysis.

Lindgren et al. (2012) generated the most comprehensive multigene phylogeny to date, based on 10 genes for over 400 cephalopod taxa (representing 42 of 47 families). In addition to providing a well-supported phylogeny, these researchers created a character matrix of habitat type and morphological characters from which to infer ancestral character states and test for correlation between habitat and morphology to study the phylogenetic and environmental forces driving character evolution. Convergent evolution was shown for all six morphological characters analysed, three of which (autogenic photophores, cornea and accessory nidamental glands) correlate with habitat and therefore environmental pressures. This study also serves to underline the importance of a combined approach in formulating robust phylogenetic hypotheses. Their results confirmed monophyly for almost all major orders and families studied except the Octopodidae *sensu lato*, which is a highly divergent group morphologically, with numerous species complexes (Norman and Hochberg, 2005; Strugnell et al., 2013).

In addition to providing insight into the phylogenetic relationships among higher-level cephalopod groups, genetic data can be used to improve natural history knowledge of individual species. For species that appear endemic to relatively small geographical areas (e.g. *Watasenia scintillans* Berry, 1911), molecular data can provide insight into population dynamics and gene flow. For others that appear to have global distributions, like the currently monospecific *Taningia*, *Vampyroteuthis* and *Spirula* (or the rarely encountered but apparently cosmopolitan *Asperoteuthis acanthoderma* Lu, 1997; see Judkins et al., 2009), genetics may reveal systematic complexities not immediately evident from morphological data alone. The genus *Taningia*, for example, has been considered monospecific for more than 150 years but may in fact have at least two species (J. Kelly, personal observation). In contrast, *Architeuthis*, a genus with 21 named species (of which recent authors have generally accepted between one and four, e.g. Jereb and Roper, 2010), now appears to contain a single cosmopolitan species, according to mitochondrial genomic work by Winkelmann et al. (2013). Moreover, the level of nucleotide diversity in *A. dux* is extremely low, suggesting a possible recent bottleneck or expansion from one region into the remainder of the temperate oceans.

4.5. Future priorities in cephalopod systematics

Despite our enhanced knowledge of deep-sea cephalopods due to recent technological advances, our ability to collect and identify cephalopods in some areas remains limited. Many developing countries lack the financial support and technology necessary for deep-sea research, which results in gaps in knowledge for particular regions. Similarly, data coverage in remote regions is limited due to logistical and financial constraints (McIntyre, 2009). In general, areas requiring the most attention are currently the southeast Pacific, southwest Atlantic, and the Indian and Arctic Oceans.

Targeting taxa that are in particular need of focused taxonomic research effort is difficult because, while the deep ocean is largely underexplored in general (Webb et al., 2010), some families are well known from some geographical regions and poorly known from others for a variety of reasons. Globally, collection and exploration efforts have been concentrated in particular areas—not necessarily areas of greatest importance or need, but largely as a result of disparate funding and resource opportunities. Furthermore, most cephalopod studies in the deep sea result from funded projects aimed at other questions/problems. For example, deep-sea benthic octopods have received some focused research effort in polar regions and Australasia but are poorly known elsewhere, although recent work in the southeastern Pacific based on specimens obtained off the shelf and slope of Chile (Ibañez et al., 2012) is now filling another knowledge gap.

Some groups, however, have notoriously unstable systematics worldwide. Foremost among them are the squid families Cranchiidae, although Voss (1980) substantially stabilized the genera, and Chiroteuthidae, which is suspected to contain several unconfirmed new genera, as well as several undescribed species in the existing genera *Chiroteuthis* and *Asperoteuthis* (Young and Roper, 2011a). One of the least-studied families appears to be the Brachioteuthidae; within *Brachioteuthis*, several species have been named but only partially described, and additional new species have been recognized (ToLweb, 2012); consequently, all named brachioteuthid taxa are in need of review and disambiguation. The validity of all taxa is now under review. Mastigoteuthids are similarly problematic, although the New Zealand fauna has recently been reviewed by Braid (2013). Octopoteuthids are similarly poorly understood and are under current review by P. Jereb, A. Guerra and J. Kelly. Families such as Cranchiidae, Opisthoteuthidae and Mastigoteuthidae, known to have high local diversity in some regions (the latter being the most abundant cephalopods in the deep sea) (M. Vecchione, personal observation), should be particular candidates for revision as such patterns suggest that additional new species may be present. Research directions can also be guided by the availability of existing preserved specimens in museum collections or the potential to obtain fresh material from active collecting programmes. Wherever possible, morphological and molecular techniques should be used together in future taxonomic revisions, with resulting sequences submitted to GenBank and BOLD and distribution records to OBIS.

5. LIFE CYCLES OF DEEP-SEA CEPHALOPODS

5.1. Reproduction

5.1.1 General reproductive system morphology

All cephalopods are dioecious (i.e. the sexes are separate). The male reproductive system is divided into an unpaired testis; a vas deferens; a spermatophoric (or spermatophore-producing) organ; a spermatophoric (or Needham's) sac, where spermatophores are stored; and a terminal organ (or "penis"), a prolongation of the spermatophoric sac. The terminal organ may have an extensor capacity and is involved in spermatophore transfer in some deep-sea cephalopods; alternatively, one or more of the male arms may bear structural and glandular modifications related to spermatophore transfer (e.g. Hoving, 2008; Nesis, 1995). Although the majority of male cephalopods bear only the left gonoduct, some deep-sea species exhibit paired (and functional) gonoducts (*Selenoteuthis scintillans* Voss, 1958, *Lycoteuthis* spp. and *Stigmatoteuthis* spp.) (Nesis, 1987a, 1995; Voss et al., 1998).

The female reproductive system is divided into an ovary, one or two oviducts and one or two oviducal glands. Paired oviducts and oviducal glands occur in Vampyromorpha, incirrate octopods and in most Oegopsida, with the oegopsid family Pyroteuthidae being an exception: *Pterygioteuthis* spp. possess only the right gonoduct, and *Pyroteuthis* spp. may bear a reduced right gonoduct. In sepiids, sepiolids, myopsids and cirrate octopods, only the left gonoduct is present. With few exceptions (i.e. Enoploteuthidae), all decapods bear a pair of nidamental glands in the anterior ventral region of the mantle cavity. All decapods except oegopsids also bear an additional pair of glands called "accessory" nidamental glands, which house symbiotic bacteria that are passed to the egg capsules (Barbieri et al., 2001). Although the exact process is largely unknown, all glands (oviducal, nidamental and accessory nidamental) are involved in egg-capsule formation, except in incirrate octopods, in which the oviducal glands form the cement material used by the female to attach eggs on the substrate (Froesch and Marthy, 1975).

5.1.2 Spermatophores

Coleoid spermatophores are considered to be one of the most elaborate reproductive structures in the animal kingdom (Mann, 1984). With the exception of cirrate octopods, which produce an unusual spermatophore (or "sperm packet"; Villanueva, 1992), coleoid spermatophores generally consist of a sperm mass, a cement body and an ejaculatory apparatus, being enveloped by the outer and middle tunics (Figure 3.7; Marian, 2012a,b). From the outside to the median axis of the spermatophore, the ejaculatory apparatus tube consists of the inner tunic, the middle and inner membranes and the spiral filament (Figure 3.7), the latter bearing minute stellate particles (Marian, 2012a,b). Hess (1987) provided an extensive overview of the spermatophores of a wide variety of cephalopods, including deep-sea species. Spermatophore sizes (relative and absolute) vary considerably among deep-sea coleoids, ranging from absolute lengths of a few millimetres in *Abralia* spp. to 20 cm in the giant squids (Hoving et al., 2004). *Heteroteuthis*

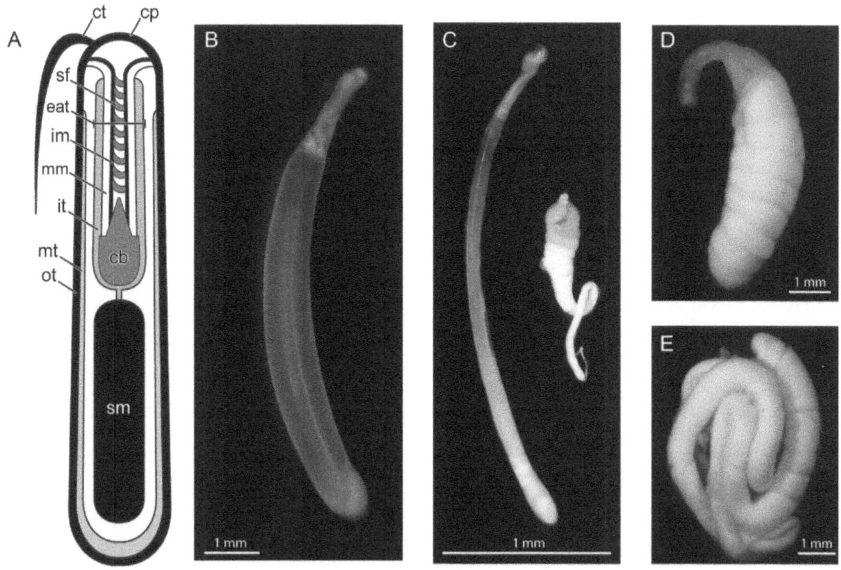

Figure 3.7 Deep-sea cephalopod spermatophores and spermatangia. (A) Diagrammatic representation of a generalized squid spermatophore. (B) *Histioteuthis miranda* spermatophore. (C) *Discoteuthis discus* spermatophore (left) and spermatangium (right). (D) and (E) *Heteroteuthis dispar* spermatophore (D) and spermatangium (E). Abbreviations: cb, cement body; cp, cap; ct, cap thread; eat, ejaculatory apparatus tube; it, inner tunic; im, inner membrane; mm, middle membrane; mt, middle tunic; ot, outer tunic; sf, spiral filament; sm, sperm mass. *(A) Originally published in Marian et al. (2012) and reproduced with kind permission from Springer Science and Business Media. (B) Originally published in Hoving et al. (2010a) and reproduced with permission. (D) and (E) Originally published in Hoving et al. (2008b) and reproduced with kind permission from Springer Science and Business Media.*

dispar (Rüppell, 1844) may have the proportionally largest spermatophores among deep-sea cephalopods (SpL = 33–43% ML; Hoving et al., 2008b). A general schematic diagram of a coleoid spermatophore is provided in Figure 3.7.

Spermatophores' intricacy is not only structural: when properly triggered, coleoid spermatophores are able to function autonomously in a process called the "spermatophoric reaction" (e.g. Hoving and Laptikhovsky, 2007; Hoving et al., 2009; Mann, 1984; Marian, 2012a,b). The spermatophoric reaction is an osmotically controlled process that comprises the evagination of the spermatophoric tunics and membranes, leading to the extrusion of the cement body and to the formation of the spermatangium (everted spermatophore containing the sperm mass). Mated females often bear spermatangia attached to their bodies in a variety of (sometimes taxon-specific) locations.

The number of spermatophores stored in the spermatophoric sac may vary widely in deep-sea squid, from approximately 80 in *Ancistrocheirus lesueurii* Orbigny, 1842 (Hoving et al., 2006), to about 250 in *Lycoteuthis lorigera* Steenstrup, 1875 (Hoving et al., 2007); 1000 in *Octopoteuthis sicula* Rüppell, 1844 (Hoving et al., 2008a); and 100–1800 in ommastrephids (e.g. Nigmatullin et al., 2003). Spermatophorogenesis (spermatophore formation) has been intensively studied in ommastrephid squids, including its ontogenetic aspects (e.g. Nigmatullin et al., 2003). These authors demonstrated that there is a distinction between physiological and functional maturity in male squid, with the production of "tentative" (devoid of sperm) and "false" spermatophores (spermatophores with very small sperm volume) during submaturity. These are constantly released through the terminal organ until the mature stage is reached and fully functional spermatophores are produced, at which point output can be considerable: a daily production rate of 10–20 spermatophores was estimated for mature males of the ommastrephid *Illex argentinus* (Arkhipkin and Laptikhovsky, 1994).

Spermatophore production apparently begins relatively early, and somatic growth continues after the production of the first spermatophores, and spermatophores may be stored in the spermatophoric sac for ca. 4.5 months (Hoving et al., 2010a). This strategy was postulated to be associated with the deep-sea environment: where encounters between mates are rare, the chances of fertilization could be maximized by beginning spermatophorogenesis earlier in the life cycle (Hoving et al., 2010a). Although this may be true, spermatophore production in coastal and epipelagic species also starts early in life and is associated with somatic growth (e.g. Nigmatullin et al., 2003).

The sperm mass of the spermatophore may contain millions of spermatozoa (Mann, 1984). Detailed ultrastructural data on the spermatozoa of one deep-sea decapod (*Rossia pacifica* Berry, 1911, Fields and Thompson, 1976) and six deep-sea octopodiform species (*Opisthoteuthis persephone* Berry, 1918, and Healy, 1993; *Graneledone gonzalezi* Guerra, Gonzalez and Cherel, 2000, and Roura et al., 2009; *Bathypolypus bairdii* Verrill, 1873 and *B. sponsalis* P. Fischer and H. Fischer, 1892, Roura et al., 2010a; *Vulcanoctopus hydrothermalis* Gonzalez and Guerra in González et al., 1998; Roura et al., 2010b; and *Vampyroteuthis infernalis* Healy, 1989, 1990) are currently available in the literature and suggest that some sperm characters may be useful for gaining insight into the systematics and phylogenetics of cephalopods, across all taxonomic levels.

Much remains to be investigated regarding sperm and spermatophore morphology in deep-sea cephalopods. In addition to contributing to the systematics and phylogenetics of the class as a whole (Hess, 1987), such detailed morphological data (such as those available for neritic squid; Marian and Domaneschi, 2012) can provide hints to sperm-transfer mechanisms (Marian, 2014).

5.1.3 Mating behaviour

Mating in deep-sea cephalopods has long been a matter of conjecture. Mating observations have been rare (e.g. Roper and Vecchione, 1996), and behaviour has often been inferred from characteristics of preserved specimens (e.g. Hoving, 2008; Nesis, 1995). As will be discussed, however, recent evidence gathered from on-board analysis of freshly caught individuals, as well as from observations performed through submersibles, has substantially improved our knowledge of the copulatory behaviour of deep-sea cephalopods. Nevertheless, much remains to be discovered, and the morphological analysis of museum specimens, particularly when combined with observations of living individuals, remains a powerful tool in providing insights into this elusive aspect of deep-sea cephalopod biology.

Spermatophores are transferred to female cephalopods by two main methods: with the aid of the hectocotylus (a male arm apparently modified specifically for spermatophore transfer) or directly through the terminal organ. Where present, the hectocotylus generally bears structural and/or histological specializations (such as the absence or modification of sucker rings and/or the presence of papillae, lobes or glandular pads, to name a few; Nesis, 1995) that differentiate it from other arms. Most of these specializations are believed to be directly involved with spermatophore transfer, that is, to facilitate holding and manipulating spermatophores. In incirrate octopods, the hectocotylus is one member of the third arm pair (Norman and Sweeney, 1997), while in decapods, one or both of the ventral arms are modified (Nesis, 1995), except in sepiolids, cranchiids and histioteuthids, in which one or both of the dorsal arms are modified (Hoving et al., 2008b; Voss, 1980; Voss et al., 1998). Spermatophore transfer through the hectocotylus appears to be the sole strategy utilized by neritic cephalopods and is also utilized by some groups of deep-sea species, such as sepiolids (Hoving et al., 2008b), incirrate octopods (Voight, 1997) and some oegopsid squids (e.g. Ommastrephidae, Enoploteuthidae, Ancistrocheiridae, Lycoteuthidae, Pyroteuthidae, Thysanoteuthidae and Cranchiidae; Nesis, 1995), including some oegopsids that reproduce in true deep-sea strata (e.g. some Cranchiidae; Nesis, 1995). This strategy is generally inferred from the presence of a hectocotylus in preserved specimens, but oceanic species from the family Ommastrephidae have also been observed to transfer spermatophores presumably by the hectocotylus: *Dosidicus gigas* (Gilly et al., 2006; Nigmatullin et al., 2001), *Todarodes pacificus* (Okutani, 1983) and *Sthenoteuthis* spp. (Zuyev et al., 2002). In these cases, the mating position was "head to head," and the spermatophores were deposited on the buccal membrane, where these species bear seminal receptacles (see "Spermatangium attachment sites"). A possible observation of mating in a deep-sea incirrate octopod *Vulcanoctopus hydrothermalis* was also reported by Rocha et al. (2002), in which four males were apparently mating with the same female simultaneously, three of them "mounting" the female and one male mating "at distance" by stretching its hectocotylus (a common strategy within shallow-water incirrate octopods). A copulatory behaviour between two males of separate deep-sea octopod genera had also previously been documented by Lutz and Voight (1994), at a depth of 2500 m along the crest of the East Pacific Rise. The authors postulated that this strategy could be related to the fact that in the deep sea, mate encounters are rare. Recently, Hoving et al. (2012) found evidence that indicate (intraspecific) same-sex sexual behaviour for the deep-sea squid *Octopoteuthis deletron* Young, 1972. In a number of males observed by ROV, implanted spermatangia could be discerned in similar body locations as those often seen on females, and such "mated" males were encountered approximately as frequently as females.

At present, the exact process of spermatophore transfer via hectocotylus in deep-sea cephalopods can only be inferred from observations of neritic species. For example, based on what is known for loliginid squids (e.g. Drew, 1919; Hanlon and Messenger, 1996), neritic sepiolids (e.g. Rodrigues et al., 2009) and idiosepiids (Sato et al., 2013), males of most deep-sea oegopsid squids and sepiolids probably retrieve a group of spermatophores from the terminal organ during mating with the aid of the hectocotylus, which then transfers them to the site of attachment on the female body. For deep-sea benthic incirrate octopods, the process of spermatophore transfer might be similar to what has been observed in shallow-water species (e.g. Wodinsky, 2008): the spermatophore is placed into the spermatophoric groove (on the hectocotylized arm) by the terminal organ; the spermatophore then travels along by peristalsis to the arm tip, which inserts the spermatophore into the distal oviduct of the female. In some pelagic incirrate octopods such as *Argonauta*,

Tremoctopus and *Ocythoe*, dwarf males possess a very elaborate hectocotylus, which remains enclosed in a protective sac until mating (Naef, 1921–1923). During copulation, the hectocotylus presumably everts from this sac, retrieves the spermatophore, autotomizes itself and relocates to inside the mantle cavity or even the reproductive system of the female (Naef, 1921–1923).

The other type of spermatophore transfer, that is, directly through the terminal organ, appears restricted to deep-sea cephalopods (e.g. Architeuthidae, Gonatidae, Octopoteuthidae, Onychoteuthidae and Pholidoteuthidae). This mode of transfer was long postulated from the presence, in several deep-sea species, of a long terminal organ (Figure 3.8) combined with the absence of a hectocotylus. In these cases, the terminal organ has recurrently been found (in preserved specimens) with its tip protruding and sometimes considerably extending beyond the mantle margin (e.g. Bolstad, 2006; Jackson and Jackson, 2004, Figure 3.8). This mode of transfer has recently been confirmed by two complementing papers. Based on a moribund specimen of *Onykia ingens* (Smith, 1881) (Onychoteuthidae), Arkhipkin and Laptikhovsky (2010) demonstrated that the terminal organ has the capacity to extend itself to almost the same length as the whole body of the animal. Then, Hoving and Vecchione (2012) made the first observations of copulation in a deep-sea squid (*Pholidoteuthis adami* Voss, 1956) by ROV, wherein the terminal organ extended considerably from the mantle cavity through the funnel to reach the female body, while the male arms (not hectocotylized) played no role beyond holding the female (Figure 3.8). In both mating pairs observed by Hoving and Vecchione (2012), the male was parallel to and positioned above the female but oriented in the opposite direction (posterior mantle above the female's head) and "upside down" (ventral side up) (Figure 3.8; ROV observations). Extending through the funnel, the terminal organ reached the dorsal surface of the female's mantle, at approximately the level of the fin insertions (Figure 3.8). This region has been observed in museum specimens as a common site of spermatangium attachment (Hoving and Vecchione, 2012).

Previously, presumed mating in *Brachioteuthis beanii* had been observed by a submersible, although alternative hypotheses (e.g. cannibalism) could also explain the observed parings (Roper and Vecchione, 1996). In this case, mating was in a "head-to-tail" position, with the presumed male grasping the posterior portion of the presumed female's mantle. Judging from the observations from Hoving and Vecchione (2012) on *P. adami*, the

Figure 3.8 Spermatophore transfer through the terminal organ. (A) Mating in *Pholidoteuthis adami* documented by ROV. Specimen size approximately 680 mm ML. This behaviour was observed during an expedition by NOAA's Ocean Exploration programme. (B) Mature male of *P. adami* from a museum collection, showing the terminal organ extending through the funnel. *(A) and (B) originally published in Hoving and Vecchione (2012) and reproduced with permission.*

presumed male could be transferring spermatophores to the female at the region of the ovary, but contrary evidence from Young (1978) and K. Bolstad (personal observation) has found spermatangia in the buccal membrane of *Brachioteuthis*.

Pholidoteuthis adami was one of the species in which spermatophore transfer was previously hypothesized to be performed by the terminal organ, based on preserved specimens observed with the terminal organ tip extending beyond the mantle margin and lacking a hectocotylus (Nesis, 1995). Following confirmation of this hypothesis, as discussed in the previous paragraph, other deep-sea squid families with similar reproductive structure morphologies (e.g. Octopoteuthidae, Onychoteuthidae, Gonatidae and Architeuthidae; Nesis, 1995) are anticipated to perform spermatophore transfer through the terminal organ as well. Several questions remain to be answered, however. For example, the mechanism enabling terminal organ extension is completely unknown. Detailed morphological investigation could help reveal whether the terminal organ of deep-sea squid function, for instance, as muscular hydrostats. Spermatophore orientation and placement are poorly understood even for species with a hectocotylus. In the case of the terminal-organ-mediated transfer, however, the mechanism by which the spermatophores are correctly oriented remains unknown yet must have great importance as they are always released with the aboral ends first (the oral end of the spermatophore being the implanting region). In some species, the tip of the terminal organ exhibits complex morphology (e.g. Nesis, 1995), suggesting a possible role in spermatophore orientation and placement, but much remains to be discovered about this enigmatic process.

5.1.4 Spermatangium attachment sites

Spermatangium attachment sites vary considerably among deep-sea cephalopods. In some species, no special receptacles are present, and the spermatangia are implanted into external surfaces of the female's body, such as the mantle, arms, tentacles and head region (e.g. Architeuthidae Hoving et al., 2004; Cranchiidae Nesis et al., 1998; Cycloteuthidae Clarke, 1980; Histioteuthidae Voss et al., 1998; Octopoteuthidae Hoving et al., 2008a, 2010b; Onychoteuthidae Bolstad and Hoving, 2011; Pholidoteuthidae Hoving and Vecchione, 2012; and Sepiolidae Hoving et al., 2008b; Figure 3.9), and/or within the mantle cavity in the vicinity of the genital opening (e.g. Cranchiidae Nesis et al., 1998; Enoploteuthidae Burgess, 1998; Histioteuthidae Voss et al., 1998; *Illex* spp. O'Dor, 1983; Onychoteuthidae Laptikhovsky et al., 2007; and Sepiolidae Hoving et al., 2008b).

In other species, specialized receptacles for spermatangium deposition/implantation are present, such as those found in the nuchal region of squids

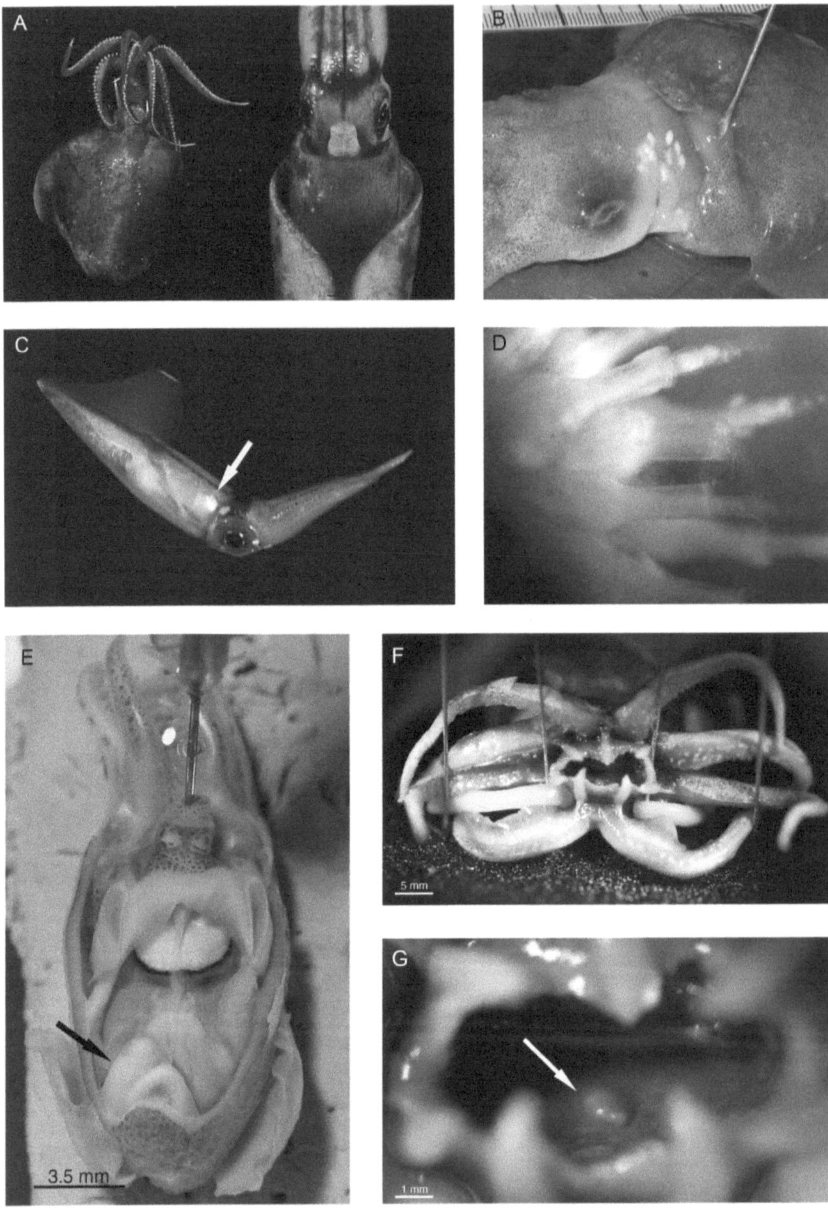

Figure 3.9 Spermatangium attachment sites. (A) and (B) Spermatangia implanted on external surfaces of the female body in (A) *Octopoteuthis deletron* (ML ca. 130 mm) and (B) *Rossia macrosoma*. (C) *Abraliopsis felis*, which has spermatangium receptacles in the nuchal region (arrow). (D) Spermatangia attached to the nuchal receptacle of *A. felis*. (E) Posterior seminal sac (arrow) containing a spermatangium in *Heteroteuthis dispar*. (F) and (G) Seminal receptacle (arrow) in the ventral buccal membrane of *Bathyteuthis berryi*. *(A) © 2006 MBARI. (B) Originally published in Hoving et al. (2009); (C) © 2010 MBARI; (E) originally published in Hoving et al. (2008b) With kind permission from Springer Science and Business Media; (F) and (G) originally published in Bush et al. (2012). All reproduced with permission.*

of the enoploteuthid families (Ancistrocheiridae, Enoploteuthidae, Lycoteuthidae and Pyroteuthidae; Burgess, 1998; Hoving et al., 2007; Hoving and Lipinski, 2014; Young and Harman, 1998; Figure 3.9). Some species of Rossinae (Sepiolidae) bear a bursa copulatrix (which involves the modification of the distal oviduct) or a ridged area near the oviduct (Hoving et al., 2008b), where spermatangia are found; in the pelagic Heteroteuthinae (Sepiolidae), a posterior seminal sac accommodates typically a single large spermatangium (Hoving et al., 2008b; Figure 3.9). Female vampire squid (*Vampyroteuthis infernalis*) bear large subcutaneous pouches located in front of the eyes, each able to store a spermatangium (Pickford, 1946, 1949). Other groups bear seminal receptacles, specialized sperm storage organs with morphological and histological specializations that enable them to store spermatozoa released from the spermatangia (Marian, 2014). For example, in some ommastrephids, numerous seminal receptacles occur in a ring on the buccal membrane (Arkhipkin et al., 1998; Ikeda et al., 1993; Nigmatullin and Markaida, 2009; Nigmatullin et al., 2002; Zuyev et al., 2002), and in bathyteuthids, a single main receptacle is located ventrally on the buccal membrane (similar to the neritic loliginids and sepiids; Bush et al., 2012; Figure 3.9). Exceptions to this general pattern (the presence of sperm and/or spermatangium receptacles) include some species in which the spermatangia are implanted within the mantle cavity in the vicinity of the genital opening (e.g. *Illex* spp. O'Dor, 1983; some Enoploteuthidae Burgess, 1998).

In incirrate octopods, seminal receptacles, when present, are located inside the oviducal glands (Froesch and Marthy, 1975). Octopods from incirrate *Eledone*, however, lack spermathecae in the oviducal glands and store spermatozoa inside the ovary, where the apical filaments of the oocytes serve as sperm storage sites (Perez et al., 1990). The sperm packets of cirrate octopods are transferred to the female oviducal gland (Aldred et al., 1983), but the presence of a seminal receptacle in this group is uncertain.

Basically, there are three types of spermatangium attachment in coleoids (Marian, 2014). One of them is typical of incirrate octopods, where the spermatangium is "plugged" into the distal oviduct, finding its way through it into the oviducal gland or into the ovary (Froesch and Marthy, 1975; Mann et al., 1970; Perez et al., 1990; Wells and Wells, 1972; Figure 3.10). Decapods may exhibit "shallow attachment," in which only the base of the spermatangium is implanted and cemented into the female tissue (e.g. Ommastrephidae, Durward et al., 1980; Sepiidae,

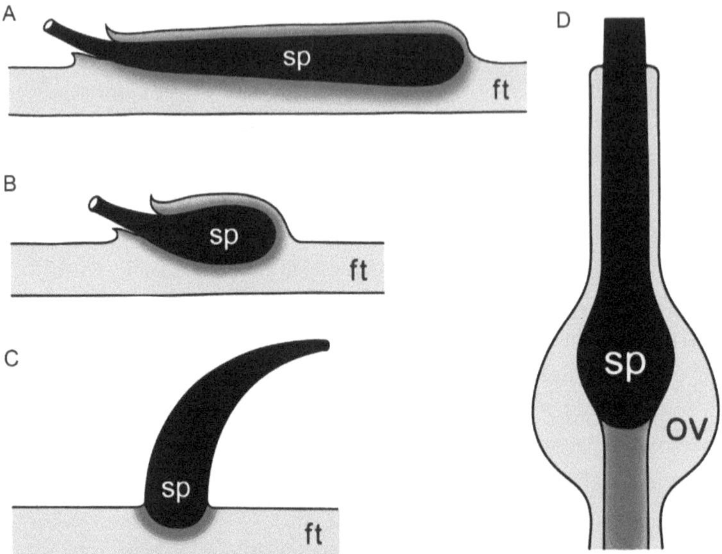

Figure 3.10 Types of spermatophore attachment in cephalopods. (A–B) Deep implantation (e.g. Architeuthidae, Cranchiidae and Octopoteuthidae). (C) Shallow implantation (e.g. Ommastrephidae, Enoploteuthidae). (D) Plugged attachment (e.g. incirrate octopods). Abbreviations: ft, female tissue; ov, oviducal gland; sp, spermatangium. *Originally published in Marian (2014) and reproduced with permission from Taylor & Francis Ltd. www.tandfonline.com.*

Hanlon et al., 1999; Figure 3.10), or "deep implantation," in which nearly the entire length of the spermatangium is implanted into female tissue (e.g. Architeuthidae Hoving et al., 2004; Cranchiidae Nesis et al., 1998; Octopoteuthidae Hoving et al., 2008a; and Sepiolidae Hoving et al., 2009; Figure 3.10). Shallow attachment is found in both neritic (e.g. Loliginidae) and deep-sea squids (e.g. Enoploteuthidae), but deep implantation appears restricted to deep-sea squid.

Until recently, the mechanism that enables deep implantation of spermatangia has been uncertain. Early hypotheses attributed the enabling of spermatangium implantation to extrinsic factors, such as a hydraulic system presumably performed by the terminal organ, or cuts made by the male hooks or beaks (Jackson and Jackson, 2004; Nesis et al., 1998; Norman and Lu, 1997). However, several recent lines of evidence (Hoving and Laptikhovsky, 2007; Hoving et al., 2009) have demonstrated that deep implantation is autonomous, that is, achieved by the spermatophore itself. Based on experimental data from neritic squid and published information from deep-sea squid, a common mechanism was proposed to explain both shallow and deep implantation (Marian, 2011, 2012a,b), involving the action of the ejaculatory apparatus and its containing spiral filament and stellate particles. This theoretical model, however, still needs to be tested (Marian, 2012b).

It is relevant to note that, among such variation in sperm-transfer mechanisms, some common patterns have been identified. Marian (2014) found evidence for convergent evolution in some cephalopod reproductive strategies: shallow implantation and seminal or spermatangium receptacles may have coevolved more than once independently (including in deep-sea lineages) and deep-spermatangium implantation multiple times independently, depending on the phylogenetic hypothesis under consideration. Arkhipkin and Laptikhovsky (2010) hypothesized that the evolution of sperm-transfer mechanisms in Decapodiformes followed two distinct pathways: transfer of spermatophores by the hectocotylus to specialized female receptacles in several neritic and epi/mesopelagic squids and transfer of spermatophores through the terminal organ with deep implantation of spermatangia on the surface of the female's body in several deep-water squids. However, this hypothesis has not yet been successfully tested. Also, because exceptions to this suggested pattern exist (e.g. bathypelagic squids having seminal receptacles, Bush et al., 2012), it remains unclear whether and how the evolution of these reproductive strategies has been influenced by the marine habitat. Further investigation requires detailed morphological and functional data on the reproductive system of the majority of the species (not currently available) and further resolution of the conflicting phylogenetic hypotheses among cephalopods (Marian, 2014).

5.1.5 Fertilization and egg (mass) production

Despite the great abundance of some oceanic cephalopods, the egg masses of many remain very poorly known or entirely unknown. Ommastrephids produce large, spherical, neutrally buoyant pelagic egg masses, which have been reported for several species (Bower and Sakurai, 1996; Durward et al., 1980; Laptikhovsky and Murzov, 1990; O'Shea et al., 2004; Staaf et al., 2008). These egg masses are spawned "on" the pycnocline (the layer between water masses of different densities) where the density gradient is the steepest, and are subsequently generally found in relatively shallow waters (up to 30 m). Encounters with egg masses are infrequent, however, and even for the well-studied, commercially valuable species *Dosidicus gigas*, of egg masses have rarely been reported (Staaf et al., 2008). Egg masses of *Thysanoteuthis rhombus* Troschel, 1857, have been repeatedly described in the literature (e.g. Guerra et al., 2002; Nigmatullin and Arkhipkin, 1998; Nigmatullin et al., 1995), and recently, pelagic egg masses encountered in South African waters were attributed to a *Lycoteuthis* species (Roberts et al., 2011).

In some deep-sea cephalopod families, females maintain contact with their eggs and embryos after spawning, a phenomenon known as post-spawning egg care or "brooding" (Okutani et al., 1995; Seibel et al., 2005). Brooding is common in benthic and pelagic octopods including deep-sea species (e.g. Voight and Grehan, 2000; Young, 1972a, 1995, 2013) but has also been reported in two deep-sea squid families: Gonatidae and Bathyteuthidae (Arkhipkin and Bjørke, 1999; Bower et al., 2012; Bush et al., 2012; Okutani et al., 1995; Seibel et al., 2000, 2005).

Fertilization mechanisms in deep-sea cephalopods remain a similar matter of conjecture and extrapolation. The species for which information on fertilization does exist all spawn in shallow oceanic waters in the upper 100 m, but summarizing the available information for these species is the only way to infer how fertilization and spawning may happen in species that spawn in the deep sea. Observations by Durward et al. (1980) and O'Dor and Balch (1985) on *Illex illecebrosus* (Lesueur, 1821) have suggested the following scenario. Thousands of eggs are released into a concentrated gel secreted from the nidamental glands. The gel is mixed with sperm from the attached spermatangia and possibly whole spermatangia that break loose from the mantle wall. The whole mixture (gel, eggs and sperm) is moved into the funnel and combined with a large quantity of water by mantle contractions (O'Dor and Balch, 1985). The occurrence of empty spermatangia inside the egg masses of *Illex illecebrosus* suggests that fertilization may also happen inside the egg mass (Durward et al., 1980).

One clue in understanding the fertilization processes in other cephalopods is the location of stored sperm. Presumably, eggs are passed along this location, wherever it may be in various taxa, so that sperm can be brought into contact with the eggs. In most ommastrephids (except *Illex*), as well as in most gonatids, sperm is stored on the buccal membrane, either in spermatangia that are superficially implanted (Gonatidae; H.J. Hoving, personal observation) or inside seminal receptacles (Ommastrephidae; e.g. Ikeda et al., 1993; Nesis, 1995). Eggs leaving the funnel are presumably subsequently brought into contact with the sperm from the buccal area. *Thysanoteuthis rhombus* is another species that stores sperm in spermatangia on the buccal membrane, but its characteristic cylindrical egg masses suggest a somewhat different process of spawning and fertilization. One proposed fertilization scenario for this species is as follows: (1) A mass of nidamental-gland secretions is moulded into a cylinder (Nigmatullin et al., 1995); (2) eggs from the oviduct are coated by jelly from the oviducal gland, forming threads that merge into one in the mantle cavity; (3) this single thread with two series of eggs is passed close to the buccal area, perhaps during exit via the funnel, and is fertilized; (4) the thread, with now-fertilized eggs, is wrapped around the cylinder of nidamental-gland secretion, forming the final columnar egg mass (e.g. Nigmatullin et al., 1995).

Species with post-spawning egg care (i.e. gonatids and bathyteuthids) also store sperm on the buccal membrane, but in these taxa, fertilization mechanisms remain unknown.

Enoploteuthids lack nidamental glands; their eggs are coated in clear oviducal-gland secretions and released as thin threads (Hayashi, 1989; Young and Harman, 1985; Young et al., 1985). These presumably pass along both sides of the nuchal cartilage, where spermatangia are stored in spermatophore receptacles (Burgess, 1998; Nesis, 1995). The collection of individual enoploteuthid eggs in plankton samples may indicate one of three things: eggs are spawned individually (Laptikhovsky, 1999); the jelly holding the eggs together in threads disintegrates after contact with water (Nesis, 1995); or the egg threads are damaged by plankton nets, releasing individual eggs (R. Young, personal communication). In other squids where females have been found with spermatangia stored in the nuchal area (*Ancistrocheirus lesueurii* Hoving and Lipinski, 2014; *Taningia danae* Hoving et al., 2010b; *Lycoteuthis lorigera* Hoving et al., 2007; and Pyroteuthidae Lindgren et al., 2013), eggs are also presumably passed along this region before fertilization, and spawning mechanisms might be similar to those of the Enoploteuthidae.

From these hypothesized scenarios, it seems that in decapod cephalopods in general, fertilization of the eggs occurs outside the body (external fertilization). However, as the eggs and sperm are first brought together in a confined space inside the mantle, nuchal area or arms, fertilization might more properly be considered "confined external" (Hanlon and Messenger, 1996; Naud and Havenhand, 2006). True internal fertilization has been hypothesized for the sepiolid *Heteroteuthis dispar* because the anatomy of this species allows for eggs to be fertilized inside the oviduct or ovary (Rüppell, 1844)—the posterior seminal receptacle in *H. dispar* is in contact with the visceropericardial coelom (Hoving et al., 2008b). Internal fertilization also happens in incirrate octopods, in which sperm is stored within the oviducal gland or within the ovary (Froesch and Marthy, 1975; Perez et al., 1990), and possibly also in the deep-sea squid *Planctoteuthis oligobessa* Young, 1972, where spermatangia have been found inside the ovary around the oocytes (Hoving, 2008; Young, 1972b). For species where sperm (or spermatangia) are implanted on external body sites, far from the oviduct openings, fertilization methods can only be hypothesized (Hoving, 2008). In *Octopoteuthis* spp., spermatangia have been reported all over the outer surface of the female's body, including the anterior and posterior mantle, (both dorsally and ventrally), on the head, on the arms and on the fins and tail (Hoving et al., 2008a, 2012; Nesis, 1995). In these species, immobile stored sperm from spermatangia may be mobilized by a chemical cue (e.g. from secretions from the nidamental or oviducal glands) produced by egg spawning. This cue could trigger the movement of the sperm from the implanted spermatangia through the water to the eggs, as has been suggested for other species (Durward et al., 1980).

5.1.6 Reproductive strategies

Coleoid cephalopods are considered semelparous (Boyle and Rodhouse, 2005; Cole, 1954): the spawning of eggs happens during one reproductive cycle, although eggs may be spawned over an extended period of time. Within semelparity, different traits (e.g. variability in egg size, ovulation and spawning patterns and patterns of somatic and reproductive growth) define the exact reproductive strategy that a cephalopod species uses. Based on the differences between these traits, it is possible to further categorize cephalopod reproductive strategies even if the overall strategy is semelparity (Hoving, 2008; Laptikhovsky et al., 2007; Rocha et al., 2001). It should be noted that recent observations on the first known mature female of *Kondakovia longimana* Filippova, 1972, have also suggested the potential for multiple reproductive cycles and, therefore, iteroparity in this species (Laptikhovsky et al., 2013). However, evidence remains inconclusive.

Various categories of reproductive strategies were recognized for some cephalopod families (e.g. Nesis, 1995; Nigmatullin and Laptikhovsky, 1994) before the comprehensive classification by Rocha et al. (2001) was published; this study defined five reproductive strategies for cephalopods based on ovulation patterns and whether or not growth occurred between spawning events. Four of these strategies can be found in coleoid cephalopods: (1) continuous spawning (as occurs in cirrate octopods), (2) simultaneous terminal spawning (spawning all eggs at once), (3) multiple spawning (eggs spawned progressively in separate batches, with growth between spawning events), and (4) intermittent terminal spawning (similar to multiple spawning, but without growth between spawning events). The classification proposed by Rocha et al. (2001) mostly applied to neritic species, because little information was available at that time on deep-sea cephalopods other than cirrates. Using new data on the deep-sea squids *Gonatus antarcticus* Lönnberg, 1898, and *Onykia ingens*, Laptikhovsky et al. (2007) divided coleoid cephalopods into two groups based on their ovulation patterns: species that ovulate (and spawn) their eggs synchronously ("simultaneous terminal spawners" sensu Rocha et al., 2001) and species that ovulate asynchronously (and spawn repeatedly). Asynchronous ovulation may involve the asynchronous ripening of individual eggs or batches of eggs. Repeated spawning includes the strategies "multiple spawning" and "intermittent terminal spawning" as defined by Rocha et al. (2001). Sample size in studies on deep-sea squids is generally small and often does not allow for identification of growth in between spawning events, making "multiple spawning" a challenging strategy to identify. However, using the division proposed by Laptikhovsky et al. (2007), it is possible to allocate squid species to a reproductive strategy based on oocyte length frequencies.

Repeated spawning is known to occur in various pelagic deep-sea cephalopods including members of the Enoploteuthidae, Lycoteuthidae, Ommastrephidae, Pyroteuthidae and Thysanoteuthidae (e.g. Hoving, 2008; Laptikhovsky et al., 2007); this strategy is also likely to occur in other species with asynchronous ovulation (e.g. Octopoteuthidae and Architeuthidae). Squids with this strategy have relatively high fecundities and release their small eggs in pelagic egg masses (strings, spheres or cylinders) in very high numbers. For example, 35,000–75,000 eggs have been estimated in egg masses of *Thysanoteuthis rhombus* (Nigmatullin et al., 1995), up to ~100,000 in *Illex illecebrosus* (Durward et al., 1980) and 200,000 in *Todarodes pacificus* (Bower and Sakurai, 1996). Based on the number of eggs observed in the oviduct, *Lycoteuthis lorigera* likely produces egg masses of 1000–3000 eggs (Hoving et al., 2014). Sepiolids also commonly have asynchronous ovary maturation, high fecundity and continuous repeated spawning, a suite of reproductive strategies that may have contributed to their evolutionary success in deep-sea habitats (Laptikhovsky et al., 2008). Moreover, benthic deep-sea sepiolids have considerably larger egg sizes (>10% ML) than shallow-water relatives (Laptikhovsky et al., 2008). Pelagic sepiolids (Heteroteuthinae) produce small eggs, have relatively high fecundity (Hoving et al., 2008b) and are thought to attach eggs to the sea floor (Okutani and Tsuchida, 2005).

Pelagic incirrates display the most elaborate reproductive strategies within cephalopods (Naef, 1921–1923; Young and Vecchione, 2008): females of *Argonauta* spp. secrete a delicate, calcareous, shell-like structure in which they brood multiple batches of eggs; females of *Tremoctopus* spp. brood multiple batches of eggs attached to elongated rods, which are held by suckers at the bases of their dorsal arms; *Ocythoe tuberculata* Rafinesque, 1814, is ovoviviparous, with the eggs developing in the oviducts until hatching. In some cirrate octopods, egg laying is apparently continuous throughout most of the lifespan, as has been suggested for *Opisthoteuthis* spp., based on observations that continuous egg production occurs in the ovary, growth continues after the onset of maturity and oviducal-gland fullness is not related to body size (Boyle and Daly, 2002; Villanueva, 1992). Moreover, females of these species are often found with a single mature egg in the distal region of the oviduct; this egg, ready to be spawned (most probably individually), is encapsulated in a sticky substance secreted by the oviducal gland that becomes a rigid shell when exposed to sea water (Vecchione et al., 1998b). Although Aldred et al. (1983) found sperm packets inside the oviducal gland of *Cirrothauma murrayi* Chun, 1911, subsequent surveys of the oviducal gland of several specimens of *Opisthoteuthis* failed to find stored sperm (Boyle and Daly, 2002), which would be expected in a continuous-spawning strategy in the deep sea. *Vampyroteuthis infernalis* apparently releases individual eggs in midwater (Nesis, 1995; Pickford, 1946, 1949).

Cephalopods with synchronous ovulation ("simultaneous terminal spawning" *sensu* Rocha et al., 2001) accumulate ripe ova in the ovary and presumably spawn all their eggs at once or within a short period of time. Females of some synchronously ovulating squids have been observed to carry their egg masses in pelagic waters; this phenomenon is known as "post-spawning egg care" or "egg brooding" and has been observed for *Gonatus onyx* Young, 1972 (Seibel et al., 2005), *Gonatus madokai* Kubodera and Okutani, 1977 (Bower et al., 2012), *Gonatopsis* spp. (Okutani et al., 1995) and *Bathyteuthis berryi* Roper, 1968 (Bush et al., 2012). Remains of egg masses have also been found on trawl-captured specimens of *Gonatus fabricii* and *Gonatus antarcticus* Lönnberg, 1898, suggesting that these species likely also brood their eggs (Arkhipkin and Bjørke, 1999; Hoving, 2008; Laptikhovsky et al., 2007). Deep-sea benthic incirrate octopods produce large eggs (10–30 mm) and have low fecundity (up to a few hundred oocytes) (e.g. Barratt et al., 2007; Laptikhovsky, 2013). Although some studies have suggested that some deep-sea incirrate octopods could possibly be multiple spawners (Barratt et al., 2007), Laptikhovsky (2013) recently demonstrated, by identifying the postovulatory follicles, that females of deep-sea incirrates probably spawn a single egg mass simultaneously or a series of several small batches over a short period of time (1–3 weeks). The same study suggested that the evolution of this strategy among deep-sea incirrates might be associated with the costs of parental care, a common strategy within these cephalopods (Voight, 2008; Voight and Grehan, 2000).

Deep-sea cephalopods produce a wide range of egg sizes and numbers (e.g. Laptikhovsky, 2013; Laptikhovsky et al., 2007, 2008), with deep-sea and polar incirrates at the high end of the size spectrum, having among the largest eggs known for any invertebrate (generally 10–30 mm, Laptikhovsky, 2013; up to 41.5 mm in length; Allcock et al., 2003). Large squids like *Architeuthis* and *Dosidicus* produce millions of small eggs (~2 mm and ~1 mm, respectively) and are repeated spawners, having the highest fecundities of all cephalopods (e.g. Hoving et al., 2004; Nigmatullin and Markaida, 2009). Among deep-sea cephalopods, pelagic brooding species (squids and octopods) and benthic cephalopods (octopods) produce larger eggs than pelagic repeated spawners. This trend also seems to be true within families. For example, *Heteroteuthis dispar*, a pelagic squid from the family Sepiolidae (which has mainly truly benthic species), produces significantly more and smaller eggs than similarly sized and even larger species in the family (Hoving et al., 2008b). Similarly, pelagic octopods like *Argonauta* produce more and smaller eggs than *Muusoctopus* spp., a genus of benthic octopod species from the same suborder (Laptikhovsky, 2001; Laptikhovsky and Salman, 2003). Cirrate octopods have intermediate sizes and number of eggs, compared to *Muusoctopus* and *Argonauta* (Laptikhovsky, 2001; Laptikhovsky and Salman, 2003), which may be related to their continuous-spawning strategy (Villanueva, 1992). Egg size and fecundity seem inversely correlated, and fecundity is related to body size, with cephalopods only capable of spawning as many eggs as the coelomic cavity can hold. Therefore, repeated spawners can increase their actual fecundity compared to synchronous spawners (Laptikhovsky et al., 2007).

5.2. Age analysis and longevity

Cephalopod longevity has implications for many other aspects of these animals' biology and evolution, but the impracticality of making direct observations in most instances has required the development of other means for inferring life-cycle length.

One technique is to keep individuals alive in the laboratory; this approach is very valuable but also very difficult, especially for deep-sea taxa. To date, to our knowledge, only one study has successfully tracked the growth rate of a living deep-sea cephalopod in the lab: O'Dor and Malacaster (1983) maintained *Bathypolypus arcturi* and reported low growth-rate estimates (0.18 % body mass/day) and a relatively high longevity (>4 years) for this species compared to other neritic octopods.

The age of an individual cephalopod may be determined postmortem by quantifying increments in its hard body structures. By determining the ages of individuals of a wide variety of sizes, it is possible to reconstruct growth curves and estimate longevity. Structures useful for age determination of individual cephalopods must meet three criteria: (1) they must contain "interpretable increment structures that are sufficiently clear to facilitate precise interpretation," (2) "the increment structure is able to be correlated with a regular and determinable time scale," and (3) the incremented parts continue "to grow at a measurable rate throughout the life of the cephalopod" (Beamish and McFarlane, 1983 from Arkhipkin, 2005). In cephalopods, statoliths, gladii, stylets, beaks and eye lenses are all structures that meet the aforementioned criteria and will be discussed in the succeeding text.

Statoliths (calcium carbonate concretions located in the statocysts, the cephalopod's organs of balance) are the most widely used structures to estimate age at size in squid. Hurley et al. (1979) and Lipinski (1979) provided the first evidence of daily deposition of statolith increments, confirming their potential as proxies for individual squid age. Their periodic growth, which is correlated with growth of the animal, results in the formation of increments, which can be counted using microscopy. However, the periodicity of statolith increment formation requires validation (e.g. Lipinski, 1986; Lipinski et al., 1998); at present, deposition is known to occur daily in several ommastrephid species (Arkhipkin, 2004), in the oceanic squid *Onychoteuthis borealijaponica* Okada, 1927 (Bigelow, 1994) and in juveniles of the oceanic squid *Abralia trigonura* Berry, 1913 (Bigelow, 1992). These species comprise a very small proportion of the known oegopsid squid taxa (24 families, 69 genera and 206 currently described species; ToLweb Young and Vecchione, 2004), but this is not surprising as validation studies, although necessary, require husbandry of living deep-sea cephalopods.

Oceanic squid age estimations based on statolith increments suggest a wide range of lifespans for different taxa (see Arkhipkin, 2004 for review). Small squids of the families Enoploteuthidae and Pyroteuthidae appear to have life cycles that are completed in less than 6 months (Arkhipkin,

1996a; Young and Mangold, 1994). Larger enoploteuthids and small ommastrephids (8–15 cm ML) also live for approximately half a year (references in Arkhipkin, 2004). Larger ommastrephids (*Sthenoteuthis pteropus* Steenstrup, 1855, and *S. oualaniensis* Lesson, 1830) and *Thysanoteuthis rhombus* each appear to have a 1-year lifespan (Arkhipkin and Mikheev, 1992; Nigmatullin et al., 1995).

Other oceanic squids seem to live longer and grow more slowly. Female *Ancistrocheirus lesueuri* mature at about 2 years and reach a mantle length of ~400 mm (Arkhipkin, 1997). On the South Atlantic continental slope of South Africa, female *Lycoteuthis lorigera* reach maturity around 11 months (ML 100 mm) and have been estimated to live for 1 year (Hoving et al., 2007), while *Histioteuthis miranda* Berry, 1918, from the same area may live for up to 2 years (Hoving and Lipinski, 2009).

While counting statolith increments can be a powerful tool for investigating the life-cycle length—and other aspects of ecology (migration, ontogeny, etc.; Arkhipkin, 2005) of some deepwater squid—their interpretation has also resulted in some questionable age estimations. If daily deposition occurs in the giant squid, *Architeuthis*, the total number of statolith increments suggests extremely high growth rates and ages of just 1–2 years for large animals of 975–2168 mm ML (e.g. Gauldie et al., 1994; Lordan et al., 1998; Roeleveld and Lipinski, 1991; review O'Shea, 2004a). Age estimations of *Architeuthis* based on other body structures (eye lenses and gladii), or using theoretical predictions, suggest a longer lifespan (Grist and Jackson, 2007; O'Shea, 2004a). The one-increment-per-day hypothesis should thus be tested for as many squid species as possible, in particular from different families.

In some cephalopods, statoliths are not suitable for quantification of increments. In *Vampyroteuthis infernalis*, for example, the statolith is not a solid concretion, but just crystals loosely clinging together (H.J. Hoving, personal observation). In octopods, the statoliths appear to have no increments and are soft and "chalklike" (Clarke, 1978), so other structures should be investigated for suitability in age determination.

The shell in squid, the gladius (or "pen"), provides quantifiable increments in both juvenile and post-juvenile phases of some species (Arkhipkin and Perez, 1998). Validation experiments have shown that these increments (in the ostracum) are also laid down daily in some oceanic squid species (e.g. *Illex*, Perez et al., 1996; Schroeder and Perez, 2013), but it has not yet been possible to calculate total age. One advantage of using the gladius for age estimation is that the structure size is directly related to the size of the squid, allowing reconstruction of individual growth rates and trajectories. These have allowed, for example, the identification of important growth shifts during the life history of *Illex illecebrosus*, as the animal transitions from oceanic to neritic environments in the Northwest Atlantic (Perez and O'Dor, 2000). Conversely, a disadvantage is that early increments are subsequently overgrown and can therefore be unreadable. The gladius is also a good alternative when statolith increments become faint and unreadable, a process that happens particularly in larger squid (e.g. *Onykia robusta* (Verrill, 1876); Bizikov and Arkhipkin, 1997). In *Architeuthis* from New Zealand, six large increments in the gladius have been found, which could potentially indicate an age of 6 years (O'Shea, 2004a).

Incirrate octopods do not have gladii; some species instead have vestigial shell remnants called stylets, which are two very fine, elongate structures in the mantle muscle at the base of the gills (Bizikov, 2008; Wells, 1978). Concentric rings have been observed in sectioned stylets, and experiments where individuals were exposed to a fluorescent marker that becomes incorporated into the stylet indicate that these, too, form daily (Doubleday et al., 2006; Hermosilla et al., 2010). However, assuming daily deposition of stylet increments, specimens of the deepwater octopod *Bathypolypus sponsalis* were estimated to be <1 year old (Barratt and Allcock, 2010), which does not agree with the longer lifespan (3–4 years) suggested by laboratory-kept individuals (O'Dor and Malacaster, 1983). A very large deepwater octopod, *Megaleledone setebos* Robson, 1932, was estimated to attain an age of 3–4 years (potentially including embryonic development) based on the number of increments in the stylets (Barratt and Allcock, 2010).

Doubleday and Semmens (2011), using the shallow-water octopod *Octopus pallidus* Hoyle, 1885, found that the age pigment lipofuscin (visible in brain tissue using standard histological techniques) is highly exponentially correlated with age in laboratory-reared animals. Because the age determined by stylets corresponded well with the age estimations from lipofuscin quantification, this method may have potential as an alternative ageing method for octopus and perhaps other deep-sea cephalopods.

Structures that may prove particularly valuable in ageing cephalopods, but which have received relatively little scientific attention thus far, are the growth increments in cephalopod beaks. Beaks are present in all cephalopods, and their species-specific morphology can be used for identification (Clarke, 1986). Daily deposition of increments in the beaks has been validated for adult *Octopus vulgaris* Cuvier, 1797 (Canali et al., 2011; Oosthuizen, 2003; Perales-Raya et al., 2010), and this method has now been successfully applied to assess the growth differences in wild populations of *O. vulgaris* (Perales-Raya et al., 2013). Quantification of increments in the beaks of squid for the purpose of age estimation has only been studied for the deep-sea squid *Onykia ingens* (Clarke, 1965). Growth increments have been observed in the beaks of the deep-sea squids *Liocranchia reinhardtii* Steenstrup, 1856, and *Chiroteuthis* cf. *veranyi* Ferussac, 1835 (Franco-Santos and Vidal, 2014). However, one potential drawback to using beaks for ageing is the fact that feeding causes structural wear, especially in voraciously predatory cephalopods.

Age estimation could also be accomplished using the cephalopod eye lens, which develops before hatching and grows throughout the individual's life, and shows visible growth increments in cross section (Baqueiro Cárdenas et al., 2011; O'Shea, 2004a). In *Architeuthis*, the total number of rings in one eye lens suggested (assuming daily formation) an age of 6.1 years in an individual of ML 1845 mm (O'Shea, 2004a). The low variation found in eye lens growth in *O. vulgaris*, contrary to highly variable mantle length/body mass relationships, suggests that it may also be a suitable structure to use for age estimations in octopods (Baqueiro Cárdenas et al., 2011).

In addition to providing insight into overall longevity, some of these structures may also allow more specific life-history inferences to be drawn. For example, microstructure of the statolith can sometimes suggest the duration of various life-cycle stages. Species of the family Cranchiidae have been suggested to spend the first part of their life (~6 months) in epi- and mesopelagic waters, after which they descend into the bathypelagic to mature (Arkhipkin, 1996b,c). *Gonatus onyx* lives at least 2 years before reproducing (Hunt and Seibel, 2000) in the bathypelagic, and then females brood the eggs (whose development is slow, due to low temperatures at bathypelagic depths) for up to 6–9 months (Seibel et al., 2005). Therefore, the brooding time, combined with maturation time, suggests a total lifespan of approximately 3 years, at least for females. Using the model on cephalopod embryogenesis developed by Laptikhovsky (2001), the development time for *Gonatus fabricii* eggs was estimated to be 12 months, based on the size of the eggs (approximately 4–6 mm) and the temperature of Arctic bathypelagic waters (−0.7 to 0.8 °C) (Arkhipkin and Bjørke, 1999). Similarly, *Bathyteuthis berryi* embryos are estimated to require 12 months to reach a TL of 5 mm at 2–4 °C (Bush et al., 2012). These findings indicate that the brooding times for some bathypelagic cephalopods are longer than the complete life cycles of some coastal and epipelagic squids (e.g. Jackson, 2004).

Other indications of deepwater cephalopods' longevity can be derived from chemical analyses of body tissues. The digestive gland of *Architeuthis* from the Bay of Biscay showed increased levels of cadmium, compared with other smaller mesopelagic squids and even higher compared to neritic species (Bustamante et al., 2008). Such levels could mean that *Architeuthis* preys upon taxa that are more contaminated than the prey of other mesopelagic squid, or that giant squid live longer than other cephalopods, allowing higher concentrations of contamination to accumulate (Bustamante et al., 2008). Isotopic analysis of body tissues from Tasmanian giant squid has suggested an age of 14 years (ML 1910–2400 cm) (Landman et al., 2004).

5.3. Trophic ecology

Oceanic squids of the families Ommastrephidae, Gonatidae and Onychoteuthidae are known for their versatile predatory feeding behaviour. Their high metabolic requirements, the generally short life cycle and the semelparous life-history strategy demand a high nutritional intake (Rodhouse and Nigmatullin, 1996). Therefore, it is not surprising that these animals feed on a wide variety of prey from pelagic crustaceans to fish to other cephalopods, which they may catch at night in surface waters after undertaking diel migrations from deeper strata (Table 3.5). This opportunism often includes cannibalism, which sometimes forms a significant dietary component in species such as *Dosidicus gigas* (Ibañez and Keyl, 2010). Cannibalism is often a response to stress (Ibañez et al., 2008), particularly during fishing operations (Markaida and Sosa-Nishizaki, 2003), and so may be overestimated in some cases, but it also appears to emerge as a natural strategy when food is scarce (Maurer and Bowman, 1985). Smaller examples of opportunistic predators likely include members of the enoploteuthid families. For example, *Watasenia scintillans* is known to prey upon crustaceans and pelagic fish (Hayashi, 1995).

Since these nektonic predators may hunt within the water column as well as near (or even on) the sea floor, versatility is required in body morphology, particularly structures involved in prey capture and feeding. The agility and strong swimming abilities of many oceanic cephalopods allow the capture of fast-moving prey (O'Dor and Shadwick, 1989), and the brachial crown (arms and tentacles) in many squid species also allows for the efficient capture of a wide range of prey sizes and types (Packard, 1972). A tremendous morphological diversity has also evolved among the armature of these structures (sucker rings and hooks), likely facilitating the capture and handling of prey (Engeser and Clarke, 1988; Nixon and Dilly, 1977).

Table 3.5 Diets of selected deep-sea cephalopods

Higher classification	Species	Size (mm) (male/ female)	Dietary components	Reference
Order Octopoda				
Suborder Incirrata				
Family Amphitretidae	*Japetella diaphana*	100	Crustaceans (euphausiids, copepods)	Passarella and Hopkins (1991)
Family Alloposidae	*Haliphron atlanticus*	4000 TL	Crustaceans, cnidarians	Willassen (1986) and O'Shea (2004b)
Family Octopodidae	*Graneledone boreopacifica*	145	Molluscs, crustaceans	Voight (2000)
	Megaleledone setebos	280	Ophiuroids, amphipods, fish	Piatkowski (unpublished)
	Pareledone charcoti	70	Crustaceans mainly	Piatkowski (unpublished)
	Pareledone turqueti	120	Benthos	Piatkowski (unpublished)
	Vulcanoctopus hydrothermalis	50	Polychaetes, amphipods, other crustaceans	González et al. (1998), Rocha et al. (2002), and Voight (2005)
Suborder Cirrata				
Family Opisthoteuthidae	*Opisthoteuthis californiana*	90	Small crustaceans, polychaetes	Pereyra (1965)
	Opisthoteuthis calypso	100	Crustaceans, polychaetes, gastropods	Villanueva and Guerra (1991)
	Stauroteuthis syrtensis	350	Copepods (mostly *Calanus finmarchicus*)	Jacoby et al. (2009)
Order Sepiolida				
Family Sepiolidae	*Rossia macrosoma*	85	Benthos, shrimp	Mangold-Wirz (1963)

Continued

Table 3.5 Diets of selected deep-sea cephalopods—cont'd

Higher classification	Species	Size (mm) (male/ female)	Dietary components	Reference
Order Teuthoidea				
Family Architeuthidae	*Architeuthis dux*	<3000	Fish, squid, crustaceans	Peréz-Gándaras and Guerra (1978), Förch (1998), Lordan et al. (1998), and Bolstad and O'Shea (2004)
Family Chiroteuthidae	*Chiroteuthis veranyi*	200	Pelagic small crustaceans, mollucs, fish	Kubota et al. (1981)
	Grimalditeuthis bonplandi	250	Cephalopods, crustaceans	Hoving et al. (2013b)
Family Cranchiidae	*Galiteuthis glacialis*	<190	Zooplankton, euphausiids (Antarctic krill)	Rodhouse and Piatkowski (1995)
Family Enoploteuthidae	*Abralia redfeldi*	30	Copepods, euphausiids	Passarella and Hopkins (1991)
	Abraliopsis atlantica	33	Copepods, euphausiids	Passarella and Hopkins (1991)
	Watasenia scintillans	70/60	Shrimp, planktonic crustaceans, pelagic fish	Hayashi (1995)
Family Gonatidae	*Berryteuthis magister*	430	Fish, squid, pelagic crustaceans, large zooplankton	Kuznetsova and Fedorets (1987) and Nesis (1997)
	Berryteuthis anonychus	150	Crustacean zooplankton, fish	Uchikawa et al. (2004)
	Gonatopsis borealis	300/280	Pelagic crustaceans, fish, squid, zooplankton	Reviewed in Roper et al. (2010a,b)

Table 3.5 Diets of selected deep-sea cephalopods—cont'd

Higher classification	Species	Size (mm) (male/female)	Dietary components	Reference
	Gonatus fabricii	385	Fish, pteropods, chaetognaths, cephalopods, crustaceans	Kristensen (1984) and Wiborg et al. (1984)
Family Histioteuthidae	*Histioteuthis bonnellii*	330	Mysids, shrimp, zooplankton, mesopelagic fish	Quetglas et al. (2010) and Fanelli et al. (2012)
	Histioteuthis reversa	200	Myctophid fish, pelagic crustaceans, cephalopods	Quetglas et al. (2010)
	Histioteuthis miranda	260/270	Crustaceans	Clarke (1980)
	Histioteuthis corona corona	190/170	Copepods and euphausiids	Passarella and Hopkins (1991)
	Histioteuthis celetaria pacifica	280/230	Crustaceans and fish	Voss (1962) from Voss et al. (1998)
	Histioteuthis arcturi	130/210	Copepods and euphausiids	Passarella and Hopkins (1991)
Family Lycoteuthidae	*Lycoteuthis lorigera*	190	Pelagic crustaceans and myctophids	Voss (1962) and Lipinski (1992)
Family Ommastrephidae	*Dosidicus gigas*	1200	Epi- and mesopelagic fish, squid, shrimp, euphausiids, cannibalism	Braid et al. (2012) Nigmatullin et al. (2001), Markaida (2006, 2008), and Hoving et al. (2013a)
	Hyaloteuthis pelagica	105	Juvenile pelagic fish and squid, crustaceans	reviewed in Roper et al. (2010a,b)

Continued

Table 3.5 Diets of selected deep-sea cephalopods—cont'd

Higher classification	Species	Size (mm) (male/female)	Dietary components	Reference
	Illex argentinus	400	Fish, squid, crustaceans (mostly euphausiids)	Ivanovic and Brunetti (1994) and Laptikhovsky (2002)
	Illex coindetii	379/279	Fish, crustaceans, cephalopods, gastropods	Rasero et al. (1996) and Sánchez et al. (1998)
	Illex illecebrosus	340/270	Mostly fish and pelagic crustaceans	Froerman (1984) and O'Dor and Shadwick (1989)
	Martialia hyadesi	400	Euphausiids, mesopelagic fish	González et al. (1997) and Dickson et al. (2004)
	Nototodarus sloanii	420	Crustaceans (euphausiids), fish, squid	reviewed in Roper et al. (2010a,b)
	Ommastrephes bartramii	420/900	Squid, fish (mostly myctophids), shrimp, cannibalism	Lipinski and Linkowski (1988), Seki (1993), and Watanabe et al. (2004)
	Ornithoteuthis antillarum	300	Crustaceans, fish, chaetognaths, squid	Arkhipkin et al. (1998)
	Sthenoteuthis oualaniensis	650	Mostly pelagic fish (myctophids), squid, shrimp	Shchetinnikov (1992), Snyder (1998), and Parry (2006)
	Sthenoteuthis pteropus	650	Mostly pelagic fish (myctophids), squid, shrimp	Reviewed in Zuyev et al. (2002)

Table 3.5 Diets of selected deep-sea cephalopods—cont'd

Higher classification	Species	Size (mm) (male/female)	Dietary components	Reference
	Todarodes angolensis	433	Mostly fish	Reviewed in Roper et al. (2010a,b)
	Todarodes filippovae	540	Fish, crustaceans, cephalopods	Reviewed in Roper et al. (2010a,b)
	Todarodes pacificus	500	Planktonic crustaceans, fish, squid	Okiyama (1965), Okutani (1983) and Song et al. (2006)
	Todarodes sagittatus	750/640	Fish, crustaceans, squid	Piatkowski et al. (1998), Quetglas et al. (1999), Lordan et al. (2001), and Fanelli et al. (2012)
Family Onychoteuthidae	*Ancistroteuthis lichtensteinii*	300	Fish and crustaceans (euphausiids)	reviewed in Roper and Jereb (2010) and Fanelli et al. (2012)
	Onychoteuthis banksii	370	Fish and squid	Arkhipkin and Nigmatullin (1997)
	Onychoteuthis borealijaponica	370/300	Small fish, squid	Okutani and Murata (1983)
	Onykia ingens	520	Euphausiids (Antarctic krill), mesopelagic fish, squid	Jackson et al. (1998) and Phillips et al. (2003a,b)
	Onykia robusta	2300	Benthic fish, echinoderms, jellyfish	Reviewed in Roper and Jereb (2010)

Continued

Table 3.5 Diets of selected deep-sea cephalopods—cont'd

Higher classification	Species	Size (mm) (male/female)	Dietary components	Reference
Family Octopoteuthidae	*Taningia danae*	1700	Fish crustaceans	Santos et al. (2001) and González et al. (2003)
Family Pyroteuthidae	*Pterygioteuthis gemmata*	40	Copepods, euphausiids	Passarella and Hopkins (1991)
	Pyroteuthis margaritifera	50	Copepods, euphausiids	Passarella and Hopkins (1991)
Family Thysanoteuthidae	*Thysanoteuthis rhombus*	1300	Fish, squid	Nigmatullin and Arkhipkin (1998)
Order Vampyromorphida				
Family Vampyroteuthidae	*Vampyroteuthis infernalis*	22–210	(Remains of) gelatinous zooplankton, crustaceans (copepods, amphipods), detritus, remains of fish and squid	Hoving and Robison (2012)

Opportunistic feeding behaviour in some oceanic squids has also become apparent from stable isotope analysis (SIA) of body tissues. In SIA, an organism's carbon and nitrogen stable isotope signatures are assumed to be influenced by what it has consumed (e.g. Kelly, 2000). In the marine environment, stable isotope ratios of carbon ($\delta^{13}C$) and nitrogen ($\delta^{15}N$) are indicators of the main feeding areas and trophic levels of consumers, respectively (Cherel and Hobson, 2005). Stable isotope analysis can also provide information on the habitat in which the individual has grown and hence yields clues to migration behaviour and habitat utilization (e.g. Cherel and Hobson, 2005; Jackson et al., 2007; Ruiz-Cooley et al., 2013).

In cephalopods, SIA has been performed on various body structures, including the mantle, beaks, gladii, arms, eye lenses and digestive glands (e.g. Cherel and Hobson, 2005; Jackson et al., 2007). Using SIA on *Dosidicus gigas* muscle tissue and gladii, for example, small individuals have been found to migrate into the northern California Current System from different regions to feed and grow (Ruiz-Cooley et al., 2013). High-resolution sampling of metabolically inactive body structures like gladii ("with no metabolic turnover after synthesis" as per Lorrain et al., 2011) may provide other information on dietary preferences and trophic position throughout ontogeny. In another study on *D. gigas*, stable isotope signatures were analysed along the gladius, revealing that this species does not necessarily increase its trophic level with age, showing that individual squid have a highly varied diet, are flexible in their feeding strategy, and migrate actively (Lorrain et al., 2011). A potentially unusual trend was identified using cephalopod beaks retrieved from stomachs of dead emperor penguin chicks at Pointe Géologie, Terre Adélie, Antarctica: varying $\delta^{15}N$ signatures in the onychoteuthid *Kondakovia longimana* showed that larger individuals of this squid tended to source prey from lower in the food web than smaller individuals, suggesting a dietary shift from fish to euphausiids during growth (Zimmer et al., 2007).

Fatty-acid analysis is another technique that can reveal valuable information on trophic position and diet but that has rarely been used for deep-sea cephalopods to date (Rosa et al., 2013b). Fatty-acid analysis utilizes the fact that certain fatty acids can only be biosynthesized by phytoplankton and macroalga species, which have distinct ratios in fatty acids that influence the ratios of fatty acids in consumers and thus give information on position in the food web (Graeve et al., 2002). Individual fatty acids or groups of fatty acids may also be related to prey type or species. The combination of fatty-acid analysis and stomach-content analysis can enable detection of temporal and spatial variations in feeding and also recognition of specific prey items that cause temporal differences in uptake of dietary lipids. In the onychoteuthid *Onykia ingens*, fatty acids analysed from the digestive gland have revealed dietary differences among regions, seasons and years, which were attributed to a dietary change from myctophids to crustaceans (Phillips et al., 2003a,b). In another study, distribution and lipid composition were studied in the Antarctic cranchiid squid *Galiteuthis glacialis* Chun, 1906 (Piatkowski and Hagen, 1994). Storage lipids (triacylglycerols) were found to be low, comprising 18–26% of total lipid, which in turn made; the latter made up only 8–11% of the dry weight, which reflected the early developmental stage of the specimens examined.

In many oceanic cephalopod families, some or all species accumulate ammonium in their tissues to increase buoyancy (see Voight et al., 1994 for review); these are called "ammoniacal cephalopods." Although some ammoniacal cephalopods are obviously very strong swimmers (e.g. Octopoteuthidae), many ammoniacal species have reduced metabolic rates and swimming capabilities (e.g. Seibel and Drazen, 2007; Seibel et al., 1997, 2004). Histioteuthids, an ammoniacal family, are known to ingest typical cephalopod prey like crustaceans, fishes and cephalopods (Table 3.5) but overall diet information on ammoniacal squid is virtually absent.

Species that certainly have a less active lifestyle are the cephalopods that have adopted an ambush-predatory (or "sit-and-wait") feeding strategy. Squids of the family Chiroteuthidae have large, robust ventral arms, with sheaths into which the very long tentacles that can be retracted (Young and Roper, 2011c). The large tentacle club is equipped with many small suckers and a distal photophore; photophores are also present along the length of the tentacle stalk. In *Chiroteuthis calyx* Young, 1972, the tentacle has been observed to be deployed and retracted, presumably to capture midwater fish (Robison, 2004).

In the same family, *Grimalditeuthis bonplandi* has tentacles that presently appear to be unique among cephalopods: the stalks are very thin and the clubs bear no armature or photophores (Hoving et al., 2013b; Young et al., 1998). ROV observations show that the club swims out from the brachial crown by undulating movements and, when positioned at some distance from the squid, is manoeuvred in such a way that it resembles the movement of a small organism. This apparently lures other cephalopods and crustaceans to within reach of the arms (Hoving et al., 2013b), demonstrating a feeding strategy known as aggressive mimicry or luring (Wickler, 1965). Mastigoteuthids, the "whiplash" squids, have been observed by ROV to hover above the sea floor with their tentacles extended just millimetres off the bottom (Roper and Vecchione, 1997). The tentacles of most species bear numerous tiny (sometimes microscopic) suckers, which are presumably used to capture copepods or perhaps larger prey (Roper and Vecchione, 1997). Magnapinnids, the "bigfin" squids, are an elusive family of deep-sea cephalopods that have only been observed below 2000 m. These squids possess exceedingly long, slender extensions of all arms, which are deployed below the squid and may be several metres long (Vecchione et al., 2001), and appear to be highly adhesive. The function of this modification is unknown but may well be related to feeding.

Deep-sea octopods also show a great variety of feeding strategies. A comprehensive review of the main prey found in stomachs of cirrate octopods was provided by Collins and Villanueva (2006, their table 7). Small crustaceans, such as gammarid amphipods, mysids, copepods, isopods and decapods, and polychaetes and bivalves are the most abundant prey items (e.g. Collins, 2003; Collins and Villanueva, 2006; Vecchione and Young, 1997; Villanueva and Guerra, 1991). Most of these studies concluded that epibenthic and benthopelagic feeding seem to be very common (Vecchione, 1987; Villanueva and Guerra, 1991). Additionally, fatty-acid analysis has revealed a benthic detrital feeding strategy in the cirrate *Opisthoteuthis calypso* Villanueva, Collins, Sanchez and Voss, 2002 (Rosa et al., 2013b). The only direct observations on the feeding behaviour of cirrate octopods were reported from laboratory studies of *Grimpoteuthis* sp. (Hunt, 1999), which exhibited three distinct feeding modes in the laboratory: envelopment of the prey (*Artemia* nauplii), entrapment and current feeding by movement of the cirri. The large pelagic incirrate octopus *Haliphron atlanticus* apparently feeds on crustaceans and cnidarians (O'Shea, 2004b; Willassen, 1986). Among the bottom-dwelling incirrate deep-sea octopods, *Graneledone* cf. *boreopacifica* and *Vulcanoctopus hydrothermalis* prey upon gastropods, polychaetes and crustaceans (González et al., 1998; Rocha et al., 2002; Voight, 2000), with the latter species also feeding on swarming amphipods by engulfing with the arms, a strategy apparently specialized to target one particular species of amphipod *Halice hesmonectes* (Voight, 2005). Another deep-sea octopod that seems to specialize in a single prey item is *Stauroteuthis syrtensis* Verrill, 1879, which was found to prey solely on the copepod *Calanus finmarchicus* (Jacoby et al., 2009) (Table 3.5).

A particularly peculiar feeding behaviour has recently been described for *Vampyroteuthis infernalis*. This species was observed (by ROV) to extend its retractile filament, which may be up to 9 × the body length (Hoving and Robison, 2012). Under laboratory conditions, food particles were found to adhere to this filament, which is equipped with microscopic hairs; the food-laden filament was then withdrawn between the arms and the food was removed and later observed as a clump in the beak (Hoving and Robison, 2012). Additional ROV observations repeatedly showed vampire squid with clumps of food in their beaks. Stomach-content analysis revealed a variety of zooplankton, crustacean moults, faecal pellets from copepods, amphipods and larvaceans, gelatinous material and incidental fragments of fish and squid that appeared to have been scavenged, based on isolated occurrence (Hoving and Robison, 2012). Larger macerated prey items such as fishes or crustaceans, which are typical of many other deep-sea cephalopods and require active hunting, were never observed. It therefore appears that vampire squid feed on already-dead, decomposing material from the water column and may opportunistically ingest copepods and other crustaceans, either dead or alive. Recent studies suggest that the ram's horn squid *Spirula spirula* is another cephalopod that ingests detrital material and marine snow (Ohkouchi et al., 2013); as mentioned in the preceding text, a third instance of benthic detrital feeding has been reported in the cirrate *Opisthoteuthis calypso* (Rosa et al., 2013b).

Although SIA and fatty-acid analysis have proven to be valuable tools in dietary analysis, they may lead to confusing results when used without other lines of evidence. For example, SIA on cephalopod beaks sourced from stomach contents of various predators from the subantarctic Crozet and Kerguelen archipelagos and Adélie Land, Antarctica, revealed a cephalopod diversity of 18 species spanning three trophic levels (Cherel and Hobson, 2005). Interestingly, the largest cephalopods did not necessarily occupy the highest trophic position, that is, body size was independent from the stable isotope ratios of nitrogen. For example, *Chiroteuthis veranyi* was positioned higher in the food web than the giant squid *Architeuthis dux* (Cherel and Hobson, 2005). SIA should therefore preferably be used in combination with stomach-content analysis. Visual identification of stomach contents, however, is time consuming and difficult; cephalopods macerate their food considerably, sometimes making identification nearly impossible.

DNA sequencing of prey tissue recovered from stomachs may help with this problem. Although this method is rather elaborate and a library of potential prey items needs to be available for comparison, when used, it can reveal interesting insights into deep-sea cephalopod diet, for example, cannibalism in giant squid (Deagle et al., 2005, confirming earlier reports by Bolstad and O'Shea, 2004) and *Vampyroteuthis infernalis* ingesting pieces of *Gonatopsis borealis* (Hoving and Robison, 2012). The advantage of direct identification (visual or via DNA sequencing) is that the results provide an instant insight into recently consumed food items. Stomach-content analysis, however, may be biased by the phenomenon of "net feeding," where cephalopods under stress begin consuming any nearby organism (Markaida and Sosa-Nishizaki, 2003; Rodhouse and Nigmatullin, 1996). Therefore, stomach-content analysis of deep-sea cephalopods that have been captured by ROV (e.g. Hoving and Robison, 2012) may provide a more unbiased source of dietary data and should be investigated where possible in future studies.

The diets of many species are likely to change ontogenetically, but relatively few studies have compared diets across different life stages to date.

The limited available data have shown interesting and dramatic changes in dietary preferences and morphology of feeding-associated structures. In *Architeuthis*, δ^{15}N profiles of beaks have indicated an early-life dietary shift, although the exact diet composition of early life stages of *Architeuthis* remains unknown (Guerra et al., 2010). Stable isotope analysis on eye lenses of the gonatid *Berryteuthis magister* (Berry, 1913) revealed an increase by one trophic level from juvenile to adult stages (Hunsicker et al., 2010). In addition to these dietary shifts, major morphological modifications have also been observed in some species. For example, in the cranchiid *Galiteuthis glacialis*, probable adaptive allometry of the brachial crown has been discussed as a response to the general increase of pelagic prey size with depth (Rodhouse and Piatkowski, 1995). Doratopsid paralarvae of the Chiroteuthidae differ radically from the subadult and adult forms (Sweeney et al., 1992; Vecchione et al., 1992), and the mechanics and function of this metamorphosis remain unknown. Interestingly, one exception seems to be the vampire squid, whose feeding apparatus in the juvenile form is almost identical to that of the adult (Young and Vecchione, 1999).

Forage locations (benthic vs. pelagic or geographical area) adopted during different cephalopod life stages may also be detected via SIA. Pelagic cephalopod species (e.g. histioteuthids and *Vampyroteuthis infernalis*) are reported to have carbon-depleted tissues and structures, while benthic cephalopods (e.g. *Bathypolypus sponsalis*) are generally enriched in carbon (Cherel et al., 2009; Fanelli et al., 2012). Two of the 18 species examined in Kerguelen waters (*Architeuthis* and the brachioteuthid *Slosarczykovia circum-antarctica* (Lipinski, 2001) were found to have spent considerable time in waters outside Kerguelen (Cherel and Hobson, 2005).

Large deep-sea squid are rarely collected or encountered in adequate numbers to achieve a good representative idea of their feeding strategies. However, the piecemeal insights that has been reported provides some interesting information. One individual of the poorly known mastigoteuthid *Idioteuthis cordiformis* (Chun, 1908) had remains of snapper (*Lutjanus* sp.) and birdbeak dogfish (*Deania calcea*) in its stomach (Braid, 2013). *Onykia robusta* has been reported to feed on benthic fish and echinoderms, as well as on the pleustonic jellyfish *Velella velella* (from Jereb and Roper, 2010). Examined stomach contents of the giant squid *Architeuthis dux* have contained benthic prey such as octopods and the Norway lobster *Nephrops norvegicus*, among other pelagic prey such as whiting and squid (probably remains of histioteuthids, onychoteuthids and ommastrephids; Bolstad and O'Shea, 2004; Förch, 1998; Lordan et al., 1998; Peréz-Gándaras and

Guerra, 1978) (Table 3.5). Several specimens of the "colossal" squid *Mesonychoteuthis hamiltoni* have been captured by longliners targeting Patagonian toothfish (*Dissostichus eleginoides*), but whether this is its natural prey remains uncertain.

6. DISCUSSION AND CONCLUSIONS

Deep-sea cephalopods have been recognized and studied for hundreds of years, and although much information remains to be gathered, many techniques for researching these animals have been developed. Great advances have already been made in the twenty-first century, with a number of these species having been observed alive in their natural habitats for the first time within just the past decade. However, deep-sea cephalopods remain difficult to collect and observe, even with large deep-sea sampling gear like double-warped trawls and advanced observation equipment such as ROVs. Every sampling device has advantages and disadvantages, and collection tools should be used in combination where possible to obtain a more complete understanding of the true community composition of any deep-sea habitat. Tools that can only observe cannot replace nets as samples are required for validation of observations and also for further biological analysis. On the other hand observational tools are crucial for providing insight into behaviour and microhabitat and also for the collection of living cephalopods for experiments. Nets cannot do this. Ideal cephalopod survey methods use a combination of observational tools that can also collect (e.g. ROVs with collection devices) and nets that can also observe (e.g. cameras integrated onto nets). Predators can provide useful corroborative data on deep-sea cephalopods (e.g. distribution, role in the food web and the location and capture of large and sexually mature specimens in particular), but studies on teuthophagous animals' feeding ecology require input from experts on cephalopod systematics, especially when using beaks. Existing museum collections of cephalopods are another important resource for the research community, but only if there is awareness of their holdings and location. Rare specimens housed in these institutions may be used to substantiate otherwise small data sets, for example, for stomach-content analysis (e.g. *Vampyroteuthis infernalis* Hoving and Robison, 2012; *Grimalditeuthis bonplandi* Hoving et al., 2013b) and morphological and taxonomic studies. Museum specimens also have great value for testing hypotheses based on in-situ observations and vice versa (Hoving and Vecchione, 2012; Hoving et al., 2012). Archived in situ observations on deep-sea cephalopods via different initiatives are now complementing the vast number of museum specimens housed in worldwide collections, and it is important to note that the value of each is greatly increased by the other. Utilizing these two data sources in concert can greatly increase the knowledge gained from both, and this interplay of resources can only become stronger and more productive in the future.

The paucity of fossil cephalopods means that much information about the phylogeny and evolution of this class must be inferred from studies of extant groups. Once consistent systematic information is gained, processed and accepted for many recent deep-sea cephalopod taxa, researchers may be able to interpret fossil records more successfully and better understand some of the many unresolved evolutionary questions. To this end, scientific cruises using a variety of collection and monitoring techniques are needed in the areas where major gaps exist worldwide, such as the Arctic, southwest Atlantic, Indian and Pacific Oceans. Material from these regions would yield additional morphological and genetic information for known taxa (and would probably reveal new taxa), adding important data to our still-incomplete understanding of cephalopod biogeography and systematics. With a relatively small number of species, compared to other animal groups, it may even be feasible for the whole class to be eventually examined using both morphology and DNA sequencing—techniques that should be used in concert for systematic work wherever possible. Improvements in DNA extraction from formalin-fixed tissue, particularly the buccal mass, may also enable new data to be gathered from the vast museum holdings of recently and historically collected cephalopod specimens worldwide.

Like many habitats on Earth, the deep-sea environment is changing physically and chemically. Understanding the phylogenetic linkages among cephalopods is imperative in order to understand the physiological and behavioural patterns observed within family groups and to predict how they may react to large-scale global events such as climate change.

One of many exciting challenges in deep-sea cephalopod research is exploring the pelagic or "deep water." Only a fraction of this, the largest biome on Earth, has been explored (Ramirez-Llodra et al., 2010). In particular, the bathypelagic (the ocean layer between 1000 and 3000 m) may provide new insights into cephalopod biology. This habitat may hold the key to encountering and understanding mature life stages and egg masses of deep-sea cephalopods; brooding female squids have been observed here (Bush et al., 2012; Seibel et al., 2005) and, as so few egg masses have ever been observed, other nonbrooding species may spawn here too. However, fully understanding the biology of many deep-sea taxa requires an understanding of photic-zone communities as well, given the commonly observed life-history strategies of ontogenetic descent and diel vertical migration in these animals.

Recent insights into reproductive strategies, tactics and behaviours of deep-sea cephalopods have shown interesting patterns that are radically different from those of well-studied coastal species. One of the key remaining questions is whether (and to what extent) the evolution of reproductive strategies in deep-sea cephalopods has been driven by habitat-related pressures. Understanding the interplay of distinct selective pressures that potentially influence the evolution of cephalopod reproductive biology will certainly require an integrative approach, including studies on phylogeny, life history and genetics as well as experimentation.

Methods for investigating trophic position, longevity and age–size comparisons in neritic species should be applied to deep-sea cephalopods in order to investigate how their life cycles differ from their shallow-water counterparts. An open question in deep-sea cephalopod biology is whether (and in what taxa) growth increments in various hard body structures may provide a proxy for age. Basic data on size-at-age trends are lacking for the majority of squid species, in part because validation (which requires the husbandry of deep-sea cephalopods) is necessary for trustworthy age estimations. Fortunately, advances in collecting living cephalopods using ROVs have allowed for improvements in husbandry and experimental designs for investigating deep-sea cephalopods (e.g. Hoving and Robison, 2012; Hunt, 1999; Robison et al., 2003). Significant progress has also been made in techniques for culturing cephalopods in the last three decades, which could serve as the foundation for technological improvements in rearing deep-sea cephalopods. If successful, captive observations could greatly enhance our ability to gain information on these animals' biology and life cycles (Vidal et al., 2014, this volume).

Although proof is accumulating that deep-sea cephalopods occupy a wide range of trophic levels, for many species and in some cases whole families, dietary information remains completely unknown. One particular combination of techniques that should be more widely applied is the traditional identification of stomach contents (visually or using molecular tools) and analysis of stable isotopes and fatty acids of body tissues. From the data that are accumulating, it seems that deep-sea cephalopods have evolved diverse feeding strategies and range from the typical, opportunistically feeding strong swimmers, to sit-and-wait predators (which in some cases may even deploy lures), to planktivores to detritivores.

While future research efforts on deep-sea habitats will likely result in exciting discoveries on deep-sea cephalopods, there is also reason for concern. Examples of change in the ocean environment and anthropogenic impacts that may influence deep-sea cephalopods include expanding hypoxia, ocean acidification, pollution, deep-ocean warming, low-frequency anthropogenic noise and deep-sea fishing (e.g. Balmaseda et al., 2013; Gilly et al., 2013; Morato et al., 2006; Seibel et al., 2004; Solé et al., 2011, 2013; Unger et al., 2008). With increasing fishing pressure on deep-sea fish stocks, high numbers of nontarget species may also be captured, including cephalopods; for example, deep-sea bottom trawling may have resulted in regional extinction for some cephalopods in New Zealand waters (Freeman et al., 2010). Another consequence of deep-sea fishing may be that cephalopods expand into the niches of exploited and removed fish, a phenomenon that has been hypothesized for several squid species (e.g. Zeidberg and Robison, 2007). Deep-sea cephalopods are also vectors of persistent anthropogenic pollutants: relatively high levels of flame retardants have been measured in deep-sea cephalopods from >2500 m (Unger et al., 2008) and mercury concentrations in the tissue of some specimens of the giant squid *Architeuthis* have been found to be higher than has been reported in any coastal cephalopod species (Bustamante et al., 2008). In addition, low-frequency anthropogenic noise similar to that produced in the marine environment by shipping, acoustic exploration and deep-sea drilling has been shown to affect neritic and oceanic cephalopods in the lab (Solé et al., 2011, 2013) and has been hypothesized to impact deep-sea species in their natural environment also (Guerra et al., 2004). Most recently, Golikov et al. (2013) examined the distribution of cephalopods in the Arctic region and found extensive geographical range extensions of deep-sea cephalopods and attributed this to warming of the Arctic.

Understanding the biology of these animals under both current and potential future conditions will require ongoing dedication, innovation and collaboration. Much of the research will need to be interdisciplinary in nature, linking cephalopod researchers with colleagues from other fields, such as predator ecologists, oceanographers and fisheries biologists. Many major gaps remain in our knowledge of deep-sea cephalopods, including (but not limited to) basic biology (reproduction and life cycle), basic ecology (distribution, abundance and availability to predators) and trophic roles, at various temporal and spatial scales. These gaps can eventually be filled, but only by using the collective expertise and resources of the greater cephalopod research community and by keeping abreast of technological advances that may improve sampling and observational capabilities.

ACKNOWLEDGMENTS

The authors would like to thank the organizers of CIAC 2012 in Florianópolis for hosting the workshop as it enabled the assembly of an international group of scientists and resulted in productive discussions on deep-sea cephalopods. H.J.T.H. wants to thank Marek Lipinski for introducing H.J.T.H. to the research field of (deep sea) cephalopod biology, and Bruce Robison is thanked for his mentorship during H.J.T.H.'s MBARI postdoctoral fellowship, during which this workshop and review was initiated. We want to thank the Monterey Bay Aquarium Research Institute for providing images for Figures 3.1, 3.2, 3.4 and 3.9. We are grateful to Erica Vidal, Vladimir Laptikhovsky and Angel Guerra for reviewing this manuscript and for providing valuable feedback. Stephanie Bush and Susan von Thun are thanked for comments on parts of section 3.4 and Sasha Arkhipkin is thanked for comments on section 5.2.

H.J.T.H. was funded by the David and Lucile Packard Foundation, the Netherlands Organization for Scientific Research (NWO) through a Rubicon grant (#825.09.016), and by a grant (CP1218) of the Cluster of Excellence 80 "The Future Ocean," which is funded within the framework of the Excellence Initiative by the Deutsche Forschungsgemeinschaft (DFG) on behalf of the German federal and state governments. J. E. A. R. M. acknowledges funding provided by the São Paulo Research Foundation—FAPESP (grants # 2013/02653-1 and 2012/14140-6) and CNPq (grant # 477233/2013-9)—and the support of the Research Center for Marine Biodiversity (NP-BioMar). K. S. R. B., A. B. E., H. E. B. and J. T. K. received support from Auckland University of Technology, with J. T. K. additionally supported by the New Zealand Marine Sciences Society.

APPENDIX A. WORKING LIST OF CEPHALOPOD COLLECTIONS

Acronym	Institution (* indicates online database available)	Location
AM	Australian Museum*	Sydney, Australia
ANSP	Academy of Natural Sciences*	Philadelphia, PA, the United States
ARC	Atlantic Reference Centre*	St. Andrews, Canada
BMNH	Natural History Museum*	London, the United Kingdom
BPBM	Bernice Pauahi Bishop Museum*	Honolulu, HI, the United States
CASIZ	California Academy of Sciences, Invertebrate Zoology*	San Francisco, CA, the United States
CBR	Colecciones Biológicas de Referencia*, Institut de Ciències del Mar	Barcelona, Spain

CMN	Canadian Museum of Nature	Ottawa, Canada
FMHU	Hokkaido University Museum	Hakodate, Japan
FMNH	Field Museum of Natural History	Chicago, IL, the United States
NTM (formerly FMT)	National Taiwan Museum	Taipei, Taiwan
FURG	Universidade Federal do Rio Grande	Rio Grande, Brazil
FWRI	Florida Wildlife Research Institute*	St. Petersburg, FL, the United States
IMAS (formerly IASOS)	Institute for Marine and Antarctic Studies	Hobart, Australia
IRSNB	Institut Royal des Sciences Naturelles de Belgique*	Brussels, Belgium
JAMSTEC	Japan Agency for Marine-Earth Science and Technology*	Yokosuka, Japan
LIPI	Indonesian Institute of Sciences	Jakarta, Indonesia
LKCNHM (formerly NMS, RMBR)	Lee Kong Chian Natural History Museum, National University of Singapore	Singapore
LS	Linnaean Society	London, the United Kingdom
MCZ	Museum of Comparative Zoology, Harvard University*	Cambridge, MA, the United States
MfN (formerly ZMB)	Museum für Naturkunde*	Berlin, Germany
MFRDB (formerly MFDT)	Marine Fisheries Research and Development Bureau, Department of Fisheries	Bangkok, Thailand
MHNG	Muséum d'Histoire Naturelle de la Ville de Genève	Geneva, Switzerland
MMI	Museum of Marine Invertebrates, University of Puerto Rico, Mayagüez*	Isla Magüeyes, Puerto Rico
MNCN	Museo Nacional de Ciencias Naturales*	Madrid, Spain
MNHN	Muséum National d'Histoire Naturelle*	Paris, France

MOM	Musée Océanographique de Monaco	Monte Carlo, Monaco
MSUT (formerly FUMT)	University Museum, University of Tokyo*	Tokyo, Japan
MV	Museum Victoria*	Melbourne, Australia
MZUSP	Museu de Zoologia da Universidade de São Paulo	São Paulo, Brazil
MZUS	Musée Zoologique de la Ville de Strasbourg, University of Strasbourg	Strasbourg, France
NCB (merger of RMNH, ZMA)	Naturalis Biodiversity Center	Leiden, the Netherlands
NIWA	National Institute of Water and Atmospheric Research, Ltd*	Wellington, New Zealand
NMNH (formerly USNM)	National Museum of Natural History, Smithsonian Institution*	Washington, DC, the United States
NMNZ	National Museum of New Zealand Te Papa Tongarewa*	Wellington, New Zealand
NMSIB (formerly NMSZ)	National Museums Scotland, Invertebrate Biology*	Edinburgh, the United Kingdom
NRM	Naturhistoriska Riksmuseet (Swedish Museum of Natural History)	Stockholm, Sweden
NMNS (formerly NSMT)	National Museum of Nature and Science*	Tokyo, Japan
PIFSC	Pacific Islands Fisheries Science Center, National Marine Fisheries Service	Honolulu, HI, the United States
RSMAS (formerly UMML)	Invertebrate Museum, Rosenstiel School of Marine & Atmospheric Science, University of Miami	Miami, FL, the United States
SAM	South African Museum	Cape Town, South Africa
SBMNH	Santa Barbara Museum of Natural History*	Santa Barbara, CA, the United States
SIO	Scripps Institute of Oceanography	La Jolla, CA, the United States
SMF	Senckenberg Forschungsinstitut und Naturmuseum (Senckenberg Museum, Frankfurt)	Frankfurt, Germany

SZN	Stazione Zoologica Anton Dohrn di Napoli*	Naples, Italy
UMUB (formerly ZMUB)	University Museum, University of Bergen	Bergen, Norway
WAM	Western Australian Museum	Perth, Australia
ZIN	Zoological Institute of the Russian Academy of Sciences	St. Petersburg, Russia
ZMH	Zoologisches Museum, Universität Hamburg	Hamburg, Germany
ZMMGU	Zoological Museum, Moscow State University	Moscow, Russia
ZMUC	Zoological Museum, University of Copenhagen	Copenhagen, Denmark
ZSI	National Zoological Collections, Zoological Survey of India	Kolkata, India

APPENDIX B. RECENT AND ONGOING SURVEYS

Ongoing surveys open to collection requests

Nature and location	Contact
Live Hawaiian bobtail squid (*Euprymna*) collected via dipnet in Oahu, Hawaii, shipped live to Wisconsin, the United States, and maintained	N. Bekiares squidroom@gmail.com
Caribbean cephalopods collected from by-catch; specimens retained at MMI	N. Schizas n_ schizas@cima.uprm.edu
Atlantis Project: targets cephalopods as well as recovery from by-catch, Falkland Islands and SW Atlantic; retained at the Spanish Institute of Oceanography, Vigo	J. Portela julio.portela@ vi.ieo.es
Trawl, by-catch, ROV capture (via *Alvin*) and donation programmes from the eastern Pacific off North and Central America; specimens retained at FMNH	J. Voight jvoight@fieldmuseum.org
Hawaii Institution of Marine Biology targets *Octopus oliveri* and *O. cyanea*, intertidal and shallow reef by hand; specimens retained at HIMB	H. Ylitalo hyw@hawaii. edu

Instituto Português do Mar e da Atmosfera collects cephalopod by-catch from Portugese Groundfish Survey; specimens retained at IPMA	J. Pereira jpereira@ipma.pt
Consejo Superior de Investigaciones Científicas works with biodiversity surveys, deep-sea sampling, groundfish by-catch and donations from the northeastern Atlantic; specimens retained at CSIC	R. Villanueva roger@icm.csic.es
Museum of Aquatic Organisms (FURG) targets cephalopods via diving, trawl net and recovery from by-catch and predators, southwestern Atlantic	T. Silva Leite leite_ts@yahoo.com.br
LIPI, CASIZ undertake a variety of collection programmes, via trawl, ROV, hand lines and by-catch recovery, equatorial Pacific and Southern Ocean; specimens retained at LIPI, CASIZ, SBMNH and BPBM. Openness to sampling depends on organization and cruise	Various

Recently completed surveys

Nature and location	Contact
Cephalopod paralarvae targeted via bongo and neuston nets, 0–200 m, Gulf of California and eastern Pacific off Mexico, 1996–1998, 2003–2007, 2010; specimens retained at the Departamento de Plancton y Ecología Marina, Centro Interdisciplinario de Ciencias Marinas (CICIMAR-IPN), La Paz, México	R. De Silva Dávila rdesilva@ipn.mx
Mesopelagic and bathypelagic cephalopods targeted via midwater trawl, Gully Marine Protected Area, northwestern Atlantic, 2007–2010; specimens retained at Delaware Museum of Natural History.	E. Kenchington ellen.kenchington@dfo-mpo.gc.ca
Pelagic and benthic cephalopods targeted via trawl during the MAR-ECO cruises along the Mid-Atlantic Ridge, North Atlantic, 2004; specimens deposited at UMUB	E. Willassen endre.willassen@um.uib.no

REFERENCES

Adam, W., 1944. Revision de l' "Etude monographique de la famille des Sepiadae" d' A.T. de Rochebrune (1844). Memo. Museu. Nation. d'Hist. Naturel. 18 (6), 219–241 (nouvelle serie).

Adam, W., Rees, W.J., 1966. A review of the cephalopod family Sepiidae. John Murray Exped. Sci. Rep. 11 (1), 1–165 (46 plates).

Aldred, R.G., Nixon, M., Young, J.Z., 1983. *Cirrothauma murrayi* Chun, a finned octopod. Philos. Trans. R. Soc. Lond. B 301, 1–54.
Alexeyev, D.O., 1992. The systematic postion of bioluminescent squids of the family Loliginidae (Cephalopoda, Myopsida). Zool. Z. 71 (11), 12–23.
Allcock, A.L., 2005. On the confusion surrounding *Pareledone charcoti* (Joubin, 1905) (Cephalopoda: Octopodidae): endemic radiation in the Southern Ocean. Zool. J. Linn. Soc. 143, 75–108.
Allcock, A.L., Piatkowski, U., Rodhouse, P.G.K., Thorpe, J.P., 2001. A study on octopodids from the eastern Weddell Sea, Antarctica. Polar Biol. 24, 832–838.
Allcock, A.L., Hochberg, F.G., Stranks, T.N., 2003. Re-evaluation of *Graneledone setebos* (Cephalopoda: Octopodidae) and allocation to the genus *Megaleledone*. J. Mar. Biol. Assoc. U.K. 83, 319–328.
Allcock, A.L., Collins, M.A., Piatkowski, U., Vecchione, M., 2004. *Thaumeledone* and other deep water octopodids from the Southern Ocean. Deep Sea Res. II 51, 1883–1901.
Allcock, A.L., Barratt, I., Eleaume, M., Linse, K., Norman, M.D., Smith, P.J., Steinke, D., Stevens, D.W., Strugnell, J.M., 2011. Cryptic speciation and the circumpolarity debate: a case study on endemic Southern Ocean octopuses using the COI barcode of life. Deep Sea Res. II 58, 242–249.
Arbuckle, N.S.M., Wormuth, J.H., 2014. Trace elemental patterns in Humboldt squid statoliths from three geographic regions. Hydrobiologia 725, 115–123.
Arkhipkin, A.I., 1996a. Age and growth of the squid *Abraliopsis pfefferi* (Oegopsina: Enoploteuthidae) from the Central-East Atlantic based on statolith microstructure. Sci. Mar. 60, 325–330.
Arkhipkin, A.I., 1996b. Age and growth of planktonic squids *Cranchia scabra* and *Liocranchia reinhardti* (Cephalopoda, Cranchiidae) in epipelagic waters of the central-east Atlantic. J. Plankton Res. 18, 1675–1683.
Arkhipkin, A.I., 1996c. Statolith microstructure and age of early life stages of planktonic squids *Galiteuthis phyllura* and *Belonella borealis* (Oegopsida, Cranchiidae) from the northern North Pacific. J. Plankton Res. 18, 123–132.
Arkhipkin, A.I., 1997. Age and growth of the mesopelagic squid *Ancistrocheirus lesueurii* (Oegopsida: Ancistrocheiridae) from the central-east Atlantic based on statolith microstructure. Mar. Biol. 129, 103–111.
Arkhipkin, A.I., 2004. Diversity in growth and longevity in short-lived animals: squid of the suborder Oegopsina. Mar. Freshw. Res. 55, 341–355.
Arkhipkin, A.I., 2005. Statoliths as 'black boxes' (life recorders) in squid. Mar. Freshw. Res. 56, 573–583.
Arkhipkin, A.I., Bjørke, H., 1999. Ontogenetic changes in morphometric and reproductive indices of the squid *Gonatus fabricii* (Oegopsida, Gonatidae) in the Norwegian Sea. Polar Biol. 22, 357–365.
Arkhipkin, A., Laptikhovsky, V., 1994. Seasonal and interannual variability in growth and maturation of winter-spawning *Illex argentines* (Cephalopoda, Ommastrephidae) in the Southwest Atlantic. Aquat. Living Resour. 7 (4), 221–232.
Arkhipkin, A.I., Laptikhovsky, V.V., 2010. Observation of penis elongation in *Onykia ingens*: implications for spermatophore transfer in deep-water squid. J. Mollus. Stud. 76, 299–300.
Arkhipkin, A.I., Mikheev, A., 1992. Age and growth of the squid *Sthenoteuthis pteropus* (Oegopsida: Ommastrephidae) from the Central-East Atlantic. J. Exp. Biol. Ecol. 163, 261–276.
Arkhipkin, A.I., Nigmatullin, C.M., 1997. Ecology of the oceanic squid *Onychoteuthis banksi* and the relationship between the genera *Onychoteuthis* and *Chaunoteuthis* (Cephalopoda: Onychoteuthidae). J. Mar. Biol. Assoc. U.K. 77, 839–869.

Arkhipkin, A.I., Perez, J.A.A., 1998. Life-history reconstruction. In: Rodhouse, P.G., Dawe, E.G., O'Dor, R.K. (Eds.), Squid recruitment dynamics. The genus *Illex* as a model, the commercial *llex* species and influences on variability. FAO Fisheries Technical Paper, vol. 376. FAO, Rome, pp. 157–180.

Arkhipkin, A.I., Laptikhovsky, V.V., Nigmatullin, C.M., Bespyatykh, A.V., Murzov, S.A., 1998. Growth, reproduction and feeding of the tropical squid *Ornithoteuthis antillarum* (Cephalopoda, Ommastrephidae) from the central-east Atlantic. Sci. Mar. 62, 273–288.

Arkhipkin, A.I., Bizikov, V.A., Fuchs, D., 2012. Vestigial phragmocone in the gladius points to a deep water origin of squid (Mollusca: Cephalopoda). Deep Sea Res. Oceanogr. Res. Pap. I 61, 109–122.

Arocha, F., 2003. A new species of *Nematolampas* (Cephalopoda: Oegopsida) from the western Atlantic with an overview of the family Lycoteuthidae. Bull. Mar. Sci. 72 (3), 941–953.

Balmaseda, M.A., Trenberth, K.E., Källén, E., 2013. Distinctive climate signals in reanalysis of global ocean heat content. Geophys. Res. Lett. 40 (9), 1754–1759.

Baqueiro Cárdenas, E.R., Medrano Correa, S., Contreras Guzman, R., Barahona, N., Briceño, F., Villegas, M.J., Paredes, R., 2011. Eye lens structure of the octopus *Enteroctopus megalocyathus*: evidence of growth. J. Shellfish Res. 30 (2), 199–204.

Barbieri, E., Paster, B.J., Hughes, D., Zurek, L., Moser, D.P., Teske, A., Sogin, M.L., 2001. Phylogenetic characterization of epibiotic bacteria in the accessory nidamental gland and egg capsules of the squid *Loligo pealei* (Cephalopoda: Loliginidae). Environ. Microbiol. 3 (3), 151–167.

Barratt, I.M., Allcock, A.L., 2010. Ageing octopods from stylets: development of a technique for permanent preparations. ICES J. Mar. Sci. 67, 1452–1457.

Barratt, I.M., Johnson, M.P., Allcock, A.L., 2007. Fecundity and reproductive strategies in deep-sea incirrate octopuses (Cephalopoda: Octopoda). Mar. Biol. 150, 387–398.

Barrett, R.T., Camphusysen, C.J., Anker-Nilssen, T., Chardine, J.W., Furness, R.W., Garthe, S., Hüppop, O., Leopold, M.F., Montevecchi, W.A., Veit, R.R., 2007. Diet studies of seabirds: a review and recommendations. ICES J. Mar. Sci. 64, 1675–1691.

Beamish, R.J., McFarlane, G.A., 1983. The forgotten requirement for age validation in fisheries biology. Trans. Am. Fish. Soc. 112, 735–743.

Beatson, E., 2007. The diet of pygmy sperm whales, Kogia breviceps, stranded in New Zealand: Implications for conservation. Reviews in Fish Biology & Fisheries 17 (2–3), 295–303.

Beaumont, N., Townsend, M., Mangi, S., Austen, M.C., 2006. Marine biodiversity: an economic valuation. In: *Building the evidence base for a Marine Bill*, DEFRA (Department for Environment Food and Rural Affairs), 73 pp.

Benoit-Bird, K.J., Gilly, W.F., 2012. Coordinated nocturnal behavior of foraging jumbo squid *Dosidicus gigas*. Mar. Ecol. Prog. Ser. 455, 211–228.

Berrow, S.D., Croxall, J.P., 1999. The diet of white-chinned petrels *Procellaria aequinoctialis*, Linnaeus 1758, in years of contrasting prey availability at South Georgia. Antarct. Sci. 11, 283–292.

Berry, S.S., 1921. A review of the cephalopod genera *Sepioloidea*, *Sepiadarium*, and *Idiosepius*. Rec. South Aust. Mus. 1 (4), 347–364.

Berry, S.S., 1932. Cephalopods of the genera *Sepioloidea*, *Sepiadarium* and *Idiosepius*. Philipp. J. Sci. 47, 39–55.

Bigelow, K.A., 1992. Age and growth in paralarvae of the mesopelagic squid *Abralia trigonura* based on daily growth increments in statoliths. Mar. Ecol. Prog. Ser. 82, 31–40.

Bigelow, K.A., 1994. Age and growth of the oceanic squid *Onychoteuthis borealijaponica* in the north Pacific. Fish. Bull. 92, 13–25.

Bizikov, V.A., 2004. The shell in Vampyropoda (Cephalopoda): morphology, functional role and evolution. Ruthenica Suppl. 3, 1–88.

Bizikov, V.A., 2008. Evolution of the Shell in Cephalopoda. VNIRO Publishing, Moscow.

Bizikov, V.A., Arkhipkin, A.I., 1997. Morphology and microstructure of the gladius and statolith from the Boreal Pacific giant squid *Moroteuthis robusta* (Oegopsida: Onychoteuthidae). J. Zool. 241, 475–492.
Blainville, H.D.d', 1823. Memoire sur les especes du genre Calmar (*Loligo*, Lamarck). J. Phys. 96, 116–135.
Block, B.A., Dewar, H., Blackwell, S.B., Williams, T.D., Prince, E.D., Farwell, C.J., Boustany, A., Teo, S.L.H., Seitz, A., Walli, A., Fudge, D., 2001. Migratory movements, depth preferences, and thermal biology of Atlantic bluefin tuna. Science 293, 1310–1314.
Block, B.A., Jonsen, I.D., Jorgensen, S.J., Winship, A.J., Shaffer, S.A., Bograd, S.J., Hazen, E.L., Foley, D.G., Breed, G.A., Harrison, A.L., Ganong, J.E., Swithenbank, A., Castleton, M., Dewar, H., Mate, B.R., Shillinger, G.L., Schaefer, K.M., Benson, S.R., Weise, M.J., Henry, R.W., Costa, D.P., 2011. Tracking apex marine predator movements in a dynamic ocean. Nature 475, 86–90.
Bolstad, K.S., 2006. Sexual dimorphism in the beaks of *Moroteuthis ingens* Smith, 1881 (Cephalopoda: Oegopsida: Onychoteuthidae). N. Zeal. J. Zool. 33, 317–327.
Bolstad, K.S., 2010. Systematics of the Onychoteuthidae Gray, 1847 (Cephalopoda: Oegopsida). Zootaxa 2696, 1–186.
Bolstad, K.S., Hoving, H.J.T., 2011. Spermatangium structure and implantation sites in onychoteuthid squid (Cephalopoda: Oegopsida). Mar. Biodivers. Rec. 4, e5. http://dx.doi.org/10.1017/S175526721000120X.
Bolstad, K.S., O'Shea, S., 2004. Gut contents of a giant squid *Architeuthis dux* (Cephalopoda: Oegopsida) from New Zealand waters. N. Zeal. J. Zool. 31, 15–21.
Bolstad, K.S.R., Vecchione, M., Young, R.E., 2010. Onychoteuthidae Gray, 1847. Hooked Squids. http://tolweb.org/Onychoteuthidae/19419/2010.12.23 (accessed 15.10.2013).
Bolstad, K.S., Perez, J.A.A., Strugnell, J.M., Vidal, E.A.G., 2014. Cranchiids of the South Atlantic Mid-Oceanic Ridge: results from the first southern MAR-ECO expedition. J. Nat. Hist. http://dx.doi.org/10.1080/00222933.2013.867375.
Borri, M., Mannini, P., Volpi, C., Biagi, V., 1986. Cephalopod molluscs from the Mediterranean in the Adolfo Targioni Tozzetti Collection. Oebalia 14, 1–14.
Bower, J.R., Sakurai, Y., 1996. Laboratory observations on *Todarodes pacificus* (Cephalopoda: Ommastrephidae) egg masses. Am. Malacol. Bull. 13 (1–2), 65–71.
Bower, J.R., Seki, K., Kubodera, T., Yamamoto, J., Nobetsu, T., 2012. Brooding in a gonatid squid off Northern Japan. Biol. Bull. 223, 259–262.
Boyle, P.R., Daly, H.I., 2002. Fecundity and spawning in a deep-water cirromorph octopus. Mar. Biol. 137 (2), 317–324.
Boyle, P.R., Rodhouse, P., 2005. Cephalopods: Ecology and Fisheries. Blackwell Science Ltd, Oxford.
Braid, H.E., 2013. Systematics and Ecology of the New Zealand Mastigoteuthidae (Cephalopoda, Oegopsida). Auckland University of Technology (M. App. Sc. thesis).
Braid, H.E., Deeds, J., DeGrasse, S.L., Wilson, J.J., Osborne, J., Hanner, R.H., 2012. Preying on commercial fisheries and accumulating paralytic shellfish toxins: a dietary analysis of invasive *Dosidicus gigas* (Cephalopoda Ommastrephidae) stranded in Pacific Canada. Mar. Biol. 159 (1), 25–31.
Brakoniecki, T.F., 1996. A revision of the genus *Pickfordiateuthis* Voss, 1953 (Cephalopoda; Myopsida). Bull. Mar. Sci. 58 (1), 9–28.
Burgess, L.A., 1998. A survey of seminal receptacles in the Enoploteuthidae. Smithson. Contr. Zool. 586, 271–276.
Bush, S.L., Caldwell, R.L., Robison, B.H., 2007. Ink utilization by mesopelagic squid. Mar. Biol. 152 (3), 485–494.
Bush, S.L., Robison, B.H., Caldwell, R.L., 2009. Behaving in the dark: locomotor, chromatic, postural, and bioluminescent behaviors of the Deep-Sea squid *Octopoteuthis deletron* Young 1972. Biol. Bull. 216 (1), 7–22.

Bush, S.L., Hoving, H.J., Huffard, C.L., Robison, B.H., Zeidberg, L.D., 2012. Brooding and sperm storage by the deep-sea squid *Bathyteuthis berryi* (Cephalopoda: Decapodiformes). J. Mar. Biol. Assoc. U.K. 92, 1629–1636.

Bustamante, P., González, A.F., Rocha, F., Miramand, P., Guerra, A., 2008. Metal and metalloid concentrations in the giant squid *Architeuthis dux* from Iberian waters. Mar. Environ. Res. 66, 278–287.

Caddy, J.F., Rodhouse, P.G., 1998. Cephalopod and groundfish landings: evidence for ecological change in global fisheries? Rev. Fish Biol. Fish. 8 (4), 431–444.

Canali, E., Ponte, G., Belcari, P., Rocha, F., Fiorito, G., 2011. Evaluating age in *Octopus vulgaris*: estimation, validation and seasonal differences. Mar. Ecol. Prog. Ser. 441, 141–149.

Carlini, D.B., Kunkle, L.K., Vecchione, M., 2006. A molecular systematic evaluation of the squid genus *Illex* (Cephalopoda: Ommastrephidae) in the North Atlantic Ocean and Mediterranean Sea. Mol. Phylogenet. Evol. 41 (2), 496–502.

Census of Marine Life Programme, 2000–2010. A Decade of Discovery. http://coml.org/ (accessed 15.12.2013).

Cherel, Y., Duhamel, G., 2004. Antarctic jaws: cephalopod prey of sharks in Kerguelen waters. Deep Sea Res. I 51, 17–31.

Cherel, Y., Hobson, K.A., 2005. Stable isotopes, beaks and predators: a new tool to study the trophic ecology of cephalopods, including giant and colossal squids. Proc. R. Soc. B Biol. Sci. 272, 1601–1607.

Cherel, Y., Ridoux, V., Spitz, J., Richard, P., 2009. Stable isotopes document the trophic structure of a deep-sea cephalopod assemblage including giant octopod and giant squid. Biol. Lett. 5, 364–367.

Childress, J.J., Barnes, A.T., Quetin, L.B., Robison, B.H., 1978. Thermally protecting cod ends for the recovery of living deep-sea animals. Deep Sea Res. 25, 419–422.

Clarke, M.R., 1962. Significance of cephalopod beaks. Nature 193, 560–561.

Clarke, M.R., 1965. Growth rings in the beaks of the squid *Moroteuthis ingens* (Oegopsida: Onychoteuthidae). Malacologia 3, 287–307.

Clarke, M.R., 1966. A review of the systematics and ecology of oceanic squids. Adv. Mar. Biol. 4, 91–300.

Clarke, M.R., 1967. A deep-sea squid, *Taningia danae* Joubin, 1931. Symp. Zool. Soc. Lond. 19, 127–143.

Clarke, M.R., 1977a. Beaks, nets and numbers. Symp. Zool. Soc. Lond. 38, 89–126.

Clarke, M.R., 1977b. A brief review of sampling techniques and tools of marine biology. In: Angel, M.V. (Ed.), A Voyage of Discovery. Pergammon Press, Oxford, pp. 439–469.

Clarke, M.R., 1978. The cephalopod statolith an introduction to its form. J. Mar. Biol. Assoc. U.K. 58, 701–712.

Clarke, M.R., 1980. Cephalopoda in the diet of sperm whales of the Southern Hemisphere and their bearing on sperm whale biology. Discov. Rep. 37, 1–324.

Clarke, M.R., 1986. A Handbook for the Identification of Cephalopod Beaks. Clarendon Press, Oxford.

Clarke, M.R., 1996. Cephalopods in the World's Oceans: cephalopods as prey. III. Cetaceans. Philos. Trans. R. Soc. Lond. B 351, 1053–1065.

Clarke, M.R., 2006. Oceanic cephalopod distribution and species diversity in the Eastern North Atlantic. Arquipélago—Life Mar. Sci. 23A, 27–46.

Clarke, M.R., 2008. Seamounts and cephalopods. In: Pitcher, T.J., Morato, T., Hart, P.J.B., Clark, M.R., Haggan, N., Santos, R.S. (Eds.), Seamounts: Ecology, Fisheries & Conservation. Blackwell Publishing Ltd, Oxford, UK. http://dx.doi.org/10.1002/9780470691953.ch11.

Clarke, M.R., Goodall, N., 1994. Cephalopods in the diets of three Odontoceti cetacean species stranded at Tierra del Fuego, *Globicephala melaena* (Traill, 1809), *Hyperoodon planifrons* Flower, 1882 and *Cephalorhynchus commersonii* (Lacepede, 1804). Antarct. Sci. 6, 149–154.

Clarke, M.R., Lu, C.C., 1975. Vertical distribution of cephalopods at 18 N 25 W in the North Atlantic. J. Mar. Biol. Assoc. U.K. 55, 165–182.

Clarke, M.R., MacLeod, N., 1982. Cephalopod remains in the stomachs of eight Weddell Seals. Br. Antarct. Surv. Bull. 57, 33–40.

Clarke, M.R., Martins, H.R., Pascoe, P., 1993. The diet of sperm whales (*Physeter macrocephalus* Linnaeus 1758) off the Azores. Philos. Trans. R. Soc. Lond. B 339, 67–82.

Cole, L.C., 1954. The population consequences of life history phenomena. Q. Rev. Biol. 29, 103–137.

Collins, M.A., 2003. The genus *Grimpoteuthis* (Octopoda: Grimpoteuthidae) in the north-east Atlantic, with descriptions of three new species. Zool. J. Linn. Soc. 139, 93–127.

Collins, M.A., Henriques, C., 2000. A revision of the family Stauroteuthidae (Octopoda: Cirrata) with redescriptions of *Stauroteuthis syrtensis* and *S. gilchristi*. J. Mar. Biol. Assoc. U.K. 80, 685–697.

Collins, M.A., Villanueva, R., 2006. Taxonomy, ecology and behaviour of the cirrate octopods. Oceanogr. Mar. Biol. Annu. Rev. 44, 277–322.

Collins, M.A., Yau, C., Allcock, L., Thurston, M.H., 2001. Distribution of deep-water benthic and bentho-pelagic cephalopods from the north-east Atlantic. J. Mar. Biol. Assoc. U.K. 81, 105–117.

Collins, M., Young, R.E., Vecchione, M., 2008. Stauroteuthidae Grimpe 1916. *Stauroteuthis* Verrill, 1879. http://tolweb.org/Stauroteuthis/20092/ (accessed 28.04.2008).

Croxall, J.P., Silk, J.R.D., Phillips, R.A., Afanasyev, V., Briggs, D.R., 2005. Global circumnavigations: tracking year-round ranges of nonbreeding albatrosses. Science 307, 249–250.

Dai, L., Zheng, X., Kong, L., Li, Q., 2012. DNA barcoding analysis of Coleoidea (Mollusca: Cephalopoda) from Chinese waters. Mol. Ecol. Resour. 12, 437–447.

Daneri, G.A., Carlini, A.R., Rodhouse, P.G.K., 2000. Cephalopod diet of the southern elephant seal, *Mirounga leonina*, at King George Island, South Shetland Islands. Antarct. Sci. 12, 16–19.

Daniel, T.L., Helmuth, B.S., Saunders, W.B., Ward, P.D., 1997. Septal complexity in ammonoid cephalopods increased mechanical risk and limited depth. Paleobiology 23, 470–481.

Davies, J., Howell, K., Guillaumont, B., Hourigan, T., 2012. Cataloguing the Deep Sea. In: Program and Abstracts for the 13th International Deep-Sea Biology Symposium, Wellington, New Zealand, 3–7 December 2012, p. 217.

Davis, R.W., Jaquet, N., Gendron, D., Markaida, U., Bazzino, G., Gilly, W., 2007. Diving behavior of sperm whales in relation to behavior of a major prey species, the jumbo squid, in the Gulf of California, Mexico. Mar. Ecol. Prog. Ser. 333, 291–302.

Deagle, B.E., Jarman, S.N., Pemberton, D., Gales, N.J., 2005. Genetic screening for prey in the gut contents from a giant squid (*Architeuthis* sp.). J. Hered. 96 (4), 417–423.

De Forest, L., Drazen, J., 2009. The influence of a Hawaiian seamount on mesopelagic micronekton. Deep Sea Res. I 56, 232–250.

Dickson, J., Morley, S.A., Mulvey, T., 2004. New data on *Martialia hyadesi* feeding in the Scotia Sea during winter; with emphasis on seasonal and annual variability. J. Mar. Biol. Assoc. U.K. 84, 785–788.

Doubleday, Z.A., Semmens, J.M., 2011. Quantification of the age-pigment lipofuscin in known-age octopus (*Octopus pallidus*): a potential tool for age determination. J. Exp. Mar. Biol. Ecol. 397 (1), 8–12.

Doubleday, Z., Semmens, J.M., Pecl, G., Jackson, G., 2006. Assessing the validity of stylets as ageing tools in *Octopus pallidus*. J. Exp. Mar. Biol. Ecol. 338, 35–42.

Doyle, P., 1990. The biogeography of the Aulacocerida (Coleoidea). In: Pallini, G. (Ed.), Fossili Evoluzione Ambiente. Atti II Convegno Piccinni. Technostampa, Rome, pp. 263–271.

Drew, G.A., 1919. Sexual activities of the squid *Loligo pealii* (Les.) II. The spermatophore; its structure, ejaculation and formation. J. Morphol. 32, 379–435.

Dunning, M., 1998. A review of the systematics, distribution and biology of the arrow squid genera *Ommastrephes* Orbigny, 1835, *Sthenoteuthis* Verrill, 1880, and *Ornithoteuthis* Okada, 1927 (Cephalopoda, Ommastrephidae). In: Voss, N.A., Vecchione, M., Toll, R.B., Sweeney, M.J. (Eds.), Systematics and Biogeography of Cephalopods. In: Smithsonian Contributions to Zoology, vol. 586. pp. 425–433, I-II.

Dunning, M., Förch, E.C., 1998. A review of the systematics, distribution and biology of arrow squids of the genus *Nototodarus* Pfeffer, 1912 (Cephalopoda, Ommastrephidae). In: Voss, N.A., Vecchione, M., Toll, R.B., Sweeney, M.J. (Eds.), Systematics and Biogeography of Cephalopods. In: Smithsonian Contributions to Zoology, vol. 586. pp. 393–404, I–II.

Dunning, M., Wormuth, J.H., 1998. The ommastrephid squid genus *Todarodes*: a review of systematics, distribution and biology (Cephalopoda, Teuthoidea). In: Voss, N.A., Vecchione, M., Toll, R.B., Sweeney, M.J. (Eds.), Systematics and Biogeography of Cephalopods. In: Smithsonian Contributions to Zoology, vol. 586. pp. 385–391, I–II.

Durward, R.D., Vessey, E., O'Dor, R.K., Amaratunga, T., 1980. Reproduction in the squid, *Illex illecebrosus*: first observations in captivity and implications for the life cycle. ICNAF Select. Pap. 6, 7–13.

Engeser, T.S., Clarke, M.R., 1988. Cephalopod hooks, both recent and fossil. In: Clarke, M.R., Trueman, R.R. (Eds.), The Mollusca. Volume 12. Paleontology and Neontology of Cephalopods. Academic Press, San Diego, pp. 133–151.

Fanelli, E., Cartes, J.E., Papiol, V., 2012. Assemblage structure and trophic ecology of deep-sea demersal cephalopods in the Balearic basin (NW Mediterranean). Mar. Freshw. Res. 63, 264–274.

Fang, S.G., Wan, Q.H., Fujihara, N., 2002. Formalin removal from archival tissue by critical point drying. Biotechniques 33, 604–611.

FAO, 2012. The State of World Fisheries and Aquaculture 2012. FAO, Rome, 209 pp.

Feulner, P.G.D., Kirschbaum, F., Mamonekene, V., Ketmaier, V., Tiedemann, R., 2007. Adaptive radiation in African weakly electric fish (Teleostei: Mormyridae: *Campylomormyrus*): a combined molecular and morphological approach. J. Evol. Biol. 20 (1), 403–414.

Fields, W.G., Thompson, K.A., 1976. Ultrastructure and functional morphology of spermatozoa of *Rossia pacifica* (Cephalopoda. Decapoda). Canad. J. Zool. 54, 908–932.

Förch, E.C., 1998. The marine fauna of New Zealand: Cephalopoda: Oegopsida: Architeuthidae (giant squid). NIWA Biodivers. Memo. 110, 1–113.

Franco-Santos, R.M., Vidal, E.A.G., 2014. Beak development of early squid paralarvae (Cephalopoda: Teuthoidea) may reflect an adaptation to a specialized feeding mode. Hydrobiologia 725, 85–103.

Frankham, R., Ballou, J.D., Dudash, M.R., Eldridge, M.D.B., Fenster, C.B., Lacy, R.C., Mendelson III, J.R., Porton, I.J., Ralls, K., Ryder, O.A., 2012. Implications of different species concepts for conserving biodiversity. Biol. Conserv. 153, 25–31.

Freeman, D.J., Marshall, B.A., Anyong, S.T., Wing, S.R., Hitchmough, R.A., 2010. Conservation status of New Zealand marine invertebrates, 2009. N. Zeal. J. Mar. Freshw. Res. 44, 129–148.

Fries, A., 2010. A Gap Analysis of the Distributions of Cephalopod Species Worldwide with a Focus on Commercially Important Species. Duke University Libraries, 63 pp.

Froerman, Y.M., 1984. Feeding spectrum and trophic relationships of short-finned squid (*Illex illecebrosus*) in the Northwest Atlantic. NAFO Sci. Council Stud. 7, 67–75.

Froesch, D., Marthy, H.J., 1975. The structure and function of the oviducal gland in octopods (Cephalopoda). Proc. R. Soc. Lond. B 188, 95–101.

Fuchs, D., 2006. Fossil erhaltungsfähige Merkmalskomplexe der Coleoidea (Cephalopoda) und ihre phylogenetische Bedeutung. Berl. Paläobiol. Abhandl. 8, 1–115.

Fuchs, D., 2009. Octobrachia—a diphyletic taxon? Berl. Paläobiol. Abhandl. 10, 182–192.

Fuchs, D., 2012. Neotenous Origin of *Spirula* Lamarck, 1799?—evidence from a Morphological Series of Fossils. CIAC, Florianopolis, p.60 (Abstracts).

Fuchs, D., Larson, N.L., 2011a. Diversity, morphology and phylogeny of coleoid cephalopods from the Upper Cretaceous Plattenkalks of Lebanon—part II: Teudopseina. J. Paleontol. 85, 815–834.

Fuchs, D., Larson, N.L., 2011b. Diversity, morphology, and phylogeny of coleoid cephalopods from the Upper Cretaceous Plattenkalks of Lebanon—part I: Prototeuthidina. J. Paleontol. 85, 234–249.

Fuchs, D., Weis, R., 2008. Taxonomy, morphology and phylogeny of Lower Jurassic loligosepiid coleoids (Cephalopoda). N. Jb. Geol. Paläont. 249, 93–112.

Fuchs, D., Weis, R., 2010. Taxonomy, morphology and phylogeny of Lower Jurassic teudopseid coleoids (Cephalopoda). Neues Jb. Geol. Paläontol. Abh. 257, 351–366.

Fuchs, D., Engeser, T., Keupp, H., 2007a. Gladius shape variation in coleoid cephalopod *Trachyteuthis* from the Upper Jurassic Nusplingen and Solnhofen Plattenkalks. Acta Palaeontol. Polon. 52, 575–589.

Fuchs, D., Klinghammer, A., Keupp, H., 2007b. Taxonomy, morphology and phylogeny of plesioteuthidid coleoids from the Upper Jurassic (Tithonian) Plattenkalks of Solnhofen. Neues Jb. Geol. Paläontol. Abh. 245, 239–252.

Fuchs, D., Bracchi, G., Weis, R., 2009. New records of octopods (Cephalopoda: Coleoidea) from the Late Cretaceous (Upper Cenomanian) of Hakel and Hadjoula (Lebanon). Palaeontology 52, 56–81.

Fuchs, D., Keupp, H., Trask, P., Tanabe, K., 2012. Taxonomy, morphology and phylogeny of Late Cretaceous spirulid coleoids (Cephalopoda) from Greenland and Canada. Palaeontology 55, 285–303.

Fuchs, D., Iba, Y., Ifrim, C., Nishimura, T., Kennedy, J., Keupp, H., Stinnesbeck, W., Tanabe, K., 2013. *Longibelus* gen. nov., a new Cretaceous coleoid genus linking Belemnoidea and Decabrachia. Palaeontology 56, 1081–1106.

Gardiner, K., Dick, T., 2010. Arctic cephalopod distributions and their associated predators. Polar Res. 29, 209–227.

Gauldie, R.W., West, I.F., Förch, E., 1994. Statocyst, statolith, and age estimation of the giant squid, *Architeuthis kirki*. Veliger 37, 93–109.

Gervais, F., 2012. Using new technologies to identify deep-sea species: Marine Life Field Guide iBook. In: Program and Abstracts for the 13th International Deep-Sea Biology Symposium, Wellington, New Zealand, 3–7 December 2012, p. 211.

Gilly, W.F., Elliger, C.A., Salinas, C.A., Camarilla-Coop, S., Bazzino, G., Beman, M., 2006. Spawning by jumbo squid *Dosidicus gigas* in San Pedro Mártir Basin, Gulf of California, Mexico. Mar. Ecol. Prog. Ser. 313, 125–133.

Gilly, W.F., Beman, J.M., Litvin, S.Y., Robison, B.H., 2013. Oceanographic and biological effects of shoaling of the oxygen minimum zone. Annu. Rev. Mar. Sci. 5, 393–420.

Gleadall, I.G., 2013. A molecular sequence proxy for *Muusoctopus januarii* and calibration of recent divergence among a group of mesobenthic octopuses. J. Exp. Mar. Biol. Ecol. 447, 106–122.

Glover, A.G., Higgs, N., Horton, T., 2013. World register of Deep-Sea species. http://www.marinespecies.org/deepsea (accessed 07.02.2013).

Golikov, A.V., Sabirov, R.M., Lubin, P.A., Jørgensen, L.L., 2013. Changes in distribution and range structure of Arctic cephalopods due to climatic changes of the last decades. Biodiversity 14, 28–35.

Gómez-Villota, F., 2007. Sperm Whale Diet in New Zealand. Auckland University of Technology (M.App.Sc. Thesis) University1–231 (Dissertation).

González, Á.F., Trathan, P.N., Yau, C., Rodhouse, P.G., 1997. Interactions between oceanography, ecology and fishery biology of the ommastrephid squid *Martialia hyadesi* in the South Atlantic. Mar. Ecol. Prog. Ser. 152 (1–3), 205–215.

González, Á.F., Guerra, Á., Pascual, S., Briand, P., 1998. *Vulcanoctopus hydrothermalis* gen. et sp. nov. (Mollusca, Cephalopoda): an octopod from a deep-sea hydrothermal vent site. Cahier Biol. Mar. 39, 169–184.

González, Á.F., Guerra, Á., Rocha, F., 2003. New data on the life history and ecology of the deep-sea hooked squid *Taningia danae*. Sarsia 88, 297–301.

Graeve, M., Kattner, G., Wiencke, C., Karsten, U., 2002. Fatty acid composition of Arctic and Antarctic macroalgae: indicator of phylogenetic and trophic relationships. Mar. Ecol. Prog. Ser. 231, 67–74.

Grist, E.P., Jackson, G.D., 2007. How long would it take to become a giant squid? Rev. Fish Biol. Fish. 17, 385–399.

Guerra, Á., Villanueva, R., Nesis, K.N., Bedoya, J., 1998. Redescription of the deep-sea cirrate octopod *Cirroteuthis magna* Hoyle, 1885, and consideration on the genus *Cirroteuthis* (Mollusca: Cephalopoda). Bull. Mar. Sci. 63 (1), 51–81.

Guerra, Á., González, Á.F., Rocha, F.J., Sagarminaga, R., Cañadas, A., 2002. Planktonic egg masses of the diamond-shaped squid *Thysanoteuthis rhombus* in the eastern Atlantic and the Mediterranean Sea. J. Plankton Res. 24 (4), 333–338.

Guerra, Á., González, Á.F., Rocha, F., 2004. A Review of Records of Giant Squid in the North-Eastern Atlantic and Severe Injuries in *Architeuthis dux* Stranded After Acoustic Exploration: ICES CM 2004. p. CC: 29.

Guerra, Á., Rodríguez-Navarro, A.B., González, Á.F., Romanek, C.S., Álvarez-Lloret, P., Pierce, G.J., 2010. Life-history traits of the giant squid *Architeuthis dux* revealed from stable isotope signatures recorded in beaks. ICES J. Mar. Sci. 67, 1425–1431.

Guerra, Á., Portela, J.M., Río, J.L.D., 2001. Cephalopods caught in the outer Patagonian shelf and its upper and medium slope in relation to the main oceanographic features. Fish. Res. 1091, 179–186.

Guerra, Á., González, Á.F., Pascual, S., Dawe, E.G., 2011. The giant squid *Architeuthis*: an emblematic invertebrate that can represent concern for the conservation of marine biodiversity. Biol. Conserv. 144 (7), 1989–1997.

Guerra, Á., Portela, J.M., Roura, A., del Río, J.l., Vecchione, M., 2012. Morphological variability of the rare bush-club squid *Batoteuthis* (Cephalopoda, Batoteuthidae)? Neues Jb. Geol. Paläontol. Abh. 266, 77–83.

Haas, W., 2002. The evolutionary history of the eight-armed Coleoidea. In: Summesberger, H., Histon, K., Daurer, A. (Eds.), Cephalopods—Present & Past. Abhandlungender Geologischen Bundesanstalt, vol. 57. pp. 341–351.

Haimovici, M., Perez, J.A., 1991. Coastal cephalopod fauna of southern Brazil. Bull. Mar. Sci. 49 (1–2), 1–2.

Haimovici, M., Santos, R.A., Fischer, L.G., 2009. Class Cephalopoda. In: Rios, E.C. (Ed.), Compendium of Brazilian Seashells. Evangraf, Rio Grande, pp. 610–649.

Hanlon, R.T., Messenger, J.B., 1996. Cephalopod Behaviour. Cambridge University Press, Cambridge.

Hanlon, R.T., Ament, S.A., Gabr, H., 1999. Behavioral aspects of sperm competition in cuttlefish, *Sepia officinalis* (Sepioidea: Cephalopoda). Mar. Biol. 134, 719–728.

Harman, R.F., Young, R.E., 1985. The larvae of ommastrephid squids (Cephalopoda, Teuthoidea) from Hawaiian waters. Vie Milieu 35, 211–222.

Hayashi, S., 1989. Spawning behavior and development of the firefly squid, *Watasenia scintillans*. Bull. Toyama Prefect. Fish. Exper. Stat. 1, 1–14.
Hayashi, S., 1995. Fishery biological studies of the firefly squid, *Watasenia scintillans* (Berry), in Toyama Bay. Bull. Toyama Prefect. Fish. Exper. Stat. 7, 1–128.
Healy, J.M., 1989. Spermatozoa of the deep-sea cephalopod *Vampyroteuthis infernalis* Chun: ultrastructure and possible phylogenetic significance. Philos. Trans. R. Soc. Lond. B 323, 589–600.
Healy, J.M., 1990. Ultrastructure of spermiogenesis in *Vampyroteuthis infernalis* Chun, a relictic cephalopod mollusc. Hel. Meer. 44, 95–107.
Healy, J.M., 1993. Sperm and spermiogenesis in *Opisthoteuthis persephone* (Octopoda: Cirrata): ultrastructure, comparison with other cephalopods and evolutionary significance. J. Mollus. Stud. 59, 105–115.
Hebert, P.D.N., Gregory, T.R., 2005. The promise of DNA barcoding for taxonomy. Syst. Biol. 54, 852–859.
Helm, R.C., 1984. Rate of digestion in three species of pinnipeds. Can. J. Zool. 62, 1751–1756.
Hermosilla, C.A., Rocha, F., Fiorito, G., González, Á.F., Guerra, Á., 2010. Age validation in common octopus *Octopus vulgaris* using stylet increment analysis. ICES J. Mar. Sci. 67, 1458–1463.
Herring, P., 2002. The Biology of the Deep Ocean. Oxford University Press, Oxford.
Hess, S.C., 1987. Comparative Morphology, Variability, and Systematic Applications of Cephalopod Spermatophores (Teuthoidea and Vampyromorpha). University of Miami, Coral Gables, FL (Dissertation. Available from: University Microfilms, Ann Arbor, MI; AAT 8716160).
Hewitt, R.A., Westermann, G.E.G., 1988. Stress and strain in Nautilus shells: some limitations on the bouyancy control and vertical migration of Ectocochliates. In: Wiedmann, J.K. (Ed.), Cephalopods—Present and Past. Schweizerbart´sche Verlagsbuchhandlung, Stuttgart, pp. 705–712.
Hewitt, R.A., Dokainish, M.A., Aghoury, M.E., Westermann, G.E.G., 1989. Bathymetric limits of a carboniferous orthoconic nautiloid deduced by finite element analysis. PALAIOS 4, 157–167.
Hewitt, R.A., Yoshike, T., Westermann, G.E.G., 1991. Shell microstructure and ecology of the Cretaceous coleoid cephalopod *Naefia* from the Santonian of Japan. Cretaceous Res. 12, 47–54.
Higgs, N., Horton, T., Glover, A., 2012. The World Register of Deep-Sea Species. In: Program and Abstracts for the 13th International Deep-Sea Biology Symposium, Wellington, New Zealand, 3–7 December 2012, p. 216.
Hilton, G.M., Houston, D.C., Furness, R.W., 1998. Which components of diet quality affect retention time of digesta in seabirds. Funct. Ecol. 12, 929–939.
Hindell, M.A., Bradshaw, C.J.A., Sumner, M.D., Michael, K.J., Burton, H.R., 2003. Dispersal of female southern elephant seals and their prey consumption during the austral summer: relevance to management and oceanographic zones. J. Appl. Ecol. 40, 703–715.
Hochberg, F.G., Huffard, C., Mangold, K.M., 2009. Octopodidae Orbigny, 1839. Octopods, octopuses, devilfishes. http://tolweb.org/Octopodidae/20194/ (accessed 29.12.2009).
Horstkotte, J., 2008. The systematics and ecological role of *Histioteuthis* (Cephalopoda: Histioteuthidae) in New Zealand Waters. Auckland University of Technology, M. App. Sc. Thesis.
Houghton, J.D., Doyle, T.K., Davenport, J., Wilson, R.P., Hays, G.C., 2008. The role of infrequent and extraordinary deep dives in leatherback turtles (*Dermochelys coriacea*). J. Exp. Biol. 211 (16), 2566–2575.

Hoving, H.J.T., 2008. Reproductive biology of oceanic decapodiform cephalopods. University of Groningen, pp.1–184 (Ph.D. thesis).

Hoving, H.J.T., Laptikhovsky, V., 2007. Getting under the skin: autonomous implantation of squid spermatophores. Biol. Bull. 212, 177–179.

Hoving, H.J.T., Lipinski, M.R., 2009. Female reproductive biology, and age of deep-sea squid *Histioteuthis miranda* from southern Africa. ICES J. Mar. Sci. 66, 1868–1872.

Hoving, H.J.T., Lipinski, M.R., 2014. Observations on age and reproduction of the oceanic squid *Ancistrocheirus lesueurii* (d'Orbigny, 1842). J. Nat. Hist. http://dx.doi.org/10.1080/00222933.2013.840748.

Hoving, H.J.T., Robison, B.H., 2012. Vampire squid: detritivores in the oxygen minimum zone. Proc. R. Soc. B Biol. Sci. 279, 4559–4567.

Hoving, H.J.T., Vecchione, M., 2012. Mating behavior of a deep-sea squid revealed by *in-situ* videography and the study of archived specimens. Biol. Bull. 223, 263–267.

Hoving, H.J.T., Roeleveld, M.A.C., Lipinski, M.R., Melo, Y., 2004. Reproductive system of the giant squid *Architeuthis* in South African waters. J. Zool. 264, 153–169.

Hoving, H.J.T., Roeleveld, M.A.C., Lipinski, M.R., Videler, J.J., 2006. Nidamental glands in males of the oceanic squid *Ancistrocheirus lesueurii* (Cephalopoda: Ancistrocheiridae)—sex change or intersexuality? J. Zool. 269 (3), 341–348.

Hoving, H.J.T., Lipinski, M.R., Roeleveld, M.A.C., Durholtz, M.D., 2007. Growth and mating of southern African *Lycoteuthis lorigera* (Steenstrup, 1875) (Cephalopoda; Lycoteuthidae). Rev. Fish Biol. Fish. 17, 259–270.

Hoving, H.J.T., Lipinski, M.R., Videler, J.J., 2008a. Reproductive system and the spermatophoric reaction of the mesopelagic squid *Octopoteuthis sicula* (Rüppell, 1844) (Cephalopoda: Octopoteuthidae) from southern African waters. Afr. J. Mar. Sci. 30, 603–612.

Hoving, H.J.T., Laptikhovsky, V., Piatkowski, U., Önsoy, B., 2008b. Reproduction in *Heteroteuthis dispar* (Rüppell, 1844) (Mollusca: Cephalopoda): a sepiolid reproductive adaptation to an oceanic lifestyle. Mar. Biol. 154, 219–230.

Hoving, H.J.T., Nauwelaerts, S., Van Genne, B., Stamhuis, E.J., Zumholz, K., 2009. Spermatophore implantation in *Rossia moelleri* Steenstrup, 1856 (Sepiolidae; Cephalopoda). J. Exp. Mar. Biol. Ecol. 372, 75–81.

Hoving, H.J.T., Lipiński, M.R., Dam, L., 2010a. The male reproductive strategy of a deep-sea squid: sperm allocation, continuous production, and long-term storage of spermatophores in *Histioteuthis miranda*. ICES J. Mar. Sci. 67 (7), 1478–1486.

Hoving, H.J.T., Lipinski, M.R., Videler, J.J., Bolstad, K.S., 2010b. Sperm storage and mating in the deep-sea squid *Taningia danae* Joubin, 1931 (Oegopsida: Octopoteuthidae). Mar. Biol. 157, 393–400.

Hoving, H.J.T., Bush, S.L., Robison, B.H., 2012. A shot in the dark: same-sex sexual behaviour in a deep-sea squid. Biol. Lett. 8 (2), 287–290.

Hoving, H.J.T., Gilly, W.F., Markaida, U., Benoit-Bird, K.J., West-Brown, Z., Daniel, P., Field, J., Paressenti, L., Liu, B., Campos, B., 2013a. Extreme plasticity in life history strategy allows a migratory predator (jumbo squid) to cope with a changing climate. Glob. Chang. Biol. 19 (7), 2089–2103.

Hoving, H.J.T., Zeidberg, L.D., Benfield, M.C., Bush, S.L., Robison, B.H., Vecchione, M., 2013b. First *in situ* observations of the deep-sea squid *Grimalditeuthis bonplandi* reveal unique use of tentacles. Proc. R. Soc. B Biol. Sci. 280 (1769), 1–5.

Hoving, H.J.T., Laptikhovsky, V.V., Lipinski, M.R., Jürgens, E., 2014. Fecundity, oogenesis, and ovulation pattern of southern African *Lycoteuthis lorigera* (Steenstrup, 1875). Hydrobiologia 725 (1), 23–32.

Hoyle, W.E., 1886. Report on the Cephalopoda collected by H.M.S. Challenger during the years 1873–1876. Rep. Scient. Res. Voyage of H.M.S. Challe. 1873–76 16, 1–245.

Hunsicker, M.E., Essington, T.E., Aydin, K.Y., Ishida, B., 2010. Predatory role of the commander squid *Berryteuthis magister* in the Bering Sea: insights from stable isotopes and food habits. Mar. Ecol. Prog. Ser. 415, 91–108.
Hunt, J.C., 1996. The Behavior and Ecology of Midwater Cephalopods from Monterey Bay: Submersible and Laboratory Observations. University of California, Los Angeles, pp.1–231 (Ph.D. dissertation).
Hunt, J.C., 1999. Laboratory observations of the feeding behaviour of the cirrate octopod, *Grimpoteuthis* sp.: one use of cirri. The Veliger 42, 152–156.
Hunt, J.C., Seibel, B.A., 2000. Life history of *Gonatus onyx* (Cephalopoda: Teuthoidea): ontogenetic changes in habitat, behavior and physiology. Mar. Biol. 136, 543–552.
Hurley, G.V., Beck, P., Drew, J., Radtke, R.L., 1979. A preliminary report on validating age readings from statoliths of the short-finned squid (*Illex illecebrosus*). In: International Commission for the Northwest Atlantic Fisheries, Research Document 79/II/26, pp. 1–6.
Ibañez, C.M., Keyl, F., 2010. Cannibalism in cephalopods. Rev. Fish Biol. Fish. 20, 123–136.
Ibañez, C.M., Arancibia, H., Cubillos, L.A., 2008. Biases in determining the diet of jumbo squid Dosidicus gigas (DÒrbigny 1835) (Cephalopoda: Ommastrephidae) off southern-central Chile (34°S–40°S). Helgoland Mar. Res. 62, 331–338.
Ibañez, C.M., Camus, P.A., Rocha, F.J., 2009. Diversity and distribution of cephalopod species off the coast of Chile. Mar. Biol. Res. 5, 374–384.
Ibañez, C., Peña, F., Pardo-Gandarillas, M.C., Poulin, E., Sellanes, J., 2012. Deep-sea octopuses from southeast Pacific Ocean. In: Program and Abstracts for the 13th International Deep-Sea Biology Symposium, Wellington, New Zealand, 3–7 December 2012, p. 217.
Ikeda, Y., Sakurai, Y., Shimazaki, K., 1993. Maturation process of the Japanese common squid *Todarodes pacificus* in captivity. In: Okutani, T., O'Dor, R.K., Kubodera, T. (Eds.), Recent Advances in Fisheries Biology. Tokai University Press, Tokyo, pp. 179–187.
INDEEP (International Network for Scientific Investigations of Deep Sea Ecosystems). http://www.indeep-project.org/ (accessed 12.02.2014).
Irigoien, X., Klevjer, T.A., Røstad, A., Martinez, U., Boyra, G., Acuña, J.L., Bode, A., Echevarria, F., Gonzalez-Gordillo, J.I., Hernandez-Leon, S., Agusti, S., Aksnes, D.L., Duarte, C.M., Kaartvedt, S., 2014. Large mesopelagic fishes biomass and trophic efficiency in the open ocean. Nat. Commun. (5 article 3271).
Ivanovic, M.L., Brunetti, N.E., 1994. Food and feeding of *Illex argentinus*. Antarct. Sci. 6 (2), 185–193.
Jackson, G.D., 2004. Advances in defining life histories of myopsid squid. Mar. Freshw. Res. 55, 357–365.
Jackson, G.D., Jackson, C.H., 2004. Mating and spermatophore placement in the onychoteuthid squid *Moroteuthis ingens*. J. Mar. Biol. Assoc. U.K. 84, 783–784.
Jackson, S., Ryan, P.G., 1986. Differential digestion rates of prey by white-chinned petrels (*Procellaria aequinoctialis*). Auk 103, 617–619.
Jackson, G.D., McKinnon, J.F., Lalas, C., Ardern, R., Buxton, N.G., 1998. Food spectrum of the deepwater squid *Moroteuthis ingens* (Cephalopoda: Onychoteuthidae) in New Zealand Waters. Polar Biol. 20, 56–65.
Jackson, G.D., Bustamante, P., Cherel, Y., Fulton, E.A., Grist, E.P.M., Jackson, C.H., Nichols, P.D., Pethybridge, H., Phillips, K., Ward, R.D., Xavier, J.C., 2007. Applying new tools to cephalopod trophic dynamics and ecology: perspectives from the Southern Ocean Cephalopod Workshop, February 2–3, 2006. Rev. Fish Biol. Fish. 17 (2), 79–99.
Jacoby, C.A., Youngbluth, M.J., Frost, J.R., Flood, P.R., Uiblein, F., Bamstedt, U., Pages, F., Shale, D., 2009. Vertical distribution, behavior, chemical composition and metabolism of *Stauroteuthis syrtensis* (Octopoda: Cirrata) in the northwest Atlantic. Aquat. Biol. 5, 13–22.

Jahn, W., 1971. Deepest photographic evidence of an abyssal cephalopod. Nature 232, 487–488.

Jeletzky, J.A., 1966. Comparative Morphology, Phylogeny and Classification of Fossil Coleoidea. The Paleontological Institute, University of Kansas, Kansas, pp. 1–166, University of Kansas Paleontological Contributions, Mollusca 7.

Jereb, P., Roper, C.F.E. (Eds.), 2010. Cephalopods of the world. An annotated and illustrated catalogue of cephalopod species known to date. In: FAO Species Catalogue for Fishery Purposes, vol. 2. FAO, Rome, Myopsidae and Oegopsidae.

Jorgensen, E.M., 2009. Field guide to squids and octopods of the eastern North Pacific and Bering Sea, Alaska Sea Grant College Program Report, SG-ED-65. Alaska Sea Grant College Program, University of Alaska Fairbanks, Fairbanks.

Joubin, L., 1902. Observations sur divers Cephalopodes. 7e note: *Heteroteuthis weberi* n. sp. Bull. Soc. Scient. Med. l'Ouest 11, 401–404.

Judkins, H., 2009. Cephalopods of the Broad Caribbean: Distribution, Abundance and Ecological Importance. University of South Florida (Dissertation, 156 pp.).

Judkins, H., Ingrao, D., Roper, C.F.E., 2009. First records of *Asperoteuthis acanthoderma* (Lu, 1977) Cephalopoda: Oegopsidae: Chiroteuthidae), from the North Atlantic Ocean, Straits of Florida. Proc. Biol. Soc. Wash. 122 (2), 162–170.

Karnovsky, N.J., Hobson, K.A., Iverson, S.J., 2012. From lavage to lipids: estimating diets of seabirds. Mar. Ecol. Prog. Ser. 451, 263–284.

Kelly, J.F., 2000. Stable isotopes of carbon and nitrogen in the study of avian and mammalian trophic ecology. Can. J. Zool. 78, 1–27.

Kemp, K.M., Jamieson, A.J., Bagley, P.M., McGrath, H., Bailey, D.M., Collins, M.A., Priede, I.G., 2006. Consumption of large bathyal food fall, a six month study in the NE Atlantic. Mar. Ecol. Prog. Ser. 310, 65–76.

Khromov, D.N., 1987a. A new species of *Sepia* (Cephalopoda: Sepiidae) from the northwestern South China Sea. Asian Mar. Biol. 4, 35–40.

Khromov, D.N., 1987b. System[atics] and phylogeny of the cuttlefish family Sepiidae (Cephalopoda). Zool. Z. 66 (8), 1164–1176.

Khromov, D.N., Lu, C.C., Guerra, Á., Dong, Z., Boletzky, S.V., 1998. A synopsis of Sepiidae outside Australian waters (Cephalopoda: Sepioidea). In: Voss, N.A., Vecchione, M., Toll, R.B., Sweeney, M.J. (Eds.), Systematics and Biogeography of Cephalopods. In: Smithsonian Contributions to Zoology, vol. 586. pp. 77–139, I-II.

Klages, N.T.W., 1996. Cephalopods as prey 2. Seals. Philos. Trans. R. Soc. Lond. B Biol. Sci. 351 (1343), 1045–1052.

Kristensen, T.K., 1984. Biology of the squid *Gonatus fabricii* (Lichtenstein, 1818) from West Greenland waters. Meddelelser om Gronland, Bioscience 13, 3–20.

Kröger, B., Vinther, J., Fuchs, D., 2011. Cephalopod origin and evolution: a congruent picture emerging from fossils, development and molecules. Bioessays 33, 602–613.

Kubodera, T., Jefferts, K., 1984a. Distribution and abundance of the early life stages of squid, primarily Gonatidae (Cephalopoda, Oegopsida), in the northern North Pacific (part 1). Bull. Natl. Sci. Mus. (Jpn.) A 10, 91–106.

Kubodera, T., Jefferts, K., 1984b. Distribution and abundance of the early life stages of squid, primarily Gonatidae (Cephalopoda, Oegopsida), in the northern North Pacific (part 2). Bull. Natl. Sci. Mus. (Jpn.) A 10, 165–193.

Kubodera, T., Mori, K., 2005. First-ever observations of a live giant squid in the wild. Proc. R. Soc. Lond. B 272, 2583–2586.

Kubodera, T., Okutani, T., 1994. Eledonine octopods from the Southern Ocean: systematics and distribution. Antarct. Sci. 6 (2), 205–214.

Kubodera, T., Piatkowski, U., Okutani, T., Clarke, M.R., 1998. Taxonomy and zoogeography of the family Onychoteuthidae (Cephalopoda: Oegopsida.). In: Voss, N.A.,

Vecchione, M., Toll, R.B., Sweeney, M.J. (Eds.), Systematics and Biogeography of Cephalopods. In: Smithsonian Contributions to Zoology, vol. 586. pp. 277–291, I-II.

Kubodera, T., Koyama, Y., Mori, K., 2007. Observations of wild hunting behaviours *and* deep-sea, eight-armed squid, *Taningia danae*. Proc. R. Soc. B 274, 1029–1034.

Kubodera, T., Hochberg, F.G., Young, R.E., Vecchione, M., 2013. Gonatidae Hoyle, 1886. http://tolweb.org/Gonatidae/19414/ (accessed 08.01.2013).

Kubota, T., Koshiga, M., Okutani, T., 1981. Rare and interesting squid from Japan—VII. Some biological data on *Chiroteuthis imperator* from Suroga Bay, Japan (Cephalopoda: Chiroteuthidae). Venus: Jap. J. Malacol. 40, 150–159.

Kuznetsova, N.A., Fedorets, Y., 1987. Feeding of the commander squid *Berryteuthis magister*. Biol. Moriya (USSR) 1, 71–73 (in Russian with English abstract).

Landman, N.H., Cochran, J.K., Cerrato, R., Mak, J., Roper, C.F.E., Lu, C.C., 2004. Habitat and age of the giant squid (*Architeuthis sanctipauli*) inferred from isotopic analyses. Mar. Biol. 144, 685–691.

Laptikhovsky, V.V., 1999. Fecundity and spawning in squid of families Enoploteuthidae and Ancistrocheiridae (Cephalopoda: Oegopsida). Sci. Mar. 63 (1), 1–7.

Laptikhovsky, V.V., 2001. Fecundity, egg masses and hatchlings of *Benthoctopus* spp. (Octopodidae) in the Falkland waters. J. Mar. Biol. Assoc. U.K. 81, 267–270.

Laptikhovsky, V.V., 2002. Diurnal feeding rhythm of the short-fin squid *Illex argentinus* (Cephalopoda: Ommastrephidae) in the Falkland waters. Fish. Res. 59, 233–237.

Laptikhovsky, V.V., 2013. Reproductive strategy of deep-sea and Antarctic octopods of the genera *Graneledone*, *Adelieledone* and *Muusoctopus* (Mollusca: Cephalopoda). Aquat. Biol. 18 (1), 21–29.

Laptikhovsky, V.V., Murzov, S.A., 1990. Epipelagic egg mass of the squid *Sthenoteuthis pteropus* collected in the tropical eastern Atlantic. Biol. Moriya 3, 62–63.

Laptikhovsky, V.V., Salman, A., 2003. On reproductive strategies of the epipelagic octopods of the superfamily Argonautoidea (Cephalopoda: Octopoda). Mar. Biol. 142 (2), 321–326.

Laptikhovsky, V.V., Arkhipkin, A.I., Hoving, H.J.T., 2007. Reproductive biology in two species of deep-sea squids. Mar. Biol. 152, 981–990.

Laptikhovsky, V.V., Nigmatullin, C.M., Hoving, H.J.T., Onsoy, B., Salman, A., Zumholz, K., Shevtsov, G.A., 2008. Reproductive strategies in female polar and deep-sea bobtail squid genera *Rossia* and *Neorossia* (Cephalopoda: Sepiolidae). Polar Biol. 31 (12), 1499–1507.

Laptikhovsky, V.V., Collins, M.A., Arkhipkin, A.I., 2013. First case of possible iteroparity among coleoid cephalopods: the giant warty squid *Kondakovia longimana*. J. Mollus. Stud. 79 (3), 270–272.

Laptikhovsky, V.V., Boersch-Supan, P., Kemp, K., Letessier, T., Rogers, A.D., Bolstad, K.S.R. Cephalopods of the Southwest Indian Ocean Ridge: a hotspot of extreme biological diversity and absence of endemism.

Lefkaditou, E., Politou, C.-Y., Papaconstantinou, C., 2000. Notes on *Brachioteuthis riisei* (Steenstrup, 1882) and *Ancistroteuthis lichtensteini* (Orbigni, 1839) (Cephalopoda: Teuthoidea) found in the Aegean Sea. Belg. J. Zool. 130 (Suppl.), 71–75.

Lindgren, A., Pankey, M., Hochberg, F., Oakley, T., 2012. A multi-gene phylogeny of Cephalopoda supports convergent morphological evolution in association with multiple habitat shifts in the marine environment. BMC Evol. Biol. 12, 129.

Lindgren, A., Young, R.E., Mangold, K.M., 2013. Pyroteuthidae Pfeffer 1912. The Fire Squid. http://tolweb.org/Pyroteuthidae/19637/ (accessed 07.01.2013).

Lipinski, M.R., 1979. The information concerning research upon ageing procedure of squids. In: International Commission for the Northwest Atlantic Fisheries, Working Paper 79/II/40, pp. 1–2.

Lipinski, M.R., 1986. Methods for the validation of squid age from statoliths. J. Mar. Biol. Assoc. U.K. 66, 505–526.

Lipinski, M.R., 1992. Cephalopods and the Benguela ecosystem: trophic relationships and impact. South Afr. J. Mar. Sci. 12, 791–802.

Lipinski, M.R., 2001. Preliminary description of two new species of cephalopods (Cephalopoda: Brachioteuthidae) from South Atlantic and Antarctic waters. Bull. Sea Fish. Inst. 152, 1–14.

Lipinski, M.R., Linkowski, T.B., 1988. Food of the squid *Ommastrephes bartramii* Lesueur, (1821) from the south-west Atlantic Ocean. South Afr. J. Mar. Sci. 6, 43–46.

Lipinski, M.R., Young, R.E., 2008. Brachioteuthidae Pfeffer 1908. http://tolweb.org/Brachioteuthidae/19409/ (accessed 06.07.2008).

Lipinski, M.R., Durholtz, M.D., Underhill, L.G., 1998. Field validation of age readings from the statoliths of chokka squid (*Loligo vulgaris reynaudii* d'Orbigny, 1845) and an assessment of associated errors. ICES J. Mar. Sci. 55, 240–257.

Lordan, C., Collins, M.A., Peralez-Raya, C., 1998. Observations on morphology, age and diet of three *Architeuthis* caught off the west coast of Ireland in 1995. J. Mar. Biol. Assoc. U.K. 78, 903–917.

Lordan, C., Collins, M.A., Key, L.N., Browne, E.D., 2001. The biology of the ommastrephid squid, *Todarodes sagittatus*, in the north-east Atlantic. J. Mar. Biol. Assoc. U.K. 81, 299–306.

Lorrain, A., Argüelles, J., Alegre, A., Bertrand, A., Munaron, J.-M., Richard, P., Cherel, Y., 2011. Sequential isotopic signature along gladius highlights contrasted individual foraging strategies of jumbo squid (*Dosidicus gigas*). PLoS One 6 (7), e22194.

Lu, C.C., 1998. A synopsis of Sepiidae in Australian waters (Sepioidea: Cephalopoda). In: Voss, N.A., Vecchione, M., Toll, R.B., Sweeney, M.J. (Eds.), Systematics and Biogeography of Cephalopods. In: Smithsonian Contributions to Zoology, vol. 586. pp. 159–190, I-II.

Lu, C.C., Boucher-Rodoni, R., 2006. A new genus and species of sepiolid squid from the waters around Tonga in the central South Pacific (Mollusca: Cephalopoda: Sepiolidae). Zootaxa 1310, 37–51.

Lu, C.C., Clarke, M.R., 1975. Vertical distribution of cephalopods at 11°N, 20°W in the North Atlantic. J. Mar. Biol. Assoc. U.K. 55, 369–389.

Lu, C.C., Ickeringill, R., 2002. Cephalopod beak identification and biomass estimation techniques: tools for dietary studies of southern Australian finfishes. Museum Victoria Science Reports 6, 1–65.

Lu, C.C., Stranks, T.N., 1994. Synopsis of *Pareledone* and *Megaleledone* species, with description of two new species from East Antarctica (Cephalopoda: Octopodidae). Mem. Mus. Vict. 54, 221–242.

Lu, C.C., Roper, C., Tait, R.W., 1985. A revision of *Loliolus* (Cephalopoda; Loliginidae), including *L. noctiluca*, a new species of squid from Australian waters. Proc. R. Soc. Vict. 97 (2), 59–85.

Lutz, R.A., Voight, J.R., 1994. Close encounter in the deep. Nature 371, 563.

Maddison, D.R., Schulz, K.-S., Maddison, W.P., 2007. The tree of life web project. Zootaxa 1668, 19–40.

Mangold, K.M., Vecchione, M., Young, R.E., 2010. Tremoctopodidae Tryon, 1879. *Tremoctopus* Chiaie 1830. Blanket Octopus. http://tolweb.org/Tremoctopus/20202/ (accessed 15.08.2010).

Mangold-Wirz, K., 1963. Biologie des Céphalopodes benthiques et nectoniques de la Mer Catalane. Vie et Milieu 13 (Suppl.), 1–285.

Mann, T., 1984. Spermatophores: development, structure, biochemical attributes and role in transfer of spermatozoa. Zoophysiology, vol. 15. Springer-Verlag, Berlin.

Mann, T., Martin, A.W., Thiersch, J.B., 1970. Male reproductive tract, spermatophores and spermatophoric reaction in the giant octopus of the North Pacific, *Octopus dofleini martini*. Proc. R. Soc. B 175, 31–61.

Marian, J.E.A.R., 2011. Perforating potential of loliginid spermatophores. J. Mollus. Stud. 77, 98–100.

Marian, J.E.A.R., 2012a. Spermatophoric reaction reappraised: novel insights into the functioning of the loliginid spermatophore based on *Doryteuthis plei* (Mollusca: Cephalopoda). J. Morphol. 273, 248–278.

Marian, J.E.A.R., 2012b. A model to explain spermatophore implantation in cephalopods (Mollusca: Cephalopoda) and a discussion on its evolutionary origins and significance. Biol. J. Linn. Soc. 105, 711–726.

Marian, J.E.A.R., Shiraki, Y., Kawai, K., Kojima, S., Suzuki, Y., Ono, K., 2012. Revisiting a medical case of "stinging" in the human oral cavity caused by ingestion of raw squid (Cephalopoda: Teuthida): new data on the functioning of squid's spermatophores. Zoomorphology 131, 293–301.

Marian, J.E.A.R., 2014. Evolution of spermatophore transfer mechanisms in cephalopods. J. Nat. Hist. http://dx.doi.org/10.1080/00222933.2013.825026.

Marian, J.E.A.R., Domaneschi, O., 2012. Unraveling the structure of squids' spermatophores: a combined approach based on *Doryteuthis plei* (Blainville, 1823) (Cephalopoda: Loliginidae). Acta Zool. 93, 281–307.

Markaida, U., 2006. Food and feeding of jumbo squid *Dosidicus gigas* in the Gulf of California and adjacent waters after the 1997–98 El Niño event. Fish. Res. 79, 16–27.

Markaida, U., Sosa-Nishizaki, O., 2003. Food and feeding habits of jumbo squid *Dosidicus gigas* (Cephalopoda: Ommastrephidae) from the Gulf of California, Mexico. J. Mar. Biol. Assoc. U.K. 83, 507–522.

Maurer, R.O., Bowman, R.E., 1985. Food consumption of squid (*Illex illecebrosus* and *Loligo pealei*) off the Northeastern United States. NAFO Sci. Council Stud. 9, 117–124.

McIntyre, A. (Ed.), 2009. Life in the World's Oceans: Diversity, Distribution and Abundance. Wiley-Blackwell, Census of Marine Life, Chichester, UK.

Milinkovitch, M.C., LeDuc, R., Tiedman, R., Dizon, A., 2002. Applications of molecular data in cetacean taxonomy and population genetics with special emphasis on defining species boundaries in cetations. In: Evans, P.G.H., Raga, J.A. (Eds.), Marine Mammals: Biology and Conservation. Kluwer Academic, New York, pp. 325–359.

Moiseev, S.I., 1991. Observation of the vertical distribution and behavior of nektonic squids using manned submersibles. Bull. Mar. Sci. 49 (1–2), 446–456.

Morato, T., Watson, R., Pitcher, T.J., Pauly, D., 2006. Fishing down the deep. Fish Fish. 7 (1), 24–34.

Naef, A., 1912a. Teuthologische Notizen. 1. Die Familien der Myopsiden. Zool. Anz. 39 (7), 241–244.

Naef, A., 1912b. Teuthologische Notizen. 2. Die Gattungen der Sepioliden. Zool. Anz. 39 (7), 244–248.

Naef, A., 1912c. Teuthologische Notizen. 3. Die Arten der Gattungen *Sepiola* und *Sepietta*. Zool. Anz. 39 (7), 262–271.

Naef, A., 1912e. Teuthologische Notizen. 7. Zur Morphologie und Systematik der *Sepiola*- und *Sepietta*-Arten. Zool. Anz. 40 (2), 78–85.

Naef, A., 1921–1923. Die Cephalopoden (Systematik). Fauna e Flora del Golfo di Napoli 35, 1–863.

Naud, M.J., Havenhand, J.N., 2006. Sperm motility and longevity in the giant cuttlefish, *Sepia apama* (Mollusca: Cephalopoda). Mar. Biol. 148 (3), 559–566.

Nesis, K.N., 1972. A review of the squid genera *Taonius* and *Belonella* (Oegopsida, Cranchiidae). Zool. Z. 51 (3), 341–350 (in Russian).

Nesis, K.N., 1974. A revision of the squid genera *Corynomma*, *Megalocranchia*, *Sandalops*, and *Liguriella* (Oegopsida, Cranchiidae). Trans. Inst. Oceanol. USSR 96, 6–22.

Nesis, K.N., 1978. The subfamily Ancistrocheirinae (Enoploteuthidae). Zool. Z. 57 (3), 446–449 (in Russian).

Nesis, K.N., 1986. Cephalopods of seamounts in the western Indian Ocean. Oceanology 26, 91–96.
Nesis, K.N., 1987a. Cephalopods of the World. TFH Publications, Neptune City.
Nesis, K.N., 1987b. Redescription of *Abralia steindachneri* and revision of subgenus systematics of the genus *Abralia* (Cephalopoda, Enoploteuthidae). Zool. Z. 66 (11), 1693–1705 (in Russian).
Nesis, K.N., 1995. Mating, spawning, and death in oceanic cephalopods: a review. Ruthenica Suppl. 6, 23–64.
Nesis, K.N., 1997. Gonatid squids in the Subarctic North Pacific: ecology, biogeography, niche diversity and role in the ecosystem. Adv. Mar. Biol. 32, 243–324.
Nesis, K.N., 2000. Squids of the family Onychoteuthidae: phylogeny, biogeography, and way of life. Zool. Z. 79 (3), 272–281 (in Russian).
Nesis, K.N., 2001. West-Arctic and East-Arctic distributional ranges of cephalopods. Sarsia 86, 1–11.
Nesis, K.N., Nikitina, I.V., 1990. Revision of the squid family Lepidoteuthidae. Zool. J. 69 (10), 38–49.
Nesis, K.N., Nigmatullin, C.M., Nikitina, I.V., 1998. Spent females of deepwater squid *Galiteuthis glacialis* under the ice at the surface of the Weddell Sea (Antarctic). J. Zool. 244, 185–200.
Nigmatullin, C.M., 1992. Systematics, phylogeny and morpho-functional evolution of squids of the family Ommastrephidae. Bull. Mar. Sci. 49 (1–2), 666 (Abstract).
Nigmatullin, C.M., Arkhipkin, A.I., 1998. A review of the biology of the Diamondback Squid, *Thysanoteuthis rhombus* (Oegopsida, Thysanoteuthidae). In: Okutani, T. (Ed.), Large Pelagic Squids. Japan Marine Fishery Resources Research Center, Tokyo, pp. 156–181.
Nigmatullin, C.M., Laptikhovsky, V.V., 1994. Reproductive strategies in the squids of the family Ommastrephidae (preliminary report). Ruthenica Suppl. 4 (1), 79–82.
Nigmatullin, C.M., Markaida, U., 2009. Oocyte development, fecundity and spawning strategy of large sized jumbo squid *Dosidicus gigas* (Oegopsida: Ommastrephinae). J. Mar. Biol. Assoc. U.K. 89, 789–801.
Nigmatullin, C.M., Arkhipkin, A.I., Sabirov, R.M., 1995. Age, growth and reproductive biology of diamond-shaped squid *Thysanoteuthis rhombus* (Oegopsida: Thysanoteuthidae). Mar. Ecol. Prog. Ser. 124 (1), 73–87.
Nigmatullin, C.M., Nesis, K.N., Arkhipkin, A.I., 2001. A review on the biology of the jumbo squid *Dosidicus gigas*. Fish. Res. 54, 9–19.
Nigmatullin, C.M., Laptikhovsky, V.V., Moustahfid, H., 2002. Brief review on the ecology in the North African population of arrow squid *Todarodes sagittatus* (Cephalopoda: Ommastrephidae). Bull. Mar. Sci. 71, 581–590.
Nigmatullin, C.M., Sabirov, R.M., Zalygalin, V.P., 2003. Ontogenetic aspects of morphology, size, structure and production of spermatophores in ommastrephid squids: an overview. Berl. Paläobiol. Abhandl. 3, 225–240.
Nixon, M., Dilly, P.N., 1977. Sucker surfaces and prey capture. Symp. Zool. Soc. Lond. 38, 447–511.
Norman, M.D., Hochberg, F.G., 2005. The current state of octopus taxonomy. Phuket Mar. Biol. Center Res. Bull. 66, 127–154.
Norman, M.D., Lu, C.C., 1997. Sex in giant squid. Nature 389, 683–684.
Norman, M.D., Sweeney, M.J., 1997. The shallow-water octopuses (Cephalopoda: Octopodinae) of the Philippine Islands. Invertebr. Taxon. 11, 89–140.
Norman, M.D., Hochberg, F.G., Boucher-Rodoni, R., 2005. A revision of the deep-water octopus genus *Scaeurgus* (Cephalopoda: Octopodidae) with description of three new species from the southwest Pacific Ocean. J. Mollus. Stud. 71, 319–337.
O'Dor, R.K., 1983. *Illex illecebrosus*. In: Boyle, P.R. (Ed.), Cephalopod Life Cycles. Volume I: Species Accounts. Academic Press, London, pp. 175–199.

O'Dor, R.K., Balch, N., 1985. Properties of *Illex illecebrosus* egg masses potentially influencing larval oceanographic distribution. NAFO Sci. Council Stud. 9, 69–76.
O'Dor, R.K., Malacaster, E.G., 1983. *Bathypolypus arcticus*. In: Boyle, P.R. (Ed.), Cephalopod Life Cycles. Academic Press, London, pp. 401–410.
O'Dor, R.K., Shadwick, R.E., 1989. Squid, the Olympian Cephalopods. J. Cephalopod Biol. 1 (1), 33–55.
Ohkouchi, N., Tsuda, R., Chikaraishi, Y., Tanabe, K., 2013. A preliminary estimate of the trophic position of the deep-water ram's horn squid *Spirula spirula* based on the nitrogen isotopic composition of amino acids. Mar. Biol. 160 (4), 773–779.
Okiyama, M., 1965. On the feeding habit of the common squid, *Todarodes pacificus* Steenstrup, in the off shore region of the Japan Sea. Bull. Jpn. Sea Region. Fish. Res. Lab. 14, 31–41.
Okutani, T., 1974. Epipelagic decapod cephalopods collected by micronekton tows during the EASTROPAC expeditions, 1967–1968 (systematic part). Bull. Tokai Region. Fish. Res. Lab. 80, 29–118.
Okutani, T., 1983. *Todarodes pacificus*. In: Boyle, P.R. (Ed.), Cephalopod Life Cycles. Volume I: Species Accounts. Academic Press, London, pp. 201–214.
Okutani, T., Kubota, T., 1976. Cephalopods eaten by lancetfish, *Alepisaurus ferox* Lowe, in Suruga Bay, Japan. Bull. Tokai Region. Fish. Res. Lab. 84, 1–9.
Okutani, T., McGowan, J.A., 1969. Systematics, distribution, and abundance of the epiplanktonic squid (Cephalopoda, Decapoda) larvae of the California Current, April 1954–March 1957. Bull. Scripps Inst. Oceanogr. 14 97 pp.
Okutani, T., Murata, M., 1983. A review of the biology of the oceanic squid *Onychoteuthis borealijaponica*. Mem. Natl. Mus. Vict. 44, 189–195.
Okutani, T., Tsuchida, S., 2005. Occurrence and living habit of bathyal bobtail squid, *Heteroteuthis hawaiiensis* (Cephalopoda: Sepioliidae) from off the Ogasawara Islands, Japan. Venus 63, 125–133.
Okutani, T., Nakamura, I., Seki, K., 1995. An unusual egg-brooding behavior of an oceanic squid in the Okhotsk Sea. Venus 54, 237–239.
Oosthuizen, A., 2003. A Development and Management Framework for a New Octopus Vulgaris Fishery in South Africa. Rhodes University (Ph.D. thesis).
O'Shea, S., 2004a. Estimating Age and Growth Rate in Architeuthis *dux*. http://www.tonmo.com/community/pages/architeuthis-age/ (accessed 13.02.2014).
O'Shea, S., 2004b. The giant octopus *Haliphron atlanticus* (Mollusca: Octopoda) in New Zealand waters. N. Zeal. J. Zool. 31, 7–13.
O'Shea, S., Bolstad, K.S., Ritchie, P.A., 2004. First records of egg masses of *Nototodarus gouldi* McCoy, 1888 (Mollusca: Cephalopoda: Ommastrephidae), with comments on egg-mass susceptibility to damage by fisheries trawl. N. Zeal. J. Zool. 31 (2), 161–166.
Packard, A., 1972. Cephalopods and fish, the limits of convergence. Biol. Rev. 47, 241–307.
Paiva, V.H., Xavier, J.C., Geraldes, P., Ramirez, I., Ramos, J.A., Garthe, S., 2010. Foraging ecology of Cory's shearwaters in different ecological environments of the North Atlantic. Mar. Ecol. Prog. Ser. 410, 257–268.
Parry, M., 2006. Comparision of the feeding behaviors of two ommastrephid squids, *Ommastrephes bartramii* and *Sthenoteuthis oualaniensis* in the vicinity of Hawaii. Mar. Ecol. Prog. Ser. 318, 229–235.
Passarella, K.C., Hopkins, T.L., 1991. Species composition and food habits of the micronektonic cephalopod assemblage in the eastern Gulf of Mexico. Bull. Mar. Sci. 49 (1–2), 638–659.
Pecl, G.T., Jackson, G.D., 2008. The potential impacts of climate change on inshore squid: biology, ecology and fisheries. Rev. Fish Biol. Fish. 18 (4), 373–385.
Perales-Raya, C., Bartolomé, A., García-Santamaría, M.T., Pascual-Alayón, P., Almansa, E., 2010. Age estimation obtained from analysis of octopus (*Octopus vulgaris* Cuvier, 1797) beaks: improvements and comparisons. Fish. Res. 106, 171–176.

Perales-Raya, C., Jurado-Ruzafa, A., Bartolomé, A., Duque, V., Carrasco, M.N., Fraile-Nuez, E., 2013. Age of spent *Octopus vulgaris* and stress mark analysis using beaks of wild individuals. Hydrobiologia 725 (1), 105–114.

Pereyra, W., 1965. New records and observations on the flapjack devilfish *Opistoteuthis californiana* Berry. Pacific Sci. 19, 427–441.

Perez, J.A.A., O'Dor, R.K., 2000. Critical transitions in early life histories of short-finned squid, *Illex illecebrosus* as reconstructed from gladius growth. J. Mar. Biol. Assoc. U.K. 80, 1–6.

Perez, J.A.A., Haimovici, M., Cousin, J.C.B., 1990. Sperm storage mechanisms and fertilization in females of two South American eledonids (Cephalopoda: Octopoda). Malacologia 32, 147–154.

Perez, J.A.A., O'Dor, R.K., Beck, P., Dawe, E., 1996. Evaluation of gladius dorsal surface structure for age and growth studies of the short-finned squid, *Illex illecebrosus* (Teuthoidea: Ommastrphidae). Can. J. Fish. Aquat. Sci. 53, 2837–2846.

Perez, J.A.A., Alves, E., Clark, M., Bergstad, O.A., Gebruk, A.V., Cardoso, I.A., Rogacheva, A., 2012. Patterns of life on the southern Mid-Atlantic Ridge. Compiling what is known and addressing future research. Oceanography 25, 14–29.

Peréz-Gándaras, G., Guerra, Á., 1978. Nueva cita de *Architeuthis* (Cephalopoda: Teuthoidea): Descripcion y alimentacion. Investig. Pesq. 42 (2), 401–414.

Phillips, K.L., Jackson, G.D., Nichols, P.D., 2003a. Temporal variations in the diet of the squid *Moroteuthis ingens* at Macquarie Island: stomach contents and fatty acid analyses. Mar. Ecol. Prog. Ser. 256, 135–149.

Phillips, K.L., Nichols, P.D., Jackson, G.D., 2003b. Dietary variation of the squid *Moroteuthis ingens* at four sites in the Southern Ocean: stomach contents, lipid and fatty acid profiles. J. Mar. Biol. Assoc. U.K. 83, 523–534.

Phillips, R.A., Croxall, J.P., Silk, J.R.D., Briggs, D.R., 2008. Foraging ecology of albatrosses and petrels from South Georgia: two decades of insights from tracking technologies. Aquat. Conserv.: Mar. Freshw. Ecosyst. 17, S6–S21.

Piatkowski, U., 1991. On the distribution of pelagic cephalopods in the Arabian Sea. Bull. Mar. Sci. 49 (1–2), 186–198.

Piatkowski, U., 2011. Psychroteuthidae Thiele 1920. *Psychroteuthis glacialis* Thiele 1920. http://tolweb.org/Psychroteuthis_glacialis/19783/ (accessed 20.03.2011).

Piatkowski, U., Hagen, W., 1994. Distribution and lipid composition of early life stages of the cranchiid squid *Galiteuthis glacialis* (Chun) in the Weddell Sea, Antarctica. Antarct. Sci. 6, 235–239.

Piatkowski, U., Pütz, K., 1994. Squid diet of emperor penguins (*Aptenodytes forsteri*) in the eastern Weddell Sea, Antarctica during late summer. Antarct. Sci. 6, 241–247.

Piatkowski, U., Pütz, K., Heinemann, H., 2001. Cephalopod prey of king penguins (*Aptenodytes patagonicus*) breeding at Volunteer Beach, Falkland Islands, during austral winter 1996. Fish. Res. 52, 79–90.

Piatkowski, U., Rodhouse, P.G., White, M.G., Bone, D.G., Symon, C., 1994. Nekton community of the Scotia Sea as sampled by the RMT 25 during austral summer. Mar. Ecol. Prog. Ser. 112, 13–28.

Piatkowski, U., Hernández-García, V., Clarke, M.R., 1998. On the biology of the European flying squid *Todarodes sagittatus* (Lamarck, 1798) (Cephalopoda, Ommastrephidae) in the central eastern Atlantic. South Afr. J. Mar. Sci. 20, 375–383.

Pickford, G.E., 1946. *Vampyroteuthis infernalis* Chun an archaic dibranchiate cephalopod. I. Natural history and distribution. Dana-Report 29, 1–40.

Pickford, G.E., 1949. *Vampyroteuthis infernalis* Chun an archaic dibranchiate cephalopod. II. External anatomy. Dana-Report 32, 1–132.

Potier, M., Ménard, F., Cherel, Y., Lorrain, A., Sabatié, R., Marsac, F., 2007. Role of pelagic crustaceans in the diet of the longnose lancetfish *Alepisaurus ferox* in the Seychelles waters. Afr. J. Mar. Sci. 29, 113–122.

Quetglas, A., Alemany, F., Carbonell, A., Merella, P., Sánchez, P., 1998. Some aspects of the biology of *Todarodes sagittatus* (Cephalopoda: Ommastrephidae) from the Balearic Sea (Western Mediterranean). Sci. Mar. 62 (1–2), 73–82.

Quetglas, A., Alemany, F., Carbonell, A., Merella, P., Sánchez, P., 1999. Diet of the European flying squid *Todarodes sagittatus* (Cephalopoda: Ommastrephidae) in the Balearic Sea (western Mediterranean). J. Mar. Biol. Assoc. U.K. 79, 479–486.

Quetglas, A., González, M., Carbonell, A., Sánchez, P., 2001. Biology of the deep-sea octopus *Bathypolypus sponsalis* (Cephalopoda: Octopodidae) from the western Mediterranean Sea. Mar. Biol. 138, 785–792.

Quetglas, A., de Mesa, A., Ordines, F., Grau, A., 2010. Life history of the deep-sea cephalopod family Histioteuthidae in the western Mediterranean. Deep Sea Res. I Oceanogr. Res. Pap. 57, 999–1008.

Ramirez-Llodra, E., Brandt, A., Danovaro, R., De Mol, B., Escobar, E., German, C.R., Levin, L.A., Arbizu, P.M., Menot, L., Buhl-Mortensen, P., et al., 2010. Deep, diverse and definitely different: unique attributes of the world's largest ecosystem. Biogeosciences 7 (9), 2851–2899.

Ramos, R., Gonzalez-Solis, J., 2012. Trace me if you can: the use of intrinsic biogeochemical markers in marine top predators. Front. Ecol. Environ. 10, 258–266.

Rasero, M., González, A.F., Castro, B.G., Guerra, Á., 1996. Predatory relationships of two sympatric squid, *Todaropsis eblanae* and *Illex coindetii* (Cephalopoda: Ommastrephidae) in Galician waters. J. Mar. Biol. Assoc. U.K. 76, 73–87.

Ratnasingham, S., Hebert, P.D.N., 2007. BOLD: the barcode of life data system. Mol. Ecol. Notes 7 (3), 355–364, http://www.barcodinglife.org.

Ratnasingham, S., Hebert, P.D.N., 2013. A DNA-based registry for all animal species: the Barcode Index Number (BIN) System. PLoS ONE 8 (7), e66213.

Reid, A., 2001. *Sepia hedleyi* Berry, 1918 (Cephalopoda: Sepiidae): a complete description and clarification of the status of *S. dannevigi* Berry, 1918 and *S. rex* (Iredale, 1926). Rec. South Aust. Mus. 34 (2), 79–97.

Reid, A., 2009. *Sepioloidea magna* sp. nov.: a new bottletail squid (Cephalopoda: Sepiadariidae) from northern Australia. Beagle, Rec. Mus. Art Gal. North. Territ. 25, 103–109.

Reid, A., Jereb, P., Roper, C.F.E., 2005. Cuttlefishes. In: Jereb, P., Roper, C.F.E. (Eds.), Cephalopods of the world. An annotated and illustrated catalogue of species known to date. FAO Species Catalogue for Fishery Purposes, No. 4, vol. 1. FAO, Rome, pp. 56–152.

Rex, M.A., Crame, J.A., Sturart, C.T., Clarke, A., 2005. Large-scale biogeographic patterns in marine molluscs: a confluence of history and productivity? Ecology 86, 2288–2297.

Roberts, M.J., Zemlak, T., Connell, A., 2011. Cyclonic eddies reveal Oegopsida squid egg balloon masses in the Agulhas Current, South Africa. Afr. J. Mar. Sci. 33 (2), 239–246.

Robison, B.H., 1989. Depth of occurrence and partial chemical composition of a giant squid, *Architeuthis*, off southern California. Veliger 32 (1), 39–42.

Robison, B.H., 2000. The coevolution of undersea vehicles and deep-sea research. Mar. Technol. Soc. J. 33, 65–73.

Robison, B.H., 2004. Deep pelagic biology. J. Exp. Mar. Biol. Ecol. 300, 253–272.

Robison, B.H., Reisenbichler, K.R., Hunt, J.C., Haddock, S.H.D., 2003. Light production by the arm tips of the deep-sea cephalopod *Vampyroteuthis infernalis*. Biol. Bull. 205, 102–109.

Rocha, F., Guerra, Á., Gonzalez, A.F., 2001. A review of reproductive strategies in cephalopods. Biol. Rev. 76 (3), 291–304.

Rocha, F., González, A.F., Segonzac, M., Guerra, Á., 2002. Behavioural observations of the cephalopod *Vulcanoctopus hydrothermalis*. Cah. Biol. Mar. 43, 299–302.

Rochebrune, A.T., 1884a. Etude monographique sur la famille des Loligopsidae. Bull. Soc. Philomath. Paris 8 (1), 7–28, series 7.

Rochebrune, A.T., 1884b. Etude monographique de la famille des Sepiidae. Bull. Soc. Philomath. Paris 8 (2), 74–122, series 7.

Rochebrune, A.T., 1884c. Etude monographique de la famille des Eledonidae. Bull. Soc. Philomath. Paris 8 (3), 152–163, series 7.

Rodhouse, P.G., 1990. Cephalopod fauna of the Scotia Sea at South Georgia: potential for commercial exploitation and possible consequences. In: Kerry, K.R., Hempel, G. (Eds.), Antarctic Ecosystems: Ecological Change and Conservation. Springer-Verlag, Berlin.

Rodhouse, P.G., 2013. Role of squid in the Southern Ocean pelagic ecosystem and the possible consequences of climate change. Deep Sea Res. II 95, 129–138.

Rodhouse, P.G., Nigmatullin, C.M., 1996. Role as consumers. Philos. Trans. R. Soc. Lond. B Biol. Sci. 351, 1003–1022.

Rodhouse, P.G., Piatkowski, U., 1995. Fine-scale distribution of juvenile cephalopods in the Scotia Sea and adaptive allometry of the brachial crown. Mar. Biol. 124 (1), 111–117.

Rodhouse, P.G., Arnbom, T.R., Fedak, M.A., Yeatman, J., Murray, A.W.A., 1992. Cephalopod prey of the southern elephant seal, *Mirounga leonina* L. Can. J. Zool. 70, 1007–1015.

Rodrigues, M., Garcí, M.E., Guerra, Á., Troncoso, J.S., 2009. Mating behavior of the atlantic bobtail squid *Sepiola atlantica* (Cephalopoda: Sepiolidae). Vie et Milieu 59 (3–4), 271–275.

Roeleveld, M.A.C., Lipinski, M.R., 1991. The giant squid *Architeuthis* in southern African waters. J. Zool. (Lond.) 224, 431–477.

Roeleveld, M.A.C., Lipinski, M.R., Augustyn, C.J., Stewart, B.A., 1992. The distribution and abundance of cephalopods on the continental slope of the eastern south Atlantic. South Afr. J. Mar. Sci. 12, 739–752.

Roper, C.F.E., 1969. Systematics and zoogeography of the worldwide bathypelagic squid *Bathyteuthis* (Cephalopoda: Oegopsida). U. S. Natl. Mus. Bull. 291, 1–210.

Roper, C.F.E., 1974. Vertical and seasonal distribution of pelagic cephalopods in the Mediterranean Sea. Preliminary report. Bull. Am. Malacol. Union 39, 27–30.

Roper, C.F.E., 2012. Bathyteuthidae Pfeffer 1900. *Bathyteuthis* Hoyle 1885. http://tolweb.org/Bathyteuthis/19429/ (accessed 2012.11.10).

Roper, C.F.E., Jereb, P., 2010. Family Onychoteuthidae. In: Jereb, P., Roper, C.F.E. (Eds.), Cephalopods of the world. An annotated and illustrated catalogue of species known to date. Myopsid and Oegopsid Squids. FAO Species Catalogue for Fishery Purposes, No. 4, vol. 2. FAO, Rome, pp. 348–369.

Roper, C.F.E., Vecchione, M., 1996. In situ observations on *Brachioteuthis beanii* (Verrill): paired behavior, probably mating (Cephalopoda, Oegopsida). Am. Malacol. Bull. 13, 55–60.

Roper, C.F.E., Vecchione, M., 1997. In-situ observations test hypotheses of functional morphology in *Mastigoteuthis* (Cephalopoda, Oegopsida). Vie Milieu 47, 87–93.

Roper, C.F.E., Young, R.E., 1968. The family Promachoteuthidae (Cephalopoda: Oegopsida). I. A re-evaluation of its systematic position based on new material from Antarctic and adjacent waters. Antarct. Res. Ser. 11, 203–214.

Roper, C.F.E., Young, R.E., 1975. Vertical distribution of pelagic cephalopods. Smithson. Contr. Zool. 209, 1–51.

Roper, C.F.E., Young, R.E., 2013. *Chiroteuthis* Orbigny, 1841. http://tolweb.org/Chiroteuthis/19462/ (accessed 08.01.2013).

Roper, C.F.E., Lu, C.C., Vecchione, M., 1998. A revision of the systematics and distribution of *Illex* species. In: Voss, N.A., Vecchione, M., Toll, R.B., Sweeney, M.J. (Eds.), Systematics and Biogeography of Cephalopods. In: Smithsonian Contributions to Zoology, vol. 586. pp. 405–423, I-II.

Roper, C.F.E., Jorgensen, E.M., Katugin, O.N., Jereb, P., 2010a. Family Gonatidae. In: Jereb, P., Roper, C.F.E. (Eds.), Cephalopods of the world. An annotated and illustrated catalogue of species known to date. Myopsid and Oegopsid Squids. FAO, Rome.

Roper, C.F.E., Nigmatullin, C., Jereb, P., 2010b. Family Ommastrephidae. In: Jereb, P., Roper, C.F.E. (Eds.), Cephalopods of the world. An annotated and illustrated catalogue of species known to date. Myopsid and Oegopsid Squids. FAO Species Catalogue for Fishery Purposes, No. 4, vol. 2. FAO, Rome, pp. 269–347.

Ropert-Coudert, Y., Kato, A., Wilson, R.P., Cannell, B., 2006. Foraging strategies and prey encounter rate of free-ranging Little Penguins. Mar. Biol. 149, 139–148.

Rosa, R., Dierssen, H.M., Gonzalez, L., Seibel, B.A., 2008. Large-scale diversity patterns of cephalopods in the Atlantic open ocean and deep sea. Ecology 89 (12), 3449–3461.

Rosa, R., Pierce, G., O'Dor, R.K., 2013a. Advances in Squid Biology, Ecology and Fisheries. Part II—Oegopsid Squids. Nova Science Publishers, New York.

Rosa, R., Pimentel, M.S., Baptista, M., Trübenbach, K., Calado, R., Nunes, M.L., Moreno, A., Pereira, J., 2013b. Trophic and reproductive biochemistry of a deep-sea gelatinous octopus, *Opisthoteuthis calypso*. Mar. Biol. 160, 263–275.

Roura, A., Guerra, Á., Gonzalez, A.F., Pascual, S., 2009. Sperm ultrastructural features of the bathyal octopod *Graneledone gonzalezi*. Vie et Milieu 59, 301–305.

Roura, A., Guerra, Á., Gonzalez, A.F., Pascual, S., 2010a. Sperm ultrastructure in *Bathypolypus bairdii* and *B. sponsalis* (Cephalopoda: Octopoda). J. Morphol. 271, 143–151.

Roura, A., Guerra, Á., González, A.F., Pascual, S., 2010b. Sperm ultrastructure of the hydrothermal vent octopod *Vulcanoctopus hydrothermalis*. J. Morphol. 271 (8), 932–936.

Ruiz-Cooley, R.I., Ballance, L.T., McCarthy, M.D., 2013. Range expansion of the jumbo squid in the NE Pacific: delta N-15 decrypts multiple origins, migration and habitat use. PLoS One 8 (3), e59651.

Sánchez, P., González, A.F., Jereb, P., Laptikhovsky, V.V., Mangold, K.M., Nigmatullin, C.M., Ragonese, S., 1998. *Illex coindetii*. In: Rodhouse, P.G., Dawe, E.G., O'Dor, R.K. (Eds.), Squid Recruitment Dynamics. The Genus *Illex* as a Model, the Commercial *Illex* Species and Influences on Variability. FAO Fisheries Technical Paper, vol. 376. FAO, Rome, pp. 59–76.

Santos, M.B., Pierce, G.J., Boyle, P.R., Reid, R.J., Ross, H.M., Patterson, I.A.P., Kinze, C.C., Tougaard, S., Lick, R., Piatkowski, U., Hernandez-Garcia, V., 1999. Stomach contents of sperm whales *Physeter macrocephalus* stranded in the North Sea 1990–1996. Mar. Ecol. Prog. Ser. 183, 281–294.

Santos, M.B., Clarke, M.R., Pierce, G.J., 2001. Assessing the importance of cephalopods in the diets of marine mammals and other top predators: problems and solutions. Fish. Res. 52, 121–139.

Sato, N., Yoshida, M.A., Fujiwara, E., Kasugai, T., 2013. High-speed camera observations of copulatory behaviour in *Idiosepius paradoxus*: function of the dimorphic hectocotyli. J. Mollus. Stud. 79 (2), 183–186.

Schroeder, R., Perez, J.A.A., 2013. Individual growth of the squid *Illex argentinus* off Brazil as reconstructed from the gladius microstructure. J. Mar. Biol. Assoc. U.K. 93, 1653–1662.

Seibel, B.A., Drazen, J.C., 2007. The rate of metabolism in marine animals: environmental constraints, ecological demands and energetic opportunities. Philos. Trans. R. Soc. Lond. B 362 (1487), 2061–2078.

Seibel, B.A., Thuesen, E.V., Childress, J.J., Gorodezky, L.A., 1997. Decline in pelagic cephalopod metabolism with habitat depth reflects differences in locomotory efficiency. Biol. Bull. 192, 262–278.

Seibel, B.A., Hochberg, F.G., Carlini, D.B., 2000. Life history of *Gonatus onyx* (Cephalopoda: Teuthoidea): deep-sea spawning and post-spawning egg care. Mar. Biol. 137, 519–526.

Seibel, B.A., Goffredib, S.K., Theusen, E.V., Childress, J.J., Robison, B.H., 2004. Ammonium content and buoyancy in midwater cephalopods. J. Exp. Mar. Biol. Ecol. 313, 375–387.

Seibel, B.A., Robison, B.H., Haddock, S.H.D., 2005. Post spawning egg-care by a squid. Nature 438, 929.

Seki, M., 1993. Trophic relationships of *Ommastrephes bartramii* during winter migrations to subtropical waters north of the Hawaiian Islands. In: Okutani, T., O'Dor, R.K., Kubodera, T. (Eds.), Recent Advances in Cephalopod Fisheries Biology. Tokai University Press, Tokyo, pp. 523–529.

Shchetinnikov, A.S., 1992. Feeding spectrum of squid, *Sthenoteuthis oualaniensis*, in the eastern Pacific. J. Mar. Biol. Assoc. U.K. 72 (4), 849–860.

Sherrard, K.M., 2000. Cuttlebone morphology limits habitat depth in eleven species of *Sepia* (Cephalopoda: Sepiidae). Biol. Bull. 198 (3), 404–414.

Sims, D.W., Southall, E.J., Humphries, N.E., Hays, G.C., Bradshaw, C.J.A., Pitchford, J.W., James, A., Ahmed, M.Z., Brierley, A.S., Hindell, M.A., Morritt, D., Musyl, M.K., Righton, D., Shepard, E.L.C., Wearmouth, V.J., Wilson, R.P., Witt, M.J., Metcalfe, J.D., 2008. Scaling laws of marine predator search behaviour. Nature 451, 1098–1102.

Sinclair, E.H., Stabeno, P.J., 2002. Mesopelagic nekton and associated physics of the southeastern Bering Sea. Deep Sea Res. II 49, 6127–6145.

Smale, M.J., 1996. Cephalopods as prey 2. Seals. Philos. Trans. R. Soc. Lond. B Biol. Sci. 351 (1343), 1067–1081.

Snyder, R., 1998. Aspects of the biology of the giant form of *Sthenoteuthis oualaniensis* (Cephalopoda: Ommastrephidae) from the Arabian Sea. J. Mollus. Stud. 64, 21–34.

Solé, M., Lenoir, M., Durfort, M., Quero, C., Mas, A., Lombarte, A., van der Schaar, M., López-Bejar, M., Morell, M., Zaugg, S., Houégnigan, L., 2011. Low-frequency sounds induce acoustic trauma in cephalopods. Front. Ecol. Environ. 9, 489–493.

Solé, M., Lenoir, M., Durfort, M., López-Bejar, M., Lombarte, A., Andre, M., 2013. Ultrastructural damage of *Loligo vulgaris* and *Illex coindetii* statocysts after low frequency sound exposure. PLoS One 8 (10), e78825. http://dx.doi.org/10.1371/journal.pone.0078825.

Söller, R., Warnke, K., Saint-Paul, U., Blohm, D., 2000. Sequence divergence of mitochondrial DNA indicates cryptic biodiversity in *Octopus vulgaris* and supports the taxonomic distinctiveness of *Octopus mimus* (Cephalopoda: Octopodidae). Mar. Biol. 136, 29–35.

Song, H., Beack, G.W., Kim, S., Huh, S.H., 2006. Feeding habits of *Todarodes pacificus* (Cephalopods: Ommastrephidae) in the coastal waters of Busan, Korea. J. Korean Fish. Soc. 39 (1), 42–48.

Staaf, D.J., Camarillo-Coop, S., Haddock, S.H.D., Nyack, A.C., Payne, J., Salinas-Zavala, C.A., Seibel, B.A., Trueblood, L., Widmer, C., Gilly, W.F., 2008. Natural egg mass deposition by the Humboldt squid (*Dosidicus gigas*) in the Gulf of California and characteristics of hatchlings and paralarvae. J. Mar. Biol. Assoc. U.K. 88, 759–770.

Stark, K.E., 2008. Ecology of the Arrow Squid (Nototodarus gouldi) in Southeastern Australian Waters: A Multi-Scale Investigation of Spatial and Temporal Variability. University of Tasmania, Doctoral dissertation.

Staudinger, M.D., Juanes, F., 2010. A size-based approach to quantifying predation on longfin inshore squid *Loligo pealeii* in the northwest Atlantic. Mar. Ecol. Prog. Ser. 399, 225–241.

Staudinger, M.D., Juanes, F., Salmon, B., Teffer, A.K., 2013. The distribution, diversity, and importance of cephalopods in top predator diets from offshore habitats of the Northwest Atlantic Ocean Deep Sea Research Part II: Topical Studies in Oceanography, 6/2012, (Issue 95), p. 182–192.

Steenstrup, J., 1881. *Sepiadarium* og *Idiosepius* to nye Slaegter af Sepiernes Familie. Med Bemaerkninger om de to beslaegtede Former *Sepioloidea* D'Orb. og *Spirula* Lmk. Danske Videnskabernes Selskabs Skrifter, 6 Raekke, Naturvidenskabelig og Mathematisk 1 (3), 211–242.

Stephen, S.J., 1985. The Systematics of the Pelagic Squid Genus *Octopoteuthis* Ruppell, 1844 (Cephalopoda; Teuthoidea) with Emphasis on Species in the North Atlantic. Memorial University of Newfoundland, M.Sc. Thesis.

Stewart, J.S., Field, J.C., Markaida, U., Gilly, W.F., 2013. Behavioral ecology of jumbo squid (*Dosidicus gigas*) in relation to oxygen minimum zones. Deep Sea Res. II: Topic. Stud. Oceanogr. 95, 197–208.

St-Onge, M., LaRue, B., Charpentier, G., 2008. A molecular revision of the taxonomic status of mermithid parasites of black flies from Quebec (Canada). J. Invertebr. Pathol. 98, 299–306.

Strugnell, J.M., Lindgren, A.R., 2007. A barcode of life database for the Cephalopoda? Considerations and concerns. Rev. Fish Biol. Fish. 17 (2–3), 337–344.

Strugnell, J., Nishiguchi, M.K., 2007. Molecular phylogeny of coleoid cephalopods (Mollusca: Cephalopoda) inferred from three mitochondrial and six nuclear loci: a comparison of alignment, implied alignment and analysis methods. J. Mollus. Stud. 73 (4), 399–410.

Strugnell, J., Norman, M., Drummond, A.J., Cooper, A.A., 2004. Neotenous origin for pelagic octopuses. Curr. Biol. 14, 300–301.

Strugnell, J.M., Lindgren, A., Allcock, A.L., 2009. Cephalopod mollusks (Cephalopoda). In: Hedges, S.B., Kumar, S. (Eds.), The Tree of Life. Oxford, University Press, pp. 242–246.

Strugnell, J., Norman, M.D., Vecchione, M., Guzic, M., Allcock, A.L., 2013. The ink sac clouds octopod evolutionary history. Hydrobiologia 725 (1), 215–235.

Sweeney, M.J., Roper, C.F.E., Mangold, K.M., Clarke, M.R., Boletzky, S.V. (Eds.), 1992. "Larval" and juvenile cephalopods: a manual for their identification. Smithsonian Contributions to Zoology 513 viii + 282 pp.

Tang, E.P.Y., 2006. Path to Effective Recovering of DNA From Formalin-Fixed Biological Samples in Natural History Collections: Workshop Summary. National Academies Press, Washington, DC.

Thistle, D., 2003. The deep-sea floor: an overview. In: Tyler, P.A. (Ed.), Ecosystems of the Deep Oceans. Elsevier, Amsterdam, pp. 5–37.

Thomas, R.F., 1977. Systematics, distribution and biology of cephalopods of the genus *Tremoctopus* (Octopoda: Tremoctopodidae). Bull. Mar. Sci. 27 (3), 353–392.

ToLweb, 2012. *Brachioteuthis*. Version 10 November 2012 (under construction). http://tolweb.org/Brachioteuthis/24101/2012.11.10 in The Tree of Life Web Project, http://tolweb.org/.

Tsuchiya, K., Okutani, T., 1998. *Abraliopsis pacificus*, a new species of the squid family Enoploteuthidae from the Northwest Pacific. Bull. Natl. Sci. Mus. (Tokyo) A 16 (2), 47–60.

Tsuchiya, K., Young, R.E., 2008. Enoploteuthidae Pfeffer 1900. http://tolweb.org/Enoploteuthidae/19634/ (accessed 23.04.2008).

Uchikawa, K., Bower, J.R., Sato, Y., Sakurai, Y., 2004. Diet of the minimal armhook squid (*Berryteuthis anonychus*) (Cephalopoda: Gonatidae) in the northeast Pacific during spring. Fish. Bull. 102, 733–739.

Unger, M.A., Harvey, E., Vadas, G.G., Vecchione, M., 2008. Persistent pollutants in nine species of deep-sea cephalopods. Mar. Pollut. Bull. 56 (8), 1498–1500.

Vecchione, M., 1987. A multispecies aggregation of cirrate octopods trawled from north of the Bahamas. Bull. Mar. Sci. 40, 78–84.

Vecchione, M., Pohle, G., 2002. Midwater cephalopods in the western North Atlantic Ocean off Nova Scotia. Bull. Mar. Sci. 71 (2), 883–892.

Vecchione, M., Roper, C.F.E., 1991. Cephalopods observed from submersibles in the Western North Atlantic. Bull. Mar. Sci. 49 (1–2), 433–445.

Vecchione, M., Young, R.E., 1997. Aspects of the functional morphology of cirrate octopods: locomotion and feeding. Vie et Milieu. 47, 101–110.

Vecchione, M., Young, R.E., 1999. Lycoteuthidae Pfeffer, 1908. http://tolweb.org/Lycoteuthidae/19636/ (accessed 01.01.1999).

Vecchione, M., Young, R.E., 2003a. *Neoteuthis* Naef, 1921. *Neoteuthis thielei* Naef, 1921. http://tolweb.org/Neoteuthis_thielei/19926/ (accessed 01.01.2003).

Vecchione, M., Young, R.E., 2003b. *Nototeuthis* Nesis & Nikitina, 1986. *Nototeuthis dimegacotyle* Nesis & Nikitina, 1986. http://tolweb.org/Nototeuthis_dimegacotyle/19927/ (accessed 01.01.2003).

Vecchione, M., Young, R.E., 2011. Illicinae Posselt 1891. *Illex* Steenstrup 1880. http://tolweb.org/Illex/19940/ (accessed 13.01.2011).

Vecchione, M., Young, R.E., 2012. Pholidoteuthidae Adam, 1950. *Pholidoteuthis* Adam, 1950. http://tolweb.org/Pholidoteuthis/19835/ (accessed 04.07.2012).

Vecchione, M., Young, R.E., 2013. Magnapinnidae Vecchione and Young, 1998. *Magnapinna* Vecchione and Young, 1998. Bigfin Squid. http://tolweb.org/Magnapinna/52203/ (accessed 08.01.2013).

Vecchione, M., Robison, B.H., Roper, C.F.E., 1992. A tale of two species: tail morphology in paralarval Chiroteuthis. Proc. Biol. Soc. Wash. 105, 683–692.

Vecchione, M., Brakoniecki, T.F., Natsukari, Y., Hanlon, R.T., 1998a. A provisional generic classification of the family Loliginidae. In: Voss, N.A., Vecchione, M., Toll, R.B., Sweeney, M.J. (Eds.), Systematics and Biogeography of Cephalopods. In: Smithsonian Contributions to Zoology, vol. 586. pp. 215–222, I-II.

Vecchione, M., Piatkowski, U., Allcock, A.L., 1998b. Biology of the cirrate octopod *Grimpoteuthis glacialis* (Cephalopoda; Opisthoteuthididae) in the south Shetland Islands, Antarctica. South Afr. J. Mar. Sci. 20 (1), 421–428.

Vecchione, M., Young, R.E., Guerra, Á., Lindsay, D.J., Clague, D.A., Bernhard, J.M., Sager, W.W., Gonzalez, A.F., Rocha, F.J., Segonzac, M., 2001. Worldwide observations of remarkable deep-sea squids. Science 294 (5551), 2505–2506.

Vecchione, M., Allcock, A.L., Piatkowski, U., 2005. Unusual incirrate octopods from the South Shetland Islands, Antarctica, including *Bathypurpurata profunda*, a newly discovered genus and species of deepwater pygmy octopod (Cephalopoda). Phuket Mar. Biol. Center Res. Bull. 66, 109–116.

Vecchione, M., Young, R.E., Lindgren, A., 2007. Mastigoteuthidae Verrill, 1881. *Mastigoteuthis* Verrill, 1881. Whip-Lash Squid. http://tolweb.org/Mastigoteuthis/19453/ (accessed 19.11.2007).

Vecchione, M., Mangold, K.M., Young, R.E., 2008a. Opisthoteuthidae Verrill, 1896. http://tolweb.org/Opisthoteuthidae/20089/ (accessed 21.05.2008).

Vecchione, M., Young, R.E., Mangold, K.M., 2008b. Cirroteuthidae Keferstein, 1866. http://tolweb.org/Cirroteuthidae/20091/ (accessed 28.04.2008).

Vecchione, M., Piatkowski, U., Allcock, A.L., Jorgensen, E., Barratt, I., 2009a. Persistent elevated abundance of octopods in an overfished Antarctic area. In: Krupnik, I. et al., (Ed.), Smithsonian at the Poles: Contributions to International Polar Year Science. Smithsonian Institution Scholarly Press, Washington, DC, pp. 197–203.

Vecchione, M., Allcock, L., Piatkowski, U., Strugnell, J., 2009b. *Benthoctopus rigbyae*, n. sp., a new species of cephalopod (Octopoda; Incirrata) from near the Antarctic peninsula. Malacologia 51 (1), 13–28.

Vecchione, M., Kubodera, T., Young, R.E., 2010a. *Taningia* Joubin, 1931. *Taningia danae* Joubin 1931. http://tolweb.org/Taningia_danae/19840/ (accessed 22.08.2010).

Vecchione, M., Young, R.E., Piatkowski, U., 2010b. Cephalopods of the northern mid-Atlantic Ridge. Mar. Biol. Res. 6 (1), 25–52.

Vecchione, M., Bergstad, O.A., Byrkjedal, I., Falkenhaug, T., Gebruk, A.V., Godø, O.R., Gislason, A., Heino, M., Høines, Å.S., Menezes, G.M.M., Piatkowski, U., Priede, I.G., Skov, H., Søiland, H., Sutton, T., de Wenneck, T. de L., 2010c. Biodiversity patterns and processes on the Mid-Atlantic Ridge. In: McIntyre, A.D. (Ed.), Life in the World's Oceans. Blackwell Publishing Ltd, pp. 103–121.

Victor, B.C., Hanner, R.H., Shivji, M., Hyde, J., Caldow, C., 2009. Identification of the larval and juvenile stages of the Cubera Snapper, *Lutjanus cyanopterus*, using DNA barcoding. Zootaxa 2215, 24–36.

Vidal, E.A.G., Villanueva, R., Andrade, J.P., Gleadall, I.G., Iglesias, J., Koueta, N., Rosas, C., Segawa, S., Grasse, B., Franco-Santos, R.M., Albertin, C.B., Caamal-Monsreal, C., Chimal, M.E., Edsinger-Gonzales, E., Gallardo, P., Le Pabic, C., Pascual, C., Roumbedakis, K., Wood, J., 2014. Cephalopod culture: current status on main biological models and research priorities. Adv. Mar. Biol. 67, 1–98.

Villanueva, R., 1992. Continuous spawning in the cirrate octopods *Opisthoteuthis agassizii* and *O. vossi*: features of sexual maturation defining a reproductive strategy in cephalopods. Mar. Biol. 114, 265–275.

Villanueva, R., Guerra, Á., 1991. Food and prey detection in two deep-sea cephalopods: *Opisthoteuthis agassizii* and *O. vossi* (Octopoda: Cirrata). Bull. Mar. Sci. 49, 288–299.

Voight, J.R., 1997. Cladistic analysis of the octopods based on anatomical characters. J. Mollus. Stud. 63, 311–325.

Voight, J.R., 2000. A deep-sea octopus (*Graneledone* cf. *boreopacifica*) as a shell-crushing hydrothermal vent predator. Symp. Zool. Soc. Lond. 252, 335–341.

Voight, J.R., 2005. Hydrothermal vent octopuses of *Vulcanoctopus hydrothermalis*, feed on bathypelagic amphipods of *Halice hesmonectes*. J. Mar. Biol. Assoc. U.K. 85 (4), 985–988.

Voight, J.R., 2008. Observations of deep-sea octopodid behavior from undersea vehicles. Am. Malacol. Bull. 24 (1), 43–50.

Voight, J.R., Grehan, A.J., 2000. Egg brooding by deep-sea octopuses in the North Pacific Ocean. Biol. Bull. 198, 94–100.

Voight, J.R., Pörtner, H.O., O'Dor, R.K., 1994. A review of ammonia-mediated buoyancy in squids (Cephalopoda: Teuthoidea). Mar. Freshw. Behav. Physiol. 25, 193–203.

Voss, G.L., 1962. A monograph of the cephalopoda of the North Atlantic I: the family Lycoteuthidae. Bull. Mar. Sci. 12 (2), 264–305.

Voss, N.A., 1969. A monograph of the cephalopoda of the North Atlantic: the family Histioteuthidae. Bull. Mar. Sci. 19, 713–867.

Voss, N.A., 1980. A generic revision of the Cranchiidae (Cephalopoda; Oegopsida). Bull. Mar. Sci. 30 (2), 365–412.

Voss, N.A., 1985. Systematics, biology and biogeography of the cranchid cephalopod genus *Teuthowenia* (Oegopsida). Bull. Mar. Sci. 36, 1–85.

Voss, G.L., 1988. Evolution and phylogenetic relationships of deep-sea octopods (Cirrata and Incirrata). In: Clarke, M.R., Trueman, E.R. (Eds.), The Mollusca—Palaeontology and Neontology of Cephalopods, vol. 12. Academic Press, San Diego, pp. 253–276.

Voss, N.A., Nesis, K.N., Rodhouse, P.G., 1998. The cephalopod family Histioteuthidae (Oegopsida): systematics, biology, and biogeography. Smithson. Contr. Zool. 586, 293–372.

Ward, P., 1984. Is *Nautilus* a living fossil? In: Eldridge, N. (Ed.), Living Fossils. Springer, New York, pp. 247–256.

Watanabe, H., Kubodera, T., Ichii, T., Kawahara, S., 2004. Feeding habits of neon flying squid *Ommastrephes bartramii* in the transitional region of the central North Pacific. Mar. Ecol. Prog. Ser. 266, 173–184.

Watanabe, H., Kubodera, T., Moku, M., Kawaguchi, K., 2006. Diel vertical migration of squid in the warm core ring and cold water masses in the transition region of the western North Pacific. Mar. Ecol. Prog. Ser. 315, 187–197.

Webb, T.J., Vanden Berghe, E., O'Dor, R., 2010. Biodiversity's big wet secret: the global distribution of marine biological records reveals chronic underexploration of the deep pelagic ocean. PLoS One 5 (8), e10223. http://dx.doi.org/10.1371/journal.pone.0010223.

Wells, M.J., 1978. Octopus: Physiology and Behaviour of an Advanced Invertebrate. Chapman and Hall, London.

Wells, M.J., Wells, J., 1972. Sexual displays and mating of *Octopus vulgaris* Cuvier and *O. cyanea* Gray and attempts to alter performance by manipulating the glandular condition of the animals. Anim. Behav. 20, 293–308.

Wenneck, T. de L., Falkenhaug, T., Bergstad, O.A., 2008. Strategies, methods and technologies adopted on the R.V. *G.O. Sars* MAR-ECO expedition to the Mid-Atlantic Ridge in 2004. Deep Sea Res. II 55 (1–2), 6–28.

Werry, J.M., Planes, S., Berumen, M.L., Lee, K.A., Braun, C.D., Clua, E., 2014. Reef-fidelity and migration of Tiger Sharks, *Galeocerdo cuvier*, across the Coral Sea. PLoS One 9 (1), e83249.

Westermann, G.E.G., 1973. Strength of concave septa and depth limits of fossil cephalopods. Lethaia 6, 383–403.

Wiborg, K.F., Gjosaeter, J., Beck, I.M., 1984. The squid *Gonatus fabricii* (Lichtenstein): investigations in the Norwegian Sea and western Barents Sea 1982–1983: ICES C.M. 1984/K:19. 14 pp.

Wickler, W., 1965. Mimicry and evolution of animal communication. Nature 208, 519.

Widder, E.A., 2007. Sly eye for the shy guy: peering into the depths with new sensors. Oceanography 20, 46–51.

Widder, E., 2013. The Kraken revealed. The technology behind the first video recordings of live giant squid. Sea Technol. 54 (8), 49–54.

Wiebe, P.H., Morton, A.W., Bradley, A.M., Backus, R.H., Craddock, J.E., Barber, V., Cowles, T.J., Flierl, G.D., 1985. New development in the MOCNESS, an apparatus for sampling zooplankton and micronekton. Mar. Biol. 87, 313–323.

Willassen, E., 1986. *Haliphron atlanticus* Steenstrup (Cephalopoda, Octopoda) from the coast of Norway. Sarsia 71, 35–40.

Wilson, R.P., La Cock, G.D., Wilson, M.P., Mollagee, F., 1985. Differential digestion of fish and squid in jackass penguins *Spheniscus demersus*. Ornis Scand. 16, 77–79.

Wilson, R.P., Steinfurth, A., Ropert-Coudert, Y., Kato, A., Kurita, M., 2002. Lip-reading in remote subjects: an attempt to quantify and separate ingestion, breathing and vocalisation in free-living animals using penguins as a model. Mar. Biol. 140, 17–27.

Winkelmann, I., Campos, P.F., Strugnell, J., Cherel, Y., Smith, P.J., Kubodera, T., Allcock, L., Kampmann, M.-L., Schroeder, H., Guerra, Á., Norman, M., Finn, J., Ingrao, D., Clarke, M., Gilbert, M.T.P., 2013. Mitochonrial genome diversity and population structure of the giant squid, *Architeuthis*: genetics sheds new light on one of the most enigmatic marine species. Proc. R. Soc. Lond. B 280 (1759), 1–9.

Wodinsky, J., 2008. Reversal and transfer of spermatophores by *Octopus vulgaris* and *O. hummelincki*. Mar. Biol. 155, 91–103.

WoRMS Editorial Board (2014). World Register of Marine Species. Available from http://www.marinespecies.org at VLIZ (accessed 26.02.2014).

Wormuth, J.H., Roper, C.F.E., 1983. Quantitative sampling of oceanic cephalopods by nets: problems and recommendations. Biol. Oceanogr. 2, 357–377.

Xavier, J.C., Cherel, Y., 2009. Cephalopod Beak Guide for the Southern Ocean. British Antarctic Survey, Cambridge, UK.

Xavier, J.C., Rodhouse, P.G., Trathan, P.N., Wood, A.G., 1999. A geographical information system (GIS) atlas of cephalopod distribution in the Southern Ocean. Antarct. Sci. 11 (1), 61–62.

Xavier, J.C., Rodhouse, P.G., Purves, M.G., Daw, T.M., Arata, J., Pilling, G.M., 2002. Distribution of cephalopods recorded in the diet of Patagonian toothfish (*Dissostichus eleginoides*) around South Georgia. Polar Biol. 25, 323–330.

Xavier, J.C., Croxall, J.P., Trathan, P.N., Rodhouse, P.G., 2003a. Inter-annual variation in the cephalopod component of the diet of wandering albatrosses *Diomedea exulans* breeding at Bird Island, South Georgia. Mar. Biol. 142, 611–622.

Xavier, J.C., Croxall, J.P., Trathan, P.N., Wood, A.G., 2003b. Feeding strategies and diets of breeding grey-headed and wandering albatrosses at South Georgia. Mar. Biol. 143, 221–232.

Xavier, J.C., Croxall, J.P., Cresswell, K.A., 2005. Boluses: an effective method to assess the proportions of cephalopods in the diet of albatrosses. Auk 122, 1182–1190.

Xavier, J.C., Geraint, G.A., Croxall, J.P., 2006. Determining large scale distribution of pelagic cephalopods, fish and crustaceans in the South Atlantic from wandering albatross (*Diomedea exulans*) foraging data. Ecography 29, 260–272.

Xavier, J.C., Cherel, Y., Assis, C.A., Sendão, J., Borges, T.C., 2010. The feeding ecology of Conger eels (*Conger conger*) in the North-east Atlantic. J. Mar. Biol. Assoc. U.K. 90, 493–501.

Xavier, J.C., Phillips, R.A., Cherel, Y., 2011. Cephalopods in marine predator diet assessments: why identifying upper and lower beaks is important. ICES J. Mar. Sci. 68, 1857–1864.

Xavier, J.C., Cherel, Y., Roberts, J., Piatkowski, U., 2013. How do cephalopods become available to seabirds: can fish gut contents from tuna fishing vessels be a major food source of deep-dwelling cephalopods? ICES J. Mar. Sci. 70, 46–49.

Young, R.E., 1972a. Brooding in a bathypelagic octopus. Pacific Sci. 26, 400–404.

Young, R.E., 1972b. The systematics and areal distribution of pelagic cephalopods from the seas off southern California. Smithson. Contr. Zool. 97, 1–108.

Young, R.E., 1978. Vertical distribution and photosensitive vesicles of pelagic cephalopods from Hawaiian waters. Fish. Bull. 76, 583–615.

Young, R.E., 1995. Aspects of the natural history of pelagic cephalopods of the Hawaiian Mesopelagic-Boundary region. Pacific Sci. 49, 143–155.

Young, R.E., 2008. Bolitaenidae Chun, 1911. http://tolweb.org/Bolitaenidae/20193/ (accessed 28.04.2008).

Young, R.E., 2009. Joubiniteuthidae Naef, 1922. *Joubiniteuthis portieri* Joubin, 1916. http://tolweb.org/Joubiniteuthis_portieri/19450/ (accessed 11.09.2009).

Young, R.E., 2010. Nautiloidea. Nautilidae Blainville 1825. Pearly Nautiluses. http://tolweb.org/Nautilidae/19397/ (accessed 15.08.2010).

Young, R.E., 2012a. Cycloteuthidae Naef, 1923. http://tolweb.org/Cycloteuthidae/19412/ (accessed 10.11.2012).

Young, R.E., 2012b. Spirulida Haeckel, 1896. Spirulida Linnaeus, 1758. http://tolweb.org/Spirulida/19989/ (accessed 04.07.2012).

Young, R.E., 2012c. Vampyroteuthidae Thiele, in Chun, 1915. *Vampyroteuthis infernalis* Chun, 1903. The Vampire Squid. http://tolweb.org/Vampyroteuthis_infernalis/20084/ (accessed 04.07.2012).

Young, R.E., 2013. Alloposidae Verrill 1881. *Haliphron atlanticus* Steenstrup 1861. http://tolweb.org/Haliphron_atlanticus/20200/ (accessed 08.01.2013).

Young, R.E., Harman, R.F., 1985. Early life history stages of enoploteuthin squids (Cephalopoda: Teuthoidea: Enoploteuthidae) from Hawaiian waters. Vie et Milieu 35 (3–4), 181–201.

Young, R.E., Harman, R.F., 1998. Phylogeny of the "Enoploteuthid" families. Smithson. Contr. Zool. 586, 257–270.

Young, R.E., Mangold, K.M., 1994. Growth and reproduction in the mesopelagic-boundary squid *Abralia trigonura*. Mar. Biol. 119, 413–421.

Young, R.E., Mangold, K.M., 1996. *Sandalops* Chun, 1906. *Sandalops melancholicus* Chun, 1906. The Sandal-Eye Squid. http://tolweb.org/Sandalops_melancholicus/19551/ (accessed 01.01.1996).

Young, R.E., Mangold, K.M., 2008. *Helicocranchia* Massy, 1907. Piglet Squid. http://tolweb.org/Helicocranchia/19550/ (accessed 27.10.2008).
Young, R.E., Mangold, K.M., 2009. *Bathothauma* Chun, 1906. *Bathothauma lyromma* Chun, 1906. http://tolweb.org/Bathothauma_lyromma/19549/ (accessed 29.12.2009).
Young, R.E., Mangold, K.M., 2010. *Teuthowenia megalops* Prosch 1947. http://tolweb.org/Teuthowenia_megalops/19613/ (accessed 15.08.2010).
Young, R.E., Mangold, K.M., 2011. *Taonius* Steenstrup, 1861. http://tolweb.org/Taonius/19558/ (accessed 22.11.2011).
Young, Richard E., Michael Vecchione, and Uwe Piatkowski. 1999. Alluroteuthis Odhner, 1923. Alluroteuthis antarcticus Odhner, 1923. Version 01 January 1999 (under construction). http://tolweb.org/Alluroteuthis_antarcticus/19925/1999.01.01 in The Tree of Life Web Project, http://tolweb.org/
Young, R.E., Roper, C.F.E., 1969. A monograph of the Cephalopoda of the North Atlantic: the family Cycloteuthidae. Smithson. Contr. Zool. 5, 1–24.
Young, R.E., Roper, C.F.E., 2011a. *Asperoteuthis* Nesis, 1980. http://tolweb.org/Asperoteuthis/19461/ (accessed 22.11.2011).
Young, R.E., Roper, C.F.E., 2011b. *Grimalditeuthis* Joubin, 1898. *Grimalditeuthis bonplandi* (Verany, 1839). http://tolweb.org/Grimalditeuthis_bonplandi/19463/ (accessed 22.11.2011).
Young, R.E., and Roper, C.F.E. (2011c). Chiroteuthidae Gray, 1849. Version 22 November 2011 (under construction). http://tolweb.org/Chiroteuthidae/19451/2011.11.22 in the Tree of Life Web Project, http://tolweb.org/.
Young, R.E., Vecchione, M., 1996. Analysis of morphology to determine primary sister-taxon relationships within coleoid cephalopods. Am. Malacol. Bull. 12, 91–112.
Young, R.E., Vecchione, M., 1999. Morphological observations on a hatchling and a paralarva of the vampire squid, *Vampyroteuthis infernalis* Chun (Mollusca: Cephalopoda). Proc. Biol. Soc. Wash. 112, 661–666.
Young, R.E., Vecchione, M., 2003. Promachoteuthidae Naef, 1912. *Promachoteuthis* Hoyle, 1885. http://tolweb.org/Promachoteuthis/19454/ (accessed 01.01.2013).
Young, R.E., Vecchione, M., 2004. Oegopsida Orbigny, 1845. http://tolweb.org/Oegopsida/19407/2004.08.18 (accessed 15.11.2013).
Young, R.E., Vecchione, M., 2005. *Narrowteuthis* Young and Vecchione, 2005. *Narrowteuthis nesisi* Young and Vecchione, 2005. http://tolweb.org/Narrowteuthis_nesisi/19928/ (accessed 12.09.2005).
Young, R.E., Vecchione, M., 2007. Histioteuthidae Verrill, 1881. http://tolweb.org/Histioteuthidae/19782/ (accessed 21.05.2007).
Young, R.E., Vecchione, M., 2008. Argonautoidea Naef, 1912. http://tolweb.org/Argonautoidea/20192/ (accessed 21.10.2008).
Young, R.E., Vecchione, M., 2009. *Ommastrephes* Orbigny 1834. *Ommastrephes bartramii* (Lesueur 1821). Red Squid. http://tolweb.org/Ommastrephes_bartramii/19947/ (accessed 29.11.2009).
Young, R.E., Vecchione, M., 2010a. Chtenopterygidae Grimpe 1922. *Chtenopteryx* Appellof 1890. Combfin Squid. http://tolweb.org/Chtenopteryx/19430/ (accessed 15.08.2010).
Young, R.E., Vecchione, M., 2010b. *Hyaloteuthis* Gray 1849. *Hyaloteuthis pelagica* (Bosc 1802). http://tolweb.org/Hyaloteuthis_pelagica/19950/ (accessed 15.08.2010).
Young, R.E., Vecchione, M., 2010c. Thysanoteuthidae Keferstein, 1866. *Thysanoteuthis rhombus* Troschel, 1857. The Diamondback Squid. http://tolweb.org/Thysanoteuthis_rhombus/19420/ (accessed 15.08.2010).
Young, R.E., Vecchione, M., 2012. Lepidoteuthidae Pfeffer 1912. *Lepidoteuthis grimaldii* Joubin, 1895. The Scaled Squid. http://tolweb.org/Lepidoteuthis_grimaldii/19833/ (accessed 10.11.2012).

Young, R.E., Vecchione, M., 2013a. *Octopoteuthis* Rüppell 1844. http://tolweb.org/Octopoteuthis/19839/ (accessed 08.01.2013).
Young, R.E., Vecchione, M., 2013b. *Dosidicus* Steenstrup 1857. *Dosidicus gigas* (Orbigny 1835). Jumbo Squid. http://tolweb.org/Dosidicus_gigas/19945/ (accessed 08.01.2013).
Young, R.E., Harman, R.F., Mangold, K.M., 1985. The common occurrence of oegopsid eggs in nearsurface oceanic waters. Pacific Sci. 39, 359–366.
Young, R.E., Vecchione, M., Donovan, D.T., 1998. The evolution of coleoid cephalopods and their present biodiversity and ecology. South Afr. J. Mar. Sci. 20, 393–420.
Young, R.E., Roper, C.F.E., Vecchione, M., 2007. Heteroteuthinae Appellöf, 1898. http://tolweb.org/Heteroteuthinae/20027/ (accessed 26.12.2007).
Young, R.E., Roper, C.F.E., Mangold, K.M., 2008a. *Heteroteuthis* Gray, 1849. http://tolweb.org/Heteroteuthis/20030/ (accessed 04.04.2008).
Young, R.E., Vecchione, M., Roper, C.F.E., 2008b. *Planctoteuthis* Pfeffer, 1912. http://tolweb.org/Planctoteuthis/19464/ (accessed 21.04.2008).
Young, R.E., Mangold, K.M., Vecchione, M., 2010. Vitreledonellidae Robson, 1932. *Vitreledonella richardi* Joubin, 1918. http://tolweb.org/Vitreledonella_richardi/20195/ (accessed 04.03.2010).
Young, R.E., Mangold, K.M., Vecchione, M., 2013. Amphitretidae Hoyle 1886. *Amphitretus* Hoyle, 1885. http://tolweb.org/Amphitretus/20191/ (accessed 07.01.2013).
Zeidberg, L.D., Robison, B.H., 2007. Invasive range expansion by the Humboldt squid, *Dosidicus gigas*, in the eastern North Pacific. Proc. Natl. Acad. Sci. 104, 12948–12950.
Zhang, J., 2010. Exploiting formalin-preserved fish specimens for resources of DNA barcoding. Mol. Ecol. Resour. 10, 935–941.
Zheng, X., Lin, X., Lu, C., Ma, R., 2012. A new species of *Cistopus* Gray, 1849 (Cephalopoda: Octopodidae) from the East and South China Seas and phylogenetic analysis based on the mitochondrial COI gene. J. Nat. Hist. 46, 355–368.
Zimmer, I., Piatkowski, U., Brey, T., 2007. The trophic link between squid and the emperor penguin *Aptenodytes forsteri* at Pointe Geologie, Antarctica. Mar. Biol. 152, 1187–1195.
Zimmer, I., Wilson, R.P., Beaulieu, M., Ropert-Coudert, Y., Kato, A., Ancel, A., Plötz, J., 2010. Dive efficiency versus depth in foraging emperor Penguins. Aquat. Biol. 8, 269–277.
Zuyev, G., Nigmatullin, C.M., Chesalin, M., Nesis, K., 2002. Main results of long-term worldwide studies on tropical nektonic oceanic squid genus *Sthenoteuthis*: an overview of the soviet investigations. Bull. Mar. Sci. 71, 1019–1060.

CHAPTER FOUR

Transitions During Cephalopod Life History: The Role of Habitat, Environment, Functional Morphology and Behaviour

Jean-Paul Robin[*,†,1], **Michael Roberts**[‡,§], **Lou Zeidberg**[¶], **Isobel Bloor**[||], **Almendra Rodriguez**[#], **Felipe Briceño**[**], **Nicola Downey**[††,‡‡], **Maite Mascaró**[§§], **Mike Navarro**[¶¶,||||], **Angel Guerra**[##], **Jennifer Hofmeister**[***], **Diogo D. Barcellos**[†††], **Silvia A.P. Lourenço**[‡‡‡], **Clyde F.E. Roper**[§§§], **Natalie A. Moltschaniwskyj**[¶¶¶], **Corey P. Green**[||||||], **Jennifer Mather**[###]

[*]Université de Caen Basse-Normandie, UMR BOREA: Biologie des ORganismes et des Ecosystèmes Aquatiques, Esplanade de la paix, CS 14032, 14032 Caen, France
[†]UMR BOREA, UMR CNRS7208, IRD207, UPMC, MNHN, UCBN, 14032 Caen, France
[‡]Rhodes University, Grahamstown, South Africa
[§]Oceans & Coasts Research, Victoria & Alfred Waterfront, Cape Town, South Africa
[¶]Department of Ecology and Evolutionary Biology, University of California, Los Angeles, California, USA
[||]School of Ocean Sciences, Bangor University, Menai Bridge, Anglesey, United Kingdom
[#]El Colegio de la Frontera Sur, Colonia Casasano, Cuautla, Morelos, Mexico
[**]Institute for Marine and Antarctic Studies (IMAS), University of Tasmania, Hobart, Tasmania, Australia
[††]Department of Ichthyology and Fisheries Science, Rhodes University, Grahamstown, South Africa
[‡‡]Bayworld Centre for Research & Education, Constantia, Cape Town, South Africa
[§§]Unidad Multidisciplinaria de Docencia e Investigación, Universidad Nacional Autónoma de México, Puerto de Abrigo s/n, Sisal, Yucatán, México
[¶¶]Integrative Oceanography Division, Scripps Institution of Oceanography, University of California, San Diego, La Jolla, California, USA
[||||]Center for Marine Biodiversity and Conservation, Scripps Institution of Oceanography, University of California, San Diego, La Jolla, California, USA
[##]Instituto de Investigaciones Marinas (CSIC), Vigo, Spain
[***]Caldwell Laboratory, Department of Integrative Biology, University of California, Berkeley, California, USA
[†††]Laboratório de Ecossistemas Pesqueiros (LabPesq), Universidade de São Paulo, Instituto Oceanográfico Praça do Oceanográfico, Butantã, São Paulo, SP, Brazil
[‡‡‡]IPMA Div. RP, Departamento do Mar e Recursos marinhos, Lisboa, Portugal
[§§§]Smithsonian Institution, National Museum of Natural History, Washington, District of Columbia, USA
[¶¶¶]School of Environmental & Life Sciences, University of Newcastle, Ourimbah, New South Wales, Australia
[||||||]Department of Environment and Primary Industries, Fisheries Victoria, Queenscliff, Victoria, Australia
[###]Psychology Department, University of Lethbridge, Lethbridge, Alberta, Canada
[1]Corresponding author: e-mail address: jean-paul.robin@unicaen.fr

Advances in Marine Biology, Volume 67
ISSN 0065-2881
http://dx.doi.org/10.1016/B978-0-12-800287-2.00004-4

Contents

Abstract

Cephalopod life cycles generally share a set of stages that take place in different habitats and are adapted to specific, though variable, environmental conditions. Throughout the lifespan, individuals undertake a series of brief transitions from one stage to the next. Four transitions were identified: fertilisation of eggs to their release from the female (1), from eggs to paralarvae (2), from paralarvae to subadults (3) and from subadults to adults (4). An analysis of each transition identified that the changes can be radical (i.e. involving a range of morphological, physiological and behavioural phenomena and shifts in habitats) and critical (i.e. depending on environmental conditions essential for cohort survival). This analysis underlines that transitions from eggs to paralarvae (2) and from paralarvae to subadults (3) present major risk of mortality, while changes in the other transitions can have evolutionary significance. This synthesis suggests that more accurate evaluation of the sensitivity of cephalopod populations to environmental variation could be achieved by taking into account the ontogeny of the organisms. The comparison of most described species advocates for studies linking development and ecology in this particular group.

Keywords: Cephalopod ontogeny, Life stages, Morphological changes, Acquisition of behaviours, Habitat shifts, Environmental variability, Cohort survival, Paralarvae, Juvenile, Subadult, Adult

1. INTRODUCTION

A very common statement about cephalopod populations is that their abundance and dynamics are very variable and that this variability is due to the sensitivity of these organisms to environmental parameters and to their fluctuations. In fished species, the high productivity of cephalopod stocks is limited by variable recruitment (Boyle and Rodhouse, 2005; Rodhouse et al., 1998), but many noncommercial species are also spatially and temporally variable in their patterns of occurrence or distribution (Collins et al., 2001; Voss, 1988). A number of environmental parameters explain population variability; however, comparisons among species are often impaired by the fact that different life stages are considered. Although most cephalopods share common life traits, such as a relatively short life span and a fast growth rate, this group of molluscs is very diverse. Cephalopods have invaded almost all marine ecosystems, and there is diversity in life histories, from "big bang" spawners to "prolonged multiple" spawners (Boyle and Boletzky, 1996; Hoving, 2008; Hoving et al., 2013; Laptikhovsky et al., 2007, 2009; Rocha et al., 2001). A series of stages in an organism's life represents in itself a range of adaptations to variable environments, and it is worth investigating in cephalopod life stages how the animal ecology changes between tiny eggs and large adults (Ebenman and Persson, 1988).

The main features of cephalopod life cycles were reviewed and described in detail three decades ago (Boyle, 1983b, 1987) and updated for some groups since Rosa et al. (2013). Our use of the terms embryo, paralarvae, juvenile, subadult and adult is consistent with those previously described by these authors and with Young and Harman (1988). From egg to spawning adult, a series of stages are undertaken (Figure 4.1) and are common to all coleoid cephalopods, with the exception of the "paralarvae", which occur only in species in which the hatchlings show either discrete morphological differences from the adults (such as the rhynchoteuthis stage

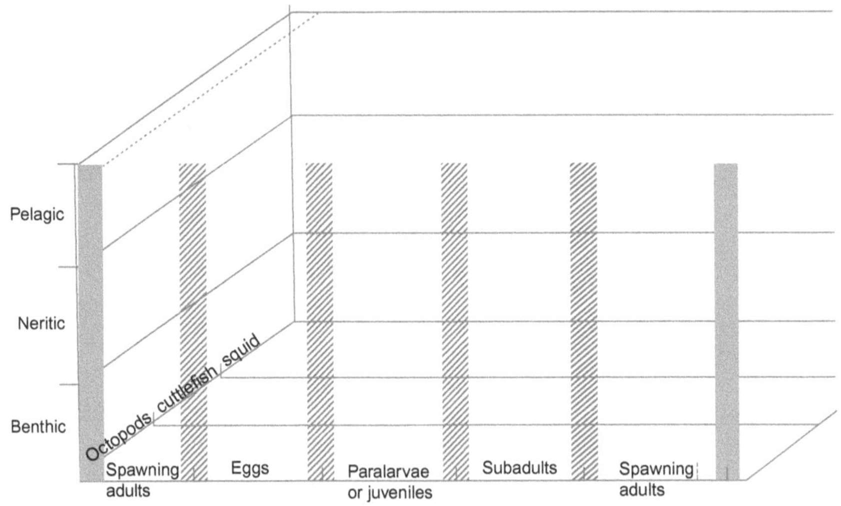

Figure 4.1 General framework for the transitional phases (shaded bars) between life stages for octopods, cuttlefish and squid for benthic, neritic and pelagic species. The scheme underlines common stages in cephalopod life cycles in spite of complex situations (change of habitats) and species differences.

of ommastrephids) or behavioural differences with planktonic posthatching versus nektonic or benthic adults (Boletzky, 1977). Life history theory (Stearns, 1992) suggests that the schedule and duration of key events in an organism's lifetime are shaped by natural selection to produce the largest possible number of surviving offspring and that these events depend on the ecological environment of the organism. The analysis of key events guided this chapter to explore how significant changes during ontogeny correspond to increased risks for a cohort's survival. While the life of an organism is a continuum of change, marine species often undergo a series of changes in life stage that can be characterised by either dramatic changes in shape and environment or subtle changes in behaviour and growth and development of internal organs. Each cephalopod life stage has its strengths and weaknesses and provides a response that fits best to the average environmental conditions. However, in a species life cycle, transitions need to be examined because they involve radical changes both in the organism and in its relationship to the environment. Transitions are critical to survival of cephalopod cohorts, and comparisons among groups of species or between ecosystems will be facilitated by looking at common transitions.

A transition refers to the time over which individuals move from one stage of their life cycle to the next (Figure 4.1). The rationale for introducing the concept of "transitions" between cephalopod life stages lies in the idea that ontogenic stages are not artificial entities defined by the observer within a continuum, but rather, they result from interactions between the organism and its environment. Each life stage corresponds to a temporary, stable interaction of the organism with the environment that lasts for a measurable duration, and during which time, the individual adopts a mode of life. As individuals move from one stage to the next, there is a transition, usually short relative to the length of the stages they are transiting between, and the start and end of each transition are defined by the change in mode of life. During the transition, individuals undergo a suite of changes that permit a change in mode of life and that contribute to cohort success, if they occur under the best possible biotic and abiotic conditions. Some of these changes may be prepared during the previous stages as the whole development is governed by the activation of a set of genes. A series of four transitions common to most cephalopod groups is presented in Figure 4.1 and will be referred to with the following headings:

1. Adult to egg: it is the transition from the mature egg inside the female to the release of the eggs, which is change in mode of life from inside the female to outside the female.
2. Egg to paralarvae: this transition is defined as the change in mode of internal to external sources of food and includes hatching that occurs during this transition.
3. First feeding to subadult: this transition is from the plankton mode of life to entering the nektonic or benthic mode of life, which involves adopting the adult lifestyle (will apply differently to species without a paralarval stage).
4. Subadult to adult: it is the transition from immature to mature adult mode of life.

This chapter examines the morphological, physiological, ecological and behavioural changes associated with each of the four transitions and their consequences at the population level. The objective of this analysis is not only synthesis but also extraction (when possible) unifying principles explaining which transitions are most critical and how transition success is related to habitats (or influenced by environmental variation). With this aim, this chapter does not intend to make a comprehensive list of all species whose life cycle changes were described, but rather to highlight phenomena of general interest. The scope of knowledge about each transition is inevitably limited by our capacity to sample and observe specimens at any stage of the life cycle. The bias related to sampling issues is mentioned in many reviews (Boyle and Rodhouse, 2005). One should keep in mind that the species most studied are either abundant, supporting intensive fisheries, or easy to observe under rearing conditions.

2. TRANSITION #1: ADULT TO EGG (COMPLETE FERTILISATION OF EGG CLUTCH TO EGG LAYING)

2.1. Introduction

As stated by Young and Harman (1988), the adult phase in cephalopods is defined by individuals who are sexually mature and of adult size. Reproduction and spawning punctuate the final stage of the adult phase, which is terminated by senescence and death and results in the transfer of adult genetic material to a new generation. The transition from adult to egg (transition 1) is defined here as the time from complete fertilisation of the egg clutch to the release of the eggs, as a drifting mass, either attached to the benthos or retained within the arms of the female. This transition involves the change from the relatively well-protected adult stage to the relatively vulnerable stage of newly deposited or released eggs. Unlike in most marine invertebrates, this transition is critical for survival, both at the individual level (e.g. transfer of genes) and at the population level (e.g. successful annual recruitment), because the short lifetime of most cephalopods (Boyle, 1983a,b, 1987; Boyle and Rodhouse, 2005; Rodhouse, 2010) means that the majority of populations consist of only one generation and thus failure to spawn will result in recruitment failure and a subsequent population crash.

Cephalopods are not broadcast spawners; instead, the female fertilises each egg, using stores of sperm provided by the male, as she releases the egg from her oviduct and encases it in a protective layer. All cephalopod species provide a protective coat for the developing eggs, but there is a diversity of habitats that the protected developing embryos are released into and the degree to which the female uses the habitat to protect the eggs also varies. Eggs can be released into the water column to move with the currents, attached to sand or macroalgae or hidden within the matrix of corals or boulder habitats; it is only in octopuses and a few oceanic squid that protection of the developing embryos by the female is provided. This section will examine the fertilisation process, the timing of egg release and the changes that occur as the egg is released, the behaviours associated with spawning and postspawning and the habitats used for spawning.

2.2. Morphological, suborganismal and physiological changes during transition

The first transition begins with the fertilisation process, which is well described in *Sepia officinalis*, where sperm are stored in the mantle cavity or in the copulatory pouch and encapsulated eggs are fertilised by sperm externally (Boletzky, 1989). The sperm-attracting peptide SepSAP produced by eggs diffuses through the gelatinous layers, directing spermatozoa so that they can reach the egg while the female is holding the egg in her arms before attaching it to the substrate (Tosti et al., 2001; Zatylny et al., 2002). It is hypothesised that once fertilised, polymerisation of the egg capsule stops, and release of SepSAP from the oocyte prevents polyspermy (Zatylny et al., 2002). For species that store sperm in the ovary, for example, *Eledone*, fertilisation occurs as eggs leave the ovary and before encapsulation (Boyle and Rodhouse, 2005; Cuomo et al., 2005). Octopods are the only group of cephalopods where fertilisation is internal, occurring in most species in the oviducal gland (Hanlon and Messenger, 1998; Mangold, 1987). The only known exception is the genus *Eledone* in which fertilisation occurs in the ovary (Boyle, 1983a,b). Once fertilised, the egg undergoes cell division and embryonic development proceeds, which usually is after deposition, although eggs can go through cleavage stages while still inside the oviduct (*Argonauta* and *Tremoctopus*) (Boletzky, 1989). In the squid species *Doryteuthis opalescens*, an individual female was observed while laying an egg mass in the laboratory. During the process, a capsule sheath was produced within the mantle cavity of the female and then ejected out slowly through the siphon. The sheath was then manipulated within her arms so that the anchor strand was distal to her arms and the free-floating end was below her head. The female then pumped embryos into the capsule sheath from the top portion towards the end that would become the anchor (video: http://gilly.stanford.edu/video/; Zeidberg, 2009). Orientation of the capsule suggests that both locations of spermatophore placement, within the mantle of guarding males or in the seminal receptacle on the buccal membrane of guarding males, provide opportunities for fertilisation.

The physiological changes and processes that occur as eggs are released and leave the ovary are best described for *S. officinalis* but are unknown for other cephalopods. Storage of ovulated eggs in the oviduct is controlled by a suite of ovarian peptides, which prevent oviducal contractions (Bernay et al., 2004, 2005; Zatylny et al., 2000a,b); this allows eggs to be ready for deposition once mating occurs. Once ready for release, oocytes of *S. officinalis* move from the oviduct to the funnel via oviduct contractions, modulated by neuropeptides, for example, FMRFamide-related peptides and APGWamide-related peptides (Henry et al., 1997, 1999), and ovarian peptides, for example, ILME and SepOvotropin (Bernay et al., 2006; Zatylny et al., 2000a,b).

The degree and nature of the encapsulation of the eggs vary among the cephalopod species, but in all cases, it is the role of the oviducal gland to produce the substances that surround the eggs, and without these gelatinous substances, fertilised eggs die several days later (Ikeda et al., 1993). The single (octopus) or paired (decapods) oviducal gland secretes the first gelatinous envelope enclosing the egg, while the cuttlefish and myopsid squid secrete a second layer of encapsulation provided by the nidamental gland (Boletzky, 1989). The cylindrical egg of *Octopus* lacks a gelatinous envelope, and the follicular tissue around the egg forms a stalk that thins out proximally (Boletzky, 1998; Naef, 1928). The eggs are coated with the encapsulation substances in the cavity of the oviducal gland after fertilisation and subsequently enter the distal oviduct before being released in a cluster (Froesch and Marthy, 1975). Like *Octopus*, the cuttlefish species deposit eggs singularly within a jelly-like substance; *S. officinalis* inserts a small quantity of ink into the egg capsule giving it the distinctive black colour. The loliginid squid deposit egg capsules that contain multiple eggs: three to eight eggs in the case of *Sepioteuthis australis* (Steer et al., 2003) to hundreds, for example, *D. opalescens* (Roper, 1965). Myopsid squid also encapsulate eggs in a sheath created by the accessory nidamental gland that is intercalated with bacteria transferred from the female (Biggs and Epel, 1991; Kaufman et al., 1998). The ommastrephid squids, for example, *Illex illecebrosus* and *Todarodes pacificus*, release the eggs within a fragile gelatinous matrix mass (O'Dor, 1983).

2.3. Ecological changes

2.3.1 Habitat and egg brooding

The first transition is complete when the adult female has deposited or released her eggs. The habitats used for egg deposition or release are diverse and range from inshore sites to open water. The squid egg masses of *Thysanoteuthis* are neustonic and thus are the most commonly observed oegopsid eggs. Eggs of the other non-deep-sea squid are found in mixed layer depths, for example, a *Dosidicus gigas* egg mass observed at 17 m depth (Staaf et al., 2008). Subsurface egg masses are laid by the female in the mixed layer, but the egg mass has a greater ion content than sea water and is rapidly inflated via osmosis: the infusion of a sea water of a certain specific density enables the inflated egg mass to be neutrally buoyant at the depth of the water where it was laid (Rodhouse et al., 1998). The exact location and depth of egg mass release are unknown for many species. Hatching experiments determined that *D. gigas* eggs have a thermal limit of 15–25 °C (Staaf et al., 2011), suggesting that *D. gigas* could spawn 500 km offshore of California as far as 45°N assuming that egg masses float at the pycnocline (Staaf et al., 2011). Ommastrephid egg masses were first observed in the laboratory in 1981 (*I. illecebrosus*; O'Dor and Balch, 1985) and more recently *in situ*, for example, *D. gigas* in the Gulf of California (Staaf et al., 2008), *Nototodarus gouldi* in New Zealand (O'Shea et al., 2004) and *Thysanoteuthis rhombus* in the Sea of Japan (Miyahara et al., 2006a).

Egg capsules of the loliginid squid are negatively buoyant, and the female attaches the egg capsules to the substrate using a mucilaginous adhesive substance (Boletzky, 1998). The nature of the substrate used varies with species; for example, *D. opalescens* attaches eggs to sandy substrates in 10–12 °C water (Roper, 1965); *Doryteuthis gahi* attaches capsules to frondless kelp stipes 0.5–2.5 m from the benthos (Arkhipkin et al., 2000); *Doryteuthis pealeii* anchor eggs to seaweed, pilings or rocks (Summers, 1983); *Loligo vulgaris* eggs are found in 20–25 m on fixed or floating supports (Worms, 1983); *Loligo reynaudii* deposit eggs on sand or low-profile rock (Sauer et al., 1993); and *S. australis* attaches eggs onto the base of the sea grass *Amphibolis antarctica* (Moltschaniwskyj et al., 2002). Cuttlefish species tend to spawn relatively few, large eggs that the females attach in clusters to structures on the seabed including plants, submerged debris and rocks (Roper et al., 1984). The requirement to attach eggs to suitable structures may be a driving force for the inshore spawning migration (Hall and Fowler, 2003). *Sepia latimanus* attaches the eggs between the corallites of dead *Lobophyllia* colonies with females temporarily compressing and elongating the egg between their arms to deposit them through the opening (Corner and Moore, 1980). The eggs of *Sepia apama* (approximately 2 cm in diameter; Cronin and Seymour, 2000) are attached to the undersides of rocks, ledges and caves (Hall and Hanlon, 2002). Aggregative spawning activity may be facilitated by β-microseminoprotein found in the capsule sheath that stimulates males to engage in aggressive mating behaviours (Cummins et al., 2011). *S. officinalis* females use their tentacles to manipulate the basal ring of the egg's gelatinous envelope around a support (e.g. *Zostera marina*, *Sabella pavonina*, Porifera sp. and *Nemertesia* sp.) to fix it in place (Blanc, 1998;

Bloor, 2012; Boletzky, 1983). The majority of squid and cuttlefish species attach their eggs to the benthos and do not provide parental care to their offspring. However, brooding behaviour has been observed in at least one genus of oceanic squid *Gonatopsis* (Gonatidae) (e.g. Okutani et al., 1995; Seibel et al., 2000).

Female octopods are known to undertake parental care of their eggs, which they brood in a rocky or coralline overhang or within an enclosed space (Boyle and Rodhouse, 2005; Mangold, 1987; Voight and Grehan, 2000). Octopod eggs are deposited by two methods in the den: cemented individually in hard substrate and cemented in clusters where individual eggs are intertwined with each other (Hanlon and Messenger, 1998; Rocha et al., 2001). Alternatively, female octopods that occur in habitats without shelters or dens (e.g. sand or mud) will carry the developing eggs in the web of their arms, e.g. *Octopus burryi* and *Wunderpus photogenicus* (Miske and Kirchhauser, 2006), *Hapalochlaena* (Norman and Reid, 2000) and *Amphioctopus* (Huffard and Hochberg, 2005). Pelagic octopus females carry eggs in a range of different ways: attached to mineralised rods (family Tremoctopodidae), within elongated distal oviducts (family Ocythoidae), within the arm crown (families Amphitretidae, Bolitaenidae, Vitreledonellidae and Alloposidae) and within a thin calcareous chamber (family Argonautidae) (Boletzky, 1998). Bathybenthic cirrate octopuses such as *Opisthoteuthis* may represent an exception, since the female deposits her eggs freely and individually onto the seabed (Boletzky, 1982).

2.3.2 Mortality

The first transition involves the change from a relatively well-protected adult stage to the relatively vulnerable stage of newly deposited or released eggs. The eggs of cephalopods, compared with most marine invertebrates, are relatively well protected by encapsulation that provides a physical and chemical buffer (Boletzky, 2003). In some cephalopod species, for example, *Octopus*, parental care (Section **2.3.1**) also helps increase the rate of survival and thus the success of the first transition, which is critical for survival of the population.

Incirrate benthic octopuses deposit their eggs in close spaces to protect them during brooding. Bathybenthic cirrate octopuses deposit their eggs individually on the seabed. Female pelagic octopuses carry their eggs in their arm crown while brooding (Boyle and Rodhouse, 2005). Predation pressure from conspecific and congeneric males is considerable (Anderson, 1997; Ibáñez and Keyl, 2010), although there is no relationship between octopus and potential reefal predators (eels) or reefal prey (crayfish) species

(Anderson, 1997). However, the consequences of predation on brooding females on the survival of the egg clutch are not known. No major predation pressures have been reported for the relatively large and well-protected, benthic egg capsules of cuttlefish (Guerra, 2006). The infusion of ink into the outer envelope of the egg capsule by female *S. officinalis* may aid in protection of embryos by camouflaging the eggs (Guerra and Gonzalez, 2011) and also potentially acting as a chemical deterrent to predators (Derby et al., 2007). Despite the intrinsic protection afforded to cuttlefish eggs, predation does occur in several species; for example, in Guam, butterfly fish (*Chaetodon ulietensis*) predate on and eat accessible eggs of *S. latimanus* (Corner and Moore, 1980). Predation on *S. officinalis* eggs does occur by the tompot blenny (*Parablennius gattorugine*) (Guerra and Gonzalez, 2011). While predators are seen on the grounds of spawning squid, it appears that the eggs are not the target, but spent females are targeted (Sauer et al., 1992). There is little in the literature to suggest that predation is a major source of mortality for either benthic squid or cuttlefish eggs.

2.3.3 Environmental variables

In general, the reproductive cycle of marine molluscs, such as cephalopods, is associated with nutrient storage and environmental parameters such as water temperature and food availability (Pliego-Cárdenas et al., 2011). Spawning sites are related to specific temperature zones, mainly because temperature regimes affect rates of embryonic development and success of hatching (Boletzky, 1974; Ortiz et al., 2011).

The eggs of most sepiolids, many octopods and myopsid squid are attached to fixed structures or substrates, and so, their successful embryonic development and survival are dictated almost entirely by biological and physical conditions encountered at the fixed spawning site. Eggs deposited at different times or locations will experience a variety of different environmental conditions including temperature, light intensity, oxygen saturation, pollution, salinity, photoperiod and predation. Bloor et al. (2013a) presented a detailed review of *S. officinalis* spawning, highlighting the importance of regulation of both temporal (e.g. timing of reproduction) and spatial (e.g. site and habitat selection) reproductive outputs of adult females to ensure successful regeneration of the spawning stock. Oceanic circulation can govern the distribution of spawning octopuses with pelagic paralarvae, for example, spawning *Octopus vulgaris* in the Strait of Sicily concentrate at the margin of eddies, favouring site retention and benthic settlement of the paralarvae (Garofalo et al., 2010).

2.4. Behavioural changes

2.4.1 Spawning patterns

Spawning patterns of cephalopods differ among species. Squid commonly form spawning aggregations with large groups of individuals at the surface or in the mid-water and smaller groups at the sea floor (Vaughan and Recksiek, 1978). Egg masses are deposited close to one another, for example, *S. australis* (Moltschaniwskyj et al., 2002), *L. reynaudii* (Sauer et al., 1992), *D. opalescens* (McGowan, 1954), *Loligo gahi* and *Loligo sanpaulensis* (Barón, 2001). Cuttlefish are generally considered solitary animals that do not shoal, but interaction between individuals does occur particularly during the reproductive period (Naud et al., 2004). In contrast, octopods are generally solitary animals with spawning also observed to be a solitary event with individuals displaying little or no courtship behaviour prior to mating (Hanlon and Messenger, 1998; Voight and Feldheim, 2009).

Female *D. opalescens* spawn 36% of their potential fecundity on the first day and 9% on each subsequent day (Macewicz et al., 2004). A small proportion of the population will continue to spawn for up to 6 days, depositing 78% of their total estimated potential fecundity, with, on average, females depositing eggs for 1.67 days on the spawning grounds before being lost from the population by natural or fishing mortality (Macewicz et al., 2004). In many squid species, the length of spawning events remains unknown. *L. reynaudii* tracked with acoustic tags were observed to perform daily migrations on and off the spawning grounds for up to 5 days (Sauer et al., 1997) and to migrate between spawning grounds for a mean of 43 km over 2 weeks (Sauer et al., 2000). This squid shows indications of atresia (degeneration and resorption of an oocyte and its follicle) that are absent in *D. opalescens* (Macewicz et al., 2004; Melo and Sauer, 1998; Sauer et al., 1999). However, serial spawning in *L. reynaudii* may occur quite rapidly, over several days to weeks (Melo and Sauer, 1999). Furthermore, individual *D. pealeii* produced multiple clutches in the laboratory over 3–6 weeks, often with several days between ovipositions (Maxwell and Hanlon 2000). Squids demonstrate a wide variety of reproductive patterns, simultaneous terminal spawning, multiple spawning, intermittent terminal spawning and continuous spawning (Rocha et al., 2001). *D. opalescens* has been described as simultaneous terminal spawner, but it seems the term "simultaneous" could range between 4 min (Hanlon et al., 2004) and 6 days (Macewicz et al., 2004) or possibly 42 days (Yang et al., 1986).

Several species of cuttlefish (e.g. *S. latimanus* and *S. officinalis*) are solitary for the majority of the year before aggregating in small loose associations (up to five individuals) in shallow coastal waters during the reproductive period to compete for mates (Boletzky, 1983; Corner and Moore, 1980). Polypeptidic waterborne pheromones expressed and secreted by accessory sex glands may facilitate aggregative egg laying behaviour in *S. officinalis* (Boal et al., 2010; Enault et al., 2012). In the Spencer Gulf, Australia, the giant cuttlefish, *S. apama*, forms the only known large spawning aggregation of cuttlefish in the world, with up to 170,000 individuals recorded (Hall and Fowler, 2003). Adults migrate inshore to this area between April and July to mate and spawn over a restricted area of rocky reef (e.g. Hall, 2002; Naud et al., 2004). The cues or mechanisms responsible for the formation of these large spawning groups in cuttlefish remain largely unknown; however, the requirement for a specific rock-based spawning substrate, the availability of which is restricted within the area, may explain the large cuttlefish spawning aggregations that occur at this specific location (Naud et al., 2004).

Cuttlefish exhibit a high degree of flexibility in their spawning patterns (e.g. *S. officinalis*; Boletzky, 1986, 1987a), including spawning once (simultaneous terminal spawning) and spawning more than once, with multiple repeated spawning events over a period of several months (intermittent terminal spawning) (Rocha et al., 2001). In several species of *Sepia* (e.g. *S. officinalis*, *Sepia pharaonis*, *S. apama* and *Sepia dollfusi*), a range of developmental stages and egg sizes are found within the ovaries of sexually mature females during the spawning season, indicating that not all eggs mature simultaneously and that eggs are produced in multiple batches suggestive of an intermittent terminal spawning strategy in these species (Boletzky, 1987a; Gabr et al., 1998; Hall and Fowler, 2003). Telemetry studies of both *S. officinalis* (Bloor et al., 2013b) and *S. apama* (Hall and Fowler, 2003) indicate that females are present in the aggregation area over an extended period during which they are not continuously involved in oviposition, but are often relatively mobile with a high degree of plasticity in reproductive behaviour and general movement patterns. Postspawning mortality is substantial in all cuttlefish species (Roper et al., 1984).

Several species of octopods perform inshore–offshore migration associated with seasonal changes in temperature such as reported for *Enteroctopus dofleini* in Hokkaido waters (Rigby and Sakurai, 2004); others such as *O. vulgaris* and *Eledone cirrhosa* made onshore shallow-water migrations in spring to match with the breeding season (Mangold, 1987). Migration of medium and large octopuses to deeper water during periods of strong thermocline (e.g. summer) is suggested to be related with a reduction of energetic cost at colder temperature in *O. vulgaris* (Katsanevakis and

Verriopoulos, 2004a,b). Thus, suitable conditions for embryonic development under stable temperature may constrain breeding in octopods (Boletzky, 1987b). Spawning sites may vary within the same species depending on the region. Before reproduction, adult *O. vulgaris* from the Atlantic coast of Spain migrate to coastal waters. Breeding occurs mainly in summer; however, mature males can be found throughout the year (Rodríguez-Rúa et al., 2005). Shallow-water cephalopods tend to spawn in cold months, so the offspring hatch at warmer temperatures. In *Octopus pallidus*, maturation is faster when exposed to decreasing temperatures so that spawning occurs in cold months. However, it has been suggested that the spawning period for this species occurs in late summer–early autumn (Semmens et al., 2011). In *Octopus hubbsorum*, there have been reports of two different spawning seasons: for one region in autumn and for another region in winter–spring. Differences in spawning seasons within the same species may be due to the differences in regional environmental conditions (Pliego-Cárdenas et al., 2011).

2.4.2 Female choice

Copulation is often assumed the final criterion for female choice, but in animals that adopt internal fertilisation strategies, copulation seldom results in direct or inevitable fertilisation (Eberhard, 1985). Prior to the start of the first transition, processes that occur both before (e.g. direct female mate choice) and after (e.g. cryptic female choice) copulation can affect the chance of a copulation ending in fertilisation (Eberhard, 1996). The classic view of mate choice is that males compete for copulations and fertilisation but that females choose which males to mate with through assessment of some male secondary sexual trait or ornamentation, potentially preferring healthier males or those with good genes (Bateman, 1948).

2.4.2.1 Direct female mate choice

Mate choice by females is known to occur in cuttlefish (e.g. *S. officinalis*; Boal, 1997), but the secondary traits or ornamentations used are not well defined. In the laboratory, Boal (1997) showed that while female *S. officinalis* did consistently prefer some males over others, these choices were not based on characteristics known to correlate with male dominance (e.g. body patterning or body size). In fact, Boal (1997) argued that female mate choice was more likely to be based on olfactory cues rather than visual cues. In *S. apama*, direct mate choice by females has also been observed, with females actively rejecting unwanted mating attempts. A total of 70% of female rejection attempts were observed to be successful with only 3% resulting in forced copulations (Hall and Fowler, 2003). Females were not receptive to mating attempts while an egg was held in the arms, and one-third of the observed rejections occurred while females were holding an egg; the remaining two-thirds of the rejections were considered a result of direct choice by the female (Hall and Fowler, 2003).

Octopuses are presumed to be solitary and polygamous animals (however, see Huffard et al., 2008 for *Abdopus aculeatus*) resulting in no clear female choices (Hanlon and Messenger, 1998).

2.4.2.2 Cryptic female choice

Observational studies of loliginids indicate that, despite arriving on the spawning ground with filled spermathecae, females copulate with multiple males (e.g. Hanlon et al., 2002) resulting in multiple paternity of eggs within the egg masses produced by loliginid squid (Buresch et al., 2001; Shaw and Boyle, 1997; Shaw and Sauer, 2004). DNA fingerprinting of *L. reynaudii* embryos indicates that four to five males can contribute their paternity with an emerging picture of complex genetic mating systems in squid that indicates potential for overt and cryptic female choice (Shaw and Sauer, 2004). Female cuttlefish are also polyandrous, accepting and storing sperm packages from multiple males. Females thus have access to sperm from two different sources: (1) sperm packages from recently mated males, which are deposited onto the buccal area (Naud et al., 2005), and (2) sperm from previous matings that are stored in their internal sperm-storage receptacles (copulatory pouch) located under the buccal mass (Hanlon et al., 1999). In *S. apama*, a study by Naud et al. (2005) confirmed that these two sperm stores contained sperm from between two and more than five males, indicating a high probability of sperm competition prior to fertilisation and highlighting a potential for stored sperm to be manipulated by females during fertilisation. This process could reflect the "cryptic female choice hypothesis" proposed by Eberhard (1996) that suggests that by actively influencing which sperm is utilised for fertilising eggs after copulation, polyandrous females who have mated with multiple males can still control which male ultimately sires her offspring. Naud et al. (2005) used microsatellite DNA analyses to determine the genetic diversity of the stored sperm in *S. apama* females and then compared it with that of their offspring. Their results indicated that fertilisation patterns with respect to the sperm stores used were not random (Naud et al., 2005). While these results represent the first account in cuttlefish of the use of sperm from multiple stores, it remains to be determined whether this fertilisation bias results from active female manipulation of stored sperm (postcopulatory choice) or from simple physical constraints on sperm access to eggs (Naud et al., 2005).

While multiple paternity has also recently been confirmed in some species such as *Graneledone boreopacifica* (Voight and Feldheim, 2009) and *O. vulgaris* (Quinteiro et al., 2011), the hypothesis of cryptic female choice has yet to be tested for cephalopods. In a study by Quinteiro et al. (2011), analysis of two microsatellite loci of four females and their respective progeny revealed that two male contributors were inferred for all assayed offspring. The authors conclude that this reproductive pattern could originate from the differential use of sperm depositions (Shaw and Sauer, 2004; Walker et al., 2006) either through a process of sperm competition (Cigliano, 1995; Hanlon and Messenger, 1998) or mediated by a cryptic female choice (Quinteiro et al., 2011).

2.4.3 Sperm competition

In a promiscuous mating system, like that exhibited by many cephalopod species, the sperm from multiple males may compete for fertilisation of a given set of ova at the start of this first transition; this is known as sperm competition (Birkhead and Parker, 1997). To try and reduce the influence of sperm competition on their fertilisation success, male cuttlefish have developed a variety of mechanisms including agonistic behaviour (male–male aggression), pre- and postcopulatory mate guarding, sperm removal (flushing) and sneaker male mating and female mimicry.

2.4.3.1 Male–male aggression

Aggression between males in competition for females has often been observed in cuttlefish species (e.g. *S. latimanus* (Corner and Moore, 1980) and *S. apama* (Hall and Hanlon, 2002; Norman et al., 1999)) and in octopus (e.g. *A. aculeatus* (Huffard et al., 2008)). In field observations of *S. latimanus*, aggression between males was often observed in competition for females (Corner and Moore, 1980). In *S. latimanus*, incidents of male–male aggression were observed to increase in frequency and intensity during the precopulatory patterns of females; typically, smaller males lost these interactions (Corner and Moore, 1980). Males dashed towards each other repeatedly; if neither backed away, then both males flared their arms upwards and outwards. If neither retreated, then the arms were extended to an umbrella pattern and the mouth exposed, which lead to submission of one combatant in six out of nine observed encounters. If the larger male was unable to keep an opponent at bay, they resorted to an exaggerated umbrella pattern dashing in a side position towards the intruder with his arms twisted over the intruders head as if to bite him; this generally resulted in the intruder paling his body and retreating several metres (Corner and Moore, 1980). Ritualistic agonistic displays have also been observed to be used by large male *S. apama* in order to challenge consorts for female matings (Hall and Hanlon, 2002). In contests of disparate sized consorts, the agonistic display typically results in the loser submitting without physical contact, but in more even contests, battles can lead to biting and many males display scars or missing arm tips from such encounters (Norman et al., 1999). Male–male aggression has also been observed in the octopus species *A. aculeatus* almost exclusively while defending a female (26 of 29 male–male combinations) both at and away from the den (Huffard et al., 2008). Male aggression included behaviours such as chasing, touching and whipping with one straight arm and grappling (Huffard et al., 2008).

2.4.3.2 Male mate guarding

Mate guarding is a form of resource defence in which males reduce the probability that a female will remate with other males, thereby increasing the chance of a newly mated male siring most of the next-laid eggs (Hall and Hanlon, 2002; Hanlon and Messenger, 1998; Naud et al., 2004). However, mate guarding can be time-consuming and reduces the available time that males have to seek out new mates (Huffard et al., 2008). Mate guarding has been observed in cuttlefish (e.g. *S. officinalis* (Boal, 1996; Hanlon et al., 1999) and *S. apama* (Hall and Hanlon, 2002)), benthic octopods (e.g. *A. aculeatus* (Huffard et al., 2008)) and squid (e.g. *D. opalescens* (Hurley, 1978)).

In cuttlefish, temporary postcopulatory mate guarding may serve to prevent the removal of the newly mated male's spermatangia on the buccal membrane during the brief period during which the spermatangia may be susceptible to removal or dilution by flushing (Hanlon et al., 1999). During this postcopulatory period, the sperm from several males may compete for entrance to the two pores of the seminal receptacle (Hanlon et al., 1999). A study by Naud et al. (2004) noted a correlation between the time that the egg was laid after mating and fertilisation success with eggs laid 20–40 min following copulation more often fertilised by the mated male than those laid before or after this time, and these results may explain the strong postcopulatory mate-guarding behaviour observed, such that by preventing the female from mating with another male for at least 40 min, the newly mated male would gain more fertilisations (Naud et al., 2004). However, postcopulatory mate guarding has been observed to last for up to 90 min (*S. apama*; Hall and Hanlon, 2002), thereby also delaying their search for additional female copulations. Mate guarding has also been observed in squid in their natural environment; for example, large consort *Loligo* males were observed to guard their paired female mates as they descend towards the egg mass to deposit an egg capsule (Hanlon, 1998). During egg guarding, male squid were observed to touch the eggs with their arms, hover over the egg mass and display red coloration on their arms and pale coloration on their mantle and head to any other male that approached the egg mass (Hurley, 1978). If the second male did not retreat, the two males would line up parallel and perform agonistic behaviours including arm touching, pushing and occasionally biting that could result in severed arms and tentacles or scars on the mantle. The original guarding male would usually win these bouts, but if the challenger won, he would take up guarding the eggs. Artificial eggs made of silicon could stimulate females to lay eggs, but male agonistic behaviour could not be elicited until natural eggs were present. Contact with *Loligo* β-microseminoprotein (*Loligo* β-msp) immediately alters male squid, causing them to compete with agonistic behaviours even in the absence of females. *Loligo* β-msp is secreted by female exocrine glands and is embedded in egg capsule sheaths (Cummins et al., 2011). Large male and female pigmy octopuses (*A. aculeatus*) were found to occupy adjacent dens with copulation between the pair occurring frequently in mate-guarding situations over successive days (Huffard et al., 2008). Mate guarding in *A. aculeatus* was used by males as an effective tactic to temporarily monopolise mating with females (Huffard et al., 2008).

2.4.3.3 Sneaker mating and female mimicry

Behaviours like sneaker mating and female mimicry have in theory evolved as a reaction to the other males with superior competitive abilities (Parker, 1990). Sneaker males and female mimics adjust their behaviour to procure copulations while reducing aggression from other males as a result of the prior knowledge they have of their (in)abilities compared with other individuals in a group (Parker, 1990).

Sneaker mating has been observed in cuttlefish, in squid and more recently in octopus. In the large spawning aggregations of the giant Australian cuttlefish (*S. apama*), sex ratios are skewed and can be as high as 11 males to 1 female (Hall and Hanlon, 2002). This leads to intense competition between males for copulation, and mate guarding occurs almost continuously (Hanlon et al., 2005). In these aggregations, consort males were found to obtain approximately 64% of matings; the remainder are obtained by small, unpaired males, known as "sneakers", which obtain extra pair copulations using open stealth (e.g. approaching a guarded female while the male is distracted), hidden stealth (e.g. meeting females under rocks) or female mimicry (e.g. mimicking the appearance and behaviour of females) (Hall and Hanlon, 2002; Hanlon et al., 2005; Norman et al., 1999). In aggregations of *S. apama*, smaller males have been observed to adopt a strategy of switching their appearance from that of a male to that of a female in order to foil the mate-guarding attempts of larger males. Single small males accompanied mating pairs while assuming the body shape and patterns of a female, thereby avoiding agonistic attacks by larger mate-guarding males. On more than 20 occasions, Norman et al. (1999) then observed these smaller males to change body pattern and behaviour to that of a male in mating display and attempt to mate with the female, often with success, while the larger male was distracted by another male intruder. Hanlon et al. (2005) verified the success of female mimicry by smaller *S. apama* males to obtain extra pair copulation (EPCs) and to successfully fertilise eggs. The authors observed five initiations of mating by mimics: One attempt was rejected by the female and another was interrupted by the consort male, and the remaining three attempts however all resulted in successful spermatophore transfer (Hanlon et al., 2005). Using DNA fingerprinting, they were able to verify that two sneaker EPCs by a small mimic male resulted in successful fertilisation of eggs (Hanlon et al., 2005), providing the first confirmed report of sneaker fertilisation in a cephalopod.

In squid, multiple paternity has been detected in the egg capsules of *L. reynaudii* (Shaw and Sauer, 2004), *Loligo forbesii* (Shaw and Boyle, 1997), *D. pealeii* (Buresch et al., 2001) and *Heterololigo bleekeri* (Iwata and Sakurai, 2007) with both guard males and sneaker males providing sperm for eggs within the same capsule. Guard or consort males clasp the mantle of the female from the ventral side and remain with her for minutes during egg laying. Guard males place their spermatophores inside the mantle cavity near the oviduct (Hanlon et al., 2002). Larger males are more likely to mate via guarding, and smaller males are more often sneakers (Hanlon et al., 2002; Iwata et al., 2005). In *H. bleekeri*, sperm size has been found to correlate to mating strategy. Sneaker males place their spermatophores in the seminal receptacle below the mouth of the female during very brief head-to-head contact. Spermatophores of sneaker males are larger but contained less spermatozoa. *In vitro* artificial fertilisation rates ranged from 70% to 98% in sneaker males and 41% to 98% in consort males (Iwata et al., 2011). Egg capsule paternity tests suggest extra pair mating is successful <10% of the time (Iwata et al., 2005). Sperm differed in size and morphology in large and small mature *H. bleekeri*, and smaller sneaker males inserted smaller drop-like sperm in the seminal receptacle, and larger guard males inserted larger rope-like sperm in the mantle cavity (Iwata et al., 2008). Guard males had a greater success in paternity (Iwata et al., 2005). Conversely, in *L. reynaudii*, consort males sired 0–48% of the embryos examined in four egg capsules (Shaw and Sauer, 2004); in one capsule, a sneaker male or a prior consort male had 100% paternity, and the consort male captured at the time of egg deposition had zero paternity.

Sneaker mating is not well documented in octopus, a single example of sneaker mating has been reported in *Octopus cyanea* in the wild (Tsuchiya and Takashi, 1997) and more recently sneaker mating has also been observed in *A. aculeatus* during focal studies in the wild (Huffard et al., 2008). These studies demonstrate that natural history observations remain necessary for testing behavioural paradigms. A concerted effort is required to undertake further natural history observations of octopus in the wild in order to test behavioural paradigms and to document the mating behaviours of additional octopus populations (Huffard et al., 2008).

2.4.3.4 Sperm removal or flushing

Sperm removal and flushing are behaviours that males of several species of cephalopods can incorporate into their reproductive repertoire in order to reduce sperm competition. These behaviours are known to occur in several cuttlefish species (e.g. *S. apama* and *S. officinalis*). Prior to sperm transfer, male cuttlefish repeatedly flush strong jets of water through their funnels towards the buccal area of the female; this action is thought to result in the removal of spermatangia from previous matings, thereby increasing the newly mated males' chance of reproductive success (e.g. *S. apama* (Hall and Hanlon, 2002) and *S. officinalis* (Hanlon et al., 1999)); however, to date, no significant relationship between flushing duration and fertilisation success has been found (Naud et al., 2004). However, it is possible that the effectiveness of this flushing mechanism may be dependent on the time elapsed since the prior mating (Hanlon et al., 1999). In octopus, the spoon-shaped tip of the male's hectocotylised arm, known as the ligula, could function in the removal of stored sperm, from previously mated males and from within the spermatheca of the female's oviducal gland. This would function to try and ensure that the most frequently occurring paternal alleles would be provided from the last copulation event (Hanlon and Messenger, 1998; Iwata et al., 2005; Quinteiro et al., 2011). However, to date, sperm removal and flushing behaviours in octopuses have not conclusively been demonstrated (Huffard et al., 2008). A study by Sato et al. (2013) using the Japanese pygmy squid (*Idiosepius paradoxus*) observed a spermatangium-removing behaviour by eating and blowing, which may be used by females for postcopulatory sexual selection. Using this process, females were able to remove the spermatangium following matings with unpreferred males before completing sperm transfer into their seminal receptacle.

2.5. Summary

During the transition from the fertilisation of eggs by spawning adults to egg release in the wild, a series of structures and phenomena are involved, which provide protection and shelter to the offspring. Fecundation and egg release are under the control of hormones and pheromones. The eggs are well protected either by encapsulation that provides a physical and chemical buffer or by parental care. Adult cephalopods have developed a wide array of behaviours, which contribute to the success of reproduction: Migrations to suitable habitats for egg release, courtship and mating optimise egg fertilisation and sneaker males maintain genetic diversity in size at maturity.

3. TRANSITION #2: EGG TO THE EXHAUSTION OF YOLK RESERVES

3.1. Introduction

The transition from egg to the exhaustion of maternally derived yolk reserves and first feeding is a critical phase within the early life history of cephalopods. This critical phase ends when individuals have survived the posthatching stage and are capable of obtaining and using external sources of energy in such a way it contributes to individual growth (Vidal et al., 2002a). It, thus, involves changes from the relatively protected environment within the egg to occupying the hatchling habitat and those occurring from dependence on maternally derived energy reserves to active predation.

Newly hatched cephalopods resemble miniature adults (Boletzky, 1974; Nesis, 1973) but do generally differ behaviourally, ecologically and morphologically from adults (Bartol et al., 2008). Across the cephalopod taxa, three different early life modes occur: (1) squid and a number of octopod (merobenthic) hatchlings are planktonic, referred to as paralarvae (Young and Harman, 1988); (2) the hatchlings of a number of benthic octopods display nektonic–benthic behaviour before settling and adopting the benthic mode of life (Boletzky, 1977; Boletzky and Boletzky, 1973); and (3) all sepiids, most sepiolids and a number of octopod (holobenthic) species have hatchlings that immediately adopt the benthic mode of life (Boletzky, 2003; Young and Harman, 1988) observed in adults of these species. These large benthic hatchlings are referred to as juveniles (Naef, 1928).

This section will examine environmental factors, both within the egg capsule and in the surrounding oceanographic environment, that affect embryo survival; it will describe the conditions that determine hatching success and major changes occurring posthatching, from dependence on yolk reserves to a life stage of exclusive prey capture (Figure 4.3).

3.2. Hatching environment and key developmental milestones among cephalopod taxa

3.2.1 The egg capsule environment and links to hatch initiation

As a further result of ecological species adaptations, the size, structure and consistency of egg capsules vary considerably across taxonomic levels (Boletzky, 1989). Cephalopod embryos are encased in either an eggshell (cuttlefish) or a chorion, and within their encasings, all are bathed with perivitelline fluid (PVF). As such, PVF is in direct contact with the embryo and is considered important for the hatching process (Marthy et al., 1976). Perivitelline fluid enables a selective exchange between sea water and the embryo (Boletzky, 1987b). The buffering function of PVF implies a limited and selective exchange with sea water, specifically in terms of water, molecules (Gomi et al., 1986) and pollutants (Bustamante et al., 2004). Embryos therefore develop under abiotic conditions differing from those encountered as hatchlings. PVF is slightly hypertonic compared to sea water, enabling the swelling of cephalopods eggs, increasing surface area and decreasing capsule thickness (Cronin and Seymour, 2000). It has also been proposed that the osmoregulatory function of PVF could be closely coupled to physiological processes during embryonic development (Dorey et al., 2012). This fluid has also been suggested to act as a natural tranquilliser for embryos of loliginid squid, preventing premature hatching (Marthy et al., 1976).

The capsules and chorions act as a barrier to the diffusion of dissolved gases such as oxygen, resulting in dissolved oxygen levels within the egg capsules becoming a limiting factor towards the end of embryonic development. This has been observed in both benthic cuttlefish eggs (Cronin and Seymour, 2000; Gutowska and Melzner, 2009) and octopod eggs (Parra et al., 2000; Uriarte et al., 2012). Low oxygen levels result from the metabolic demands during embryonic development leading to an accumulation of metabolic CO_2 and a simultaneous drop in pO_2 (reviewed recently by Marquez et al., 2013). Gutowska and Melzner (2009) found that PVF pCO_2 increases as a function of embryo mass in cuttlefish. Both the postorganogenetic embryonic growth phase (Lacoue-Labarthe et al., 2010) and the hatching process (Parra et al., 2000) have been found to have a high energetic cost resulting in the rapid increase of oxygen consumption before and during hatching. The progressive depletion of oxygen at late embryonic stages may act as a hatching trigger in some species (Rosa et al., 2012).

Sea water properties such as pH and temperature also affect abiotic conditions of PVF, with important implications for cuttlebone calcification of embryonic and juvenile stages of the cephalopod *S. officinalis* (Dorey et al., 2012). Decreasing of sea water pH by 0.25–0.5 units resulted in hypercalcification by 17–80% of cuttlebone of cuttlefish embryos and juveniles (Dorey et al., 2012).

Elemental concentrations and requirements differ between eggshell-protected embryos (e.g. cuttlefish) and chorionic embryos (e.g. octopuses). The eggshell acts as a partially selective permeable barrier, depending on developmental stage, to both dissolved essential and nonessential trace elements (Bustamante et al., 2004; Lacoue-Labarthe et al., 2012), whereas in eggs surrounded by a chorion only, the uptake of elements is nonselective (further details in Villanueva and Bustamante, 2006). Despite the lack of information, particularly with regard to implications on the hatching process, a number of studies suggest that the bioaccumulation of trace elements influences embryonic development. Calcium reserves are needed for metabolic process being expected to be contained in yolk of cuttlefish eggs (Boletzky, 1989). Thus, calcium concentrations increased over the oocyte development of *O. vulgaris* suggesting the importance of sea water intake (Villanueva and Bustamante, 2006). Sea water properties such as pCO_2 may affect this property (Lacoue-Labarthe et al., 2011). Abnormal hatchlings resulted from using artificial water without strontium in *O. vulgaris*, *L. vulgaris* and *S. officinalis* (Hanlon et al., 1989a). In addition, a high requirement of copper for hatchlings and juveniles was found in these species, particularly in octopus paralarvae, and is suggested to be related with haemocyanin requirements for oxygen transport (Villanueva and Bustamante, 2006). Not only is the environment within the egg capsule important, but also the conditions immediately surrounding the egg clutch also influence mortality. The position of embryos within an egg clutch may lead to differences in hatchling traits such as developmental rate, survival and growth (Strathmann and Strathmann, 1995). Steer et al. (2003) found differences in the embryonic development rate within single egg strands of *S. australis*, with eggs situated at the fixed or proximal end developing more slowly than those situated at the free or distal end. It was also noted that embryos developing at the proximal end suffered higher mortalities. Differences in hatchling mantle length (ML) were found in *Illex coindetii* embryos incubated at different eggs densities during an *in vitro* experiment. Emergent hatchlings were larger at lower egg densities (≤ 5 eggs ml^{-1}) (Villanueva et al., 2011). Asynchronous development rates and variation in the size of hatchlings within a single egg strand or cluster have also been recorded for *Sepioteuthis sepioidea* and *Sepioteuthis lessoniana* (Chung, 2003; Ikeda et al., 1999; LaRoe, 1971). Strathmann and Strathmann (1995) suggested that the proximal embryos were not sufficiently oxygenated and/or accumulated excess excretory products and responded by either retarding or arresting development. Chung (2003) observed that large egg clusters tended to block the current causing less circulation within the clusters. This would indeed influence oxygenation. Pelagic eggs are extruded in a continuous fragile gelatinous mass, which, as a result of water absorption, has a neutral buoyancy. The absorption process increases the space between embryos and so allowing proper ventilation throughout the egg mass (Lee and Strathmann, 1998). The variability of hatchlings from the same egg strand can also be a result of multiple paternities (Buresch et al., 2001).

3.2.2 The changing oceanographic environment and effects on embryo survival and hatching

Few studies have assessed how natural embryonic mortality results. Steer et al. (2002) found that natural mortalities of embryos of the southern calamari (*S. australis*) can vary between 4% and 20% in late November, decreasing to 10% until late December. In *D. gahi*, Arkhipkin et al. (2000) reported higher mortalities in winter, resulting in four times spring and summer mortalities. There is no clarity whether seasonal differences in embryo mortalities are related to thermal variations or other environmental conditions (e.g. rainfall (Steer et al., 2002) and salinity (Şen, 2005)). Recent studies in thermal tolerance in benthic octopus embryos have shown a higher thermal sensitivity in embryos than in preadults and adults (Uriarte et al., 2012). Cephalopod embryos are very vulnerable to thermal fluctuations, especially to extreme temperatures as those used in laboratory-based studies

(Oosthuizen et al., 2002a; Rosa et al., 2012; Uriarte et al., 2012). Furthermore, from embryo position studies, we know that oxygenation is important for embryo development and survival (Strathmann and Strathmann, 1995). Ocean pH effects on survival are also unknown for most species.

Cephalopods are well studied in terms of temperature effects on embryonic development and hatching. Results of those studies have important implications for the prediction of impacts of climate change (e.g. warming process). Thermal niche classification throughout embryonic stages has resulted in the potential to understand cephalopod distribution and the influence of oceanographic and physical conditions (e.g. *D. gigas* (Staaf et al., 2011)). A study on the effects of temperature on the embryonic development of this species found the upper temperature limit for successful embryonic development through hatching to be 25 °C (Staaf et al., 2011). Embryos experiencing temperatures outside the optimal thermal range resulted in mortalities by abnormal embryonic development in *L. reynaudii* (Oosthuizen et al., 2002a). A study by Rosa et al. (2012) investigated the effects of ocean warming (+2 °C) on *L. vulgaris* embryonic development. A metabolic suppression of embryos incubated at +2 °C of the optimal thermal range resulted in premature hatching under hypoxia.

Most cephalopod species are stenohaline with some exceptions (i.e. *Lolliguncula brevis*; Hendrix et al., 1981; Laughlin and Livingston, 1982) and are osmoconformers. Salinity was shown to affect the hatching rate, eggs size, first hatching time and dorsal ML of *L. vulgaris* under controlled conditions (Şen, 2005). Low salinity levels have been reported to have more severe effects on *S. lessoniana* embryonic development, in terms of abnormalities, compared to temperature (Chung, 2003). Very low salinity levels (20‰ and 25‰) were found to cause the death of *S. lessoniana* embryos, even at optimal temperatures (Chung, 2003). Similarly, the survival of loliginid embryos appears to be more affected by heavy rainfall and consequent reduction in salinity levels, rather than temperature changes (*S. australis*, Steer et al., 2002). Areas where salinity is expected to rapidly change (e.g. places with more frequent rain events) may negatively affect cephalopod inhabitants.

The global decrease of pO_2 concentration in the ocean, deoxygenation, is caused by global warming and increased stratification (Keeling et al., 2010). Increasing declines in pO_2 availability are predicted to affect cephalopod physiology, behaviour and demographics (Gilly et al., 2013; Pauly, 2010; Pörtner et al., 2005). The beginning of the transition, from hatch initiation to the exhaustion of yolk reserves, is especially susceptible to deoxygenation as critical *p*O2 levels can be reached just prior to hatching in "normal" environmental conditions (Cronin and Seymour, 2000; Gutowska and Melzner, 2009). Although the metabolic demands differ among taxa, Pimentel et al. (2012) found that *L. vulgaris* had significantly higher demand than *S. officinalis*. They attributed this demand to reflect taxa differences in locomotion (Aitken and O'Dor, 2004), activity (Aitken et al., 2005), buoyancy (Denton and Gilpin-Brown, 1961) and oxygen extraction capabilities (Gutowska et al., 2010b; Strobel et al., 2012; Wells et al., 1988). In general, teuthiods have faster metabolic rates than octopods or sepiids throughout life stages including during the hatching process (Wells et al., 1988). Taxa with greater metabolic demands during embryogenesis may be especially vulnerable to asphyxiation or premature hatching in areas intensely impacted by deoxygenation.

To keep pO_2 levels above critical low levels, hypoxia-induced metabolic suppression occurs until the embryo is mature enough to hatch, and it is suspected that this continues until a threshold of low pO_2 ultimately triggers hatching (Pimentel et al., 2012). For loliginids, the PVF likely plays a key role during the transition by tranquillising near-hatch embryos, suppressing metabolic demand and allowing the embryo to fully develop (Marthy et al., 1976). *L. vulgaris* metabolic rate is more than doubled after hatching (Pimentel et al., 2012).

From what is known about cephalopods, teuthoids may be the most sensitive to deoxygenation with a broad spectrum of effects across species. These effects can be categorised as one of two types: lethal and sublethal. Lethal effects include underdevelopment or behavioural impairment preventing embryos from hatching, whereas sublethal effects include underdevelopment and/or malformations that carry over negative effects to later life stages (i.e. carry-over effects). Sublethal effects lower survivorship rates for paralarvae, reducing their abilities for feeding and/or for defence, and overall lower their fitness.

Ongoing anthropogenic pCO_2 uptake by the ocean increases acidification of sea water and decreases global ocean pH (ocean acidification; Doney et al., 2009; Feely et al., 2004, Orr et al., 2005). Molluscs are vulnerable to acidification during larval stages (Bryne, 2011). Many invertebrate embryos exposed to global warming experience a mortality "bottleneck" leaving larval stages (and presumably the transition from hatch initiation to yolk exhaustion) vulnerable to acidification (Bryne, 2011). Gastropod embryos exposed to a pH of 7.6 exhibited "subtle" effects such as a slower heartbeats and reduced locomotion implicating lower survival rates in later life stages including the larval stage (Ellis et al., 2009). Abalone and oyster were also found to be susceptible to acidification during the larval stage (Parker et al., 2013). Most mollusc ocean acidification studies have not focused on cephalopod embryos or young hatchlings (Bryne, 2011). Gutowska and Melzner (2009) showed that cuttlefish embryos experience low levels of pH naturally (i.e. 7.2). However, it is an open question whether low environmental pH interacts with PVF pH and, if so, whether the PVF is lowered by environmental pH. Large gaps remain in the study of ocean acidification effects on cephalopods; the effects of organisms exposed to increased concentrations of hydrogen ion and pCO_2 (i.e. lower pH) are in their infancy.

Several burgeoning fields are developing in ecotoxicology and comparative physiology. Bioaccumulation of pollutants under low-pH conditions has been found for some species of sepiids (Lacoue-Labarthe et al., 2009, 2012) and teuthiods (Lacoue-Labarthe et al., 2011). Sepiids have been shown to calcify at higher rates having thicker densities possibly leading to negative effects on locomotion (Gutowska et al., 2008, 2010b). This likely impacts sensory abilities necessary for locomotion as has been found in fish. Sepiids could perform "normal" acid–base regulation under environmental pH below 7.2 (Gutowska et al., 2008, 2010a). Physiological study on gill membranes showed that late-stage embryos did experience significant reduction in somatic growth and incomplete development (Hu et al., 2011). Although some have suggested that cephalopods are "unscathed" by ocean acidification (Branch et al., 2013), we caution that many more studies across cephalopod taxa are needed to arrive at a clear understanding of the effects of high environmental pCO_2 and associated low-pH exposure to cephalopods.

3.3. Hatching

Hatching conditions in cephalopods are flexible suggesting a phase of hatching competence rather than a well-defined hatching stage by morphological and/or physiological features (Boletzky, 2003). This flexibility is suspected to optimise posthatching survival (Boletzky, 2003). The timing of hatching, or hatching period, varies greatly across systematic groups, ranging from 2 days in *Octopus laqueus* (Kaneko et al., 2006) to 78 days in *E. dofleini* (Gabe, 1975). Villanueva and Norman (2008) undertook an extensive review of hatching times for many merobenthic octopods. They suggest that reported periods may be underestimated since observations likely come from a single major hatching event. In *Octopus maya*, an observed variation in hatching time and size of emergent hatchlings from a single female implied a source of variation in initial conditions (e.g. initial size) (Briceño et al., 2010).

Mechanical stimulation, as well as biological and environmental factors, is suggested to trigger hatching (under wild and laboratory conditions). The mechanical stimulation of eggs, such as agitation or turbulence generated by a brooding octopus female, has also been suggested to promote or regulate the timing of hatching in benthic octopods (Villanueva and Norman, 2008). Octopus females are able to expel water forcibly over the eggs, aiding the ejection of hatchlings out of the den (Boletzky, 2003; Hanlon and Messenger, 1998). This behaviour has been largely observed at night, possibly enhancing hatchling survival (Hanlon and Messenger, 1998). The agitation of eggs incubated under artificial conditions (without parental care) was observed to promote hatching in *E. dofleini* (Snyder, 1986), *Octopus tetricus* (Joll, 1976) and *O. maya* (C. Rosas, personal observation).

Light has been proposed to modulate the hatching process in cephalopods, with most hatching events occurring during sunset and at night, which possibly minimises predatory risk (Rodrigues et al., 2011; Summers, 1985; Villanueva and Norman, 2008). Fields (1965) suggested that by emerging in darkness, hatchlings will be transported by tidal drift to clear spawning areas before light returns sight to any predators present in the vicinity of the spawning grounds. Under laboratory conditions, a majority of night-time hatchings were recorded for *O. vulgaris* (Villanueva and Norman, 2008), *Octopus briareus* (Hanlon and Wolterding, 1989), *S. pharaonis* (Nair et al., 1986) and *S. officinalis* (Paulij et al., 1991). In loliginid squid, such as *L. vulgaris*, hatching is influenced by light–dark transition, which is suggested to act as "zeitgeber" or synchroniser (Paulij et al., 1990). Other factors such as tidal and lunar rhythms, as well as external synchronisers and circadian rhythms in adults, are suggested to be associated with the hatching process (Villanueva and Norman, 2008). Variation in light regimes, the degree of exposure of study animals, observer behaviour, use of flash photography, mechanical vibrations and temperature fluctuations have been proposed as hatching stimuli under laboratory conditions (Villanueva and Norman, 2008).

Hatching success depends partially on the degree to which the chorion membrane swells (Villanueva et al., 2011). For example, the swelling process in *O. vulgaris* eggs has been found to occur in two distinct phases: first, an initial slow swelling from Naef's stages 1 to 10–15 and, second, a subsequent accelerated swelling of the eggs (Marquez et al., 2013). Marquez et al. (2013) compared these two chorion swelling phases to a number of other studies and reported a second accelerated swelling process for loliginids, sepiolids, cuttlefish and octopods. Chorion expansion is essential to allow space for embryos to move and respire (Sakurai et al., 1995) and is said to be essential for successful hatching (Villanueva et al., 2011).

The hatching process begins with increased mantle contractions and fin movement by the embryo (Lee et al., 2009). Hoyle's organ, located at the distal tip of the mantle (Boletzky, 2003), and terminal spine are pushed directly against the inner wall of the chorion (Lee et al., 2009). This is thought to activate the secretion of enzymes from the hatching gland cells (Boletzky, 1987b; Villanueva and Norman, 2008). These enzymes digest the chorion membrane, while the terminal spine perforates the chorion and the enzyme-resistant outer capsule (Lee et al., 2009). Using mantle contractions and fin undulations, the embryo expels itself out of the opening (Lee et al., 2009).

3.4. From dependence on yolk reserves to exclusive prey capture

3.4.1 Changes from an internal to external energy source

The yolk mass of late-stage cephalopod embryos is made up of the outer yolk sac, remaining outside of the embryo, and the inner yolk, the portion lying inside the embryo (Boletzky, 2002). The yolk neck connects the outer and inner yolk sacs (Boletzky, 2002). Towards the end of embryonic development, a shift in partial pressure, as a result of further development of the visceral mass, causes the yolk to "flow" from the outer to the inner sac (Boletzky, 2002). In cuttlefish *S. officinalis*, the transfer of yolk has been noted to take place during the final 2 weeks of embryonic development (Lacoue-Labarthe et al., 2010). The yolk volume inside the mantle cavity would peak around hatching, as has been observed in *T. rhombus* embryos (Miyahara et al., 2006b). These inner yolk reserves are essential to the survival of hatchlings during the first few days posthatching (Vidal et al., 2002a). As the outer yolk sac is normally shed after hatching (Boletzky, 2002), hatching before all the yolk has been transferred into the mantle cavity would limit the energy available (Boyle and Rodhouse, 2005).

After the yolk reserves are completely exhausted, hatchlings need to catch prey in order to survive (Vidal et al., 2002a). The switch from yolk utilisation to exclusive prey capture is a critical phase in the early life history of cephalopods, and hatchlings are extremely sensitive to starvation during this period (Boletzky, 2003; Vidal et al., 2002a, 2006). The high mortality rates shortly after hatching, observed in numerous cephalopod rearing experiments (Boletzky and Hanlon, 1983; Vidal et al., 2002b; Villanueva, 2000; Yang et al., 1983), confirm this.

The rates of yolk utilisation in fed and starved laboratory-reared loliginid paralarvae suggest that yolk utilisation is dependent on feeding conditions, with fed paralarvae having lower utilisation rates (Vidal et al., 2002a). Initially, loliginid paralarvae undergo a no-net-growth phase as a result of the exponential rate of yolk utilisation to support the metabolic costs of maintenance (Vidal et al., 2002a, 2005). The weight lost is regained by feeding and the end of the no-net-growth phase reached when hatching weight is achieved (Vidal et al., 2002a, 2005).The no-net-growth phase shortly after hatching corresponds to the high mortality rates observed during this period (Vidal et al., 2002a, 2005). Loliginid statolith growth studies too indicate a period of no growth (Villanueva et al., 2007). This no-net-growth phase has also been observed in *O. maya* juveniles (Moguel et al., 2010). Analysis of RNA/DNA ratios in *D. opalescens* paralarvae confirms the poor nutritional condition of paralarvae during this period and their extreme sensitivity to starvation (Vidal et al., 2006). To survive the transition from yolk utilisation to exclusively prey capture, cephalopod hatchlings must not only initiate exogenous feeding but also find sufficient food to fuel metabolism and growth and improve prey-capture efficiency (Vidal et al., 2002a).

3.4.2 Hatchling morphology and prey capture

O. maya (Moguel et al., 2010) and *S. officinalis* (Darmaillacq et al., 2006) hatchlings do not immediately respond to or attack prey within the first few days posthatching. This lack of predatory behaviour has been attributed to the use of yolk reserves as the main energy source immediately after hatching (Moguel et al., 2010). Although hatchlings can survive on yolk reserves alone for a number of days (Miyahara et al., 2006b; Vidal et al., 2002a), yolk energy is used to fuel standard metabolism and activity and is quickly depleted (Vidal et al., 2005, 2006). The inner yolk reserve in cephalopods is separated from the digestive gland and is directly connected to the venous system (Boletzky, 2002). This enables simultaneous endogenous (yolk utilisation) and exogenous (prey-capture) feeding (Boletzky, 2002) and an opportunity to develop and improve prey-capture skills (Kier, 1996; Shea, 2005; Vidal et al., 2005). The capture and ingestion of prey at the first feeding induce the digestive process and the secretory activity of the digestive gland (Boletzky, 2002; Boucaud-Camou and Roper, 1995; Mangold and Young, 1998; Moguel et al., 2010). Active feeding during the yolk utilisation phase has been observed in cuttlefish (Boletzky, 1975), *O. maya* (Moguel et al., 2010) and *D. opalescens* (Vidal et al., 2002a).

The first feeding depends on the cephalopod's ability to process visual and chemosensory information to recognise the sight or odour of potential prey (Romagny et al., 2012). This ability develops differently among cephalopod taxa. Young *S. officinalis* hatchlings innately prefer and target shrimplike prey (Darmaillacq et al., 2006). However, visually exposing embryos during late-stage embryonic development (Darmaillacq et al., 2008) or new hatchlings (Darmaillacq et al., 2006) of crab results in a preference for crab when active feeding commences a number of days later (Darmaillacq et al., 2006, 2008). Darmaillacq et al. (2006) also showed that the efficiency of this familiarisation in young hatchlings was dependent on the length of time and the number of crab that were visually exposed. The ability to learn the visual characteristics of prey in ova immediately after hatching facilitates imprinting and preference for prey abundant in the immediate area (Darmaillacq et al., 2008; Guerra and Gonzalez, 2011). This could have a beneficial effect during the transition from endogenous (yolk utilisation) to exogenous feeding (active predation), during which hatchlings are extremely sensitive to starvation (Vidal et al., 2006). Guerra and Gonzalez (2011) suggested that similarly late-stage embryos can recognise potential predators in ova and avoid them after hatching.

In contrast, prey selection experiments on *O. maya* suggest that preference in early juveniles (7 and 16 days posthatching) has a strong innate component that cannot be modified through imprinting or associative learning during the first weeks of life (Portela E., unpublished data). To preferentially attack prey that follow a predetermined search image of food may result in fast and successful foraging bouts that reduce both the time spent searching and testing alternative prey and the risk of failing to identify a deadly threat. In addition, to attack prey resembling a predetermined search image whenever it is present but readily to consume unfavoured prey when these are abundant would confer flexibility in foraging and allow predators to adjust to unpredictable changes in food quantity and quality.

Newly hatched cephalopods have a saclike mantle and a circular fin shape (Sweeney et al., 1992). Whether planktonic, nektonic–benthic or benthic, hatchlings swim by way of jet propulsion. Initially, the fins of squid paralarvae are small and rudimentary, and jet propulsion is the primary means of locomotion (Bartol et al., 2008, 2009). After hatching, paralarvae immediately begin swimming within intermediate Reynolds numbers (*Re*) (Bartol et al., 2008). The viscosity at this *Re* range prevents coasting, the continuous forward momentum observed in adults after the expulsion of water from the mantle (Bartol et al., 2008). As a result, paralarvae make use of pulse jet propulsion, with more rapid and frequent mantle contractions than observed in adults (Bartol et al., 2008; Thompson and Kier, 2001). Bartol et al. (2008) also noted that squid paralarvae swim predominately along a vertical axis, with sinking occurring as the mantle is refilled. Backward swimming is considered routine swimming and has been observed in planktonic squid (*L. vulgaris*, *L. forbesii* and *I. illecebrosus*) and octopus (*O. vulgaris*) hatchlings (Villanueva et al., 1997). Forward swimming forms part of predatory swimming behaviour (Villanueva et al., 1997).

Prey capture in cephalopods is both instinctive and experience-dependent (Messenger, 1977). In *D. opalescens* paralarvae, fundamental features of basic attack have been observed as early as the day of hatching (Chen et al., 1996). A rearing study has shown that the capture of copepods by squid paralarvae is a skill that must be acquired in an experience-dependent manner (Chen et al., 1996). Over the course of numerous unsuccessful attempts and the occasional successful capture, paralarva attacks become progressively more refined and complex (Chen et al., 1996).

Cephalopod hatchlings are visual predators (Boletzky, 1977; Villanueva and Norman, 2008) with prey capture achieved by a visual feedback system involving fixating the prey binocularly, assuming an attack position and striking the prey (Nair et al., 1986). According to Nair et al. (1986), it is essential that prey must be within the visual field of the hatchling in order for an attack to be initiated. Most teuthoids do not hatch with functional feeding tentacles, and the strike behaviour of young hatchlings differs from that of adults (Kier, 1996). In contrast, most sepiolids capture prey using the rapid tentacle elongation observed in adults (Boletzky 1987b; Boyle, 1983a,b). Kier (1996) noted that, considering the longer embryonic development phase in sepiolids compared with teuthoids, it appears that a longer developmental period is required for hatchlings to emerge with functional feeding tentacles. Changes in arm size relative to mantle size in early *O. maya* hatchlings suggest variations in attack success rate and prey capture as octopuses grow into full juveniles (Moguel et al., 2010). These changes together with those in the number (Boletzky and Boletzky, 1969) and structure of suckers (Kier and Smith, 1990) are believed to be in close correspondence with the transition from nektonic–benthic to fully benthic habits in octopuses.

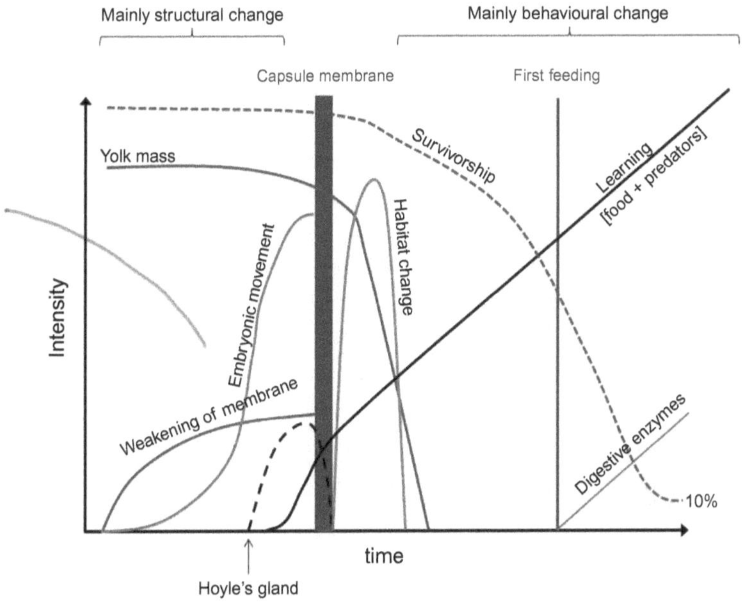

Figure 4.2 Overview of the main phenomena occurring to the organism during the transition from egg to paralarvae with qualitative indications of their intensity.

3.5. Summary

The numerous changes that occur during this transition are represented in a schematic diagram (Figure 4.2), which shows that some phenomena start before (and others continue after) the boundaries of this transition. Both the environment within the embryo and the surrounding oceanographic environment affect embryonic survival and hatching success. Once hatched, survival is largely dependent on passing through the critical transition from endogenous to exclusively exogenous feeding. Feeding and digestion of new items take part gradually. Brain development and learning start in the late embryo, and early acquisition of essential sequences such as feeding, swimming or predator escape might explain the high variability in the performance of specimens within the same cohort.

4. TRANSITION #3: FIRST FEEDING TO SUBADULT

4.1. Introduction

The transition from paralarva to subadult in cephalopods does not involve a metamorphosis as found in many other marine invertebrates (Boletzky, 1974). Paralarvae have, in general, similar morphology to subadult and adult individuals. The changes occurring are subtle and concern the variation in growth rates, changes in body proportions or changes in the structure or function of specific organs, for example, the separation of the fused tentacles (proboscis) in ommastrephid rhynchoteuthion. In this respect, the transition from paralarva to subadult should be identified for each family or even for each species since the transition into a subadult is not followed by ecological or habitat changes in all species (Shea and Vecchione, 2010). For example, morphological aspects, such as the presence of a proboscis, have been used to identify ommastrephid squid paralarva (Froerman and Dubinina, 1984; Harman and Young, 1985; Wormuth et al., 1992), while for other oegopsid squid and sepiids, the differentiation of the tentacles has been also used as marker of the end of the paralarval stage (Kier, 1996). Both examples are not linked with changes in ecological function of tentacles but are used as keystone criteria to identify the transition between paralarva and subadult.

Obvious morphological changes during growth are characterised by discontinuities in relative growth that highlight crucial limits in stages of development (Nesis, 1979), and the first discontinuity seems to coincide with the transition from paralarva to subadult (Young and Harman, 1988). These changes in growth have been documented in many families including changes in body proportions in the Cranchiidae (Voss, 1980) and changes in chromatophore patterns in the Brachioteuthidae (Young et al., 1985). Many other examples can be found in Sweeney et al. (1992) and Vecchione (1983).

Shea and Vecchione (2010) examined the timing of ecological and morphological changes in oegopsid squids (*Chtenopteryx sicula*, *Mastigoteuthis magna* and *Brachioteuthis* sp.), and their results showed that the diel vertical migration (DVM) patterns are species-specific, and the end point of the paralarval stage cannot be defined ecologically in these three species. None of the three species examined had a sudden, obvious onset of DVM, and no clear pattern was found in the size of the specimens collected deeper than 250 m by day, the main benchmark for having left the paralarval stage (Young and Harman, 1988). In their work, Shea and Vecchione (2010) proposed a revised definition of paralarvae: "newly hatched cephalopods that have a distinctly different mode of life from the adults, with an endpoint identified by ecologically significant allometric changes in morphological characters". After going through these significant changes in morphology, the subadults grow until the onset of the maturity process.

All stages of transition from one development stage to another are crucial to the survival of any animal, including cephalopods. The transition between paralarva and subadult is not an exception. Mortality rates are higher during the early paralarval stages in many small-edged octopod species, and high peaks of mortality occur at settlement. Villanueva and Norman (2008) reviewed this aspect and showed that in laboratory experiments, the survival of benthic octopus at settlement is around 10% for individuals fed with crustacean zoeae. For species for which the transition from paralarva to subadult does not involve habitat changes as in oegopsid or myopsid squid, the survival numbers grow considerably. Using indirect mortality estimates, Bigelow (1992) estimated for *Abralia trigonura* an average instantaneous mortality rate of 7%, while González et al. (2010) using the same methodology estimated for *L. vulgaris* an instantaneous mortality rate between 5% and 10% depending on biotic and abiotic factors. In their study, González et al. (2010) showed that survival of paralarvae decreases with age and this trend is stronger between 40 and 50 days after hatching. The recruitment, which is an arbitrary measure of population breeding success, is also very dependent on the growth and survival of the paralarva to subadult stage (Boyle and Rodhouse, 2005).

Nixon and Young (2003) had provided an extensive range of examples in which one can observe changes in the nervous system and sense organs as animal transitions from a paralarva to a subadult and the consequences on behaviour. Therefore, this chapter will not discuss neurological underpinnings of this transition. Additionally, environmental factors evolved in this ecologically significant change in the mode of life can be quite diverse. The importance of each specific environmental factor in the transition from paralarva to subadult depends on the habitat occupied by the adult form of the species or group of species. For example, in the ommastrephid squid such as *Illex argentinus*, a significant change in feeding habitats marks the change in life stage (Vidal, 1994; Vidal and Haimovici, 1998). In others, as *O. vulgaris*, paralarvae become subadult when they settle to the benthic habitat (Villanueva and Norman, 2008). Thus, we have attempted to create a general synthesis of the broad morphological, physiological, ecological and behavioural changes associated with the transition from paralarva to subadult, but the specific context for groups and subgroups of cephalopods is equally important in our understanding of the importance of these changes.

4.2. Morphological, suborganismal and physiological changes during transition

Paralarvae exhibit distinct growth patterns that change as the animal transitions into subadulthood. The ML, the head length, the arm lengths and the proboscis length of the paralarvae of the squids *I. illecebrosus* and *I. argentinus* grow allometrically (Froerman and Dubinina, 1984; Vidal, 1994).

The transition from paralarva to subadult often involves the modification, appearance or loss of the function of some organs. The newly hatched and paralarvae of all Ommastrephidae are known as rhynchoteuthions, in which both tentacles are fused into a "proboscis". This organ becomes separated as individuals grow. The end of the rhynchoteuthion stage is marked by the complete separation of both tentacles, but the size at which this is completed is variable both within and between species (Okutani, 1987). The squid *I. argentinus* has three stages of growth during transition 3: (1) a rapid development of arms, suckers and fins with positive allometric growth (1–14 mm ML); (2) a rapid development of tentacles and clubs with isometric growth (14–28 mm ML); and (3) a greater increase in length relative to the other body parts with negative allometric growth (>28 mm ML) (Vidal, 1994). These shifts in growth pattern seem to be related to animal activity and food acquisition in different environments (Vidal et al., 2010).

In benthic octopuses with planktonic stages, the metamorphosis from paralarva to subadult occurs at settlement. Morphological changes associated with settlement include positive allometric arm growth; chromatophore, iridophore and leucophore geneses; the loss of Kölliker organs that cover the body surface; the loss of the "lateral line system"; and the loss of the oral denticles of the beaks (Villanueva and Norman, 2008). Changes in body proportions in *O. vulgaris* during this transition are shown in Figure 4.3.

According to Young and Harman (1988), the number of chromatophores on the dorsal surface of the mantle (and others body surfaces) varies during the transition between stages of paralarva and subadult. There is an increase in number and complexity of changes in the disposition of chromatophores. An increase in number of chromatophores and change in pigmentation pattern on the mantle were observed in three *Onychoteuthis* species at 12–16 mm dorsal ML (Young and Harman, 1987), between 14 and 17 mm ML for *I. argentinus* (Vidal, 1994), at 4.5 mm ML for *Doryteuthis pealeii* (Vecchione, 1981), between 7 and 10 mm ML for *S. lessoniana* (Segawa, 1987) and at 20 mm ML for *Gonatus madokai* (Kubodera and Okutani, 1977).

The size and shape of the eyes, especially in the oegopsid squids, vary widely within the paralarvae. Given that the major causes of mortality for the paralarvae are starvation and predation and that both depend critically

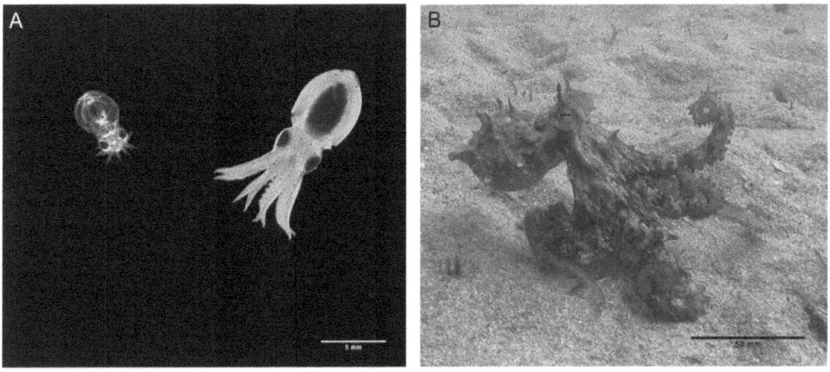

Figure 4.3 Life cycle of *Octopus vulgaris* from hatching to settlement. Recently hatched *O. vulgaris* paralarva have three suckers per arm (paralarvae on the left in A, 2.1 mm DML; photo by Jorge H. Urcera). After a period of constant swimming the planktonic octopuses undergo a transitional period from a pelagic lifestyle (juvenile on the right in A, 5.75 mm DML with 21 suckers; modified with permission from Iglesias et al., 2007) to the adult benthic life (subadult in B, 5 cm DML; photo by Manuel E. Garci). *Figure composition by Alvaro Roura and Sílvia Lourenço.*

on visual adaptations, it is likely that ocular specialisations that enhance visual efficacy have been favoured by evolution (Weihs and Moser, 1981). The two major limitations for foraging efficiency are the perception distance and the aperture of the visual field. Increasing the panoramic field of vision in cephalopod paralarvae has followed a progressive course. The first of these steps involves the possession of sessile eyes, round and firmly seated in their sockets. This type of eye is characteristic of loliginid paralarvae, paralarvae of some oegopsid squid species (e.g. the rhynchoteuthion of ommastrephids, *Onychoteuthis banksii* and/or *Gonatopsis borealis*) and species of octopuses with planktonic stages. The second stage involves the development of short eyestalks, so that the eyes are directly in line with the outer margin of the head, such as in *Helicocranchia pfefferi* paralarvae. Such eyes rotate freely, allowing the animal to redirect the visual cone without moving the body. In the third stage, the eyestalks are arranged at different lengths. These eyes, besides allowing changes in the direction of the visual axis, increase visual perception of distance, so the volume of observable water from a given point becomes a sphere. Additionally, prey can be detected from farther away and with minimal movement of the paralarva. This is the case, for example, of the cranchiid squid *Bathothauma*, which stalked eyes may be connected with plankton feeding at great depths, perhaps improving distance judgement (Young, 1970). The disappearance of these eyestalks with growth could be related to the different strategies of prey capture and predator avoidance for paralarvae and adults. For paralarvae, visual perception and immobility may be crucial, whereas in adults, these characteristics are less important (Guerra and Pérez-Gándaras, 1984).

In many cephalopods, the shape and features of the beak change during the transition from paralarva to subadult. In their planktonic stages, *L. vulgaris* and *D. pealeii* have teeth on the lower jaw, while the upper jaw is toothless (Boletzky, 1971). The beaks of the squid paralarvae also have a similar structure. Franco-Santos and Vidal (2014) observed that at hatching, the beak is rudimentary. Additionally, all beaks measured showed strong positive allometry in relation to ML, which has clear implications in a diet change as the animal grows. This character is found in both myopsid and oegopsid squids (Wakabayashi et al., 2005). In many species of incirrate octopods, such as *O. vulgaris*, *Octopus joubini*, *E. cirrhosa*, *Argonauta argo* and *Tremoctopus violaceus*, the anterior margin of both lower and upper beaks forms a row of relatively large teeth. This dentition is more important in species with planktonic paralarvae than in those whose offspring are benthic, like *O. maya* or *O. briareus*, in which the reinforced edges of the beaks seem more like the smooth cutting edges of the adult mandibles.

Although there is no direct evidence that toothed beaks are related to the planktonic life stages, this assumption is endorsed by the fact that teeth are present in common, newly hatched planktonic stages of different cephalopod groups: Octopoda with planktonic early life stages and paralarval Teuthida that are all planktonic. Whereas in all young Sepioidea, most of which are benthonic, the beaks are adult-like. The function of the teeth is possibly related to quick ingestion of planktonic prey (Boletzky, 1974). In consequence, this paralarval feature could be considered as a transitory formation in adaptation to the type of diet as the individual grows. Thus, the progressive strengthening of the jaws and teeth of planktonic larvae of cephalopods represents a change of role in the trophic web.

Photophore development can also mark the transition out of the paralarvae, especially in Myopsida and Oegopsida. Examples of photophore development have been shown in many oegopsid families, especially within Enoploteuthidae, Cranchiidae and Ommastrephidae, which are typically photophore-bearing squids (see Henning et al., 2002 and Okutani, 1987 for a review). Photophores develop on the ventral region of the eye in *Stenoteuthis oualaniensis* between the paralarval sizes of 1.8 and 6.3 mm (Ramos-Castillejos, 2007). Bykov and Dubinina (1984) examined the early ontogenic stages of the squids of the families Enoploteuthidae. They identified successive development stages on the basis of the appearance of photophores on the ventral arms (*Abraliopsis morisii*) and intramantle photophores (*Pterygioteuthis giardi*) that defines roughly the morphological changes occurring during the transition from paralarva to subadult stage. Four stages were established for *A. morisii*: (1) stage without photophores (1.5–3.1 mm ML), (2) stage with one photophore on the ends of the ventral arms (2.7–3.8 mm ML), (3) stage with two photophores on the ventral arms (3.7–5.5 mm ML) and (4) stage with three photophores on the arms of the four pair; numerous photophores appear on the ventral surface of mantle, head and eyes (over 5.5 mm). Four stages were also established for *P. giardi*: (1) stage without photophores (ML is less than 1.9 mm); (2) stage with 1 pair of anal photophores, 1 photophore on each eye (ML 1.9–2.5 mm); (3) stage of branchial photophores, 3 photophores on each eye (ML 2.9–4.0 mm); and (4) stage of ventral photophores (ML more than 4.5 mm), 8 intramantle photophores—one paired anal, one paired branchial and four unpaired ventral—and 9–10 photophores on eyes.

An excellent example of the morphological changes with growth in paralarvae can be found with the diamond squid *T. rhombus* (Wakabayashi et al., 2005). Paralarvae with ML <3 mm possessed a round mantle with many chromatophores, long tentacles and small fins. At 3–6 mm ML, paralarval arms grew rapidly, and a primordial protective membrane developed simultaneously. By 15 mm ML, the shape of the mantle has become similar to that of the subadults, and fins were present along the entire length of the lateral mantle. Although arm suckers were present, they were not prominent and were absent from the distal third of the arms in specimens smaller than 6 mm ML. The development of protective membranes and the relatively long arms in the postlarval stage may be adaptations for floating in ocean currents rather than for active swimming and suggest that *T. rhombus* might have a long planktonic phase (Wakabayashi et al., 2005).

Arm and tentacle morphologies change to accommodate new methods of food acquisition (Kier, 1996; Messenger, 1977; Shea 2005). In cephalopod families such as Onychoteuthidae, Enoploteuthidae, Pyroteuthidae, Ancistrocheiridae, Octopoteuthidae, Gonatidae and Cranchiidae, some suction rings develop during ontogeny to form chitinous hooks, which facilitate grab and tighten the prey (Engeser and Clarke, 1988; Falcon et al., 2000; Kubodera and Okutani, 1977; Vecchione et al., 2001; Young and Harman, 1988). In other occasions, the transition between paralarva and subadult stages is marked by a loss. In *Taningia danae*, the tentacles are reduced to minute appendages in young subadults, and they are absent in later stages, which influences hunting behaviour (Kubodera et al., 2007). Similar losses have been observed in *Octopoteuthis* and in *Chiroteuthis* spp., where the initial tentacular clubs are lost (Young, 1991).

Although only external morphological changes in the transition from paralarva to subadult stage have been discussed, there are also modifications in the functional morphology of some internal organs. However, these transformations mainly affect the size, growth and maturation of the gonads. No changes in the digestive tract were observed between paralarva and adult stages in many species, as cephalopods maintain carnivorous diet throughout their life (Boucaud-Camou and Roper, 1995). Nevertheless, the prey type proportion in the cephalopod diet changes ontogenically. While paralarvae fed mostly in larval stages of decapod crustaceans, subadult and adult shift their diet to larger crustacean life stages and species and to fishes and other cephalopods during growth (Rodhouse and Nigmatullin, 1996). *O. vulgaris* and *S. officinalis* clearly follow this trend (Nixon, 1985) with paralarvae feeding on large copepod and crustacean zoeae and changing gradually to bigger prey as arms grow and attack performance increases.

4.3. Ecological changes

It is widely recognised that cephalopods are highly sensitive to environmental conditions and changes at a range of spatial and temporal scales due to their short life cycles and rapid growth (Boyle and Rodhouse, 2005; Pierce et al., 2010). The specific environmental conditions that paralarvae and subadults are exposed to can affect the recruitment success and post-recruitment life history characteristics such as growth rate (Forsythe and Hanlon, 1988a), maturation (Boavida-Portugal et al., 2010; Forsythe and Hanlon, 1988b; Smith et al., 2005) and spawning (Lourenço et al., 2012; Otero et al., 2007).

Sea water temperature affects all phases of the life cycle, including paralarval growth and settlement of benthic species (Garofalo et al., 2010). Temperature directly affects growth in cephalopods (Forsythe, 1993; Forsythe, 2004; Forsythe and Hanlon, 1988a; Forsythe and Van Heukelem, 1987; Semmens et al., 2004) with increasing temperature (until reaching a threshold temperature value) positively affecting the growth rates in all cephalopod species. The initial exponential phase of growth is more strongly affected by temperature than the consequent phases of the growth curve (Briceño et al., 2010). This indicates that paralarvae experiencing higher temperatures will transition into the subadult phase sooner. In *O. vulgaris*, the duration of the planktonic period lasted 47–54 days at 21.2 °C, 30–35 days at 23 °C and 40 days at 22.5 °C after which the octopuses started setting to the bottom (Iglesias et al., 2004; Imamura, 1990; Villanueva, 1995). *Octopus bimaculoides* grew from an approximate hatching size of 0.07 g to a mean of 619 g in 404 days at 18 °C. Growth was exponential from hatching until day 156 (21.18 g; 38.64 mm ML), with a mean relative growth rate of 3.56% of body weight per day. Growth was slower after day 156, becoming logarithmic. The weight data for the warm-water population (23 °C) revealed the same pattern of growth over slightly different time periods. The faster exponential growth phase lasted until day 142 (49.70 g; 51.83 mm ML) with an average growth rate of 4.5% of body weight per day. The slower logarithmic phase lasted from day 142 to day 324 (Forsythe and Hanlon, 1988a).

Nutrition is a key factor in octopus paralarval rearing success and, therefore, transition into a subadult. Comparisons of the effect of food type on paralarva survival when raised between 19 and 25 °C demonstrated that survival was more dependent on nutrition than on temperature (Iglesias et al., 2007). Meeting metabolic requirements is also important for the development of early stages of cephalopods. For most cephalopod species studied, the energetic requirement to grow depends on the protein degradation metabolic pathway (Lee, 1995), as several studies show on wild and captivity species where the availability or choice of a diet rich in protein or with a high ratio of protein/lipid content benefits the paralarva growth and survival (e.g. Rosas et al., 2013). Additionally, essential fatty acids also play an important role in the transition from paralarva to subadult. For *O. vulgaris*, prey rich in DHA (docosahexaenoic acid $22{:}6n-3$), EPA (eicosapentaenoic acid $20{:}5n-3$) and copper are required for optimal growth of the paralarvae (Navarro and Villanueva, 2000, 2003; Villanueva and Bustamante, 2006; Villanueva et al., 2004).

Changes in ocean currents, mixing, deepwater production and coastal upwelling will directly affect the distribution and abundance of plankton and of many fish and cephalopod species, as well as their migrations (e.g. González et al., 2005; Roberts and van den Berg, 2002; Waluda and Rodhouse, 2006). This, in turn, affects the temperature and nutritional options available to paralarvae as they transition to the subadult stage. Although oceanographic conditions are of particular significance for mobile pelagic species such as the ommastrephid squid, the less widely ranging demersal and benthic species may depend more on other physical habitat characteristics (e.g. substrata and bathymetry). Coastal species may be affected by variations in water quality and salinity (related to rainfall and river flow) (Sobrino et al., 2002). In northwestern Atlantic Iberian upwelling system, the wind stress structure during the spring–summer (prior to the hatching peak) and autumn–winter (during the planktonic stage) was found to affect the early life phase of *O. vulgaris* and explains up to 85% of the total variance of the year-to-year variability of the adult catch (Otero et al., 2008). Despite this bottom-up modulation via environmental conditions, the results also provide evidence for a between-cohort density-dependent interaction, probably caused by cannibalism and competition for habitat (Otero et al., 2008). Moreover, Otero et al. (2009) suggested that the increase in *O. vulgaris* paralarval abundance and biomass was significantly correlated with the simultaneous decrease of water column integrated nitrate, ammonium and chlorophyll levels in the Ria de Vigo. These conditions occur during the early stage of the relaxation phase of coastal upwelling events, when nutrient salts are consumed to produce biogenic matter, which is retained in the system and transferred through the food web.

4.4. Behavioural changes

For holobenthic octopus, prey capture by paralarvae is not fundamentally different from that of a subadult. The prey captured is smaller but the prey type stays relatively constant (Hanlon and Messenger, 1998). However, other cephalopod species, like pelagic squids with very characteristic paralarvae (e.g. the rhynchoteuthion in the ommastrephids) and the benthic octopods that do have a planktonic stage, differ in form and in size from the adults, and consequently, their feeding behaviour is substantially different.

Ontogenic changes in predatory behaviour are closely related to morphological changes in arms and tentacles of hatchlings. Newly hatched *D. opalescens* exhibit only one predatory behaviour: basic attack (BA). During BA, arms and tentacles were spread apart before being thrust forwards to make first contact with the prey. Forty days after hatchling, BA was replaced by arm net (AN) and tentacular strike behaviours (Chen et al., 1996).

Changes in the diet during development have been followed in several species of squids: together, they provide evidence that prey size, prey behaviour and prey density are critical. Vovk and Khvichiya (1980) recognised four stages of feeding in *D. pealeii*. Yang et al. (1983) followed these changes in *D. opalescens* reared in the laboratory and found that during the first

70 days, squid selected copepods less than 4 mm in length, as well as *Artemia* and chaetognaths; from day 60 to day 130, they ate mysid and shrimp larvae up to 10 mm long; and from day 100, they ate various shrimps (up to 25 mm) and specially fish (up to 70 mm). Three-week-old *S. sepioidea* stopped attacking mysids, and even before this stage, the young showed signs of selectivity. Mysids and juvenile fish were preferred over copepods, amphipods, polychaetes and zoeae of various crustacean species (LaRoe, 1971).

For many social species, the onset of social behaviour can indicate a transition to the subadult stage. The appearance of schooling behaviour, which depends partly upon size and swimming strength, was observed in *L. vulgaris*, *D. opalescens*, *D. pealeii* and *I. argentinus* 4–11 mm ML and 20–60 days old (Hanlon et al., 1987; Turk et al., 1986; Vidal et al., 2010; Yang et al., 1986). The schooling behaviour is associated with the increase of the swimming ability related to the onset of specific morphological features as improved vision, development of communication skills (chromatophores patterns) and the development of functional clubs (Vidal et al., 2010). This behaviour allows squid subadults to migrate vertically and explore more food resources, enhancing growth and survival (O'Dor, 1998). As the ability to swim against the current is crucial (Hurley, 1978), the formation of schools depends on size rather than age as proved for *D. opalescens* and *I. argentinus* (Vidal et al., 2010).

Transition between paralarva and subadult stages obviously implies the development and modification of some elements (chromatophores, iridophores, internal organs, muscles, photophores, etc.), units or groups of elements contributing to components, components or constituents of a pattern and the appearance of whole animal, which constitute the hierarchical categories of any behavioural pattern (Hanlon and Messenger, 1998). This transition also implies deep changes in the central nervous systems and receptors of the animal (Nixon and Young, 2003). However, little attention has been devoted to these aspects until present. The development of body patterning has been studied, at least, in *S. officinalis* (a species lacking paralarvae) and three octopus species (*O. vulgaris*, *Octopus rubescens* and *O. briareus*). Hanlon and Messenger (1998), compiling results of several researches, showed that in many cephalopods, new elements, units and components of body patterns develop in skin as the animal grows and that some adult body patterns and display are developed or refined during ontogeny. These changes in both enoploteuthid squids represent obvious modifications in behaviour of the animals. The acquisition of a complete set of photophores means full development of behaviour in relation to intraspecific communication, between males and females, as well as the increased possibility of camouflage against predators.

4.5. Summary

Planktonic paralarvae common in oegopsid squids, merobenthic octopods with small hatchlings and the exceptional pygmy cuttlefish *Idiosepius pygmaeus* (Boletzky, 1974) are not found in benthic octopods or cuttlefish of longer embryonic development and bigger hatchlings. Nevertheless, between the first feeding after hatch and the subadult phase, several important changes occur in morphology, ecology and behaviour in pelagic, benthic and merobenthic cephalopods. In species where this transitional stage is followed by migration to different habitats, important morphological, suborganismal and physiological changes occur. Examples of these changes are the separation of the rhynchoteuthis "proboscis" in two tentacles in the ommastrephid, the allometric growth in octopus paralarvae, the changes in chromatophore pattern in *T. rhombus* and the development of photophore patterns in oegopsid paralarvae. In all cases, the transition is associated with behavioural changes in diet, predation, migration and social patterns. In this, as in other transitional stages, environmental conditions play an important role in survival success. By finding the best environmental or nutritional conditions (e.g. primary production, water temperature and water column mixing), a quick transition can be achieved to the subadult stage where chances of individual survival are exponentially greater.

5. TRANSITION #4: SUBADULT TO ADULT

5.1. Introduction

This chapter will examine the suite of changes that occur during the subadult phase and are directly associated with the transition to mature adult, including morphology, physiology, behaviour, diet and habitat. A subadult is an individual who, while appearing to be functionally and morphologically similar to an adult, is generally smaller than an adult and not yet reproductively mature, while an adult is an individual who is sexually mature and of adult size (Young and Harman, 1988). So effectively, we are examining the transition from being reproductively immature to mature; males will have spermatophores present and females will have mature ova in the ovary and are functionally mature (Arkhipkin, 1992). While there is a suite of changes when individuals transition from subadult to adult, these changes are not as dramatic and rapid as other transitions in the cephalopod life cycle, yet they are crucial to reproductive success of the population. During this protracted transition, it is difficult for biologists to identify when this transition has been completed as there are no external signs of sexual maturation, which is first evident internally with the start of ova and sperm production. Furthermore, the size and age at which individuals start maturation are highly variable, both within and between populations of species, for example, *S. australis* (Pecl et al., 2004), *S. lessoniana* (Jackson and Moltschaniwskyj, 2002), *D. pealeii* (Macy and Brodziak, 2001) and *I. argentinus* (Arkhipkin and Laptikhovsky, 1994). Age and size at completion of the transition can differ between the sexes, with males usually maturing sooner than females, for example, *O. pallidus* (Leporati et al., 2008), *I. illecebrosus* (Rodhouse and Hatfield, 1990) and *D. gigas* (Liu et al., 2013); as a result, males may have a substantially longer reproductive life.

Given the absence of external characters indicating sexual maturation for most cephalopod species, especially in females, fisheries and population biologists identify the size and age at which subadults of a species become reproductively mature adults using logistic regression models. These models estimate the age or size that a specified percentage of the population has attained sexual maturity, usually 50%, and has finished the transition from subadult to adult. Although not usually reported, it is possible to use this approach to identify the age and size of the start of the transition, for example, size or age when 5% of the population are mature. If the presence of spermatophores or ovulated eggs defines the end point of the transition, then the best external indicators of the end of this transition for males are mating behaviour and the transfer of spermatophores to females. However, mating is not a good indicator of sexual maturation in females, as she is receptive to mating even when not fully reproductively mature and will store spermatophores until ready to fertilise eggs. For female cephalopods, the start of egg deposition may be the best external identifier of the transition end point. Not all species will have a subadult phase; if the acquisition of the adult morphology and diagnostic characteristics and the attainment of sexual maturity are almost simultaneous, the transition from a juvenile to adult effectively skips the subadult phase, for example. *Leachia pacifica* (Young, 1975).

Mortality estimates during this transition are not available in the literature, but it is assumed that most cephalopod species have a type III mortality pattern, with the highest mortality rates expected during the larval phase and following spawning. Therefore, the transition from subadult to adult most probably has a relatively low mortality risks, although for species that undertake migrations (Section 5.5) during this transition, there are mortality risks associated with cannibalism and shifts in diet (Section 5.3). However, this transition is critical from an evolutionary biology perspective as the reproductive success of an individual determines its contribution to the next generation and the short life span of cephalopod means that the reproductive window is months not years. Delays in the start of the transition or extending the transition over a longer period can carry a risk of a shorter reproductive window once the adult phase begins. However, allocation of energy to reproduction and away from production of muscle tissue (Moltschaniwskyj and Carter, 2013) can influence capacity to escape predators and capture prey, affecting survival. Timing of the transition among individuals within a species is typically variable for cephalopods and is not necessarily directly a function of size or age (Pecl, 2001). Factors such as environmental and oceanographic conditions, food availability and behavioural interactions have the capacity to influence the timing of transition (Pecl et al., 2004, Pierce et al., 2005). Variability in size at maturity for squid is a function of the time of spawning and hatching occurring throughout the year, with individuals experiencing different growth rates dependent on environment conditions (Pecl et al., 2004). For *S. officinalis*, the length of time a subadult individual spends in optimal conditions (inshore summer) determines the speed and timing of maturation (Boletzky, 1983). For a few species, death occurs very shortly after completing the transition, for example, after ovulation in female *Onykia ingens* (previously *Moroteuthis ingens*; Jackson and Mladenov, 1994) and *Gonatus fabricii* (Arkhipkin and Bjorke, 1999), or shortly after sperm transfer, for example, in male epipelagic octopuses *Argonauta*, *Tremoctopus* and *Ocythoe* because they leave the hectocotylus in the female during sperm transfer (Mangold, 1987).

5.2. Morphological, suborganismal and physiological changes during the transition

Whole animal growth during the subadult stage is accompanied by allometric growth of external morphological structures, for example, *Photololigo* sp. (Moltschaniwskyj, 1995a) and *D. opalescens* (Zeidberg, 2004). Individuals reach their final body shape towards the end of the subadult phase with no evidence that further shape change occurs during the transition from subadult to adult. However, some morphological changes do occur during this transition. The males of most coleoid cephalopods develop a hectocotylus, a specialisation of one or more of the arm tip(s) used to transfer spermatozoa to females. The number of modified arms and the nature of the specialisation are species-specific but can develop early in the reproductive maturation process, for example, *I. coindetii* (Zecchini et al., 2012). However, not all species use a hectocotylus for spermatophore transfer; for example, *Nautilus* have a large erectile mass known as a spadix (Arnold, 1987; Saunders and Spinosa, 1978), or there is no specialisation for spermatophore transfer, for example, cirrate octopods (Mangold, 1987). Females of some cephalopod species will develop secondary reproductive structures during the maturation phase; for example, female *Argonauta* develop specialised dorsal arms that secrete a calcareous brood chamber for her eggs. Epi-, meso- and bathypelagic octopuses (families Tremoctopodidae, Vitrellodellidae and Bolitaenidae) brood eggs in their arms, as do some deep-sea squid from the family Gonatide (Seibel et al., 2000). However, we do not know if or when the development of the specialisations of the arms and suckers necessary to hold the eggs occurs. Additional secondary sexual characteristics that can develop during this transition are extra-large suckers on the second and third arms in mature males, for example, *O. cyanea* (Van Heukelem, 1973) and *Euprymna tasmanica* (Norman and Lu, 1997); mature male *Octopus* use them for displaying to conspecifics (Packard, 1961) and/or they may be used in chemoreception (Voight, 1991). Although *L. pacifica* has an abbreviated adult lifetime, immediately before sexual maturation, the female develops brachial photophores that may be used to attract mates (Young, 1975).

Hormones control the reproductive maturation process in most animals, and in *Octopus* spp., the optic glands secrete a small quantity of gonadotropin before the onset of sexual maturity (Wells and Wells 1959). The optic gland sits on the optic tract, which connects the optic lobes with the central regions of the brain, with its size, colour and structure changing with sexual development in a number of species, for example, *E. cirrhosa* (Boyle and Thorpe, 1984), *S. officinalis* (Koueta et al., 1995) and *S. sepioidea* (Arrieche, 1999). However, we know little about the endogenous and exogenous factors influencing optic gland secretory activity. It is possible that female *O. vulgaris* respond to cues from males, since females mature faster in the presence of males than when kept in isolation (Estefanell et al., 2010).

Attainment of sexual maturity involves growth and maturation of all the reproductive organs and is a consistent characteristic of the transition from subadult to adult. However, there is considerable variability in the nature and extent of changes to other organs, such as the digestive gland and mantle muscle, during the development and growth of reproductive tissue. If maturation, that is, transfer of sperm and egg deposition, occurs before final adult size is attained, then energy is allocated to somatic and reproductive growth simultaneously, for example, *Photololigo* sp. (Moltschaniwskyj, 1995b); *S. lessoniana* and *S. australis* (Ho et al., 2004; Pecl, 2001); *I. argentinus* (Hatfield et al., 1992); *Octopus chierchiae* (Rodaniche, 1984), *O. vulgaris* (Otero et al., 2007; Rosa et al., 2004a) and *Octopus defilippi* (Rosa et al., 2004a); *E. cirrhosa* and *Eledone moschata* (Rosa et al., 2004b); and *S. dollfusi* (Gabr et al., 1999a) and *S. pharaonis* (Gabr et al., 1999b). These species show none of the dramatic changes in size or proximal composition of muscle tissue and digestive gland that are typical when stored energy reserves are mobilised; instead, individuals fuel reproductive growth directly from ingested food. However, somatic growth is affected, with specific growth rates of many cephalopod species slowing as individuals get older and larger, for example, *Photololigo* sp. (Moltschaniwskyj, 1995b), *S. lessoniana* and *S. australis* (Ho et al., 2004; Pecl, 2001), *O. chierchiae* (Rodaniche, 1984), *I. argentinus* (Hatfield et al., 1992), *D. gigas* (Mejía-Rebollo et al., 2008) and *N. gouldi* (Jackson et al., 2003). Slowing of somatic growth is due to slowing in the rates of both protein synthesis and degradation in the mantle muscle during reproductive maturation, allowing more energy to be allocated to reproductive growth (Moltschaniwskyj and Carter, 2013).

In contrast, some cephalopod species have dramatic changes in their digestive gland and/or muscle tissue size and composition during the transition from subadult to adult. Muscle tissue mass and integrity are lost as protein in the tissue is used as an energy source to fuel maturation, for example, *M. ingens* (Jackson and Mladenov, 1994; Jackson et al., 2004) and *G. fabricii* (Arkhipkin and Bjorke, 1999). Individuals mobilise energy from the mantle during reproductive growth either because they cease feeding as they reach sexual maturation or because they cannot catch sufficient food to support metabolism, movement and maturation.

5.3. Changes in feeding and diet

For many cephalopods, the transition from subadult to adult occurs while growing, and they may undertake a change in habitat at the same time (see section 5.5), both of these allow or necessitate a change in prey items. Cephalopods that are opportunistic predators change their diet based on the availability of prey items, for example, *I. argentinus* (Rodhouse and Nigmatullin, 1996), *I. coindetii* (Castro and Hernández-García, 1995) and *D. gigas* (Markaida and Sosa-Nishizaki, 2003). Increasing body size and associated changes in shape (see section 5.2) allow larger prey items to be caught and consumed, although smaller prey are often retained in the diet, for example, *L. forbesii* (Collins and Pierce, 1996), which effectively increases the diversity of prey items consumed. A change in diet or increased prey diversity can be accompanied by a morphometric change in beak shape (Castro and Hernández-García, 1995). Differences in fatty acid profiles between mature and immature female *N. gouldi* suggest dietary differences during this transition, but it is likely that larger mature females are able to catch different prey as a function of their size, and there is no evidence that a change of diet directly contributes to the process of reproductive maturity (Pethybridge et al., 2012). Opportunistic use of prey species is critical to supporting fuelling of reproductive growth directly from ingested food (see section above) and maximises opportunities to reproduce regardless of nutritional history.

The risk of using income energy to fuel reproductive maturation in species that are relatively short-lived and have a short reproductive period is that insufficient energy is available to support all the biological demands at this time (Bonnet et al., 1998; Jonsson, 1997). One of the dietary options available, particularly to larger individuals in a population, is cannibalism, which increases during this transition (Ibáñez and Keyl, 2010). While there is a reproductive behavioural component to this (see Section 5.4), there is also evidence that this is a source of energy for individuals (Ibáñez and Keyl, 2010). Increased energy demands associated with reproduction and movement to spawning sites mean that cannibalism is important for many species, for example, *I. illecebrosus* (O'Dor, 1983) and *Sepioteuthis* spp. (Pecl, 2001).

5.4. Behavioural changes

During the transition from subadult to mature adult, it is likely that individuals will need to learn and adopt new behaviours associated with courting and/or mating and also any new behaviours needed to use new habitats (see Section 5.5). Our knowledge of the acquisition or learning of new behaviours as individuals undergo the transition from subadult to adult is extremely limited; anecdotal evidence suggests that males become more active and females less as they approach maturity (Mather, 2006). While a body of research has described the behaviours during courting, mating and egg deposition, these descriptions of behaviours are in adults that have already undertaken the transition from subadult to adult. The changes in behaviour, how they are learnt and adopted and how they are mediated hormonally warrant investigation (Hanlon and Messenger, 1998; Mather, 2006), as acquiring these behaviours is important for both survival and successful reproduction.

In some cephalopod species, the subadults are asocial and have limited interactions with conspecifics; this is particularly the case for *Octopus* species, many of the *Sepia* species and the sepiolids. Some of these species will continue a solitary existence as they transition from subadult to adult and will mate opportunistically when encountering the opposite sex. For species that show no evidence of courtship behaviours prior to copulation, for example, *I. paradoxus* (Kasugai, 2000), *Sepiola atlantica* (Rodrigues et al., 2009) and *S. officinalis* (Hanlon et al., 1999), development of courtship behaviours will be absent. However, some species have extended and complex courtship behaviours that need to be developed and potentially learned by individuals during this transition, for example, *S. apama* (Norman et al., 1999), *S. australis* (Jantzen and Havenhand, 2003) and *L. reynaudii* (Hanlon et al., 1994). Males of species that school very early in their lifetime develop intraspecific aggressive behaviours that are expressed before, during and after mating, for example, *L. forbesii* (Hanlon et al., 1989b). For some species, for example, *S. australis* and *D. gigas*, subadults typically exist in small and socially loose aggregations; however, during the transition from subadult to adult, individuals experience great densities when individuals converge at a given location for spawning (Nigmatullin et al., 2001; Pecl et al., 2006; Schoen et al., 2002; Steer et al., 2005). The effect of aggregation on the timing of the transition from subadult to adult is unknown but is likely to promote a behavioural response that facilitates courtship and copulation. During the transition from subadult to adult, the reproductive behaviours of the adults develop and become apparent. Sexually mature female cuttlefish can distinguish between conspecific males and conspecific females (Palmer et al., 2006), while obviously an important element to reproductive behaviours, when this capacity develops, is unknown. Subadult *S. sepioidea* in the presence of reproductively mature adults display colour patterns out of normal sequence or to the wrong sex (Mather, 2006), suggesting that learning of some reproductive behaviours may occur as subadults join spawning aggregations.

The few studies that have examined ontogenic changes in behaviour in cephalopods have focused on the development of behaviours and learning. Cuttlefish (*S. officinalis)* significantly improve their learning ability during their transition from subadult to adult, and this improvement is correlated with the superior frontal and vertical lobe development (Agin et al., 2006; Dickel et al., 2001). Increased quality and complexity of the surrounding environment at this time have a positive effect on the maturation of memory, highlighting the importance of habitat in the development of certain behavioural abilities (Dickel et al., 2000). Additionally, changes in individual personality occur at approximately the age that individuals start to undergo the maturation process (Sinn and Moltschaniwskyj, 2005), but there is no evidence of any further change at the time that mating and egg deposition occur. However, as this study described changes in personality in individuals held in isolation for their entire life, there is substantial scope to determine changes in behavioural interactions between males and females as individuals move from subadult to adult.

The use of visual displays and their importance changes as cephalopods transition from the subadult stage to the adult stage. Cuttlefish hatchlings are born with the ability to produce almost every adult body pattern; however, the visual displays shift from being used primarily for concealment and predator evasion to courtship and reproductive behaviours (Hanlon and Messenger, 1998). Body patterns beneficial for camouflage, such as light mottle or a disruptive pattern, are displayed more often in juvenile *S. officinalis*, and patterns associated with reproductive behaviours, like intense zebra, are developed around the time of sexual maturity (Hanlon and Messenger, 1998). The squid *S. sepioidea* present the first indications of sexual behaviour via visual displays (Mather, 2006). However, immature *S. sepioidea* males can mimic the displays of mature ones, making it difficult to use body patterns as indicators of the transition from subadult to adult. Regardless, the use and prevalence of reproductively associated visual displays are a key change during this transition, and the quality of these displays can affect the animal's fitness (Boal, 1997).

Cannibalism within a cohort, that is, among individuals of similar size and age, increases close to and during the reproductive season and changes with maturity status of individuals (Ibáñez and Keyl, 2010). This type of cannibalism appears to have both a strong element of behaviour, aggression and competition and also energy requirements (Ibáñez and Keyl, 2010). Where cannibalism is related to behaviours of aggression and competition, this occurs once maturation has occurred and often during courting and mating (see review by Ibáñez and Keyl, 2010).

5.5. Migration and habitat change during the transition

For many cephalopods, migration, large (thousands of kilometres) or small (tens–hundreds of kilometres), is a major event in the transition from subadult to adult and is commonly a prelude to mating and egg deposition.

Migrations can involve important changes in habitat, for example, from the continental slope to the shelf or vice versa, or migrations can occur along coasts or throughout oceans with little in the way of habitat change. Regardless of the magnitude of the migration, individuals will experience some changes in their physical and biological environment, including temperature, salinity, water chemistry, prey, predators and current systems. Although the nature and magnitude of migration during this transition vary among cephalopod species, migration appears to be associated with increasing the chances of finding a mate, finding and using habitats suitable for egg attachment or release. Formation of spawning aggregations, for example, *L. reynaudii* (Sauer et al., 1992), *S. apama* (Lu, 1998) and *Graneledone* sp. (Drazen et al., 2003), increases spawning potential, influences the number of offspring produced during spawning events and promotes genetic diversity within the population or stock (Naud et al., 2004).

Our knowledge of the distribution and abundance patterns of subadults is largely a function of methods of sampling and the nature of the data used to derive the pattern. If we derive our knowledge from fishery data, then the distribution of fishing effort and fishing technique will determine our capacity to track changes in the distribution patterns of subadults and adults. For *I. argentinus*, jigs catch larger squid than trawlers in similar regions (Koronkiewicz, 1995), so it is possible that the fishery catches a relatively greater proportion of larger subadults. Analyses of migration characteristics during this transition will require a sampling method, such as using fine mesh cod ends, that ensures all individuals in the population have equal chance of capture.

In some cases, the locations of the end point of migration are temporally and spatially predictable and stable, for example, southern calamari *S. australis* in Tasmania (Pecl et al., 2011) and *S. apama* in South Australia (Drazen et al., 2003). However, this is not always the case, and often, the location of the spawning grounds is strongly associated with characteristics of the physical environment (see review by Pierce et al., 2008). This association of distribution with environmental characteristics, such as temperature, may result in different migratory behaviours within a species; for example, Australian populations of *N. gouldi* migrate to the shelf waters of southern Australia to spawn (Green, 2011), but the New Zealand population does not undertake migrations to the shelf waters of New Zealand (Uozumi, 1998).

Whether cephalopods migrate between habitat types or within a habitat type will determine the magnitude of change in environmental and oceanographic factors they experience, which in turn can affect the timing of transition. Shallow coastal habitats have greater spatial and temporal variability in biotic and abiotic factors, such as in temperature, salinity, nutrient load, current speed, turbidity, bottom types and food availability, compared with offshore waters (Boyle and Rodhouse, 2005). Species whose transition from subadult to adult occurs in a similar oceanographic realm experience relatively small changes in environmental and oceanographic conditions, for example, incirrate octopus and some benthic squid *Sepiadarium austrinum* (Norman and Reid, 2000). In contrast, species are likely to experience greater changes in habitat and environmental and oceanographic variables if they migrate between inshore and offshore habitats, for example, *L. gahi* (Arkhipkin et al., 2004); move in and out current systems, for example, *I. illecebrosus* (Falkland Current; Dawe et al., 2000) and *T. pacificus* (Kuroshio and Oyashio Currents; Sugimoto and Tameishi, 1992); or have significant inshore movements, for example, *S. australis* (Pecl et al., 2006). Changes in environmental conditions experienced by individuals undertaking migration are when energy and resources are being allocated to reproductive development and, as such, will have substantial influence on the size, age, growth and condition at maturity.

5.5.1 Vertical migrations during transition: Change from well-lit, warm to cold, dark environment

We are aware that some open-water cephalopod species undergo vertical migrations during the transition from subadult to adult, for example, *L. pacifica* (Young, 1975), *Eledonella pygmaea* (Voight, 1995) and *Gonatus onyx* (Hunt and Seibel, 2000). Collections of juveniles and mature adults with eggs indicate that individuals move to find a mate, with individuals moving from well-lit warm near-surface waters to dark and cold waters at depths of >1000 m. However, there are physiological impacts associated with this thermal change; in particular, metabolic rates decrease significantly for species that undertake vertical migrations associated with reproductive activities (Seibel et al., 1997). It is not clear what the biological implications are for these species, as slowing metabolic rates will affect locomotion and in turn activities such as finding mates and capturing prey (Seibel et al., 1997).

5.5.2 Migrations between the offshore neritic and the inshore neritic

Loliginids, sepiolids and octopods have a strong association with the sea bottom, where there is a diversity of habitats in this photic zone, with nutrients provided from terrestrial sources, and rich diversity of food resources, but with large variations in environmental conditions (Boyle and Rodhouse, 2005). Many of these species do not migrate over long distances, preferring to deposit their benthic eggs in shallow waters (Boyle and Rodhouse, 2005). So, a common migration characteristic of neritic cephalopod species is for subadults to move from deeper offshore neritic waters to inshore shallower neritic waters, for example, *L. gahi* (Arkhipkin et al., 2004; Hatfield et al., 1990), *D. gigas* (Ibáñez and Cubillos, 2007; Nigmatullin et al., 2001), *D. pealeii* (Hatfield and Cadrin, 2002) and *S. officinalis* (Boucaud-Camou and Boismery, 1991; Le Goff, 1991). This may result in changes in the habitat that the individuals experience, as they move from bare low-profile sediment substrate to a more physically complex habitat, which supports marine plants and greater diversity of marine organisms, for example, *S. australis* who attaches eggs primarily to sea grass (Moltschaniwskyj and Pecl, 2003; Steer et al., 2007). In contrast, some species move inshore but remain on bare sediment and use this substrate to deposit their eggs, for example, *L. reynaudii* (Sauer et al., 1992). These shallower inshore habitats typically are warmer (Bettencourt and Guerra, 1999), which can potentially increase metabolic rates (Boucher-Rodoni and Mangold, 1995), but greater food intake needed to support this is readily available in these productive areas.

5.5.3 Migrations within habitats

There are groups of cephalopod species that display no dramatic shift in either habit or habitat during their transition from subadult to adult, but they may move within the habitat. The subadults occupy the same habitat as the adults, and it is here that they will become reproductively mature and start mating and deposit eggs, for example, *Photololigo* sp. (Moltschaniwskyj, 1995b). Some species such as *L. forbesii* have relatively short migrations where they make seasonal migrations along the coast of the United Kingdom (Waluda and Pierce, 1998), whereas *Uroteuthis edulis* and *H. bleekeri* have shown to prefer inshore continental shelf waters in the Sea of Japan (Natsukari and Tashiro, 1991, Semmens et al., 2007). For *O. vulgaris*, migration is on a very small scale from intertidal to the subtidal areas to spawn (Oosthuizen et al., 2002b). Depending on the population structure and dynamics, this transition may occur in the presence of adults, that is, overlapping generations when there are multiple spawning events or in the absence of adults when there is a single spawning event for an annual species. Fisheries stock assessment typically makes no distinction between subadult and adult, and size limits are not used as a management tool; as a result, these individuals are treated as a single group irrespective to their status as a subadult or adult or if they are going through this transition.

5.6. Summary

The start and end of this transition is very difficult to identify because there are few external signs of sexual maturation, which is first evident internally with the onset of gonadal maturation. While the transition from subadult to adult occurs at a phase when mortality risks are relatively small, this transition is critical from an evolutionary biology perspective as the reproductive success of an individual determines its contribution to the next generation. Although allometric growth leads to changes in body proportions between subadults and adults, the main morphological change is the differentiation of mating structures in males. Most male coleoid cephalopods develop a hectocotylus, a specialisation of one or more of the arm tip(s) used to transfer spermatozoa to females. The main behavioural changes around the migration to spawning grounds, where known, occurs when there is differences in habitats; and this exposes individuals to challenges associated with new environmental factors and differences in prey. Those species that are generally solitary will experience behavioural changes associated with increased interaction with other adults during spawning, although these changes may be gradual.

6. DISCUSSION

This description of four life stage transitions from morphological, ecological and behavioural perspectives gives us an opportunity to see how these transitions might be selected for, occur and vary within cephalopods. Transitions might be defined by major morphological change, habitat shift, shifts in energy allocation from growth to reproduction and internal changes such as hormone regulation (Bishop et al., 2006). In the four transitions undertaken by cephalopod species, morphological changes are present, but not extreme, for example, hatching gland development in transition 2, change in body proportions in transition 3 and growth of gonads in transition 4. Habitat shift is a common, though not universal, feature of these changes. Many cephalopod species seek specific sites for egg deposition during transition 1, they move from a benthic to pelagic habitat in transition 2, they transition from pelagic to benthic habitat in transition 3 and they may gather in specific locations during transition 4. There must be major epigenetic regulations during these changes; we know little about them, although we know that hormonal influences on the optic gland trigger the transition from subadult to adult (transition 4).

If different stages of cephalopod life cycles occupy different niches, transitions may need to be rapid (Hadfield, 2000), so that each stage occupies the niche to which it is closely adapted for as long as possible and reduces the time taken to transition between the two niches (Moran, 1994). Our knowledge about the transitions from egg to the exhaustion of the yolk reserves (transition 2) and from paralarva to subadult (transition 3) identifies the major risk of mortality during these transitions. For both transitions, there are obvious changes in niche, and both are rapid relative to the time spent in each phase. In contrast, paralarvae, resembling the subadult more closely than many larvae of other marine animals, may delay their transition to subadult for some time until a suitable habitat has been found. Likewise, the transition from subadult to adult may be much longer as the search for suitable mates and appropriate habitat and the major morphological changes of the reproductive system take longer. Understanding the scope of transitions and the time over which they occur will assist in the management of many cephalopod species, as well as their cultivation in captivity. The challenges of quantitatively sampling cephalopods in every life stage are preventing us from defining "key" life stages (Yamamura, 1999). Nevertheless, this synthesis highlights the most critical transitions and more importantly identifies the factors known to be critical during the life history of cephalopod species.

Heyland and Moroz (2006) suggested a series of changes that must underlie any such transition. The first change necessarily prepares the animal for the new competence required of it, and Figure 4.2 demonstrates that the changes during hatching show this well. First, the membrane must weaken, the embryo begins to move, and Hoyle's gland develops—all before hatching. For the transition from paralarva to subadult, chromatophores increase in number and the relative size of arms to body changes before the paralarvae settle. Hadfield's (2000) next transitional step (for larval fishes) is the differentiation into structures appropriate for the next stage. Only after this should we see loss of the structures appropriate for the earlier stage and habitat change. Such a sequence is generally known for our transitions 2 and 3, much less so for 4 and 1, where many of the changes are behavioural. Major morphological changes, such as growth of gonads, are well described from dead animals, but behavioural changes and learning by individuals during the transitions require observations of live animals, preferably in their natural habitats, and this poses us with substantial challenges in most marine habitats although there is a lot of ongoing work in this area (Hoving and Vecchione, 2012; Hunt and Lindsay, 2012).

This chapter has identified some critical questions about the transition from one life stage to the next. Early stages of development should be evolutionarily conservative, which suggests that transitions 1 and 2 should be more stereotyped than 3 and 4. Departures from the set four stages and transitions might occur in cephalopods. In stable environments, direct development, larger and fewer eggs, more yolk and longer embryonic development should occur—a pattern typical of deep benthic octopods. Such direct development should lead to fewer chromosomes, smaller geographic ranges and greater taxonomic diversity, which is a testable hypothesis. Another possible departure is neoteny, the adult retention of juvenile characteristics, which has been identified in some physiological characteristics of cephalopods (Rodhouse, 1998). Another departure from standard developmental patterns is found when larvae have accumulated enough stored energy that adults need neither feed nor grow, as is seen in many insects. Is this true for the adult cephalopod after transition 4 to the reproductive adult? Such a trend has been documented but not fully explored. Under what circumstances do cephalopods abandon semelparity?

Behind these descriptions is a deeper question, around the programming and control of saltatory development. The transitional stages are morphological, physiological and behavioural, a product of genes and environment. Development can be modelled as a sequential flow between stabilised states (stages) and relatively fast changes (transitions) (Balon, 2001); however, the induction of change must be the result of the activation of an array of genes at a particular time in the life cycle (Jackson et al., 2003). Many of the transitions may be brought on by specific environmental cues, such as temperature and light, and activated by external chemicals such as pheromones and internal ones such as hormones. Understanding the control of the transitions will take a mechanistic approach involving physiologists, behavioural ecologists, evolutionary biologists and molecular biologists with particular focus on developmental biology.

ACKNOWLEDGEMENTS

This chapter was prepared after a workshop held during the CIAC 2012 Symposium (Florianopolis, Brazil). The authors wish to thank Erica A.G. Vidal for the organisation of this meeting and long after for her advice and support when working remotely on the manuscript. The authors also wish to acknowledge Jorge H. Urcera and Manuel E. Garci for providing the original photographs used in Figure 4.3 of this review.

REFERENCES

Agin, V., Poirier, R., Chichery, R., Dickel, L., Chichery, M.P., 2006. Developmental study of multiple memory stages in the cuttlefish, *Sepia officinalis*. Neurobiol. Learn. Mem. 86, 264–269.

Aitken, J.P., O'Dor, R.K., 2004. Respirometry and swimming dynamics of the giant Austrailian cuttlefish, *Sepia apama* (Mollusca, Cephalopoda). Mar. Freshw. Behav. Physiol. 4, 217–234.

Aitken, J.P., O'Dor, R.K., Jackson, G.D., 2005. The secret life of the giant Australian cuttlefish *Sepia apama* (Cephalopoda): behavior and energetics in nature revealed through radio acoustic positioning and telemetry (RAPT). J. Exp. Mar. Biol. Ecol. 320, 77–91.

Anderson, T., 1997. Habitat selection and shelter use by *Octopus tetricus*. Mar. Ecol. Prog. Ser. 150, 137–148.

Arkhipkin, A.I., 1992. Reproductive system structure, development and function in cephalopods with a new general scale for maturity stages. J. Northw. Atl. Fish. Sci. 12, 63–74.

Arkhipkin, A.I., Bjorke, H., 1999. Ontogenetic changes in morphometric and reproductive indices of the squid *Gonatus fabricii* (Oegopsida, Gonatidae) in the Norwegian Sea. Polar Biol. 22, 357–365.

Arkhipkin, A., Laptikhovsky, V., 1994. Seasonal and interannual variability in growth and maturation of winter-spawning *Illex argentinus* (Cephalopoda, Ommastrephidae) in the Southwest Atlantic. Aquat. Living Resour. 7, 221–232.

Arkhipkin, A.I., Laptikhovsky, V.V., Middleton, D.A.J., 2000. Adaptations for the cold water spawning in squid of the family Loliginidae: *Loligo gahi* around the Falkland Islands. J. Molluscan Stud. 66, 551–564.

Arkhipkin, A.I., Grzebielec, R., Sirota, A.M., Remeslo, A.V., Polishchuk, I.A., Middleton, D.A.J., 2004. The influence of seasonal environmental changes on ontogenetic migrations of the squid *Loligo gahi* on the Falkland shelf. Fish. Oceanogr. 13, 1–9.

Arnold, J.M., 1987. Reproduction and embryology of *Nautilus*. In: Saunders, W., Landman, N. (Eds.), Nautilus: The Biology and Paleobiology of a Living Fossil. Plenum Press, New York, pp. 353–372.

Arrieche, D., 1999. Ultrastructure of the optic gland of the squid *Sepiotheutis sepioidea* (Cephalopoda: Loliginidae). Rev. Biol. Trop. 47, 831–842.

Balon, E.K., 2001. Saltatory ontogeny and the life-history model: neglected processes and patterns of evolution. J. Bioecon. 3, 1–26.

Barón, P.J., 2001. First description and survey of the egg masses of *Loligo gahi* (D'Orbigny, 1835) and *Loligo sanpaulensis* (Brakoniecki, 1984) from coastal waters of Patagonia. J. Shellfish Res. 20, 289–295.

Bartol, I.K., Krueger, P.S., Thompson, J.T., Stewart, W.J., 2008. Swimming dynamics and propulsive efficiency of squids throughout ontogeny. Integr. Comp. Biol. 48, 720–733.

Bartol, I.K., Krueger, P.S., Thompson, J.T., Stewart, W.J., 2009. Pulsed jet dynamics of squid hatchlings at intermediate Reynolds numbers. J. Exp. Biol. 212, 1506–1518.

Bateman, A.J., 1948. Intra-sexual selection in *Drosophila*. Heredity 2, 349.

Bernay, B., Gagnon, J., Henry, J., 2004. Egg capsule secretion in invertebrates: a new ovarian regulatory peptide identified by mass spectrometry comparative screening in *Sepia officinalis*. Biochem. Biophys. Res. Commun. 314, 215–222.

Bernay, B., Baudy-Floc'h, M., Zanuttini, B., Gagnon, J., Henry, J., 2005. Identification of SepCRP analogues in the cuttlefish *Sepia officinalis*: a novel family of ovarian regulatory peptides. Biochem. Biophys. Res. Commun. 338, 1037–1047.

Bernay, B., Baudy-Floc'h, M., Gagnon, J., Henry, J., 2006. Ovarian jelly-peptides (OJPs), a new family of regulatory peptides identified in the cephalopod *Sepia officinalis*. Peptides 27, 1259–1268.

Bettencourt, V., Guerra, A., 1999. Carbon-and oxygen-isotope composition of the cuttlebone of *Sepia officinalis*: a tool for predicting ecological information? Mar. Biol. 133, 651–657.

Bigelow, K., 1992. Age and growth in paralarvae of the mesopelagic squid *Abralia trigonura* based on daily growth increments in statoliths. Mar. Ecol. Prog. Ser. 82, 31–40.

Biggs, J., Epel, D., 1991. Egg capsule sheath of *Loligo opalescens* Berry: structure and association with bacteria. J. Exp. Zool. 259, 263–267.

Birkhead, T.R., Parker, G.A., 1997. Sperm competition and mating systems. In: Krebs, J.R., Davies, N.B. (Eds.), Behavioural Ecology: An Evolutionary Approach. Blackwell, Oxford.

Bishop, C.D., Erezyilmaz, D.F., Flatt, T., Georgiou, C.D., Hadfield, M.G., Heyland, A., Hodin, J., Jacobs, M.W., Maslakova, S.A., Pires, A., Reitzel, A.M., Santagata, S., Tanaka, K., Youson, J.H., 2006. What is metamorphosis? Integr. Comp. Biol. 46, 655–661.

Blanc, A., 1998. Recherches Bio-Ecologique Et Ecophysiologique De La Phase Juvenile De La Seiche *Sepia officinalis* Linne (Mollusque, Cephalopod, Sepiidae) Dans Le Golfe Du Morbihan (Sud Bretagne). PhD Thesis, University of Rennes, France.

Bloor, I.S.M., 2012. The Ecology, Distribution and Spawning Behaviour of the Commercially Important Common Cuttlefish (*Sepia officinalis*) in the Inshore Waters of the English Channel. Marine Institute and Marine Biological Association of the United Kingdom; Plymouth University, Plymouth.

Bloor, I.S.M., AttrilL, M.J., Jackson, E.L. (Eds.), 2013a. A Review of the Factors Influencing Spawning, Early Life Stage Survival and Recruitment Variability in the Common Cuttlefish (*Sepia officinalis*). Academic Press, Burlington.

Bloor, I.S.M., Wearmouth, V.J., Cotterell, S.P., Mchugh, M.J., Humphries, N.E., Jackson, E.L., Attrill, M.J., Sims, D.W., 2013b. Movements and behaviour of European common cuttlefish *Sepia officinalis* in English Channel inshore waters: first results from acoustic telemetry. J. Exp. Mar. Biol. Ecol. 448, 19–27.

Boal, J.G., 1996. Absence of social recognition in laboratory-reared cuttlefish, *Sepia officinalis* L. (Mollusca: Cephalopoda). Anim. Behav. 52, 529–537.

Boal, J.G., 1997. Female choice of males in cuttlefish (Mollusca: Cephalopoda). Behaviour 134, 975–988.

Boal, J.G., Prosser, K.N., Holm, J.B., Simmons, T.L., Haas, R.E., Nagle, G.T., 2010. Sexually mature cuttlefish are attracted to the eggs of conspecifics. J. Chem. Ecol. 36, 834–836.

Boavida-Portugal, J., Moreno, A., Gordo, L., Pereira, J., 2010. Environmentally adjusted reproductive strategies in females of the commercially exploited common squid *Loligo vulgaris*. Fish. Res. 106, 193–198.

Boletzky, S.V., 1971. Mandibules denticulées chez les larves des Teuthoidés et des Octopodes (Mollusca: Cephalopoda). C. R. Acad. Sci. Paris 23, 2904–2906.

Boletzky, S.V., 1974. The "larvae" of Cephalopoda: a review. Thalass. Jugosl. 10, 45–76.

Boletzky, S.V., 1975. The reproductive cycle of Sepiolidae (Mollusca, Cephalopoda). Pubbl. Staz. Zool. Napoli 39, 84–95.

Boletzky, S.V., 1977. Post-hatching behaviour and mode of life in cephalopods. Symp. Zool. Soc. Lond. 38, 557–567.

Boletzky, S.V., 1982. Structure of the embryonic integument and its relation to the mode of hatching in cephalopods. Bull. Soc. Zool. France 107, 475–482.

Boletzky, S.V., 1983. Sepia officinalis. In: Boyle, P.R. (Ed.), Cephalopod Life Cycles. Academic Press, London.

Boletzky, S.V., 1986. Reproductive strategies in cephalopods: variation and flexibility of life-history patterns. Adv. Invertebr. Reprod. 4, 379–389.

Boletzky, S.V., 1987a. Fecundity variation in relation to intermittent or chronic spawning in the cuttlefish, *Sepia officinalis* L. (Mollusca, Cephalopoda). Bull. Mar. Sci. 40, 382–387.

Boletzky, S.V., 1987b. Embryonic phase. In: Boyle, P.R. (Ed.), Cephalopod Life Cycles. Volume II. Comparative Reviews. Academic Press, London, pp. 5–31.
Boletzky, S.V., 1989. Recent studies on spawning, embyonic development, and hatching in the Cephalopoda. Adv. Mar. Biol. 25, 85–115.
Boletzky, S.V., 1998. Cephalopod eggs and egg masses. Oceanogr. Mar. Biol. Annu. Rev., 36, 341–371.
Boletzky, S.V., 2002. Yolk sac morphology in cephalopod embryos. In: Summesberger, H., Histon, K., Daurer, A. (Eds.), Cephalopods—Present and Past, vol. 57. Abhandlungen der Geologischen Bundesanstalt, Wein, pp. 57–68.
Boletzky, S.V., 2003. Biology of early life stages in cephalopods molluscs. Adv. Mar. Biol. 44, 144–203.
Boletzky, S.V., Boletzky, M.V.v., 1969. First results in rearing *Octopus joubini* Robson, 1929. Verh. Naturforsch. Ges. Basel 80, 56–61.
Boletzky, S.V., Boletzky, M.V.v., 1973. Observations on the embryonic and early post-embryonic development of *Rossia macrosoma* (Mollusca, Cephalopoda). Helgoländer Meeresun. 25, 135–161.
Boletzky, S.V., Hanlon, R.T., 1983. A review of the laboratory maintenance, rearing and culture of cephalopod molluscs. Mem. Natl. Mus. Victoria 44, 147–187.
Bonnet, X., Bradshaw, D., Shine, R., 1998. Capital versus income breeding: an ectothermic perspective. Oikos 83, 333–342.
Boucaud-Camou, E., Boismery, J., 1991. The migrations of the cuttlefish (*Sepia officinalis* L) in the English Channel. In: Boucaud- Camou, E. (Ed.), The Cuttlefish, First International Symposium on the Cuttlefish Sepia. Centre de Publications de l'Universite de Caen, Caen.
Boucaud-Camou, E., Roper, C.F.E., 1995. Digestive enzymes in paralarval cephalopods. Bull. Mar. Sci. 57, 313–327.
Boucher-Rodoni, R., Mangold, K., 1995. Ammonia production in cephalopods, physiological and evolutionary aspects. Mar. Freshw. Behav. Physiol. 25, 53–60.
Boyle, P.R., 1983a. Eledone cirrhosa. In: Boyle, P.R. (Ed.), Cephalopod Life Cycles. Volume I: Species Accounts. Academic Press, London, pp. 365–386.
Boyle, P.R. (Ed.), 1983b. Cephalopod Life Cycles. In: Species Accounts, vol. 1. Academic Press, London, p. 475.
Boyle, P.R. (Ed.), 1987. Cephalopod Life Cycles. In: Species Accounts, vol. 2. Academic Press, London, p. 441.
Boyle, P.R., Boletzky, S.V., 1996. Cephalopod populations: definitions and dynamics. Philos. Trans. R. Soc. B 351, 985–1002.
Boyle, P.R., Rodhouse, P., 2005. Cephalopods Ecology and Fisheries. Wiley-Blackwell, Oxford, p. 472.
Boyle, P.R., Thorpe, R., 1984. Optic gland enlargement and female gonad maturation in a population of the octopus *Eledone cirrhosa*: a multivariate analysis. Mar. Biol. 79, 127–132.
Branch, T.A., DeJoseph, B.M., Ray, L.J., Wagner, C.A., 2013. Impacts of ocean acidification on marine seafood. Trends Ecol. Evol. 28, 178–186.
Briceño, F., Mascaró, M., Rosas, C., 2010. GLMM-based modelling of growth in juvenile *Octopus maya* siblings: does growth depend on initial size? ICES J. Mar. Sci. 67, 1509–1516.
Bryne, M., 2011. Impact of ocean warming and ocean acidification on marine invertebrate life history stages: vulnerabilities and potential for persistence in a changing ocean. Oceanogr. Mar. Biol. Annu. Rev. 49, 1–42.
Buresch, K.M., Hanlon, R.T., Maxwell, M.R., Ring, S., 2001. Microsatellite DNA markers indicate a high frequency of multiple paternity within individual field-collected egg capsules of the squid Loligo pealeii. Mar. Ecol. Prog. Ser. 210, 161–165.

Bustamante, P., Teyssié, J., Danis, B., Fowler, S.W., Miramand, P., Cotret, O., 2004. Uptake, transfer and distribution of silver and cobalt in tissues of the common cuttlefish *Sepia officinalis* at different stages of its life cycle. Mar. Ecol. Prog. Ser. 269, 185–195.

Bykov, A.N., Dubinina, T.S., 1984. Some pecularities of larval development of squids of the family Enoploteuthidae from de Central-East Atlantic. Malacol. Rev. 17, 118.

Castro, J., Hernández-García, V., 1995. Ontogenetic changes in mouth structures, foraging behaviour and habitat use of *Scomber japonicus* and *Illex coindetii*. Sci. Mar. 59, 347–355.

Chen, D.S., Dykhuizen, G.V., Hodge, J., Gilly, W.F., 1996. Ontogeny of copepod predation in juvenile squid (*Loligo opalescens*). Biol. Bull. 190, 69–81.

Chung, W.S., 2003. Effects of temperature, salinity and photoperiod on the deposition of growth increments in statoliths of the oval squid *Sepioteuthis lessoniana* Lesson, 1830 (Cephalopoda: Loliginidae) during early stages. MSc thesis, National Sun Yat-sen University, 65 pp.

Cigliano, J.A., 1995. Assessment of the mating history of female pygmy octopuses and a possible sperm competition mechanism. Anim. Behav. 49, 849–851.

Collins, L.A., Pierce, G.J., 1996. Size selectivity in the diet of *Loligo forbesi* (Cephalopoda: Loliginidae). J. Mar. Biol. Assoc. U.K. 76, 1081–1091.

Collins, M., Yau, C., Allcock, L., Thurston, M., 2001. Distribution of deep-water benthic and bentho-pelagic cephalopods from the north-east Atlantic. J. Mar. Biol. Assoc. U.K. 81, 105–117.

Corner, B.D., Moore, H.T., 1980. Field observations on the reproductive behavior of *Sepia latimanus*. Micronesica 16, 235–260.

Cronin, E.R., Seymour, R.S., 2000. Respiration of the eggs of the giant cuttlefish *Sepia apama*. Mar. Biol. 136, 863–870.

Cummins, S.F., Boal, J.G., Buresch, K.C., Kuanpradit, C., Sobhan, P., Holm, J.B., Degnan, B.M., Nagle, G.T., Hanlon, R.T., 2011. Extreme aggression in male squid induced by a β-MSP-like pheromone. Curr. Biol. 21, 322–327.

Cuomo, A., Di Cristo, C., Paolucci, M., Di Cosmo, A., Tosti, E., 2005. Calcium currents correlate with oocyte maturation during the reproductive cycle in *Octopus vulgaris*. J. Exp. Zool. A Comp. Exp. Biol. 303, 193–202.

Darmaillacq, A.S., Chichery, R., Dickel, L., 2006. Food imprinting, new evidence from the cuttlefish *Sepia officinalis*. Biol. Lett. 2, 345–347.

Darmaillacq, A.S., Chichery, R., Dickel, L., 2008. Embryonic visual learning in cuttlefish, *Sepia officinalis*. Anim. Behav. 76, 131–134.

Dawe, E.G., Colbourne, E.B., Drinkwater, K.F., 2000. Environmental effects on recruitment of short-finned squid (*Illex illecebrosus*). ICES J. Mar. Sci. 57, 1002–1013.

Denton, E.J., Gilpin-Brown, J.B., 1961. The buoyancy of the cuttlefish, *Sepia officinalis*. J. Mar. Biol. Assoc. 41, 319–342.

Derby, C.D., Kicklighter, C.E., Johnson, P.E., Zhang, X., 2007. Chemical composition of inks of diverse marine molluscs suggests convergent chemical defences. J. Chem. Ecol. 33, 1105–1113.

Dickel, L., Boal, J.G., Budelmann, B.U., 2000. The effect of early experience on learning and memory in cuttlefish. Dev. Psychobiol. 36 (2), 101–110.

Dickel, L., Chichery, M.P., Chichery, R., 2001. Increase of learning abilities and maturation of the vertical lobe complex during postembryonic development in the cuttlefish, *Sepia*. Dev. Psychobiol. 39 (2), 92–98.

Doney, S.C., Fabry, V.J., Feely, R.A., Kleypas, J.A., 2009. Ocean acidification: the other CO2 problem. Annu. Rev. Mar. Sci. 1, 169–192.

Dorey, N., Melzner, F., Martin, S., Oberhänsli, F., Teyssie, J.-L., Bustamante, P., Gattuso, J.-P., Lacoue-Labarthe, T., 2012. Ocean acidification and temperature rise: effects on calcification during early development of the cuttlefish *Sepia officinalis*. Mar. Biol. 160, 2007–2022.

Drazen, J.C., Goffredi, S.K., Schlining, B., Stakes, D.S., 2003. Aggregations of egg-brooding deep-sea fish and cephalopods on the Gorda Escarpment: a reproductive hot spot. Biol. Bull. 205, 1–7.

Ebenman, B., Persson, L. (Eds.), 1988. Size-Structured Populations: Ecology and Evolution. Springer-Verlag, Berlin, Germany.

Eberhard, W.G., 1985. Sexual Selection and Animal Genitalia. Harvard University Press, Cambridge, MA.

Eberhard, W.G., 1996. Female Control: Sexual Selection by Cryptic Female Choice. Princeton University Press, Princeton, NJ.

Ellis, R.P., Bersey, J., Rundle, S.D., Hall-Spencer, J.M., Spicer, J.I., 2009. Subtle but significant effects of CO2 acidified seawater on embryos of the intertidal snail, *Littorina obtusata*. Aquat. Biol. 5, 41–48.

Enault, J., Zatylny-Gaudin, C., Bernay, B., Lefranc, B., Leprince, J., Baudy-Floc'h, M., Henry, J., 2012. A complex set of sex pheromones identified in the cuttlefish *Sepia officinalis*. PLoS One 7 (10), e46531. http://dx.doi.org/10.1371/journal.pone.0046531.

Engeser, T.S., Clarke, M.R., 1988. Cephalopod hooks, both recent and fossil. In: Clarke, M.R., Trueman, E.R. (Eds.), The Mollusca, Vol. 12: Paleontology and neontology of Cephalopods. Academic Press, London, pp. 133–151.

Estefanell, J., Socorro, J., Roo, F.J., Guirao, R., Fernández-Palacios, H., Izquierdo, M., 2010. Gonad maturation in *Octopus vulgaris* during ongrowing, under different conditions of sex ratio. ICES J. Mar. Sci. 67, 1487–1493.

Falcon, L.I., Vecchione, M., Roper, C.F.E., 2000. Paralarval gonatid squids (Cephalopoda: Oegopsida) from the Mid-North Atlantic. Proc. Biol. Soc. Wash. 113, 532–541.

Feely, R.A., Sabine, C.L., Lee, K., Berelson, W., Kleypas, J., Fabry, V.J., Millero, F.J., 2004. Impact of anthropogenic CO2 on CaCO3 system in the oceans. Science 305, 362–366.

Fields, W.G., 1965. The structure, development, food relations, reproduction, and life history of the squid *Loligo opalescens* Berry. Fish. Bull. 131, 1–108.

Forsythe, J.W., 1993. A working hypothesis on how seasonal temperature change may impact the field growth of young cephalopods. In: Okutani, T., O'Dor, R.K., Kubodera, T. (Eds.), Recent Advances in Cephalopod Fisheries Biology. Tokai University Press, Tokyo, Japan.

Forsythe, J.W., 2004. Accounting for the effect of temperature on squid growth in nature: from hypothesis to practice. Mar. Freshw. Res. 55, 331–339.

Forsythe, J.W., Hanlon, R.T., 1988a. Behaviour, body patterning and reproductive biology of *Octopus bimaculoides* from California. Malacologia 29, 40–56.

Forsythe, J.W., Hanlon, R.T., 1988b. Effect of temperature on laboratory growth, reproduction and life span of *Octopus bimaculoides*. Mar. Biol. 98, 369–379.

Forsythe, J.W., Van Heukelem, W.F., 1987. Cephalopod growth. In: Boyle, P.R. (Ed.), Cephalopod Life Cycles, Volume II, Comparative Reviews. Academic Press, London.

Franco-Santos, R.M., Vidal, E.A.G., 2014. Beak development of early squid paralarvae (Cephalopoda: Teuthoidea) may reflect an adaptation to a specialized feeding mode. Hydrobiologia 725, 85–103.

Froerman, Y.M., Dubinina, T.S., 1984. Distribution and peculiarities of allometric growth of larval *Illex* in the Northwest Atlantic. NAFO Scr. Doc. 84/IX/115, 25 pp.

Froesch, D., Marthy, H.J., 1975. The structure and function of the oviducal gland in octopods (Cephalopoda). Proc. R. Soc. Lond. B Biol. Sci. 188, 95–101.

Gabe, S.H., 1975. Reproduction in the giant octopus of the North Pacific, Octopus dofleini martini. Veliger 18, 146–150.

Gabr, H.R., Hanlon, R.T., Hanafy, M.H., El-Etreby, S.G., 1998. Maturation, fecundity and seasonality of reproduction of two commercially valuable cuttlefish, *Sepia pharaonis* and *S. dollfusi*, in the Suez Canal. Fish. Res. 36, 99–115.

Gabr, H., Hanlon, R.T., El-Etreby, S., Hanafy, M., 1999a. Reproductive versus somatic tissue growth during the life cycle of the cuttlefish *Sepia pharaonis* Ehrenberg, 1831. Fish. Bull. 97, 802–811.

Gabr, H., Hanlon, R.T., Hanafy, M., El-Etreby, S., 1999b. Reproductive versus somatic tissue allocation in the cuttlefish *Sepia dollfusi* Adam (1941). Bull. Mar. Sci. 65, 159–173.

Garofalo, G., Ceriola, L., Gristina, M., Fiorentino, F., Pace, R., 2010. Nurseries, spawning grounds and recruitment of *Octopus vulgaris* in the Strait of Sicily, central Mediterranean Sea. ICES J. Mar. Sci. 67, 1363–1371.

Gilly, W.F., Beman, J.M., Litvin, S.Y., Robison, B.H., 2013. Oceanographic and biological effects of shoaling of the oxygen minimum zone. Annu. Rev. Mar. Sci. 5, 393–420.

Gomi, F., Masamich, Y., Nakazawa, T., 1986. Swelling of egg during development of the cuttlefish, *Sepiella japonica*. Zool. Sci. 3 (4), 641–645.

González, A.F., Otero, J., Guerra, Á., Prego, R., Rocha, F., Dale, A.W., 2005. Distribution of common octopus and common squid paralarvae in a wind-driven upwelling area (Ria of Vigo, northwestern Spain). J. Plankton Res. 27, 271–277.

González, A.F., Otero, J., Pierce, G., Guerra, A., 2010. Age, growth, and mortality of *Loligo vulgaris* wild paralarvae: implications for understanding of the life cycle and longevity. ICES J. Mar. Sci. 67, 1119–1127.

Green, C.P., 2011. Influence of environmental factors on population structure of arrow squid *Nototodarus gouldi*: implications for stock assessment. PhD Thesis, University of Tasmania, Hobart, Australia.

Guerra, A., 2006. Ecology of *Sepia officinalis*. Vie Milieu 56, 97–107.

Guerra, A., Gonzalez, J.L., 2011. First record of predation by a tomplot blenny on the common cuttlefish *Sepia officinalis* eggs. Vie Milieu 61, 45–48.

Guerra, A., Pérez-Gándaras, G., 1984. Larvas de cefalópodos. Investig. Ciencia 89, 6–15.

Gutowska, M.A., Melzner, F., 2009. Abiotic condition in cephalopod (*Sepia officinalis*) eggs: embryonic development at low pH and high pCO2. Mar. Biol. 156, 515–519.

Gutowska, M.A., Pörtner, H.O., Melzner, F., 2008. Growth and calcification in the cephalopod *Sepia officinalis* under elevated seawater pCO2. Mar. Ecol. Prog. Ser. 373, 303–309.

Gutowska, M.A., Melzner, F., Langenguch, M., Bock, C., Claireaux, G., Pörtner, H.O., 2010a. Acid-base regulatory ability of the cephalopod (*Sepia officinalis*) in response to environmental hypercapnia. J. Comp. Physiol. B. 180, 323–335.

Gutowska, M.A., Melzner, F., Pörtner, H.O., Meier, S., 2010b. Cuttlebone calcification increases during exposure to elevated seawater pCO2 in the cephalopod *Sepia officinalis*. Mar. Biol. 157, 1653–1663.

Hadfield, M.G., 2000. Why and how marine invertebrate larvae metamorphose so fast. Semin. Cell Dev. Biol. 11, 437–443.

Hall, K.C., 2002. The Life History and Fishery of a Spawning Aggregation of the Giant Australian Cuttlefish *Sepia apama*. Department of Environmental Biology, University of Adelaide.

Hall, K.C., Fowler, A.J., 2003. The fisheries biology of the cuttlefish *Sepia apama* Gray, in South Australian waters: Final Report to FRDC (Project No. 98/151). Adelaide.

Hall, K.C., Hanlon, R.T., 2002. Principle features of the mating system of a large spawning aggregation of the giant Australian cuttlefish. Mar. Biol. 140, 533–545.

Hanlon, R.T., 1998. Mating systems and sexual selection in the squid *Loligo*: how might commercial fishing on spawning squids affect them? Calcofi Rep. 39, 92–100.

Hanlon, R.T., Messenger, J.B., 1998. Cephalopod Behaviour. Cambridge University Press, Cambridge, UK.

Hanlon, R.T., Wolterding, M.R., 1989. Behavior, body patterning, growth and life history of *Octopus briareus* cultured in the laboratory. Am. Malacol. Bull. 7, 21–45.

Hanlon, R.T., Turk, P.E., Yang, W.T., 1987. Laboratory rearing of the squid *Loligo pealei* to the juvenile stage: growth comparisons with fishery data. Fish. Bull. 85, 163–167.

Hanlon, R.T., Bidwell, J.P., Tait, R., 1989a. Strontium is required for statolith development and thus normal swimming behaviour of hatchling cephalopods. J. Exp. Biol. 141, 187–195.

Hanlon, R.T., Yang, W.T., Turk, P.E., Lee, P.G., Hixon, R.F., 1989b. Laboratory culture and estimated life span of the Eastern Atlantic squid, *Loligo forbesi* (Steenstrup, 1856) (Mollusca: Cephalopoda). Aquac. Fish. Manag. 20, 15–34.

Hanlon, R.T., Smale, M.J., Sauer, W.H., 1994. An ethogram of body patterning behavior in the squid *Loligo vulgaris reynaudii* on spawning grounds in South Africa. Biol. Bull. 187, 363–372.

Hanlon, R.T., Ament, S.A., Gabr, H., 1999. Behavioral aspects of sperm competition in cuttlefish, *Sepia officinalis* (Sepioidea: Cephalopoda). Mar. Biol. 134, 719–728.

Hanlon, R.T., Smale, M.J., Sauer, W.H.H., 2002. The mating system of the squid *Loligo vulgaris reynaudii* (Cephalopoda, Mollusca) off South Africa: fighting, guarding, sneaking, mating and egg laying behavior. Bull. Mar. Sci. 71, 331–345.

Hanlon, R.T., Kangas, N., Forsythe, J.W., 2004. Egg capsule deposition and how behavioral interactions influence spawning rate in the squid *Loligo opalescens* in Monterey Bay, California. Mar. Biol. 145, 923–930.

Hanlon, R.T., Naud, M.J., Shaw, P.W., Havenhand, J.N., 2005. Transient sexual mimicry leads to fertilisation. Nature 433, 212.

Harman, R.F., Young, R.E., 1985. The larvae of ommastrephid squids (Cephalopoda, Teuthoidea) from Hawaiian waters. Vie Milieu 35, 211–222.

Hatfield, E., Cadrin, S., 2002. Geographical and temporal patterns in size and maturity of the longfin inshore squid (*Loligo pealeii*) off the northwestern United States. Fish. Bull. 100 (2), 200–213.

Hatfield, E.M.C., Rodhouse, P.G., Porebski, J., 1990. Demography and distribution of the Patagonian squid (*Loligo gahi* d'Orbigny) during the austral winter. ICES J. Mar. Sci. 46, 306–312.

Hatfield, E.M.C., Rodhouse, P.G., Barber, D.L., 1992. Production of soma and gonad in maturing female *Illex argentinus* (Mollusca: Cephalopoda). J. Mar. Biol. Assoc. U.K. 72, 281–291.

Hendrix, J.P., McKinney, L.D., Salzer, R.R., Case, R.J., 1981. Salinity tolerance and the responses to hypo osmotic stress of the bay squid *Lolliguncula brevis*, a euryhaline cephalopod mollusk. Comp. Biochem. Physiol. 69A, 641–648.

Henning, G.B., Bird, C.M., Wichmann, F.A., 2002. Contrast discrimination with pulse trains in pink noise. J. Opt. Soc. Am. A 19, 1259–1266.

Henry, J., Favrel, P., Boucaud-Camou, E., 1997. Isolation and identification of a novel Ala-Pro-Gly-Trp-amide-related peptide inhibiting the motility of the mature oviduct in the cuttlefish, *Sepia officinalis*. Peptides 18, 1469–1474.

Henry, J., Zatylny, C., Boucaud-Camou, E., 1999. Peptidergic control of egg-laying in the cephalopod *Sepia officinalis*: involvement of FMRFamide and FMRFamide-related peptides. Peptides 20, 1061–1071.

Heyland, A.L., Moroz, L., 2006. Signaling mechanisms underlying metamorphic transitions in animals. Integr. Comp. Biol. 46 (6), 743–759.

Ho, J.D., Moltschaniwskyj, N.A., Carter, C.G., 2004. The effect of variability in growth on somatic condition and reproductive status in the southern calamary *Sepioteuthis australis*. Mar. Freshw. Res. 55, 423–428.

Hoving, H.J.T., 2008. Reproductive biology of oceanic decapodiform cephalopods. PhD Thesis, University of Groningen, 184 pp.

Hoving, H.J.T., Vecchione, M., 2012. Mating behavior of a deep-sea squid revealed by in situ videography and the study of archived specimens. Biol. Bull. 223, 263–267.

Hoving, H.J.T., Gilly, W.F., Markaida, U., Benoit-Bird, K.J., Brown, Z.W., Daniel, P., 2013. Extreme plasticity in life-history strategy allows a migratory predator (jumbo squid) to cope with a changing climate. Glob. Change Biol. 19, 2089–2103.

Hu, M.Y., Tseng, Y., Stumpp, M., Gutowska, M.A., Kiko, R., Lucassen, M., Melzner, F., 2011. Elevated seawater pCO2 differentially affects brancial acid-base transporters over the course of development in cephalopod *Sepia officinalis*. Am. J. Physiol. Regul. Integr. Comp. Physiol. 300, R1100–R1114.

Huffard, C.L., Hochberg, F.G., 2005. Description of a new species of the genus *Amphioctopus* (Mollusca: Octopodidae) from the Hawaiian Islands. Molluscan Res. 25, 113–128.

Huffard, C.L., Cadwell, R.L., Boneka, F., 2008. Mating behavior of *Abdopus aculeatus* (d'Orbigny 1834) (Cephalopoda: Octopodidae) in the wild. Mar. Biol. 154, 353–362.

Hunt, J.C., Lindsay, D.J., 2012. In situ behavioral observations of the mesopelagic squid *Stigmatoteuthis dofleini* (Cephalopoda: Histioteuthidae). Am. Malacol. Bull. 30, 335–338.

Hunt, J.C., Seibel, B., 2000. Life history of *Gonatus onyx* (Cephalopoda: Teuthoidea): ontogenetic changes in habitat, behavior and physiology. Mar. Biol. 136, 543–552.

Hurley, A.C., 1978. School structure of the squid *Loligo opalescens*. Fish. Bull. 76 (2), 433–442.

Ibáñez, C.M., Cubillos, L.A., 2007. Seasonal variation in the length structure and reproductive condition of the jumbo squid *Dosidicus gigas* (d'Orbigny, 1835) off central-south Chile. Sci. Mar. 71, 123–128.

Ibáñez, C.M., Keyl, F., 2010. Cannibalism in cephalopods. Rev. Fish Biol. Fish. 20, 123–136.

Iglesias, J., Otero, J.J., Moxica, C., Fuentes, L., Sánchez, F.J., 2004. The completed life cycle of the octopus (*Octopus vulgaris*, Cuvier) under culture conditions: paralarval rearing using Artemia and Zoeae, and first data on juvenile growth up to 8 months of age. Aquac. Int. 12, 481–487.

Iglesias, J., Sánchez, F.J., Bersano, J.G.F., Carrasco, J.F., Dhont, J., Fuentes, L., Linares, F., Munoz, J.L., Okumura, S., Roo, J., van der Meeren, T., Vidal, E.A.G., Villanueva, R., 2007. Rearing of *Octopus vulgaris* paralarvae: present status, bottlenecks and trends. Aquaculture 266, 1–15.

Ikeda, Y., Sakurai, Y., Shimazaki, K., 1993. Fertilizing capacity of squid (*Todarodes pacificus*) spermatozoa collected from various sperm storage sites, with special reference to the role of gelatinous substance from oviducal gland in fertilization and embryonic development. Integr. Comp. Biol. 23, 39–44.

Ikeda, Y., Wada, Y., Arai, N., Sakamoto, W., 1999. Note on size variation of body and statoliths in the oval squid *Sepioteuthis lessoniana* hatchlings. J. Mar. Biol. Assoc. U.K. 79 (4), 757–759.

Imamura, S., 1990. Larval rearing of Octopus (*Octopus vulgaris* Cuvier); the progress of technological development and some problems remained. Collect. Breed. 52, 339–343.

Iwata, Y., Sakurai, Y., 2007. Threshold dimorphism in ejaculate characteristics in the squid *Loligo bleekeri*. Mar. Ecol. Prog. Ser. 345, 141–146.

Iwata, Y., Munehara, H., Sakurai, Y., 2005. Dependence of paternity rates on alternative reproductive behaviors in the squid *Loligo bleekeri*. Mar. Ecol. Prog. Ser. 298, 219–228.

Iwata, Y., Ito, K., Sakurai, Y., 2008. Effect of low temperature on mating behavior of squid *Loligo bleekeri*. Fish. Sci. 74, 1345–1347.

Iwata, Y., Shaw, P., Fujiwara, E., Shiba, K., Kakiuchi, Y., Hirohashi, N., 2011. Why small males have big sperm: dimorphic squid sperm linked to alternative mating behaviours. BMC Evol. Biol. 11, 236.

Jackson, G.D., Mladenov, P.V., 1994. Terminal spawning in the deepwater squid *Moroteuthis ingens* (Cephalopoda: Onychoteuthidae). J. Zool. 234, 189–201.

Jackson, G.D., Moltschaniwskyj, N.A., 2002. Spatial and temporal variation in growth rates and maturity in the Indo-Pacific squid *Sepioteuthis lessoniana* (Cephalopoda: Loliginidae). Mar. Biol. 140, 747–754.

Jackson, G.D., McGrath Steer, B., Wotherspoon, S., Hobday, A.J., 2003. Variation in age, growth and maturity in the Australian arrow squid *Nototodarus gouldi* over time and space—what is the pattern? Mar. Ecol. Prog. Ser. 264, 57–71.

Jackson, G.D., Semmens, J.M., Phillips, K.L., Jackson, C.H., 2004. Reproduction in the deepwater squid *Moroteuthis ingens*, what does it cost? Mar. Biol. 145, 905–916.

Jantzen, T.M., Havenhand, J.N., 2003. Reproductive behavior in the squid *Sepioteuthis australis* from South Australia: interactions on the spawning grounds. Biol. Bull. 204, 305–317.

Joll, L.M., 1976. Mating, egg-laying and hatching of *Octopus tetricus* (Mollusca: Cephalopoda) in the Laboratory. Mar. Biol. 36, 327–333.

Jonsson, K.I., 1997. Capital and income breeding as alternative tactics of resource use in reproduction. Oikos 78, 57–66.

Kaneko, N., Oshima, Y., Ikeda, Y., 2006. Egg brooding behavior and embryonic development of *Octopus laqueus* (Cephalopoda: Octopodidae). Molluscan Res. 26, 113–117.

Kasugai, T., 2000. Reproductive behavior of the pygmy cuttlefish *Idiosepius paradoxus* in an aquarium. Jpn. J. Malacol. 59, 37–44.

Katsanevakis, S., Verriopoulos, G., 2004a. Relative abundance of *Octopus vulgaris* on soft sediment. Sci. Mar. 68, 553–560.

Katsanevakis, S., Verriopoulos, G., 2004b. Den ecology of *Octopus vulgaris* Cuvier, 1797, on soft sediment: availability and types of shelter. Sci. Mar. 68, 147–157.

Kaufman, M.R., Ikeda, Y., Patton, C., Van Dijkhausen, G., Epel, D., 1998. Bacterial symbionts colonize the accessory nidamental gland of the squid *Loligo opalescens* via horizontal transmission. Biol. Bull. 194, 36–43.

Keeling, R.F., Kortzinger, A., Gruber, N., 2010. Ocean deoxygenation in a warming world. Annu. Rev. Mar. Sci. 2, 199–229.

Kier, W.M., 1996. Muscle development in squid: ultrastructural differentiation of a specialized muscle fiber type. J. Morphol. 229, 271–288.

Kier, W.M., Smith, A.M., 1990. The morphology and mechanics of octopus suckers. Biol. Bull. 178, 126–136.

Koronkiewicz, A., 1995. Size and maturity differences between trawl and jigger caught short-finned squid *Illex argentinus* (Cephalopoda: Ommastrephidae). Acta Ichthyol. Piscat. 25, 97–111.

Koueta, N., Camou-Boucaud, E., Renou, A., 1995. Gonadotropic mitogenic activity of the optic gland of the cuttlefish, *Sepia officinalis*, during sexual maturation. J. Mar. Biol. Assoc. U.K. 75, 391–404.

Kubodera, T., Okutani, T., 1977. Descriptions of a new species of gonatid squid, *Gonatus madokai*, n. sp., from the Northwest Pacific, with notes on morphological changes with growth and distribution in immature stages (Cephalopoda: Oegopsida). Jpn. J. Malacol. 36, 123–151.

Kubodera, T., Koyama, Y., Mori, K., 2007. Observations of wild hunting behaviour and bioluminescence of a large deep-sea, eight-armed squid, *Taningia danae*. Proc. R. Soc. B 274, 1029–1034.

Lacoue-Labarthe, T., Martin, S., Oberhansli, F., Teyssie, J.L., Markich, S., Bustamante, P., 2009. Effects of increased pCO2 and temperature on trace element (Ag, Cd and Zn) bioaccumulation in the eggs of the common cuttlefish, *Sepia officinalis*. Biogeosciences 6, 2561–2573.

Lacoue-Labarthe, T., Le Bihan, E., Borg, D., Koueta, N., Bustamante, P., 2010. Acid phosphatase and cathepsin activity in cuttlefish (*Sepia officinalis*) eggs: the effects of Ag, Cd, and Cu exposure. ICES J. Mar. Sci. 67, 1517–1523.

Lacoue-Labarthe, T., Reveillac, E., Oberhansli, F., Teyssie, J.L., Jeffree, R., Gattuso, J.P., 2011. Effects of ocean acidification on trace element accumulation in the early-life stages of squid *Loligo vulgaris*. Aquat. Toxicol. 105, 166–176.

Lacoue-Labarthe, T., Martin, S., Oberhansli, F., Teyssie, J., Jeffree, R., Gattuso, J., Bustamante, P., 2012. Temperature and pCO2 effect on the bioaccumulation of radionuclides and trace elements in the eggs of the common cuttlefish, *Sepia officinalis*. J. Exp. Mar. Biol. Ecol. 413, 45–49.

Laptikhovsky, V.V., Arkhipkin, A.I., Hoving, H.J.T., 2007. Reproductive biology in two species of deep-sea squids. Mar. Biol. 152, 981–990.

Laptikhovsky, V., Pereira, J., Salman, A., Arkhipov, A., Costa, A.M., 2009. A habitat-dependence in reproductive strategies of cephalopods and pelagophile fish in the Mediterranean Sea. Boll. Malacol. 45, 95–102.

LaRoe, E.T., 1971. The culture and maintenance of the loliginid squids *Sepioteuthis sepioidea* and *Doryteuthis plei*. Mar. Biol. 9, 9–25.

Laughlin, R.A., Livingston, R.J., 1982. Environmental and trophic determinants of the spatial/temporal distribution of the brief squid (*Lolliguncula brevis*) in the Apalachicola estuary (North Florida, USA). Bull. Mar. Sci. 32, 489–497.

Le Goff, R., 1991. Biologie et Migration de la seiche *Sepia officinalis* L. (Mollusque, Cephalopode, Sepiidae) dans le secteur Morbraz-golfe du Morbihan (sud Bretagne). University of Rennes, Rennes, France.

Lee, P.G., 1995. Nutrition of cephalopods: fuelling the system. Mar. Freshw. Behav. Physiol. 25, 35–51.

Lee, C.E., Strathmann, R.R., 1998. Scaling of gelatinous clutches: effects of siblings' competition for oxygen on clutch size and parental investment per offspring. Am. Nat. 151, 293–310.

Lee, P.N., Callaerts, P., de Couet, H.G., 2009. The embryonic development of the Hawaiian bobtail squid (*Euprymna scolopes*). Cold Spring Harb. Protoc. 4 (11), 1426–1435.

Leporati, S.C., Pecl, G.T., Semmens, J.M., 2008. Reproductive status of *Octopus pallidus*, and its relationship to age and size. Mar. Biol. 155, 375–385.

Liu, B., Chen, X., Chen, Y., Tian, S., Li, J., Fang, Z., Yang, M., 2013. Age, maturation, and population structure of the Humboldt squid *Dosidicus gigas* off the Peruvian Exclusive Economic Zones. Chin. J. Oceanol. Limnol. 31, 81–91.

Lourenço, S., Moreno, A., Narciso, L., Gonzalez, A., Pereira, J., 2012. Seasonal trends of the reproductive cycle of *Octopus vulgaris* in two environmentally distinct coastal areas. Fish. Res. 127&128, 116–124.

Lu, C.C., 1998. A synopsis of Sepiidae in Australian waters (Cephalopods: Sepioidea). In: Voss, N.A., Vecchione, M., Toll, R.B., Sweeney, M.J. (Eds.), Systematics and Biogeography of Cephalopods. Smithsonian Institution Press, Washington, pp. 159–190.

Macewicz, B.J., Hunter, J.R., Lo, N.C.H., LaCasella, E.L., 2004. Fecundity, egg deposition, and mortality of market squid (Loligo opalescens). Fish. Bull. 102, 306–327.

Macy, W.K., Brodziak, J.K.T., 2001. Seasonal maturity and size at age of *Loligo pealeii* in waters of southern New England. ICES J. Mar. Sci. 58, 852–864.

Mangold, K., 1987. Reproduction. In: Boyle, P.R. (Ed.), In: Cephalopod Life Cycles, vol. 2. Academic Press, London, UK, pp. 157–200.

Mangold, K.M., Young, R.E., 1998. The systematic value of the digestive organs. In: Voss, N.A., Vecchione, M., Toll, R.B., Sweeney, M.J. (Eds.), Systematics and biogeography of cephalopods. In: Smithson. Contrib. Zool., 586, pp. 21–30.

Markaida, U., Sosa-Nishizaki, O., 2003. Food and feeding habits of jumbo squid *Dosidicus gigas* (Cephalopoda: Ommastrephidae) from the Gulf of California, Mexico. J. Mar. Biol. Assoc. U.K. 83, 507–522.

Marquez, L., Quintana, D., Lorenzo, A., Almanza, E., 2013. Biometrical relationship in developing eggs and neonates of *Octopus vulgaris* in relation to parental diet. Helgol. Mar. Res. 67, 461–470.

Marthy, H.J., Hauser, R., School, A., 1976. Natural tranquilliser in cephalopod eggs. Nature 261, 496–497.

Mather, J.A., 2006. Behaviour development: a cephalopod perspective. Int. J. Comp. Psychol. 19, 98–115.

Maxwell, M.R., Hanlon, R.T., 2000. Female reproductive output in the squid *Loligo pealeii*: multiple egg clutches and implications for spawning strategy. Mar. Ecol. Prog. Ser. 199, 159–170.

McGowan, J.A., 1954. Observations on the sexual behavior and spawning of the squid, *Loligo opalescens*, at La Jolla, California. Calif. Fish. Game Bull. 40, 47–54.

Mejía-Rebollo, A., Salinas-Zavala, C.A., Quiñónez-Velázquez, C., Markaida, U., 2008. Age, growth and maturity of jumbo squid (*Dosidicus gigas* D'orbigny, 1835) off the western coast of the Baja California Peninsula. Calif. Coop. Ocean. Fish. Invest. Rep. 49, 256–262.

Melo, Y.C., Sauer, W.H.H., 1998. Ovarian atresia in cephalopods. In: Payne, A.I.L., Lipinski, M.R., Clarke, M.R., Roeleveld, M.A.C. (Eds.), Cephalopod Biodiversity, Ecology and Evolution. In: South African Journal of Marine Science, vol. 20, pp. 143–151.

Melo, Y.C., Sauer, W.H.H., 1999. Confirmation of serial spawning in the chokka squid *Loligo vulgaris reynaudii* off the coast of South Africa. Mar. Biol. 135, 307–313.

Messenger, J.B., 1977. Prey capture and learning in the cuttlefish *Sepia*. Symp. Zool. Soc. Lond. 38, 347–376.

Miske, V., Kirchhauser, J., 2006. First record of brooding and early life cycle stages in *Wunderpus photogenicus* Hochberg, Norman and Finn, 2006 (Cephalopoda: Octopodidae). Molluscan Res. 26, 169–171.

Miyahara, K., Fukui, K., Nagahama, T., Ohatani, T., 2006a. First record of planktonic egg masses of the diamond squid, *Thysanoteuthis rhombus* Troschel, in the Sea of Japan. Plank. Benth. Res. 1, 59–63.

Miyahara, K., Fukui, K., Ota, T., Minami, T., 2006b. Laboratory observations on the early life stages of the diamond squid *Thysanoteuthis rhombus*. J. Molluscan Stud. 72, 199–205.

Moguel, C., Mascaró, M., Avila-Poveda, O.H., Caamal-Monsreal, C., Sanchez, A., Pascual, C., Rosas, C., 2010. Morphological, physiological, and behavioral changes during post-hatching development of *Octopus maya* (Mollusca: Cephalopoda) with special focus on the digestive system. Aquat. Biol. 9, 35–48.

Moltschaniwskyj, N.A., 1995a. Changes in shape associated with growth in the loliginid squid *Photololigo* sp.: a morphometric approach. Can. J. Zool. 73, 1335–1343.

Moltschaniwskyj, N.A., 1995b. Multiple spawning in the tropical squid *Photololigo* sp.: what is the cost in somatic growth? Mar. Biol. 124, 127–135.

Moltschaniwskyj, N.A., Carter, C.G., 2013. The adaptive response of protein turnover to the energetic demands of reproduction in a cephalopod. Physiol. Biochem. Zool. 86, 119–126.

Moltschaniwskyj, N.A., Pecl, G.T., 2003. Small-scale spatial and temporal patterns of egg production by the temperate loliginid squid *Sepioteuthis australis*. Mar. Biol. 142, 509–516.

Moltschaniwskyj, N.A., Pecl, G., Lyle, J., 2002. An assessment of the use of short-term closures to protect spawning southern calamary aggregations from fishing pressure in Tasmania, Australia. Bull. Mar. Sci. 71, 501–514.

Moran, N.A., 1994. Adaptation and constraint in the complex life cycles of animals. Annu. Rev. Ecol. Syst. 25, 573–600.

Naef, A., 1928. Die Cephalopoden. Embryologie. Fauna Flora Golf Neapel 35, 1–357, English translation by Boletzky, S.V., 2001. The Cephalopoda-Embryology. Smithsonian Institution Press, Washington, District of Columbia, USA.

Nair, K.P., Thomas, P.A., Gopakumar, G., Vincent, S.G., Omana, T.A., 1986. Some observations on the hatching and post-hatch- ing behaviour of the cuttlefish *Sepia pharaonis* Ehrenberg. In: Silas, E.G. (Ed.), Cephalopod Bionomics. Fisheries and Resource of the Exclusive Economic Zone of India, vol. 37. Bulletin of the Central Marine and Freshwater Research Institute, pp. 157–159.

Natsukari, Y., Tashiro, M., 1991. Neritic squid resources and cuttlefish resources in Japan. Mar. Behav. Physiol. 18, 149–226.

Naud, M.J., Hanlon, R.T., Hall, K.C., Shaw, P.W., Havenhand, J.N., 2004. Behavioural and genetic assessment of reproductive success in a spawning aggregation of the Australian giant cuttlefish, *Sepia apama*. Anim. Behav. 67, 1043–1050.

Naud, M.J., Shaw, P.W., Hanlon, R.T., Havenhand, J.N., 2005. Evidence for biased use of sperm sources in wild female giant cuttlefish (*Sepia apama*). Proc. R. Soc. Lond. B 272, 1047–1051.

Navarro, J.C., Villanueva, R., 2000. Lipid and fatty acid composition of early stages of cephalopods: an approach to their lipid requirements. Aquaculture 183, 161–177.

Navarro, J.C., Villanueva, R., 2003. The fatty acid composition of *Octopus vulgaris* paralarvae reared with live and inert food: deviation from their natural fatty acid profile. Aquaculture 219, 613–631.

Nesis, K.N., 1973. Ecological classification (life forms) of cephalopods. Itogi Nauki Tekh. (Zool. Bespozv.) 2, 8–59, VINITI (in Russian).

Nesis, K.N., 1979. Larvae of cephalopods. Biol. Morya 4, 26–37.

Nigmatullin, C.M., Nesis, K.N., Arkhipkin, A.I., 2001. A review of the biology of the jumbo squid *Dosidicus gigas* (Cephalopods: Ommastrephidae). Fish. Res. 54, 9–19.

Nixon, M., 1985. Capture of prey, diet and feeding of *Sepia officinalis* and *Octopus vulgaris* (Mollusca: Cephalopoda) from hatchling to adult. Vie Milieu 35, 255–261.

Nixon, M., Young, J.Z., 2003. The Brain and Lives of Cephalopods. Oxford University Press, Oxford, 384 pp.

Norman, M.D., Lu, C.C., 1997. Redescription of the southern dumpling squid *Euprymna tasmanica* and a revision of the genus *Euprymna* (Cephalopoda: Sepiolidae). J. Mar. Biol. Assoc. U.K. 77, 1109–1137.

Norman, M., Reid, A., 2000. Guide to Squid, Cuttlefish and Octopuses of Australasia. CSIRO Publishing, Collingwood, Vic.

Norman, M.D., Finn, J., Tregenza, T., 1999. Female impersonation as an alternative reproductive strategy in giant cuttlefish. Proc. R. Soc. B Biol. Sci. 266, 1347–1349.

O'Dor, R.K., Balch, N., 1985. Properties of *Illex illecebrosus* egg masses potentially influencing larval oceanographic distribution. NAFO Sci. Council Stud. 9, 69–76.

O'Dor, R.K., 1983. Illex illecebrosus. In: Boyle, P.R. (Ed.), Cephalopod Life Cycles. Academic Press, London, pp. 175–200.

O'Dor, R.K., 1998. Squid life history strategies. In: Rodhouse, P.G., Dawe, E.G., O'Dor, R.K. (Eds.), Squid Recritment Dynamics. FAO Fisheries Technical Paper, 376, pp. 233–254.

Okutani, T., 1987. Juvenile morphology. In: Boyle, P.R. (Ed.), Cephalopod Life Cycles. Volume II, Comparative Reviews. Academic Press, London, pp. 33–44.

Okutani, T., Nakamura, I., Seki, K., 1995. An unusual egg-brooding behaviour of an oceanic squid in the Okhotsk Sea. Venus 54, 237–239.

Oosthuizen, A., Roberts, M.J., Sauer, W.H., 2002a. Temperature effects on the embryonic development and hatching success of the squid *Loligo vulgaris reynaudii*. Bull. Mar. Sci. 71, 619–632.

Oosthuizen, A., Smale, M.J., Sauer, W.H.H., 2002b. Aspects of biology, ecology and fishery potential of *Octopus* in the Eastern Cape, South Africa. Bull. Mar. Sci. 71, 1134–1135.

Orr, J.C., Fabry, V.J., Aumont, O., Bopp, L., Doney, S.C., Feely, R.A., Gnanadesikan, A., Gruber, N., Ishida, A., Joos, F., Key, R.M., Lindsay, K., Maier-Reimer, E., Matear, R., Monfray, P., Mouchet, A., Najjar, R.G., Plattner, G., Rodgers, K.B., Sabine, C.L., Sarmiento, J.L., Schlitzer, R., Slater, R.D., Totterdell, I.J., Weirig, M., Yamanaka, Y., Yool, A., 2005. Anthropogenic ocean acidification over the twenty-first century and its impact on calcifying organisms. Nature 437, 681–686.

Ortiz, N., Ré, M.E., Márquez, F., Glembocki, N.G., 2011. The reproductive cycle of the red octopus *Enteroctopus megalocyathus* in fishing areas of Northern Patagonian coast. Fish. Res. 110, 217–223.

O'Shea, S., Bolstad, K.S., Ritchie, P.A., 2004. First records of egg masses of *Nototodarus gouldi* McCoy, 1888 (Cephalopoda: Ommastrephidae), with comments on egg-mass susceptibility to damage by fisheries trawl. N. Z. J. Zool. 31, 161–166.

Otero, J., Gonzalez, A.F., Sieiro, M., Guerra, A., 2007. Reproductive cycle and energy allocation of *Octopus vulgaris* in Galician waters, NE Atlantic. Fish. Res. 85, 122–129.

Otero, J., Álvarez-Salgado, X., González, Á.F., Miranda, A., Groom, S., Cabanas, J., Casas, G., Wheatley, B., Guerra, Á., 2008. Bottom-up control of common octopus *Octopus vulgaris* in the Galician upwelling system, northeast Atlantic Ocean. Mar. Ecol. Prog. Ser. 362, 181–192.

Otero, J., Álvarez-Salgado, X., González, Á.F., Gilcoto, M., Guerra, Á., 2009. High-frequency coastal upwelling events influence *Octopus vulgaris* larval dynamics on the NW Iberian shelf. Mar. Ecol. Prog. Ser. 386, 123–132.

Packard, A., 1961. Sucker display of *Octopus*. Nature 190, 736–737.

Palmer, M., Calve, M., Adamo, S.A., 2006. Response of female cuttlefish *Sepia officinalis* (Cephalopoda) to mirrors and conspecifics: evidence for signaling in female cuttlefish. Anim. Cogn. 9, 151–155.

Parker, G.A., 1990. Sperm competition games: sneaks and extra-pair copulations. Proc. R. Soc. Biol. Sci. B 242, 127–133.

Parker, L.M., Ross, P.M., O'Connor, W.A., Pörtner, H.O., Scanes, E., Wright, J.M., 2013. Predicting the response of molluscs to the impact of ocean acidification. Biology 2, 651–692.

Parra, G., Villanueva, R., Yufera, M., 2000. Respiration rates in late eggs and early hatchlings of the common octopus, *Octopus vulgaris*. J. Mar. Biol. Assoc. U.K. 80, 557–558.

Paulij, W.P., Herman, P.M.J., Van Hannen, E.J., Denuce, J.M., 1990. The impact of photoperiodicity on hatching of *Loligo vulgaris* and *Loligo forbesi*. J. Mar. Biol. Assoc. U.K. 70, 597–610.

Paulij, W.P., Herman, P.M.J., Roozen, M.E.F., Denuce, J.M., 1991. The influence of photoperiodicity on hatching of *Sepia officinalis*. J. Mar. Biol. Assoc. U.K. 71, 665–678.

Pauly, D., 2010. Gasping fish and panting squids: oxygen, temperature and the growth of water-breathing animals. In: Kinne, O. (Ed.), Excellence in ecology, vol. 22. International Ecology Institute, Oldendorf/Luhe.

Pecl, G., 2001. Flexible reproductive strategies in tropical and temperate *Sepioteuthis* squids. Mar. Biol. 138, 93–101.

Pecl, G.T., Moltschaniwskyj, N.A., Tracey, S., Jordan, A., 2004. Inter-annual plasticity of squid life-history and population structure: ecological and management implications. Oecologia 139, 515–524.

Pecl, G.T., Tracey, S.R., Semmens, J.M., Jackson, G.D., 2006. Use of acoustic telemetry for spatial management of southern calamary *Sepioteuthis australis*, a highly mobile inshore squid species. Mar. Ecol. Prog. Ser. 328, 1–15.

Pecl, G.T., Tracey, S.R., Danyushevsky, L., Wotherspoon, S., Moltschaniwskyj, N.A., 2011. Elemental fingerprints of southern calamary (*Sepioteuthis australis*) reveal local recruitment sources and allow assessment of the importance of closed areas. Can. J. Fish. Aquat. Sci. 68, 1351–1360.

Pethybridge, H., Virtue, P., Casper, R., Yoshida, T., Green, C.P., Jackson, G., Nichols, P.D., 2012. Seasonal variations in diet of arrow squid (*Nototodarus gouldi*): stomach content and signature fatty acid analysis. J. Mar. Biol. Assoc. U.K. 92, 187–196.

Pierce, G.J., Zuur, A.F., Smith, J.M., Santos, M.B., Bailey, N., Chen, C., Boyle, P.R., 2005. Interannual variation in life-cycle characteristics of the veined squid (*Loligo forbesi*) in Scottish (UK) waters. Aquat. Living Resour. 18, 327–340.

Pierce, G.J., Valavanis, V.D., Guerra, A., Jereb, P., Orsi-Relini, L., Bellido, J.M., Katara, I., Piatkowski, U., Pereira, J., Balguerias, E., 2008. A review of cephalopod—environment interactions in European Seas. Hydrobiologia 612, 49–70.

Pierce, G.J., Allcock, L., Bruno, I., Bustamante, P., Gonzalez, A., Guerra, A., Jereb, P., Lefkaditou, E., Malham, S., Moreno, A., Pereira, J. a, Piatkowski, U., Rasero, M., Sanchez, P., Santos, M.B., Santurtun, M., Seixas, S., Sobrino, I., Villanueva, R., 2010. Cephalopod Biology and Fisheries in Europe. Technical report, ICES Cooperative research report 303. Copenhagen, Denmark.

Pimentel, M.S., Trubenbach, K., Faleiro, F., Boavida-Portugal, J., Repolho, T., Rosa, R., 2012. Impact of ocean warmning on the early ontogeny of cephalopods: a metabolic approach. Mar. Biol. 159, 2051–2059.

Pliego-Cárdenas, García-Domínguez, R.F.A., Ceballos-Vázquez, B.P., Villalejo-Fuerte, M., Arellano-Martínez, M., 2011. Reproductive aspects of *Octopus hubbsorum* (Cephalopoda: Octopodidae) from Espiritu Santo Island, southern Gulf of California, México. Cien. Mar. 37, 23–32.

Pörtner, H.O., Langenbuch, M., Michaelidis, B., 2005. Synergistic effects of temperature extremes, hypoxia, and increases in CO2 on marine animals: from Earth history to global change. J. Geophys. Res. 110, C09S10. doi:10.1029/2004JC002561.

Quinteiro, J., Baibai, T., Oukhattar, L., Soukri, A., Seixas, P., Rey-Mendez, M., 2011. Multiple paternity in the common octopus *Octopus vulgaris* (Cuvier, 1797), as revealed by microsatellite DNA analysis. Molluscan Res. 31 (1), 15–20.

Ramos-Castillejos J.E., 2007. Análisis morfológico y genético de las paralarvas Rhynchoteuthion del calamar gigante *Dosidicus gigas* (D'Orbigny, 1835) y del calamar púrpura *Sthenoteuthis oualaniensis* (Lesson, 1830). Master's thesis. Centro de Investigaciones Biológicas del Noroeste, La Paz, B.C.S., Mexico, 135 pp.

Rigby, R.P., Sakurai, Y., 2004. Temperature and feeding related to growth efficiency of immature octopuses *Enteroctopus dofleini*. Suisanzoshoku 52, 29–36.

Roberts, M.J., van den Berg, M., 2002. Recruitment variability of chokka squid (*Loligo vulgaris reynaudii*)—role of currents on the Agulhas Bank (South Africa) in paralarvae distribution and food abundance. Bull. Mar. Sci. 71, 691–710.

Rocha, F., Guerra, A., Gonzalez, A.F., 2001. A review of reproductive strategies in cephalopods. Biol. Rev. 76, 291–304.

Rodaniche, A.F., 1984. Iteroparity in the lesser Pacific striped octopus *Octopus chierchiae* (Jatta, 1889). Bull. Mar. Sci. 35, 99–104.

Rodhouse, P.G., 1998. Physiological progenesis in cephalopod molluscs. Biol. Bull. 195, 17–20.

Rodhouse, P.G., 2010. Effects of environmental variability and change on cephalopod populations: an introduction to the CIAC '09 Symposium special issue. ICES J. Mar. Sci. 67 (7), 1311–1313.

Rodhouse, P.G., Hatfield, E.M.C., 1990. Dynamics of growth and maturation in the cephalopod *Illex argentinus* de Castellanos, 1960 (Teuthoidea: Ommastrephidae). Philos. Trans. R. Soc. Lond. 329, 229–241.

Rodhouse, P.G., Nigmatullin, C.M., 1996. Role as consumers. Philos. Trans. R. Soc. B Biol. Sci. 351, 1003–1022.

Rodhouse, P.G., Dawe, E.G., O'Dor, R.K. (Eds.), 1998. Squid recruitment dynamics. The genus *Illex* as a model. The commercial Illex species. Influences on variability. FAO Fisheries Technical Paper, 376, p. 273.

Rodrigues, M., Garci, M.E., Guerra, A., Troncoso, J.S., 2009. Mating behavior of the Atlantic bobtail squid *Sepiola atlantica* (Cephalopoda: Sepiolidae). Vie Milieu 59, 271–275.

Rodrigues, M., Guerra, Á., Troncoso, J., 2011. The embryonic phase and its implication in the hatchling size and condition of Atlantic bobtail squid *Sepiola atlantica*. Helgol. Mar. Res. 65, 211–216.

Rodríguez–Rúa, A., Pozuelo, I., Prado, M.A., Gómez, M.J., Bruzón, M.A., 2005. The gametogenic cycle of *Octopus vulgaris* (Mollusca: Cephalopoda) as observed on the Atlantic coast of Andalusia (south of Spain). Mar. Biol. 147, 927–933.

Romagny, S., Darmaillacq, A.-S., Guibé, M., Bellanger, C., Dickel, L., 2012. Feel, smell and see in an egg: emergence of perception and learning in an immature invertebrate, the cuttlefish embryo. J. Exp. Biol. 215, 4125–4130.

Roper, C.F., 1965. A note on egg deposition by *Doryteuthis plei* (Blainville, 1823) and its comparison with other North American loliginid squids. Bull. Mar. Sci. 15, 589–598.

Roper, C.F.E., Sweeney, M.J., Nauen, C.E., 1984. Cephalopods of the World: an annotated and illustrated catalogue of species of interest to fisheries. FAO Fish. Synop., 125.

Rosa, R., Costa, P.R., Nunes, M.L., 2004a. Effect of sexual maturation on the tissue biochemical composition of *Octopus vulgaris* and *O. defilippi* (Mollusca: Cephalopoda). Mar. Biol. 145, 563–574.

Rosa, R., Costa, P.R., Pereira, J., Nunes, M.L., 2004b. Biochemical dynamics of spermatogenesis and oogenesis in *Eledone cirrhosa* and *Eledone moschata* (Cephalopoda: Octopoda). Comp. Biochem. Physiol. B 139, 299–310.

Rosa, R., Pimentel, M.S., Boavida-Portugal, J., Teixeira, T., Tru¨benbach, K., Diniz, M., 2012. Ocean warming enhances malformations, premature hatching, metabolic suppression and oxidative stress in the early life stages of a keystone squid. PLoS One 7(6). http://dx.doi.org/10.1371/journal.pone.0038282.

Rosa, R., Pierce, G.J., O'Dor, R., 2013. Advances in Squid Biology, Ecology and Fisheries. Nova Science Publishers, Hauppauge, NY.

Rosas, C., Valero, A., Caamal-Monsreal, C., Uriarte, I., Farias, A., Gallardo, P., Sanchez, A., Domingues, P., 2013. Effects of dietary protein sources on growth, survival and digestive capacity of *Octopus maya* juveniles (Mollusca: Cephalopoda). Aquac. Res. 44, 1029–1044.

Sakurai, Y., Young, R.E., Hirota, J., Mangold, K., Vecchione, M., Clarke, M.R., Bower, J., 1995. Artificial fertilization and development through hatching in the oceanic squids *Ommastrephes bartramii* and *Sthenoteuthis oualaniensis* (Cephalopoda: Ommastrephidae). Veliger 38, 185–191.

Sato, N., Kasugai, T., Munehara, H., 2013. Sperm transfer or spermatangia removal: postcopulatory behaviour of picking up spermatangium by female Japanese pygmy squid. Mar. Biol. 160, 553–561.

Sauer, W.H.H., Smale, M.J., Lipinski, M.R., 1992. The location of spawning grounds, spawning and schooling behaviour of the squid *Loligo vulgaris reynaudii (*Cephalopoda: Myopsida) off the eastern Cape coast, South Africa. Mar. Biol. 114, 97–107.

Sauer, W.H.H., McCarthy, C., Smale, M.J., Koorts, A.S., 1993. An investigation of the egg distribution of the chokka squid, *Loligo vulgaris reynaudii*, in Krom bay, South Africa. Bull. Mar. Sci. 53, 1066–1077.

Sauer, W.H.H., Koberts, M.J., Lipinski, M.R., Snide, M.J., Hanlon, R.T., Webber, D.M., O'Dor, R.K., 1997. Choreography of the squid's "nuptial dance". Biol. Bull. 192, 203–207.

Sauer, W.H.H., Melo, Y., De Wet, W., 1999. Fecundity estimation in the chokka squid *Loligo vulgaris reynaudii*. Mar. Biol. 135, 315–319.

Sauer, W.H.H., Lipinski, M.R., Augustyn, C.J., 2000. Tag recapture studies of the chokka squid *Loligo vulgaris reynaudii* d'Orbigny, 1845 on inshore spawning grounds on the south-east coast of South Africa. Fish. Res. 45, 283–289.

Saunders, W.B., Spinosa, C., 1978. Sexual dimorphism in *Nautilus* from Palau. Paleobiology 4, 349–358.

Schoen, P.J., Sauer, W.H.H., Roberts, M.J., 2002. Environmental influences on spawning aggregations and jig catches of chokka squid *Loligo vulgaris reynaudii*: a 'black box' approach. Bull. Mar. Sci. 71, 783–800.

Segawa, S., 1987. Life history of oval squid *Sepioteuthis lessoniana* in Kominato and adjacent waters central Honshu. Jpn. J. Tokyo Univ. Fish. 74, 67–105.

Seibel, B.A., Thuesen, E.V., Childress, J.J., Gorodezky, L.A., 1997. Decline in pelagic cephalopod metabolism with habitat depth reflects differences in locomotory efficiency. Biol. Bull. 192, 262–278.

Seibel, B.A., Hochberg, F.G., Carlini, D.B., 2000. Life history of *Gonatus onyx* (Cephalopoda: Teuthoidea): deep-sea spawning and post-spawning egg care. Mar. Biol. 137, 519–526.

Semmens, J.M., Pecl, G.T., Villanueva, R., Jouffre, D., Sobrino, I., Wood, J.B., Rigby, P.R., 2004. Understanding octopus growth: patterns, variability and physiology. Mar. Freshw. Res. 55, 367–377.

Semmens, J.M., Pecl, G.T., Gillanders, B.M., Waluda, C.M., Shea, E.K., Jouffre, D., Ichii, T., Zumholz, K., Katugin, O.N., Leporati, S.C., Shaw, P.W., 2007. Approaches to resolving cephalopod movement and migration patterns. Rev. Fish Biol. Fish. 17, 401–423.

Semmens, J., Doubleday, Z., Hoyle, K., Pecl, G., 2011. A multilevel approach to examining cephalopod growth using *Octopus pallidus* as a model. J. Exp. Biol. 214, 2799–2807.

Şen, H., 2005. Incubation of European Squid (*Loligo vulgaris* Lamarck, 1798) eggs at different salinities. Aquac. Res. 36, 876–881.

Shaw, P.W., Boyle, P.R., 1997. Multiple paternity within the brood of single females of *Loligo forbesi* (Cephalopoda: Loliginidae), demonstrated with microsatellite DNA markers. Mar. Ecol. Prog. Ser. 160, 279–282.

Shaw, P.W., Sauer, W.H.H., 2004. Evidence for multiple paternity and complex fertilisation dynamics in the squid *Loligo vulgaris reynaudii*. Mar. Ecol. Prog. Ser. 270, 173–179.

Shea, E.K., 2005. Ontogeny of the fused tentacles in three species of ommastrephid squids (Cephalopoda, Ommastrephidae). Invertebr. Biol. 124, 25–38.

Shea, E.K., Vecchione, M., 2010. Ontogenic changes in diel vertical migration patterns compared with known allometric changes in three mesopelagic squid species suggest an expanded definition of a paralarva. ICES J. Mar. Sci. 67, 1436–1443.

Sinn, D.L., Moltschaniwskyj, N.A., 2005. Personality traits in dumpling squid (*Euprymna tasmanica*): context-specific traits and their correlation with biological characteristics. J. Comp. Psychol. 119 (1), 99.

Smith, J.M., Pierce, G.J., Zuur, A.F., Boyle, P.R., 2005. Seasonal patterns of investment in reproductive and somatic tissues in the squid *Loligo forbesi*. Aquat. Living Resour. 18, 341–351.

Snyder, S., 1986. Laboratory culture of *Octopus dofleini* from hatching to settlement. Am. Malacol. Bull. 4, 241.

Sobrino, I., Silva, L., Bellido, J.M., Ramos, F., 2002. Rainfall, river discharges and sea temperature as factors affecting abundance of two coastal benthic cephalopod species in the Gulf of Cádiz (SW Spain). Bull. Mar. Sci. 71, 851–865.

Staaf, D., Camarillo-Coop, S., Haddock, S.H.D., Nyack, A.C., Payne, J., Salinas-Zavala, C.A., Seibel, B.A., Trueblood, L., Widmer, C., Gilly, W.F., 2008. Natural egg mass

deposition by the Humboldt squid (*Dosidicus gigas*) in the Gulf of California and characteristics of hatchlings and paralarvae. J. Mar. Biol. Assoc. U.K. 88, 759–770.
Staaf, D., Zeidberg, L.D., Gilly, W.F., 2011. Effects of temperature on embryonic development of the Humboldt squid *Dosidicus gigas*. Mar. Ecol. Prog. Ser. 441, 165–175.
Stearns, S.C., 1992. The Evolution of Life Histories. Oxford University Press, Oxford, p. 249.
Steer, M.A., Moltschaniwskyj, N.A., Gowland, F.C., 2002. Temporal variability in embryonic development and mortality in the southern calamary, *Sepioteuthis australis*: a field assessment. Mar. Ecol. Prog. Ser. 243, 143–150.
Steer, M.A., Moltschaniwskyj, N.A., Jordan, A.R., 2003. Embryonic development of southern calamary (*Sepioteuthis australis*) within the constraints of an aggregated egg mass. Mar. Freshw. Res. 54, 217–226.
Steer, M.A., Fowler, A.J., Jackson, W.B., Jennings, P.R., 2005. Southern Calamary (*Sepioteuthis australis*) Fishery. SARDI Aquatic Sciences Publication, 135.
Steer, M.A., Lloyd, M.T., Jackson, W.B., 2007. Assessing the feasibility of using 'by-product' data as a pre-recruit index in South Australia's southern calamary (*Sepioteuthis australis*) fishery. Fish. Res. 88, 42–50.
Strathmann, R.R., Strathmann, M.F., 1995. Oxygen supply and limits on aggregation of embryos. J. Mar. Biol. Assoc. U.K. 75, 413–428.
Strobel, A., Hu, M.Y.A., Gutowska, M.A., Lieb, B., Lucassen, M., Melzner, F., Pörtner, H.O., Mark, F.C., 2012. Influence of temperature, hypercapnia, and development on the relative expression of different hemocyanin isoforms in the common cuttlefish *Sepia officinalis*. J. Exp. Zool. 317A, 511–523.
Sugimoto, T., Tameishi, H., 1992. Warm-core rings, streamers and their role on the fishing ground formation around Japan. Deep Sea Research Part A. Oceanographic Research Papers, 39, pp. S183–S201.
Summers, W.C., 1983. Loligo pealeii. In: Boyle, P.R. (Ed.), Cephalopod Life Cycles. Academic Press, New York, pp. 115–142.
Summers, W.C., 1985. Ecological implications of life stage timing determined from the cultivation of *Rossia pacifica* (Mollusca: Cephalopoda). Vie Milieu 35, 249–254.
Sweeney, M.J., Roper, C.F.E., Mangold, K.M., Clarke, M.R., Boletzky, S.V., 1992. "Larval" and juvenile cephalopods: a manual for their identification. Smithson. Contrib. Zool. 513, 282.
Thompson, J.T., Kier, W.M., 2001. Ontogenetic changes in mantle kinematics during escape-jet locomotion in the oval squid, *Sepioteuthis lessoniana* Lesson, 1830. Biol. Bull. 201, 154–166.
Tosti, E., Di Cosmo, A., Cumo, A., Di Cristo, C., Gragnaniello, G., 2001. Progesterone induces activation in *Octopus vulgaris* spermatozoa. Mol. Reprod. Dev. 59, 97–105.
Tsuchiya, K., Takashi, U., 1997. Sneaker male in *Octopus*. Jpn. J. Malacol. 56, 177–181.
Turk, P., Hanlon, R., Tanlon, R.T., Bradford, L.A., Yang, W.T., 1986. Aspects of feeding, growth and survival of the European squid *Loligo vulgaris* (Lamarck 1799), reared through the early growth stages. Vie Milieu 36, 9–13.
Uozumi, Y., 1998. Fishery biology of arrow squids, *Nototodarus gouldi* and *N. sloanii*, in New Zealand waters. Bull. Natl. Res. Inst. Far. Seas Fish. 35, 1–111, Shimizu.
Uriarte, I., Espinoza, V., Herrera, M., Zuñiga, O., Olivares, A., Carbonell, P., Pino, S., Farías, A., Rosas, C., 2012. Effect of temperature on embryonic development of *Octopus mimus* under controlled conditions. J. Exp. Mar. Biol. Ecol. 416 & 417, 168–175.
Van Heukelem, W.F., 1973. Growth and life-span of *Octopus cyanea* (Mollusca: Cephalopoda). J. Zool. (Lond.) 169, 299–315.
Vaughan, D.L., Recksiek, C.W., 1978. An acoustic investigation of market squid, *Loligo opalescens*. Calif. Dep. Fish Game Fish. Bull. 169, 135–147.

Vecchione, M., 1981. Aspects of early life history of *Loligo pealei* (Cephalopoda: Myopsida). J. Shellfish Res. 1, 171–180.
Vecchione, M., 1983. Proceedings of the International Squid Symposium. Trans. Am. Fish. Soc. 112, 731–732.
Vecchione, M., Young, R.E., Guerra, A., Lindsay, D.J., Clague, D.A., Bernhard, J.M., Sager, W.W., Gonzalez, A.F., Rocha, F.J., Segonaz, M., 2001. Worldwide observations of remarkable deep-sea squids. Science 294, 2505–2506.
Vidal, E.A.G., 1994. Relative growth of paralarvae and juvenile of *Illex argentinus* (Castellanos, 1960) in southern Brazil. Antart. Sci. 2 (6), 275–282.
Vidal, E.A.G., Haimovici, M., 1998. Feeding and the possible role of the proboscis and mucus cover in the ingestion of microorganisms by rhynchoteuthion paralarvae (Cephalopoda: Ommastrephidae). Bull. Mar. Sci. 63, 305–316.
Vidal, E.A.G., DiMarco, F.P., Wormuth, J.H., Lee, P.G., 2002a. Influence of temperature and food availability on survival, growth and yolk utilization in hatchling squid. Bull. Mar. Sci. 71 (2), 915–931.
Vidal, E.A.G., DiMarco, F.P., Wormuth, J.H., Lee, P.G., 2002b. Optimizing rearing conditions of hatchling loliginid squid. Mar. Biol. 140, 117–127.
Vidal, E.A.G., Roberts, M.J., Martins, R.S., 2005. Yolk utilization, metabolism and growth in reared Loligo vulgaris reynaudii paralarvae. Aquat. Living Resour. 18 (4), 385–393.
Vidal, E.A.G., DiMarco, P., Lee, P., 2006. Effects of starvation and recovery on the survival, growth and RNA/DNA ratio in loliginid squid paralarvae. Aquaculture 260 (1–4), 94–105.
Vidal, E.A.G., Haimovici, M., Hackbart, V.C., 2010. Distribution of paralarvae and small juvenile cephalopods in relation to primary production in an upwelling area off southern Brazil. ICES J. Mar. Sci. 67, 1346–1352.
Villanueva, R., 1995. Experimental rearing and growth of planktonic *Octopus vulgaris* from hatching to settlement. Can. J. Fish. Aquat. Sci. 52, 2639–2650.
Villanueva, R., 2000. Effect of temperature on statolith growth of European squid *Loligo vulgaris* during the early life. Mar. Biol. 136, 449–460.
Villanueva, R., Bustamante, P., 2006. Composition in essential and non-essential elements of early stages of cephalopods and dietary effects on the elemental profiles of *Octopus vulgaris* paralarvae. Aquaculture 261, 225–240.
Villanueva, R., Norman, M.D., 2008. Biology of the planktonic stages of benthic octopuses. Oceanogr. Mar. Biol. Annu. Rev. 46, 105–202.
Villanueva, R., Nozais, C., Boletzky, S.v., 1997. Swimming behaviour and food searching in planktonic *Octopus vulgaris* Cuvier from hatching to settlement. J. Exp. Mar. Biol. Ecol. 208, 169–184.
Villanueva, R., Riba, J., Ruiz-Capillas, C., Gonzalez, A.V., Baeta, M., 2004. Amino acid composistion of early stages of cephalopods and effect of amino acid dietary treatments on *Octopus vulgaris* paralarvae. Aquaculture 242, 455–478.
Villanueva, R., Moltschaniwskyj, N.A., Bozzano, A., 2007. Abiotic influences on embryo growth: statoliths as experimental tools in the squid early life history. Rev. Fish Biol. Fish. 17, 101–110.
Villanueva, R., Quintana, D., Petroni, G., Bozzano, A., 2011. Factors influencing the embryonic development and hatchling size of the oceanic squid *Illex coindetii* following in vitro fertilization. J. Exp. Mar. Biol. Ecol. 407, 54–62.
Voight, J.R., 1991. Enlarged suckers as an indicator of male maturity in *Octopus*. Bull. Mar. Sci. 49, 98–106.
Voight, J.R., 1995. Sexual dimorphism and niche divergence in a mid-water octopod (Cephalopoda: Bolitaenidae). Biol. Bull. 189, 113–119.
Voight, J.R., Feldheim, K.A., 2009. Microsatellite inheritance and multiple patrnity in the deep-sea octopus *Graneledone boreopacifica* (Mollusca: Cephalopoda). Invertebr. Biol. 128 (1), 26–30.

Voight, J.R., Grehan, A.J., 2000. Egg brooding by deep-sea octopuses in the North Pacific Ocean. Biol. Bull. 198, 94–100.
Voss, N.A., 1980. A generic revision of the Cranchiidae (Cephalopoda: Oegopsida). Bull. Mar. Sci. 30, 365–412.
Voss, G.L., 1988. The biogeography of the deep-sea Octopoda. Malacologia 29, 295–307.
Vovk, A.N., Khvichiya, L.A., 1980. On feeding of long-finned squid (Loligo pealei) juveniles in Subareas 5 and 6. Northwest Atlantic Fisheries Organization (NAFO), SCR Doc. 80/VI/50. 9.
Wakabayashi, T., Tsuchiya, K., Segawa, S., 2005. Morphological changes with growth in the paralarvae of the diamondback squid *Thysanoteuthis rhombus* Troschel, 1857. Phuket Mar. Biol. Center Res. Bull. 66, 167–174.
Walker, D., Power, A.J., Sweeney-Reeves, M.J., Avise, J.C., 2006. Multiple paternity and female sperm usage along egg case strings of the knobbed whelk, *Busycon carica* (Molluscal; Melongenidae). Mar. Biol. 151, 53–61.
Waluda, C.M., Pierce, G.J., 1998. Temporal and spatial patterns in the distribution of squid Loligo spp. in United Kingdom waters. S. Afr. J. Mar. Sci. 20, 323–336.
Waluda, C.M., Rodhouse, P.G., 2006. Remotely sensed mesoscale oceanography of the Central Eastern Pacific and recruitment variability in *Dosidicus gigas*. Mar. Ecol. Prog. Ser. 310, 25–32.
Weihs, D., Moser, H.G., 1981. Stalked eyes as an adaptation towards more efficient foraging in marine fish larvae. Bull. Mar. Sci. 31, 31–36.
Wells, M.J., Wells, J., 1959. Hormonal control of sexual maturity in *Octopus*. J. Exp. Biol. 36, 1–33.
Wells, M.J., Hanlon, R.T., Lee, P.G., Dimarco, F.P., 1988. Respiratory and cardiac performance in *Lolliguncula brevis* (Cephalopoda, Myopsida): the effects of activity, temperature, and hypoxia. J. Exp. Biol. 138, 17–36.
Worms, J., 1983. Loligo vulgaris. In: Boyle, P.R. (Ed.), Cephalopod Life Cycles, Volume 1. Species Accounts. Academic Press, London, pp. 143–157.
Wormuth, J.H., O'Dor, R.K., Balch, N., Dunning, M.C., Forch, E.C., Harman, R.F., Rowell, T.W., 1992. Family Ommastrephidae. Smithson. Contrib. Zool. 513, 105–119.
Yamamura, K., 1999. Key-factor/key-stage analysis for life table data. Ecology 80 (2), 533–537.
Yang, W.T., Hanlon, R.T., Krejci, M.E., Hixon, R.F., Hulet, W.H., 1983. Laboratory rearing of *Loligo opalescens*, the market squid of California. Aquaculture 31, 77–88.
Yang, W.T., Hixon, R.F., Turk, P.E., Krejci, M.E., Hulet, W.H., Hanlon, R.T., 1986. Growth, behaviour and sexual maturation of the market squid, *Loligo opalescens*, cultured through the life cycle. Fish. Bull. 84, 771–798.
Young, J.Z., 1970. The Anatomy of the Nervous System of *Octopus vulgaris*. Oxford, Clarendon, pp. 507–530.
Young, R.E., 1975. *Leachia pacifica* (Cephalopoda, Teuthoidea): spawning habitat and function of the brachial photophores. Pac. Sci. 29, 19–25.
Young, R.E., 1991a. Chiroteuthid and related paralarvae from Hawaiian waters. Bull. Mar. Sci. 49, 162–185.
Young, R.E., Harman, R.F., 1987. Three species of larval *Onychoteuthis* from Hawaiian waters. Veliger 29, 313–321.
Young, R.E., Harman, R.F., 1988. "Larva", "paralarva" and "subadult" in cephalopod terminology. Malacologia 29, 201–207.
Young, R.E., Harman, R.F., Mangold, K.M., 1985. The eggs and larvae of Brachioteuthis sp. (Cephalopoda: Teuthoidea) from Hawaiian waters. Vie Milieu 35, 203–209.
Zatylny, C., Gagnon, J., Boucaud-Camou, E., Henry, J., 2000a. The SepOvotropin: a new ovarian peptide regulating oocyte transport in *Sepia officinalis*. Biochem. Biophys. Res. Commun. 276, 1013–1018.

Zatylny, C., Gagnon, J., Boucaud-Camou, E., Henry, J., 2000b. ILME: a waterborne pheromonal peptide released by the eggs of *Sepia officinalis*. Biochem. Biophys. Res. Commun. 275, 217–222.

Zatylny, C., Marvin, L., Gagnon, J., Henry, J., 2002. Fertilization in *Sepia officinalis*: the first mollusk sperm-attracting peptide. Biochem. Biophys. Res. Commun. 296, 1186–1193.

Zecchini, F., Vecchione, M., Belcari, P., Roper, C.F.E., 2012. Development of the hectocotylus in *Illex coindetii* (Verany, 1837) (Cephalopoda: Ommastrephidae). Sci. Mar. 76, 463–472.

Zeidberg, L., 2004. Allometry measurements from *in situ* video recordings can determine the size and swimming speeds of juvenile and adult squid *Loligo opalescens* (Cephalopoda: Myopsida). J. Exp. Biol. 207, 4195–4203.

Zeidberg, L.D., 2009. First observations of sneaker mating in the California market squid *Doryteuthis opalescens* (Cephalopoda: Myopsida). J. Mar. Biol. Assoc. Marine Biodiversity Records 2, e6. http://dx.doi.org/10.1017/S1755267208000067.

SUBJECT INDEX

Note: Page numbers followed by "*f*" indicate figures and "*t*" indicate tables.

E

R

S

T

V

W

TAXONOMIC INDEX

Note: Page numbers followed by "*f*" indicate figures, and "*t*" indicate tables, and "*np*" indicate footnotes.

A

B

C

M

N

O

FSC
www.fsc.org
MIX
Papier | Fördert gute Waldnutzung
FSC® C083411

Kolibri 360 GmbH
Ferdinand-Jühlke-Straße 7
99095 Erfurt, Deutschland
produktsicherheit@kolibri360.de